Ken Brigham –

We look forward to teaming with you and your colleagues to discover useful patterns in data! God bless,

John Elder

Charlottesville Virginia
March 2010

HANDBOOK OF STATISTICAL ANALYSIS AND DATA MINING APPLICATIONS

"Great introduction to the real-world process of data mining. The overviews, practical advice, tutorials, and extra DVD material make this book an invaluable resource for both new and experienced data miners."

Karl Rexer, Ph.D.
(President and Founder of Rexer Analytics, Boston, Massachusetts,
www.RexerAnalytics.com)

"Statistical thinking will one day be as necessary for efficient citizenship as the ability to read and write."

H. G. Wells (1866 – 1946)

*"Today we aren't quite to the place that H. G. Wells predicted years ago, but society is getting closer out of necessity. Global businesses and organizations are being forced to use <u>statistical analysis and data mining applications</u> in a format that combines **art** and **science–intuition** and **expertise** in collecting and understanding data in order to make **accurate models** that realistically **predict** the **future** that lead to informed strategic **decisions** thus allowing correct **actions** ensuring **success**, before it is too late . . . <u>today, numeracy is as essential as literacy</u>. As John Elder likes to say: 'Go data mining!' It really does save enormous time and money. For those with the patience and faith to get through the early stages of business understanding and data transformation, the cascade of results can be extremely rewarding."*

Gary Miner, March, 2009

HANDBOOK OF STATISTICAL ANALYSIS AND DATA MINING APPLICATIONS

ROBERT NISBET
Pacific Capital Bankcorp N.A.
Santa Barbara, CA

JOHN ELDER
Elder Research, Inc., Charlottesville, VA

GARY MINER
StatSoft, Inc., Tulsa, Oklahoma

ELSEVIER

AMSTERDAM • BOSTON • HEIDELBERG • LONDON
NEW YORK • OXFORD • PARIS • SAN DIEGO
SAN FRANCISCO • SINGAPORE • SYDNEY • TOKYO
Academic Press is an imprint of Elsevier

Academic Press is an imprint of Elsevier
30 Corporate Drive, Suite 400, Burlington, MA 01803, USA
525 B Street, Suite 1900, San Diego, California 92101-4495, USA
84 Theobald's Road, London WC1X 8RR, UK

Library of Congress Cataloging-in-Publication Data
Nisber, Robert, 1942-
   Handbook of statistical analysis and data mining applications / Robert Nisbet, John Elder,
Gary Miner.
      p. cm.
   Includes index.
   ISBN 978-0-12-374765-5 (hardcover : alk. pager) 1. Data mining–Statistical methods. I. Elder, John F.
(John Fletcher) II. Miner, Gary. III. Title.
   QA76.9.D343N57 2009
   006.3'12–dc22   2009008997

British Library Cataloguing-in-Publication Data
A catalogue record for this book is available from the British Library.

ISBN: 978-0-12-374765-5

For information on all Academic Press publications
visit our Web site at www.elsevierdirect.com

Printed in Canada
09   10   9   8   7   6   5   4   3   2   1

Table of Contents

I

HISTORY OF PHASES OF DATA ANALYSIS, BASIC THEORY, AND THE DATA MINING PROCESS

II

THE ALGORITHMS IN DATA MINING AND TEXT MINING, THE ORGANIZATION OF THE THREE MOST COMMON DATA MINING TOOLS, AND SELECTED SPECIALIZED AREAS USING DATA MINING

III

TUTORIALS—STEP-BY-STEP CASE STUDIES AS A STARTING POINT TO LEARN HOW TO DO DATA MINING ANALYSES

Guest Authors of the Tutorials

IV

MEASURING TRUE COMPLEXITY, THE "RIGHT MODEL FOR THE RIGHT USE," TOP MISTAKES, AND THE FUTURE OF ANALYTICS

Foreword 1

This book will help the novice user become familiar with data mining. Basically, data mining is doing data analysis (or statistics) on data sets (often large) that have been obtained from potentially many sources. As such, the miner may not have control of the input data, but must rely on sources that have gathered the data. As such, there are problems that every data miner must be aware of as he or she begins (or completes) a mining operation. I strongly resonated to the material on "The Top 10 Data Mining Mistakes," which give a worthwhile checklist:

- Ensure you have a response variable and predictor variables—and that they are correctly measured.
- Beware of overfitting. With scads of variables, it is easy with most statistical programs to fit incredibly complex models, but they cannot be reproduced. It is good to save part of the sample to use to test the model. Various methods are offered in this book.
- Don't use only one method. Using only linear regression can be a problem. Try dichotomizing the response or categorizing it to remove nonlinearities in the response variable. Often, there are clusters of values at zero, which messes up any normality assumption. This, of course, loses information, so you may want to categorize a continuous response variable and use an alternative to regression. Similarly, predictor variables may need to be treated as factors rather than linear predictors. A classic example is using marital status or race as a linear predictor when there is no order.
- Asking the wrong question—when looking for a rare phenomenon, it may be helpful to identify the most common pattern. These may lead to complex analyses, as in item 3, but they may also be conceptually simple. Again, you may need to take care that you don't overfit the data.
- Don't become enamored with the data. There may be a substantial history from earlier data or from domain experts that can help with the modeling.
- Be wary of using an outcome variable (or one highly correlated with the outcome variable) and becoming excited about the result. The predictors should be "proper" predictors in the sense that (a) they are measured prior to the outcome and (b) are not a function of the outcome.
- Do not discard outliers without solid justification. Just because an observation is out of line with others is insufficient reason to ignore it. You must check the circumstances that led to the value. In any event, it is useful to conduct the analysis with the observation(s) included and excluded to determine the sensitivity of the results to the outlier.

- Extrapolating is a fine way to go broke—the best example is the stock market. Stick within your data, and if you must go outside, put plenty of caveats. Better still, restrain the impulse to extrapolate. Beware that pictures are often far too simple and we can be misled. Political campaigns oversimplify complex problems ("My opponent wants to raise taxes"; "My opponent will take us to war") when the realities may imply we have some infrastructure needs that can be handled only with new funding, or we have been attacked by some bad guys.

Be wary of your data sources. If you are combining several sets of data, they need to meet a few standards:

- The definitions of variables that are being merged should be identical. Often they are close but not exact (especially in meta-analysis where clinical studies may have somewhat different definitions due to different medical institutions or laboratories).
- Be careful about missing values. Often when multiple data sets are merged, missing values can be induced: one variable isn't present in another data set, what you thought was a unique variable name was slightly different in the two sets, so you end up with two variables that both have a lot of missing values.
- How you handle missing values can be crucial. In one example, I used complete cases and lost half of my sample—all variables had at least 85% completeness, but when put together the sample lost half of the data. The residual sum of squares from a stepwise regression was about 8. When I included more variables using mean replacement, almost the same set of predictor variables surfaced, but the residual sum of squares was 20. I then used multiple imputation and found approximately the same set of predictors but had a residual sum of squares (median of 20 imputations) of 25. I find that mean replacement is rather optimistic but surely better than relying on only complete cases. If using stepwise regression, I find it useful to replicate it with a bootstrap or with multiple imputation. However, with large data sets, this approach may be expensive computationally.

To conclude, there is a wealth of material in this handbook that will repay study.

Peter A. Lachenbruch, Ph.D.,
Oregon State University
Past President, 2008, American Statistical Society
Professor, Oregon State University
Formerly: FDA and professor at Johns Hopkins University;
UCLA, and University of Iowa, and
University of North Carolina Chapel Hill

Foreword 2

A November 2008 search on Amazon.com for "data mining" books yielded over 15,000 hits—including 72 to be published in 2009. Most of these books either describe data mining in very technical and mathematical terms, beyond the reach of most individuals, or approach data mining at an introductory level without sufficient detail to be useful to the practitioner. The *Handbook of Statistical Analysis and Data Mining Applications* is the book that strikes the right balance between these two treatments of data mining.

This volume is not a theoretical treatment of the subject—the authors themselves recommend other books for this—but rather contains a description of data mining principles and techniques in a series of "knowledge-transfer" sessions, where examples from real data mining projects illustrate the main ideas. This aspect of the book makes it most valuable for practitioners, whether novice or more experienced.

While it would be easier for everyone if data mining were merely a matter of finding and applying the correct mathematical equation or approach for any given problem, the reality is that both "art" and "science" are necessary. The "art" in data mining requires experience: when one has seen and overcome the difficulties in finding solutions from among the many possible approaches, one can apply newfound wisdom to the next project. However, this process takes considerable time and, particularly for data mining novices, the iterative process inevitable in data mining can lead to discouragement when a "textbook" approach doesn't yield a good solution.

This book is different; it is organized with the practitioner in mind. The volume is divided into four parts. Part I provides an overview of analytics from a historical perspective and frameworks from which to approach data mining, including CRISP-DM and SEMMA. These chapters will provide a novice analyst an excellent overview by defining terms and methods to use, and will provide program managers a framework from which to approach a wide variety of data mining problems. Part II describes algorithms, though without extensive mathematics. These will appeal to practitioners who are or will be involved with day-to-day analytics and need to understand the qualitative aspects of the algorithms. The inclusion of a chapter on text mining is particularly timely, as text mining has shown tremendous growth in recent years.

Part III provides a series of tutorials that are both domain-specific and software-specific. Any instructor knows that examples make the abstract concept more concrete, and these tutorials accomplish exactly that. In addition, each tutorial shows how the solutions were developed using popular data mining software tools, such as Clementine, Enterprise Miner, Weka, and *STATISTICA*. The step-by-step specifics will assist practitioners in learning not only how to approach a wide variety of problems, but also how to use these software products effectively. Part IV presents a look at the future of data mining, including a treatment of model ensembles and "The Top 10 Data Mining Mistakes," from the popular presentation by Dr. Elder.

However, the book is best read a few chapters at a time while actively doing the data mining rather than read cover-to-cover (a daunting task for a book this size). Practitioners will appreciate tutorials that match their business objectives and choose to ignore other tutorials. They may choose to read sections on a particular algorithm to increase insight into that algorithm and then decide to add a second algorithm after the first is mastered. For those new to a particular software tool highlighted in the tutorials section, the step-by-step approach will operate much like a user's manual. Many chapters stand well on their own, such as the excellent "History of Statistics and Data Mining" and "The Top 10 Data Mining Mistakes" chapters. These are broadly applicable and should be read by even the most experienced data miners.

The *Handbook of Statistical Analysis and Data Mining Applications* is an exceptional book that should be on every data miner's bookshelf or, better yet, found lying open next to their computer.

Dean Abbott
President
Abbott Analytics
San Diego, California

Preface

Data mining scientists in research and academia may look askance at this book because it does not present algorithm theory in the commonly accepted mathematical form. Most articles and books on data mining and knowledge discovery are packed with equations and mathematical symbols that only experts can follow. Granted, there is a good reason for insistence on this formalism. The underlying complexity of nature and human response requires teachers and researchers to be extremely clear and unambiguous in their terminology and definitions. Otherwise, ambiguities will be communicated to students and readers, and their understanding will not penetrate to the essential elements of any topic. Academic areas of study are not called *disciplines* without reason.

This rigorous approach to data mining and knowledge discovery builds a fine foundation for academic studies and research by experts. Excellent examples of such books are

- *The Handbook of Data Mining*, 2003, by Nong Ye (Ed.). Mahwah, New Jersey: Lawrence Erlbaum Associates.
- *The Elements of Statistical Learning: Data Mining, Inference, and Prediction*, 2nd edition, 2009, by T. Hastie, R. Tibshirani, & J. Friedman. New York: Springer-Verlag.

Books like these were especially necessary in the early days of data mining, when analytical tools were relatively crude and required much manual configuration to make them work right. Early users had to understand the tools in depth to be able to use them productively. These books are still necessary for the college classroom and research centers. Students must understand the theory behind these tools in the same way that the developers understood it so that they will be able to build new and improved versions.

Modern data mining tools, like the ones featured in this book, permit ordinary business analysts to follow a path through the data mining process to create models that are "good enough." These less-than-optimal models are far better in their ability to leverage faint patterns in databases to solve problems than the ways it used to be done. These tools provide default configurations and automatic operations, which shield the user from the technical complexity underneath. They provide one part in the crude analogy to the automobile interface. You don't have to be a chemical engineer or physicist who understands moments of force to be able to operate a car. All you have to do is learn to turn the key in the ignition, step on the gas and the brake at the right times, turn the wheel to change direction in a safe manner, and *voila*, you are an expert user of the very complex technology under the hood. The *other* half of the story is the instruction manual and the driver's education course that help you to learn how to drive.

This book provides that instruction manual and a series of tutorials to train you how to do data mining in many subject areas. We provide both the right tools and the right intuitive explanations (rather than formal mathematical definitions) of the data mining process and algorithms, which will enable even beginner data miners to understand the basic concepts necessary to understand *what* they are doing. In addition, we provide many tutorials in many different industries and businesses (using many of the most common data mining tools) to show *how* to do it.

OVERALL ORGANIZATION OF THIS BOOK

We have divided the chapters in this book into three parts for the same general reason that the ancient Romans split Gaul into three pieces—for the ease of management. The fourth part is a group of tutorials, which serve in principle as Rome served—as the central governing influence. The central theme of this book is the education and training of beginning data mining practitioners, not the rigorous academic preparation of algorithm scientists. Hence, we located the tutorials in the middle of the book in Part III, flanked by topical chapters in Parts I, II, and IV.

This approach is "a mile wide and an inch deep" by design, but there is a lot packed into that inch. There is enough here to stimulate you to take deeper dives into theory, and there is enough here to permit you to construct "smart enough" business operations with a relatively small amount of the right information. James Taylor developed this concept for automating operational decision making in the area of Enterprise Decision Management (Taylor, 2007). Taylor recognized that companies need decision-making systems that are automated enough to keep up with the volume and time-critical nature of modern business operations. These decisions should be deliberate, precise, consistent across the enterprise, smart enough to serve immediate needs appropriately, and agile enough to adapt to new opportunities and challenges in the company. The same concept can be applied to nonoperational systems for Customer Relationship Management (CRM) and marketing support. Even though a CRM model for cross-sell may not be optimal, it may enable several times the response rate in product sales following a marketing campaign. Models like this are "smart enough" to drive companies to the next level of sales. When models like this are proliferated throughout the enterprise to lift all sales to the next level, more refined models can be developed to do even better. This enterprise-wide "lift" in intelligent operations can drive a company through evolutionary rather than revolutionary changes to reach long-term goals.

When one of the primary authors of this book was fighting fires for the U.S. Forest Service, he was struck by the long-term efficiency of Native American contract fire fighters on his crew in Northern California. They worked more slowly than their young "whipper-snapper" counterparts, but they didn't stop for breaks; they kept up the same pace throughout the day. By the end of the day, they completed far more fire line than the other members of the team. They leveraged their "good enough" work at the moment to accomplish optimal success overall.

Companies can leverage "smart enough" decision systems to do likewise in their pursuit of optimal profitability in their business.

Clearly, use of this book and these tools will not make you experts in data mining. Nor will the explanations in the book permit you to understand the complexity of the theory behind the algorithms and methodologies so necessary for the academic student. But we will conduct you through a relatively thin slice across the wide practice of data mining in many industries and disciplines. We can show you how to create powerful predictive models in your own organization in a relatively short period of time. In addition, this book can function as a springboard to launch you into higher-level studies of the theory behind the practice of data mining. If we can accomplish those goals, we will have succeeded in taking a significant step in bringing the practice of data mining into the mainstream of business analysis.

The three coauthors could not have done this book completely by themselves, and we wish to thank the following individuals, with the disclaimer that we apologize if, by our neglect, we have left out of this "thank you list" anyone who contributed.

Foremost, we would like to thank Acquisitions Editor Lauren Schultz of Elsevier's Boston office; Lauren was the first to catch the vision and see the need for this book and has worked tirelessly to see it happen. Also, Leah Ackerson, Marketing Manager for Elsevier, and Tom Singer, then Elsevier's Math and Statistics Acquisitions Editor, who were the first to get us started down this road. Yet, along with Elsevier's enthusiasm came their desire to have it completed within two months of their making a final decision. . . . So that really pushed us. But Lauren and Leah continually encouraged us during this period by, for instance, flying into the 2008 Knowledge Discovery and Data Mining conference to work out many near-final details.

Bob Nisbet would like to honor and thank his wife, Jean Nisbet, Ph.D., who blasted him off in his technical career by retyping his dissertation five times (before word processing), and assumed much of the family's burdens during the writing of this book. Bob also thanks Dr. Daniel B. Botkin, the famous global ecologist, for introducing him to the world of modeling and exposing him to the distinction between viewing the world as *machine* and viewing it as *organism*. And, thanks are due to Ken Reed, Ph.D., for inducting Bob into the practice of data mining. Finally, he would like to thank Mike Laracy, a member of his data mining team at NCR Corporation, who showed him how to create powerful customer response models using temporal abstractions.

John Elder would like to thank his wife, Elizabeth Hinson Elder, for her support— keeping five great kids happy and healthy while Dad was stuck on a keyboard—and for her inspiration to excellence. John would also like to thank his colleagues at Elder Research, Inc.—who pour their talents, hard work, and character into using data mining for the good of our clients and community—for their help with research contributions throughout the book. You all make it a joy to come to work. Dustin Hux synthesized a host of material to illustrate the interlocking disciplines making up data mining; Antonia de Medinaceli contributed valuable and thorough edits; Stein Kretsinger made useful suggestions; and Daniel Lautzenheiser created the figure showing a non-intuitive type of outlier.

Co-author Gary Miner wishes to thank his wife, Linda A. Winters-Miner, Ph.D., who has been working with Gary on similar books over the past 10 years and wrote several of the tutorials included in this book, using real-world data. Gary also wishes to thank the following people from his office who helped in various ways, from keeping Gary's computers running properly to taking over some of his job responsibilities when he took days off to write this book, including Angela Waner, Jon Hillis, Greg Sergeant, Jen Beck, Win Noren, and Dr. Thomas Hill, who gave permission to use and also edited a group of the tutorials that had been written over the years by some of the people listed as guest authors in this book.

Without all the help of the people mentioned here, and maybe many others we failed to specifically mention, this book would never have been completed. Thanks to you all!

Bob Nisbet (bob2@rnisbet.com)
John Elder (elder@datamininglab.com)
Gary Miner (miner.gary@gmail.com)
October 31, 2008

General inquiries can be addressed to: handbook@datamininglab.com

References

Taylor, J. 2007. *Smart (Enough) Systems*. Upper Saddle River, NJ: Prentice-Hall.

SAS

To gain experience using SAS® Enteprise Miner™ for the Desktop using tutorials that take you through all the steps of a data mining project. visit HYPERLINK "http://www.support.sas.com/statandDMapps" www.support.sas.com/statandDMapps.

The tutorials include problem definition and data selection, and continue through data exploration, data transformation, sampling, data partitioning, modeling, and model comparison. The tutorials are suitable for data analysts, qualitative experts, and others who want an introduction to using SAS Enterprise Miner for the Desktop using a free 90–day evaluation.

STATSOFT

To gain experience using *STATISTICA* Data Miner + QC-Miner + Text Miner for the Desktop using tutorials that take you through all the steps of a data mining project, please install the free 90-day *STATISTICA* that is on the DVD bound with this book. Also, please see the "DVD Install Instructions" at the end of the book for details on installing the software and locating the additional tutorials that are only on the DVD.

SPSS

Call 1.800.543.2185 and mention offer code US09DM0430C to get a free 30-day trial of SPSS Data Mining software (PASW Modeler) for use with the HANDBOOK.

Introduction

Often, data miners are asked, "What are statistical analysis and data mining?" In this book, we will define what data mining is from a procedural standpoint. But most people have a hard time relating what we tell them to the things they know and understand. Before moving on into the book, we would like to provide a little background for data mining that everyone can relate to.

Statistical analysis and data mining are two methods for simulating the unconscious operations that occur in the human brain to provide a rationale for decision making and actions. Statistical analysis is a very directed rationale that is based on norms. We all think and decide on the basis of norms. For example, we consider (unconsciously) what the norm is for dress in a certain situation. Also, we consider the acceptable range of variation in dress styles in our culture. Based on these two concepts, the norm and the variation around that norm, we render judgments like, "That man is inappropriately dressed." Using similar concepts of mean and standard deviation, statistical analysis proceeds in a very logical way to make very similar judgments (in principle). On the other hand, data mining learns case by case and does not use means or standard deviations. Data mining algorithms build patterns, clarifying the pattern as each case is submitted for processing. These are two very different ways of arriving at the same conclusion: a decision. We will introduce some basic analytical history and theory in Chapters 1 and 2.

The basic process of analytical modeling is presented in Chapter 3. But it may be difficult for you to relate what is happening in the process without some sort of tie to the real world that you know and enjoy. In many ways, the decisions served by analytical modeling are similar to those we make every day. These decisions are based partly on patterns of action formed by experience and partly by intuition.

PATTERNS OF ACTION

A pattern of action can be viewed in terms of the activities of a hurdler on a race track. The runner must start successfully and run to the first hurdle. He must decide very quickly how high to jump to clear the hurdle. He must decide when and in what sequence to move his legs to clear the hurdle with minimum effort and without knocking it down. Then he must run a specified distance to the next hurdle, and do it all over again several times, until he crosses the finish line. Analytical modeling is a lot like that.

The training of the hurdler's "model" of action to run the race happens in a series of operations:

- Run slow at first.
- Practice takeoff from different positions to clear the hurdle.
- Practice different ways to move the legs.
- Determine the best ways to do each activity.
- Practice the best ways for each activity over and over again.

This practice trains the sensory and motor neurons to function together most efficiently.

Individual neurons in the brain are "trained" in practice by adjusting signal strengths and firing thresholds of the motor nerve cells. The performance of a successful hurdler follows the "model" of these activities and the process of coordinating them to run the race. Creation of an analytical "model" of a business process to predict a desired outcome follows a very similar path to the training regimen of a hurdler. We will explore this subject further in Chapter 3 and apply it to develop a data mining process that expresses the basic activities and tasks performed in creating an analytical model.

HUMAN INTUITION

In humans, the right side of the brain is the center for visual and aesthetic sensibilities. The left side of the brain is the center for quantitative and time-regulated sensibilities. Human intuition is a blend of both sensibilities. This blend is facilitated by the neural connections between the right side of the brain and the left side. In women, the number of neural connections between right and left sides of the brain is 20% greater (on average) than in men. This higher connectivity of women's brains enables them to exercise intuitive thinking to a greater extent than men. Intuition "builds" a model of reality from both quantitative building blocks and visual sensibilities (and memories).

PUTTING IT ALL TOGETHER

Biological taxonomy students claim (in jest) that there are two kinds of people in taxonomy—those who divide things up into two classes (for dichotomous keys) and those who don't. Along with this joke is a similar recognition that taxonomists are divided into the "lumpers" (who combine several species into one) and the "splitters" (who divide one species into many). These distinctions point to a larger dichotomy in the way people think.

In ecology, there used to be two schools of thought: autoecologists (chemistry, physics, and mathematics explain all) and the synecologists (organism relationships in their environment explain all). It wasn't until the 1970s that these two schools of thought learned that both perspectives were needed to understand ecosystems (but more about that later). In business, there are the "big picture" people versus "detail" people. Some people learn by following an intuitive pathway from general to specific (inductive). Often, we call them "big picture" people. Other people learn by following an intuitive pathway from specific to general (deductive). Often, we call them "detail" people.

This distinction is reflected in many aspects of our society. In Chapter 1, we will explore this distinction to a greater depth in regard to the development of statistical and data mining theory through time.

Many of our human activities involve finding patterns in the data input to our sensory systems. An example is the mental pattern that we develop by sitting in a chair in the middle of a shopping mall and making some judgment about patterns among its clientele. In one mall, people of many ages and races may intermingle. You might conclude from this pattern that this mall is located in an ethnically diverse area. In another mall, you might see a very different pattern. In one mall in Toronto, a great many of the stores had Chinese titles and script on the windows. One observer noticed that he was the only non-Asian seen for a half-hour. This led to the conclusion that the mall catered to the Chinese community and was owned (probably) by a Chinese company or person.

Statistical methods employed in testing this "hypothesis" would include

- Performing a survey of customers to gain empirical data on race, age, length of time in the United States, etc.;
- Calculating means (averages) and standard deviations (an expression of the average variability of all the customers around the mean);
- Using the mean and standard deviation for all observations to calculate a metric (e.g., student's t-value) to compare to standard tables;

If the metric exceeds the standard table value, this attribute (e.g., race) is present in the data at a higher rate than expected at random.

More advanced statistical techniques can accept data from multiple attributes and process them in combination to produce a metric (e.g., average squared error), which reflects how well a subset of attributes (selected by the processing method) predicts desired outcome. This process "builds" an analytical equation, using standard statistical methods. This analytical "model" is based on averages across the range of variation of the input attribute data. This approach to finding the pattern in the data is basically a deductive, top-down process (general to specific). The general part is the statistical model employed for the analysis (i.e., normal parametric model). This approach to model building is very "Aristotelian." In Chapter 1, we will explore the distinctions between Aristotelian and Platonic approaches for understanding truth in the world around us.

Both statistical analysis and data mining algorithms operate on patterns: statistical analysis uses a predefined pattern (i.e., the Parametric Model) and compares some measure of the observations to standard metrics of the model. We will discuss this approach in more detail in Chapter 1. Data mining doesn't start with a model; it *builds* a model with the data. Thus, statistical analysis uses a model to characterize a pattern in the data; data mining uses the pattern in the data to build a model. This approach uses *deductive reasoning*, following an Aristotelian approach to truth. From the "model" accepted in the beginning (based on the mathematical distributions assumed), outcomes are deduced. On the other hand, data mining methods discover patterns in data *inductively*, rather than *deductively*, following a more Platonic approach to truth. We will unpack this distinction to a much greater extent in Chapter 1.

Which is the best way to do it? The answer is ... it depends. It depends on the data. Some data sets can be analyzed better with statistical analysis techniques, and other data sets can be analyzed better with data mining techniques. How do you know which approach to use for a given data set? Much ink has been devoted to paper to try to answer that question. We will not add to that effort. Rather, we will provide a guide to general analytical theory (Chapter 2) and broad analytical procedures (Chapter 3) that can be used with techniques for either approach. For the sake of simplicity, we will refer to the joint body of techniques as *analytics*.

In Chapters 4 and 5, we introduce basic process and preparation procedures for analytics.

Chapters 6–9 introduce accessory tools and some basic and advanced analytic algorithms used commonly for various kinds of analytics projects, followed by the use of specialized algorithms for the analysis of textual data.

Chapters 10–12 provide general introductions to three common analytics tool packages and the two most common application areas for those tools (classification and numerical prediction).

Chapter 13 discusses various methods for evaluating the models you build. We will discuss

- Training and testing activities
- Resampling methods
- Ensemble methods
- Use of graphical plots
- Use of lift charts and ROC curves

Additional details about these powerful techniques can be found in Chapter 5 and in Witten and Frank (2006).

Chapters 14–17 guide you through the application of analytics to four common problem areas: medical informatics, bioinformatics, customer response modeling, and fraud.

One of the guiding principles in the development of this book is the inclusion of many tutorials in the body of the book and on the DVD. There are tutorials for SAS-Enterprise Miner, SPSS Clementine, and *STATISTICA* Data Miner. You can follow through the appropriate tutorials with *STATISTICA* Data Miner. If you download the free trials of the other tools (as described at the end of the Preface), you can follow the tutorials based on them. In any event, the overall principle of this book is to provide enough of an introduction to get you started doing data mining, plus at least one tool for you to use in the beginning of your work.

Chapters 18–20 discuss the issues in analytics regarding model complexity, parsimony, and modeling mistakes.

Chapter 18, on how to measure true complexity, is the most complex and "researchy" chapter of the book, and can be skipped by most readers; but Chapter 20, on classic analytic mistakes, should be a big help to anyone who needs to implement real models in the real world.

Chapter 21 gives you a glimpse of the future of analytics. Where is data mining going in the future? Much statistical and data mining research during the past 30 years has focused on designing better algorithms for finding faint patterns in "mountains" of data. Current directions in data mining are organized around how to link together many processing instances rather than improving the mathematical algorithms for pattern recognition. We can see these developments taking shape in at least these major areas:

- RAID (Radio Frequency Identification Technologies)
- Social networks
- Visual data mining: object identification, video and audio, and 3D scanning
- Cloud computing

It is likely that even these global processing strategies are not the end of the line in data mining development. Chapter 1 ends with the statement that we will discover increasingly novel and clever ways to mimic the most powerful pattern recognition engine in the universe: the human brain. Chapter 22 wraps up the whole discussion with a summary.

Much concern in the business world now is organized around the need for effective *business intelligence* (BI) processes. Currently, this term refers just to business reporting, and there is not much "intelligence" in it. Data mining can bring another level of "intelligence" to bear on problem solving and pattern recognition. But even the state that data mining may assume in the near future (with cloud computing and social networking) is only the first step in developing truly intelligent decision-making engines.

One step further in the future could be to drive the hardware supporting data mining to the level of nanotechnology. Powerful biological computers the size of pin heads (and smaller) may be the next wave of technological development to drive data mining advances. Rather than the sky, the atom is the limit.

References

Whitten, I. H. & Frank, E. (2006). *Data Mining: Practical Machine Learning Tools and Techniques*. San Francisco: Morgan-Kaufmann.

List of Tutorials by Guest Authors

Tutorials are located in three places:

1. Tutorials A–N are located in Part III of this book.
2. Tutorials O–Z, AA–KK are located on the DVD bound with this book.
3. Additional Tutorials are located on the book's companion Web site:
   http://www.elsevierdirect.com/companions/9780123747655

Tutorials in Part III of the printed book, with accompanying datasets and results located on the DVD that is bound with this book:

Tutorial A (*Field: General*)
How to Use Data Miner Recipe *STATISTICA* Data Miner Only
Gary Miner, Ph.D.

Tutorial B (*Field: Engineering*)
Data Mining for Aviation Safety Using Data Mining Recipe "Automatized Data Mining" from *STATISTICA*
Alan Stolzer, Ph.D.

Tutorial C (*Field: Entertainment Business*)
Predicting Movie Box-Office Receipts Using SPSS Clementine Data Mining Software
Dursun Delen, Ph.D

Tutorial D (*Field: Financial–Business*)
Detecting Unstatisfied Customers: A Case Study Using SAS Enterprise Miner Version 5.3 for the Analysis
Chamont Wang, Ph.D.

Tutorial E (*Field: Financial*)
Credit Scoring Using *STATISTICA* Data Miner
Sachin Lahoti and Kiron Mathew

Tutorial F (*Field: Business*)
Churn Analysis using SPSS-Clementine
Robert Nisbet, Ph.D.

Tutorial G (*Field: Customer Satisfaction–Business*)
Text Mining: Automobile Brand Review Using STATISTICA Data Miner and Text Miner
Sachin Lahoti and Kiron Mathew, edited by Gary Miner, Ph.D.

Tutorial H (*Field: Industry Quality Control*)
Predictive Process Control: QC-Data Mining Using *STATISTICA* Data Miner and QC-Miner
Sachin Lahoti and Kiron Mathew, edited by Gary Miner, Ph.D.

Tutorials I, J, and K
Three Short Tutorials Showing the Use of Data Mining and Particularly C&RT to Predict and Display Possible Structural Relationships among Data
edited by Linda A. Miner, Ph.D.

Tutorial I (*Field: Business Administration*)
Business Administration in a Medical Industry: Determining Possible Predictors for Days with Hospice Service for Patients with Dementia
Linda A. Miner, Ph.D., James Ross, MD, and Karen James, RN, BSN, CHPN

Tutorial J (*Field: Clinical Psychology & Patient Care*)
Clinical Psychology: Making Decisions about Best Therapy for a Client: Using Data Mining to Explore the Structure of a Depression Instrument
David P. Armentrout, Ph.D. and Linda A. Miner, Ph.D.

Tutorial K (*Field: Leadership Training–Business*)
Education–Leadership Training for Business and Education Using C&RT to Predict and Display Possible Structural Relationships
Greg S. Robinson, Ph.D., Linda A. Miner, Ph.D., and Mary A. Millikin, Ph.D.

Tutorial L (*Field: Dentistry*)
Dentistry: Facial Pain Study Based on 84 Predictor Variables (Both Categorical and Continuous)
Charles G. Widmer, DDS, MS.

Tutorial M (*Field: Financial–Banking*)
Profit Analysis of the German Credit Data using SAS-EM Version 5.3
Chamont Wang, Ph.D., edited by Gary Miner, Ph.D.

Tutorial N (*Field: Medical Informatics*)
Predicting Self-Reported Health Status Using Artificial Neural Networks
Nephi Walton, Stacey Knight, MStat, and Mollie R. Poynton, Ph.D., APRN, BC

Tutorials on the DVD bound with this book including datasets, data mining projects, and results:

Tutorial O (*Field: Demographics*)
Regression Trees Using Boston Housing Data Set
Kiron Mathew, edited by Gary Miner, Ph.D.

Tutorial P (*Field: Medical Informatics & Bioinformatics*)
Cancer Gene
Kiron Mathew, edited by Gary Miner, Ph.D.

Tutorial Q (*Field: CRM – Customer Relationship Management*)
Clustering of Shoppers: Clustering Techniques for Data Mining Modeling
Kiron Mathew, edited by Gary Miner, Ph.D.

Tutorial R (*Field: Financial–Banking*)
Credit Risk using Discriminant Analysis in a Data Mining Model
Sachin Lahoti and Kiron Mathew, edited by Gary Miner, Ph.D.

Tutorial S (*Field: Data Analysis*)
Data Preparation and Management
Kiron Mathew, edited by Gary Miner, Ph.D.

Tutorial T (*Field: Deployment of Predictive Models*)
Deployment of a Data Mining Model
Kiron Mathew and Sachin Lahoti

Tutorial U (*Field: Medical Informatics*)
Stratified Random Sampling for Rare Medical Events: A Data Mining Method to Understand Pattern and Meaning of Infrequent Categories in Data
David Redfearn, Ph.D., edited by Gary Miner, Ph.D.
[This Tutorial is not included on the DVD bound with the book, but instead is on this book's companion Web site.]

Tutorial V (*Field: Medical Informatics–Bioinformatics*)
Heart Disease Utilizing Visual Data Mining Methods
Kiron Mathew, edited by Gary Miner, Ph.D.

Tutorial W (*Field: Medical Informatics–Bioinformatics*)
Type II Diabetes Versus Assessing Hemoglobin A1c and LDL, Age, and Sex: Examination of the Data by Progressively Analyzing from Phase 1 (Traditional Statistics) through Phase 4 (Advanced Bayesian and Statistical Learning Theory) Data Analysis Methods, Including Deployment of Model for Predicting Success in New Patients
Dalton Ellis, MD and Ashley Estep, DO, edited by Gary Miner, Ph.D.

Tutorial X (*Field: Separating Competing Signals*)
Independent Component Analysis
Thomas Hill, Ph.D, edited by Gary Miner, Ph.D.

Tutorial Y (*Fields: Engineering–Air Travel–Text Mining*)
NTSB Aircraft Accident Reports
Kiron Mathew, by Thomas Hill, Ph.D., and Gary Miner, Ph.D.

Tutorial Z (*Field: Preventive Health Care*)
Obesity Control in Children: Medical Tutorial Using *STATISTICA* Data Miner Recipe—
Childhood Obesity Intervention Attempt
Linda A. Miner, Ph.D., Walter L. Larimore, MD, Cheryl Flynt, RN, and Stephanie Rick

Tutorial AA (*Field: Statistics–Data Mining*)
Random Forests Classification
Thomas Hill, Ph.D. and Kiron Mathew, edited by Gary Miner, Ph.D.

Tutorial BB (*Field: Data Mining–Response Optimization*)
Response Optimization for Data Mining Models
Kiron Mathew and Thomas Hill, Ph.D., edited by Gary Miner, Ph.D.

Tutorial CC (*Field: Industry–Quality Control*)
Diagnostic Tooling and Data Mining: Semiconductor Industry
Kiron Mathew and Sachin Lahoti, edited by Gary Miner, Ph.D.

Tutorial DD (*Field: Sociology*)
Visual Data Mining: Titanic Survivors
Kiron Mathew and Thomas Hill, Ph.D., edited by Gary Miner, Ph.D.

Tutorial EE (*Field: Demography–Census*)
Census Data Analysis: Basic Statistical Data Description
Kiron Mathew, edited by Gary Miner, Ph.D.

Tutorial FF (*Field: Environment*)
Linear and Logistic Regression (Ozone Data)
Jessica Sieck, edited by Gary Miner, Ph.D.

Tutorial GG (*Field: Survival Analysis–Medical Informatics*)
R-Integration into a Data Miner Workspace Node: R-node Competing Hazards Program
Named cmprsk from the R-Library
Ivan Korsakov and Wayne Kendal, MD., Ph.D., FRCSC, FRPCP, edited by Gary Miner,
Ph.D.

Tutorial HH (*Fields: Social Networks–Sociology & Medical Informatics*)
Social Networks among Community Organizations: Tulsa Cornerstone Assistance
Network Partners Survey of Satisfaction Derived From this Social Network by Members
of Cornerstone Partners: Out of 24 Survey Questions, Which are Important in Predicting
Partner Satisfaction?
Enis Sakirgil, MD and Timothy Potter, MD, edited by Gary Miner, Ph.D.

Tutorial II (*Field: Social Networks*)
Nairobi, Kenya Baboon Project: Social Networking among Baboon Populations in
Kenya on the Laikipia Plateau
Shirley C. Strum, Ph.D., edited by Gary Miner, Ph.D.

Tutorial JJ (*Field: Statistics Resampling Methods*)
Jackknife and Bootstrap Data Miner Workspace and MACRO for *STATISTICA* Data
Miner
Gary Miner, Ph.D.

Tutorial KK (*Field: Bioinformatics*)
Dahlia Mosaic Virus: A DNA Microarray Analysis on 10 Cultivars From a Single
Source: Dahlia Garden in Prague, Czech Republic
Hanu R. Pappu, Ph.D., edited by Gary Miner, Ph.D.

Tutorials that are on the book's companion Web site:
http://www.elsevierdirect.com/companions/9780123747655

Tutorial LL (*Field: Physics–Meteorology*)
Prediction of Hurricanes from Cloud Data
Jose F. Nieves, Ph.D. and Juan S. Lebron

Tutorial MM (*Field: Education–Administration*)
Characteristics of Student Populations in Schools and Universities of the United States
Allen Mednick, Ph.D. Candidate

Tutorial NN (*Field: Business–Psychology*)
Business Referrals based on Customer Service Records
Ronald Mellado Miller, Ph.D.

Tutorial OO (*Field: Medicine–Bioinformatics*)
Autism: Rating From Parents and Teachers Using an Assessment of Autism Measure and
Also Linkage with Genotype: Three Different Polymorphisms Affecting Serotonin
Functions Versus Behavioral Data
Ira L. Cohen, Ph.D.

Tutorial PP (*Field: Ecology*)
Human Ecology: Clustering Fiber Length Histograms by Degree of Damage
Mourad Krifa, Ph.D.

Tutorial QQ (*Field: Finance–Business*)
Wall Street: Stock Market Predictions
Gary Miner, Ph.D.

Tutorial RR (*Field: Scorecards–Business Financial*)
Developing Credit Scorecards Using SAS® Enterprise Miner™
R. Wayne Thompson, SAS, Carey, North Carolina

Tutorial SS (*Field: Astronomy*)
Astro-Statistics: Data Mining of SkyLab Project: Hubble Telescope Star and
Galaxy Data
Joseph M. Hilbe, JD, Ph.D., Gary Miner, Ph.D., and Robert Nisbet, Ph.D.

Tutorial TT (*Field: Customer Response–Business Commercial*)
Boost Profitability and Reduce Marketing Campaign Costs with PASW Modeler by
SPSS Inc.
David Vergara

HISTORY OF PHASES OF DATA ANALYSIS, BASIC THEORY, AND THE DATA MINING PROCESS

Part I focuses on the historical and theoretical background for statistical analysis and data mining, and integrates it with the data discovery and data preparation operations necessary to prepare for modeling. Part II presents some basic algorithms and applications areas where data mining technology is commonly used. Part III is not a set of chapters, but is rather a group of tutorials you can follow to learn data mining by example. In fact, you don't even have to read the chapters in the other parts at first. You can start with a tutorial in an area of your choice (if you have the tool used in that tutorial) and learn how to create a model successfully in that area. Later, you can return to the text to learn *why* the various steps were included in the tutorial and understand what happened behind the scenes when you performed them. The third group of chapters in Part IV leads you into some advanced data mining areas, where you will learn how create a "good-enough" model and avoid the most common (and sometimes devastating) mistakes of data mining practice.

1

The Background for Data Mining Practice

PREAMBLE

You must be interested in learning how to practice data mining; otherwise, you would not be reading this book. We know that there are many books available that will give a good introduction to the process of data mining. Most books on data mining focus on the features and functions of various data mining tools or algorithms. Some books do focus on the challenges of performing data mining tasks. This book is designed to give you an introduction to the *practice* of data mining in the real world of business.

One of the first things considered in building a business data mining capability in a company is the selection of the data mining tool. It is difficult to penetrate the hype erected around the description of these tools by the vendors. The fact is that even the most mediocre of data mining tools can create models that are at least 90% as good as the best tools. A 90% solution performed with a relatively cheap tool might be more cost effective in your organization than a more expensive tool. How do you choose your data mining tool?

Few reviews are available. The best listing of tools by popularity is maintained and updated yearly by KDNuggets.com. Some detailed reviews available in the literature go beyond just a discussion of the features and functions of the tools (see Nisbet, 2006, Parts 1–3). The interest in an unbiased and detailed comparison is great. We are told the "most downloaded document in data mining" is the comprehensive but decade-old tool review by Elder and Abbott (1998).

The other considerations in building a business's data mining capability are forming the data mining team, building the data mining platform, and forming a foundation of good data mining practice. This book will not discuss the building of the data mining platform. This subject is discussed in many other books, some in great detail. A good overview of how to build a data mining platform is presented in *Data Mining: Concepts and Techniques* (Han and Kamber, 2006). The primary focus of this book is to present a practical approach to building cost-effective data mining models aimed at increasing company profitability, using tutorials and demo versions of common data mining tools.

Just as important as these considerations in practice is the background against which they must be performed. We must not imagine that the background doesn't matter . . . it *does* matter, whether or not we recognize it initially. The reason it matters is that the capabilities of statistical and data mining methodology were not developed in a vacuum. Analytical methodology was developed in the context of prevailing statistical and analytical theory. But the major driver in this development was a very pressing need to provide a simple and repeatable analysis methodology in medical science. From this beginning developed modern statistical analysis and data mining. To understand the strengths and limitations of this body of methodology and use it effectively, we must understand the strengths and limitations of the statistical theory from which they developed. This theory was developed by scientists and mathematicians who "thought" it out. But this thinking was not one sided or unidirectional; there arose several views on how to solve analytical problems. To understand how to approach the solving of an analytical problem, we must understand the different ways different people tend to think. This history of statistical theory behind the development of various statistical techniques bears strongly on the ability of the technique to serve the tasks of a data mining project.

A SHORT HISTORY OF STATISTICS
AND DATA MINING

Analysis of patterns in data is not new. The concepts of average and grouping can be dated back to the 6th century BC in Ancient China, following the invention of the bamboo rod abacus (Goodman, 1968). In Ancient China and Greece, statistics were gathered to help heads of state govern their countries in fiscal and military matters. (This makes you wonder if the words *statistic* and *state* might have sprung from the same root.) In the sixteenth and seventeenth centuries, games of chance were popular among the wealthy, prompting many questions about probability to be addressed to famous mathematicians (Fermat, Leibnitz, etc.). These questions led to much research in mathematics and statistics during the ensuing years.

MODERN STATISTICS: A DUALITY?

Two branches of statistical analysis developed in the eighteenth century: Bayesian and classical statistics. (See Figure 1.1.) To treat both fairly in the context of history, we will consider both in the First Generation of statistical analysis. For the Bayesians, the probability of an event's occurrence is equal to the probability of its past occurrence times the likelihood of its occurrence in the future. Analysis proceeds based on the concept of conditional probability: the probability of an event occurring given that another event has already occurred. Bayesian analysis begins with the quantification of the investigator's existing state of knowledge, beliefs, and assumptions. These *subjective priors* are combined with observed data quantified probabilistically through an objective function of some sort. The classical statistical approach (that flowed out of mathematical works of Gauss and Laplace) considered that the joint probability, rather than the conditional probability, was the appropriate basis for analysis. The joint probability function expresses the probability that simultaneously X takes the specific values x and Y takes value y, as a function of x and y.

Interest in probability picked up early among biologists following Mendel in the latter part of the nineteenth century. Sir Francis Galton, founder of the School of Eugenics in England, and his successor, Karl Pearson, developed the concepts of regression and correlation for analyzing genetic data. Later, Pearson and colleagues extended their work to the social sciences. Following Pearson, Sir R. A. Fisher in England developed his system for inference testing in medical studies based on his concept of standard deviation. While the development of probability theory flowed out of the work of Galton and Pearson, early predictive methods followed Bayes's approach. Bayesian approaches to inference testing could lead to widely different conclusions by different medical investigators because they used different sets of subjective priors. Fisher's goal in developing his system of statistical inference was to provide medical investigators with a common set of tools for use in comparison studies of effects of different treatments by different investigators. But to make his system work even with large samples, Fisher had to make a number of assumptions to define his "Parametric Model."

FIGURE 1.1   Rev. Thomas Bayes (1702–1761).

Assumptions of the Parametric Model

1. Data Fits a Known Distribution (e.g., Normal, Logistic, Poisson, etc.)

Fisher's early work was based on calculation of the parameter *standard deviation*, which assumes that data are distributed in a *normal* distribution. The normal distribution is bell-shaped, with the mean (average) at the top of the bell, with "tails" falling off evenly at the sides. Standard deviation is simply the "average" of the absolute deviation of a value from the mean. In this calculation, however, averaging is accomplished by dividing the sum of the absolute deviations by the total – 1. This subtraction expresses (to some extent) the increased uncertainty of the result due to grouping (summing the absolute deviations). Subsequent developments used modified parameters based on the logistic or Poisson distributions. The assumption of a particular known distribution is necessary in order to draw upon the characteristics of the distribution function for making inferences. All of these *parametric* methods run the gauntlet of dangers related to force-fitting data from the real world into a mathematical construct that does not fit.

2. Factor Independency

In parametric predictive systems, the variable to be predicted (Y) is considered as a function of predictor variables (X's) that are assumed to have independent effects on Y. That is, the effect on Y of each X-variable is not dependent on effects on Y of any other X-variable. This situation could be created in the laboratory by allowing only one factor (e.g., a treatment) to vary, while keeping all other factors constant (e.g., temperature, moisture, light, etc.). But, in the real world, such laboratory control is absent. As a result, some factors that do affect other factors are permitted to have a joint effect on Y. This problem is called *collinearity*. When it occurs between more than two factors, it is termed *multicollinearity*. The multicollinearity problem led statisticians to use an interaction term in the relationship that supposedly represented the combined effects. Use of this interaction term functioned as a magnificent kluge, and the reality of its effects was seldom analyzed. Later development included a number of interaction terms, one for each interaction the investigator might be presenting.

3. Linear Additivity

Not only must the X-variables be independent, their effects on Y must be cumulative and linear. That means the effect of each factor is added to or subtracted from the combined effects of all X-variables on Y. But what if the relationship between Y and the predictors (X-variables) is not additive, but multiplicative or divisive? Such functions can be expressed only by exponential equations that usually generate very nonlinear relationships. Assumption of linear additivity for these relationships may cause large errors in the predicted outputs. This is often the case with their use in business data systems.

4. Constant Variance (Homoscedasticity)

The variance throughout the range of each variable is assumed to be constant. This means that if you divided the range of a variable into bins, the variance across all records for bin #1 is the same as the range for all the other bins in the range of that variable. If the variance throughout the range of a variable differs significantly from constancy, it is said to be *heteroscedastic*. The error in the predicted value caused by the combined heteroscedasticity among all variables can be quite significant.

5. *Variables Must Be Numerical and Continuous*

The assumption that variables must be numerical and continuous means that data must be numeric (or it must be transformable to a number before analysis) and the number must be part of a distribution that is inherently *continuous*. Integer values in a string are not continuous; they are *discrete*. Classical parametric statistical methods are not valid for use with discrete data, because the probability distributions for continuous and discrete data are different. But both scientists and business analysts have used them anyway.

In his landmark paper, Fisher (1921; see Figure 1.2) began with the broad definition of probability as the intrinsic probability of an event's occurrence divided by the probability of occurrence of all competing events (very Bayesian). By the end of his paper, Fisher modified his definition of probability for use in medical analysis (the goal of his research) as the intrinsic probability of an event's occurrence *period*. He named this quantity *likelihood*. From that foundation, he developed the concepts of standard deviation based on the normal distribution. Those who followed Fisher began to refer to *likelihood* as *probability*. The concept of likelihood approaches the classical concept of probability only as the sample size becomes very large and the effects of subjective priors approach zero (von Mises, 1957). In practice, these two conditions may be satisfied sufficiently if the initial distribution of the data is known and the sample size is relatively large (following the Law of Large Numbers).

Why did this duality of thought arise in the development of statistics? Perhaps it is because of the broader duality that pervades all of human thinking. This duality can be traced all the way back to the ancient debate between Plato and Aristotle.

FIGURE 1.2　Sir Ronald Fisher.

TWO VIEWS OF REALITY

Whenever we consider solving a problem or answering a question, we start by conceptualizing it. That means we do one of two things: (1) try to reduce it to key elements or (2) try to conceive of it in general terms. We call people who take each of these approaches "detail people" and "big picture people," respectively. What we don't consider is that this distinction has its roots deep in Greek philosophy in the works of Aristotle and Plato.

Aristotle

Aristotle (Figure 1.3) believed that the true being of things (reality) could be discerned only by what the eye could see, the hand could touch, etc. He believed that the highest level of intellectual activity was the detailed study of the tangible world around us. Only in that way could we understand reality. Based on this approach to truth, Aristotle was led to believe that you could break down a complex system into pieces, describe the pieces in detail, put the pieces together and understand the whole. For Aristotle, the "whole" was equal to the sum of its parts. This nature of the whole was viewed by Aristotle in a manner that was very *machine-like*.

Science gravitated toward Aristotle very early. The nature of the world around us was studied by looking very closely at the physical elements and biological units (species) that composed it. As our understanding of the natural world matured into the concept of the ecosystem, it was discovered that many characteristics of ecosystems could not be explained by traditional (Aristotelian) approaches. For example, in the science of forestry, we discovered that when a tropical rain forest is cut down on the periphery of its range, it may take a very long time to regenerate (if it does at all). We learned that the reason for this is that in areas of relative stress (e.g., peripheral areas), the primary characteristics necessary for the survival and growth of tropical trees are maintained by the forest itself! High rainfall leaches nutrients down beyond the reach of the tree roots, so almost all of the nutrients for tree growth must come from recently fallen leaves and branches.

FIGURE 1.3   Aristotle before the bust of Homer.

When you cut down the forest, you remove that source of nutrients. The forest canopy also maintains favorable conditions of light, moisture, and temperature required by the trees. Removing the forest removes the very factors necessary for it to exist at all in that location. These factors *emerge* only when the system is whole and functioning. Many complex systems are like that, even business systems. In fact, these emergent properties may be the major drivers of system stability and predictability.

To understand the failure of Aristotelian philosophy for completely defining the world, we must return to Ancient Greece and consider Aristotle's rival, Plato.

Plato

Plato (Figure 1.4) was Aristotle's teacher for 20 years, and they both agreed to disagree on the nature of being. While Aristotle focused on describing tangible things in the world by detailed studies, Plato focused on the world of ideas that lay behind these tangibles. For Plato, the only thing that had lasting being was an idea. He believed that the most important things in human existence were beyond what the eye could see and the hand could touch. Plato believed that the influence of ideas transcended the world of tangible things that commanded so much of Aristotle's interest. For Plato, the "whole" of reality was *greater* than the sum of its tangible parts.

The concept of the nature of being was developed initially in Western thinking upon a Platonic foundation. Platonism ruled philosophy for over 2,000 years—up to the Enlightenment. Then the tide of Western thinking turned toward Aristotle. This division of thought on the nature of reality is reflected in many of our attempts to define the nature of reality in the world, sometimes unconsciously so. We speak of the difference between "big picture people" and "detail people"; we contrast "top-down" approaches to organization versus "bottom-up" approaches; and we compare "left-brained" people with "right-brained" people. These dichotomies of perception are little more than a rehash of the ancient debate between Plato and Aristotle.

FIGURE 1.4   Plato.

THE RISE OF MODERN STATISTICAL ANALYSIS:
THE SECOND GENERATION

In the 1980s, it became obvious to statistical mathematicians that the rigorously Aristotelian approach of the past was too restrictive for analyzing highly nonlinear relationships in large data sets in complex systems of the real world. Mathematical research continued dominantly along Fisherian statistical lines by developing nonlinear versions of parametric methods. Multiple curvilinear regression was one of the earliest approaches for accounting for nonlinearity in continuous data distributions. But many nonlinear problems involved discrete rather than continuous distributions (see Agresti, 1996). These methods included the following:

- **Logit Model (including Logistic Regression):** Data is assumed to follow a logistic distribution and the dependent variable is categorical (e.g., 1:0). In this method, the dependent variable (Y) is defined as an exponential function of the predictor variables (X's). As such, this relationship can account for nonlinearities in the response of the X-variables to the Y-variable, but not in the interaction between X-variables.
- **Probit Model (including Poisson Regression):** Like the Logit Model, except data are assumed to follow a Poisson rather than a Logistic distribution.
- **The Generalized Linear Model (GLM):** The GLM expands the general estimation equation used in prediction, $Y = f\{X\}$, f is some function and X is a vector of predictor variables. The left side of the equal sign was named the *deterministic component*, the right side of the equation the *random component*, and the equal sign one of many possible *link functions.* Statisticians recognized that the deterministic component could be expressed as an exponential function (like the Logistic function), the random component accumulated effects of the X-variables and was still linear, and the link function could be any logical operator (equal to, greater than, less than, etc.). The equal sign was named the *identity link.* Now mathematicians had a framework for defining a function that could fit data sets with much more nonlinearity. But it would be left to the development of neural networks (see following text) to express functions with any degree of nonlinearity.

While these developments were happening in the Fisherian world, a stubborn group of Bayesians continued to push their approach. To the Bayesians, the practical significance (related to what happened in the past) is more significant than the statistical significance calculated from joint probability functions. For example, the practical need to correctly diagnose cancerous tumors (true-positives) is more important than the error of misdiagnosing a tumor as cancerous when it is not (false-positives). To this extent, their focus was rather Platonic, relating correct diagnosis to the data environment from which any particular sample was drawn, rather than just to data of the sample alone. To serve this practical need, they had to ignore the fact that you can consider only the probability of events that actually happened in the past data environment, not the probability of events that *could* have happened but did not (Lee, 1989).

In Fisherian statistics, the *observation* and the corresponding *alpha error* determine whether it is different from what is expected (Newton and Rudestam, 1999). The alpha error is the probability of being wrong when you think you are right, while the beta error is the probability of being right when you think you are wrong. Fisherians set the alpha error in the beginning of the analysis and referred to significant differences between data populations in terms of the alpha error that was specified. Fisherians would add a suffix to their prediction, such as "... at the 95% Confidence Level." The Confidence Level (95% in this case) is the complement of the alpha error (0.05%). It means that the investigator is willing be wrong 5% of the time. Fisherians use the beta error to calculate the "power" or "robustness" of an analytical test. Bayesians feel free to twiddle with both the alpha and beta errors and contend that you cannot arrive at a true decision without considering the alternatives carefully. They maintain that a calculated probability level of 0.023 for a given event in the sample data does not imply that the probability of the event within the entire universe of events is 0.023.

Which approach is right, Fisherian or Bayesian? The answer depends on the nature of the study, the possibility of considering priors, the relative cost of false-positive errors and false-negative errors. Before selecting one, we must bear in mind that all statistical tests have advantages and disadvantages. We must be informed about the strengths and weaknesses of both approaches and have a clear understanding of the meaning of the results produced by either one. Regardless of its problems and its "bad press" among the Fisherians, Bayesian statistics eventually did find its niche in the developing field of data mining in business in the form of Bayesian Belief Networks and Naïve Bayes Classifiers. In business, success in practical applications depends to a great degree upon analysis of all *viable* alternatives. Nonviable alternatives aren't worth considering. One of the tutorials on the enclosed DVD uses a Naïve Bayes Classifier algorithm.

Data, Data Everywhere ...

The crushing practical needs of business to extract knowledge from data that could be leveraged immediately to increase revenues required new analytical techniques that enabled analysis of highly nonlinear relationships in very large data sets with an unknown distribution. Development of new techniques followed *three* paths rather than the two classical paths described previously. The third path (machine learning) might be viewed as a blend of the Aristotelian and Platonic approach to truth, but it was not Bayesian.

MACHINE LEARNING METHODS:
THE THIRD GENERATION

The line of thinking known as machine learning arose out of the Artificial Intelligence community in the quest for the intelligent machine. Initially, these methods followed two parallel pathways of developments: artificial neural networks and decision trees.

Artificial Neural Networks. The first pathway sought to express a nonlinear function directly (the "cause") by means of assigning weights to the input variables, accumulate their effects, and "react" to produce an output value (the "effect") following some sort of decision function. These systems (artificial neural networks) represented simple analogs of the way the human brain works by passing neural impulses from neuron to neuron across synapses. The "resistance" in transmission of an impulse between two neurons in the human brain is variable. The complex relationship of neurons and their associated synaptic connections is "trainable" and could "learn" to respond faster as required by the brain. Computer scientists began to express this sort of system in very crude terms in the form of an artificial neural network that could be used to learn how to recognize complex patterns in the input variables of a data set.

Decision Trees. The second pathway of development was concerned with expressing the *effects* directly by developing methods to find "rules" that could be evaluated for separating the input values into one of several "bins" without having to express the functional relationship directly. These methods focused on expressing the rules explicitly (rule induction) or on expressing the relationship among the rules (decision tree) that expressed the results. These methods avoided the strictures of the Parametric Model and were well suited for analysis of nonlinear events (NLEs), both in terms of combined effects of the X-variables with the Y-variable and interactions between the independent variables. While decision trees and neural networks could express NLEs more completely than parametric statistical methods, they were still intrinsically linear in their aggregation functions.

STATISTICAL LEARNING THEORY: THE FOURTH GENERATION

Logistic regression techniques can account for the *combined* effects of interaction among all predictor variables by virtue of the nonlinear function that defines the dependent variable (Y). Yet, there are still significant limitations to these linear learning machines (see Minsky and Papert, 1969). Even neural nets and decision trees suffered from this problem, to some extent. One way of expressing these limitations is to view them according to their "hypothesis space." The hypothesis space is a mathematical construct within which a solution is sought. But this space of possible solutions may be highly constrained by the linear functions in classical statistical analysis and machine learning techniques. Complex problems in the real world may require much more expressive hypothesis spaces than can be provided by linear functions (Cristianini and Shawe-Taylor, 2000). Multilayer neural nets can account for much more of the nonlinear effects by virtue of the network architecture and error minimization techniques (e.g., back-propagation).

An alternative approach is to arrange data points into vectors (like rows in a customer record). Such vectors are composed of elements (one for each attribute in the customer record). The vector space of all rows of customers in a database can be characterized conceptually and mathematically as a space with the N-dimensions, where N is the number of customer attributes (predictive variables). When you view data in a customer record as a vector, you can take advantage of linear algebra concepts, one of which is that you can express all of the differences between the attributes of two customer records by calculating the *dot product* (or the inner product). The dot product of two vectors is the sum of all the products between corresponding attributes of the two vectors. Consequently, we can express our data as a series of dot products composed into an inner product space with N dimensions. Conversion of our data into inner products is referred to as "mapping" the data to inner product space.

Even classical statistical algorithms (like linear regression) can be expressed in this way. In Statistical Learning Theory, various complex functions, or "kernels," replace the inner product. When you map data into these complex kernel spaces, the range of possible solutions to your problem increases significantly. The data in these spaces are referred to as "features" rather than as attributes that characterized the original data.

A number of new learning techniques have taken advantage of the properties of kernel learning machines. The most common implementation is a Support Vector Machine. When a neural net is "trained," rows of customer data are fed into the net, and errors between predicted and observed values are calculated (an example of supervised learning). The learning function of the training and the error minimization function (that defines the best approximate solution) are closely intertwined in neural nets. This is not the case with support vector machines. Because the learning process is separated from the approximation process, you can experiment by using different kernel definitions with different learning theories. Therefore, instead of choosing from among different architectures for a neural net application, you can experiment with different kernels in a support vector machine implementation.

Several commercial packages include algorithms based on Statistical Learning Theory, notably *STATISTICA* Data Miner and KXEN (Knowledge Extraction Engine). In the future, we will see more of these powerful algorithms in commercial packages. Eventually, data mining methods may become organized around the steps that enable these algorithms to work most efficiently. For example, the KXEN tool incorporates a smart data recoder (to standardize inputs) and a smart variable derivation routine that uses variable ratios and recombination to produce powerful new predictors.

Is the Fourth Generation of statistical methods the last? Probably it is not. As we accumulate more and more data, we will probably discover increasingly clever ways to simulate more closely the operation of the most complex learning machine in the universe—the human brain.

POSTSCRIPT

New strategies are being exploited now to spread the computing efforts among multiple computers connected like the many neurons in a brain:

- **Grid Computing:** Utilizing a group of networked computers to divide and conquer computing problems.
- **"Cloud" Computing:** Using the Internet to distribute data and computing tasks to many computers anywhere in the world, but without a centralized hardware infrastructure of grid computing.

These strategies for harnessing multiple computers for analysis provide a rich new milieu for data mining. This approach to analysis with multiple computers is the next logical step in the development of artificial "brains." This step might develop into the Fifth Generation of data mining.

References

Agresti, A. (1996). *An Introduction to Categorical Data Analysis*. New York, NY: John Wiley & Sons.

Cristianini, N., & Shawe-Taylor, J. (2000). *An Introduction to Support Vector Machines*. Cambridge, UK: Cambridge University Press.

Elder, J., & Abbott, D. (1998). *A Comparison of Leading Data Mining Tools*. 4th Annual Conference on Knowledge Discovery and Data Mining, New York, NY, August 28, 1998. http://www.datamininglab.com/Resources/ TOOLCOMPARISONS/tabid/58/Default.aspx

Fisher, R. A. (1921). *On the Mathematical Foundations of Theoretical Statistics*. London: Philos. Trans. Royal Soc, A 222.

Goodman, A. F. (1968). The interface of computer science and statistica: An historical perspective. *The American Statistician, 22*, 17–20.

Han, J., & Kamber, M. (2006). *Data Mining: Concepts and Techniques*. New York: Morgan Kaufmann.

Lee, P. M. (1989). *Bayesian Sensitivity: An Introduction*. NewYork: Oxford University Press.

Minsky, M., & Papert, S. (1969). *Perceptrons: An Introduction to Computational Geometry*. Cambridge, MA: MIT Press (3rd edition published in 1988).

Newton, R. R., & Rudestam, K. E. (1999). *Your Statistical Consultant*. Thousand Oaks, CA: Sage Publishing.

Nisbet, R. A. (2006). Data mining tools: Which one is best for CRM? Part 1. *DM-Review Special Report*, January 23, 2006. http://www.dmreview.com/specialreports/20060124/1046025-1.html

Nisbet, R. A. (2006). Data mining tools: Which one is best for CRM? Part 2. *DM-Review Special Report*, February, 2006. http://www.dmreview.com/specialreports/20060207/1046597-1.html

Nisbet, R.A. (2006). Data mining tools: Which one is best for CRM? Part 3. *DM-Review Special Report*. http://www .dmreview.com/specialreports/20060321/1049954-1.html

Von Mises, R. (1957). *Probability, Statistics, and Truth*. New York: Dover Publications.

Theoretical Considerations for Data Mining

PREAMBLE

In Chapter 1, we explored the historical background of statistical analysis and data mining. Statistical analysis is a relatively old discipline (particularly if you consider its origins in China). But data mining is a relatively new field, which developed during the 1990s and coalesced into a field of its own during the early years of the twenty-first century. It represents a confluence of several well-established fields of interest:

- Traditional statistical analysis
- Artificial intelligence

- Machine learning
- Development of large databases

Traditional statistical analysis follows the *deductive method* in the search for relationships in data sets. Artificial intelligence (e.g., expert systems) and machine learning techniques (e.g., neural nets and decision trees) follow the *inductive method* to find faint patterns of relationship in data sets. Deduction (or deductive reasoning) is the Aristotelian process of analyzing detailed data, calculating a number of metrics, and forming some conclusions based (or deduced) solely on the mathematics of those metrics. Induction is the more Platonic process of using information in a data set as a "springboard" to make general conclusions, which are not wholly contained directly in the input data. The scientific method follows the inductive approach but has strong Aristotelian elements in the preliminary steps.

THE SCIENTIFIC METHOD

The scientific method is as follows:

1. Define the problem.
2. Gather existing information about a phenomenon.
3. Form one or more hypotheses.
4. Collect new experimental data.
5. Analyze the information in the new data set.
6. Interpret results.
7. Synthesize conclusions, based on the old data, new data, and *intuition.*
8. Form new hypotheses for further testing.
9. Do it again (iteration).

Steps 1–5 involve deduction, and steps 6–9 involve induction. Even though the scientific method is based strongly on deductive reasoning, the final products arise through inductive reasoning. Data mining is a lot like that. In fact, machine learning algorithms used in data mining are designed to mimic the process that occurs in the mind of the scientist. Data mining uses mathematics, but the results are *not* mathematically determined. This statement may sound somewhat contradictory until you view it in terms of the human brain. You can describe many of the processes in the human conceptual pathway with various mathematical relationships, but the *result* of being human goes far beyond the mathematical descriptions of these processes. Women's intuition, mother's wisdom regarding their offspring, and "gut" level feelings about who should win the next election are all *intuitive models* of reality created by the human brain. They are based largely on empirical data, but the mind extrapolates beyond the data to form the conclusions following a purely inductive reasoning process.

WHAT IS DATA MINING?

Data mining can be defined in several ways, which differ primarily in their focus on different aspects of data mining. One of the earliest definitions is

The *non-trivial* extraction of *implicit,* previously unknown, and potentially useful information from data (Frawley et al., 1991).

As data mining developed as a professional activity, it was necessary to distinguish it from the previous activity of statistical modeling and the broader activity of knowledge discovery. For the purposes of this handbook, we will use the following working definitions:

- **Statistical modeling:** The use of parametric statistical algorithms to group or predict an outcome or event, based on predictor variables.
- **Data mining:** The use of machine learning algorithms to find faint patterns of relationship between data elements in large, noisy, and messy data sets, which can lead to actions to increase benefit in some form (diagnosis, profit, detection, etc.).
- **Knowledge discovery:** The entire process of data access, data exploration, data preparation, modeling, model deployment, and model monitoring. This broad process includes data mining activities, as shown in Figure 2.1.

As the practice of data mining developed further, the focus of the definitions shifted to specific aspects of the information and its sources. In 1996, Fayyad et al. proposed the following:

Knowledge discovery in databases is the non-trivial process of identifying valid, novel, potential useful, and ultimately understandable patterns in data.

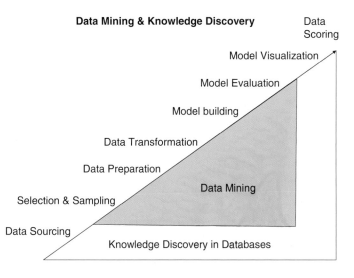

FIGURE 2.1 The relationship between data mining and knowledge discovery.

The second definition focuses on the *patterns* in the data rather than just information in a generic sense. These patterns are faint and hard to distinguish, and they can only be *sensed* by analysis algorithms that can evaluate nonlinear relationships between predictor variables and their targets and themselves. This form of the definition of data mining developed along with the rise of machine learning tools for use in data mining. Tools like decision trees and neural nets permit the analysis of nonlinear patterns in data easier than is possible in parametric statistical algorithms. The reason is that machine learning algorithms learn the way humans do—by example, not by calculation of metrics based on averages and data distributions.

The definition of data mining was confined originally to just the process of model building. But as the practice matured, data mining tool packages (e.g., SPSS-Clementine) included other necessary tools to facilitate the building of models and for evaluating and displaying models. Soon, the definition of data mining expanded to include those operations in Figure 2.1 (and some include model visualization also).

The modern Knowledge Discovery in Databases (KDD) process combines the mathematics used to discover interesting patterns in data with the entire process of extracting data and using resulting models to apply to other data sets to leverage the information for some purpose. This process blends business systems engineering, elegant statistical methods, and industrial-strength computing power to find *structure* (connections, patterns, associations, and basis functions) rather than statistical parameters (means, weights, thresholds, knots). In Chapter 3, we will expand this rather linear organization of data mining processes to describe the iterative, closed-loop system with feedbacks that comprise the modern approach to the practice of data mining.

A THEORETICAL FRAMEWORK FOR THE DATA MINING PROCESS

The evolutionary nature of the definition and focus of data mining occurred primarily as a matter of experience and necessity. A major problem with this development was the lack of a consistent body of theory, which could encompass all aspects of what information is, where it comes from, and how is it used. This logical concept is sometimes called a *model-theoretic*. Model theory links logic with algebraic expressions of structure to describe a system or complex process with a body of terms with a consistent syntax and relationships between them (semantics). Most expressions of data mining activities include inconsistent terms (e.g., attribute and predictor), which may imply different logical semantic relations with the data elements employed.

Mannila (2000) summarized a number of criteria that should be satisfied in an approach to develop a model-theoretic for data mining. These criteria include the ability to

- Model typical data mining tasks (clustering, rule discovery, classification)
- Describe data and the inductive generalizations derived from the data
- Express information from a variety of forms of data (relational data, sequences, text, Web)
- Support interactive and iterative processes

- Express comprehensible relationships
- Incorporate users in the process
- Incorporate multiple criteria for defining what is an "interesting" discovery

Mannila describes a number of approaches to developing an acceptable model-theoretic but concludes that none of them satisfy all the above criteria. The closest we can come is to combine the microeconomic approach with the inductive database approach.

Microeconomic Approach

The starting point of the microeconomic approach is that data mining is concerned with finding actionable *patterns* in data that have some *utility* to form a decision aimed at getting something done (e.g., employ interdiction strategies to reduce attrition). The goal is to find the decision that maximizes the total utility across all customers.

Inductive Database Approach

An inductive database includes all the data available in a given structure *plus* all the questions (queries) that could be asked about patterns in the data. Both stored and derived facts are handled in the same way. One of the most important functions of the human brain is to serve as a pattern recognition engine. Detailed data are submerged in the unconscious memory, and actions are driven primarily by the stored patterns.

Manilla suggests that the microeconomic approach can express most of the requirements for a model-theoretic based on stored facts, but the inductive database approach is much more facile to express derived facts. One attempt to implement this was taken in the development of the Predictive Modeling Markup Language (PMML) as a superset of the standard Extended Markup Language (XML). Most data mining packages available today store internal information (e.g., arrays) in XML format and can output results (analytical models) in the form of PMML. This combination of XML and PMML permits expression of the same data elements and the data mining process either in a physical database environment or a Web environment. When you choose your data mining tool, look for these capabilities.

STRENGTHS OF THE DATA MINING PROCESS

Traditional statistical studies use past information to *determine* a future state of a system (often called prediction), whereas data mining studies use past information to construct patterns based not solely on the input data, but also on the *logical consequences* of those data. This process is also called *prediction*, but it contains a vital element missing in statistical analysis: the ability to provide an orderly expression of *what might be* in the future, compared to *what was* in the past (based on the assumptions of the statistical method).

Compared to traditional statistical studies, which are often hindsight, the field of data mining finds patterns and classifications that look toward and even predict the future. In summary, data mining can (1) provide a more complete understanding of data by finding patterns previously not seen and (2) make models that predict, thus enabling people to make better decisions, take action, and therefore mold future events.

CUSTOMER-CENTRIC VERSUS ACCOUNT-CENTRIC: A NEW WAY TO LOOK AT YOUR DATA

Most computer databases in business were designed for the efficient storage and retrieval of account or product information. Business operations were controlled by accounting systems; it was natural that the application of computers to business followed the same data structures. The focus of these data structures was on transactions, and multiple transactions were stored for a given account. Data in early transactional business systems were held in Indexed Sequential Access Method (ISAM) databases. But as data volumes increased and the need for flexibility increased, Relational Database Management Systems (RDBMS) were developed. Relational theory developed by C. J. Codd distributed data into tables linked by primary and foreign keys, which progressively reduced data redundancy (like customer names) in (eventually) six "normal forms" of data organization. Some of the very large relational systems using NCR Teradata technology extend into the hundreds of terabytes. These systems provide relatively efficient systems for storage and retrieval of account-centric information.

Account-centric systems were quite efficient for their intended purpose, but they have a major drawback: it is difficult to manage *customers* per se as the primary responders, rather than accounts. One person could have one account or multiple accounts. One account could be owned by more than one person. As a result, it was very difficult for a company on an RDBMS to relate its business to specific customers. Also, accounts (per se) don't buy products or services; products don't buy themselves. *People* buy products and services, and our businesses operations (and the databases that serve them) should be oriented around the customer, not around accounts.

When we store data in a customer-centric format, extracts to build the Customer Analytical Record (CAR) are much easier to create (see later section for more details on the CAR). And customer-centric databases are much easier to update *in relation to the customer*.

The Physical Data Mart

One solution to this problem is to organize data structures to hold specific aspects (dimensions) of customer information. These structures can be represented by tables with common keys to link them together. This approach was championed by Oracle to hold customer-related information apart from the transactional data associated with them. The basic architecture was organized around a central (fact) table, which stored general information about a customer. This fact table formed the hub of a structure like a wheel (Figure 2.2). This structure became known as the *star-schema*.

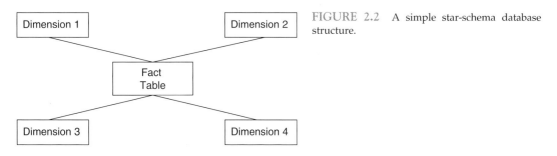

FIGURE 2.2   A simple star-schema database structure.

Another name for a star-schema is a multidimensional database. In an online store, the dimensions can hold data elements for Products, Orders, Back-orders, etc. The transactional data are often stored in another very different data structure. The customer database system is refreshed daily with summaries and aggregations from the transactional system. This smaller database is "dependent" on the larger database to create the summary and aggregated data stored in it. When the larger database is a data warehouse, the smaller dependent database is referred to as a dependent data mart. In Chapter 3, we will see how a system of dependent data marts can be organized around a relational data warehouse to form the Corporate Information Factory.

The Virtual Data Mart

As computing power and disk storage capacity increased, it became obvious in the early 1990s that a business could appeal to customers directly by using characteristics and historical account information, and Customer Relationship Management (CRM) was born. One-to-one marketing appeals could be supported, and businesses became "smarter" in their ability to convince customers to buy more goods and services. This success of CRM operations changed the way some companies looked at their data. No longer must companies view their databases in terms of just accounts and products, but rather they could view their customers directly, in terms of all accounts, products and demographic data associated with each customer. These "logical" data marts could even be implemented as "views" in an RDBMS.

Householded Databases

Another way to gain customer-related insights is to associate all accounts to the customers who own them and to associate all individual members of the same household. This process is called *householding*. The householding process requires some fuzzy matching to aggregate all accounts to the same customer. The reason is that the customer names may not be spelled exactly the same way in all records. An analogous situation occurs when trying to gather all individuals into the same household, because not all addresses are listed in exactly the same format. This process of fuzzy matching can be performed by a number of data integration and data quality tools available in the market today (DataFlux, Trillium, Informatica Data Quality, IBM Quality Stage).

The householded data structure could consist of the following tables:

- Accounts
- Individuals
- Households

Historical data could be combined with each of the preceding hierarchical levels of aggregation. Alternatively, the preceding tables could be restricted to current data, and historical data could be installed in historical versions of the same tables (e.g., Accounts_Hst), linked together with common keys. This compound structure would optimize speed of database queries and simplify data extraction for most applications requiring only current data. Also, the historical data would be available for trending in the historical tables.

THE DATA PARADIGM SHIFT

The organization of data structures suitable for data mining requires a basic shift in thinking about data in business. Data do not serve the account; data should be organized to serve the customer who buys goods and services. To directly serve customers, data must be organized in a customer-centric data structure to permit the following:

- Relationship of all data elements must be relevant to the customer.
- Data structures must make it relatively easy to convert all required data elements into a form suitable for data mining: the Customer Analytical Record (CAR).

CREATION OF THE CAR

All input data must be loaded into the CAR (Accenture Global Services, 2006). This process is similar to preparing for a vacation by automobile. If your camping equipment is stored in one place in your basement, you can easily access it and load it into the automobile. If it is spread throughout your house and mixed in with noncamping equipment, access will be more difficult because you have to separate (extract) it from among other items. Gathering data for data mining is a lot like that. If your source data is a data warehouse, this process will *denormalize* your data. Denormalization is the process of extracting data from normalized tables in the relational model of a data warehouse. Data from these tables must be associated with the proper individuals (or households) along the way. Data integration tools (like SAS DataFlux or Informatica) are required to extract and transform data from the relational database tables to build the CAR. See any one of a number of good books on relational data warehousing to understand what this process entails. If your data are already in a dimensional or householding data structure, you are already halfway there. The CAR includes the following:

1. All data elements are organized into one record per customer.
2. One or more "target" (Y) variables are assigned or derived.

The CAR is expressed as a textual version of

$$\text{An equation: } Y = X_1 + X_2 + X_3 + \ldots X_n$$

This expression represents a computerized "memory" of the information about a customer. These data constructs are analyzed by either statistical or machine learning "algorithms," following specific methodological operations. Algorithms are mathematical expressions that describe relationships between the variable predicted (Y or the customer response) and the predictor variables ($X_1 + X_2 + X_3 + \ldots X_n$). Basic and advanced data mining algorithms are discussed in Chapters 7 and 8.

The CAR is analyzed by parametric statistical or machine learning algorithms, within the broader process of Knowledge Discovery in Databases (KDD), as shown in Figure 2.1. The data mining aspect of KDD consists of an ordered series of activities aimed at training and evaluating the best patterns (for machine learning) or equations (for parametric statistical procedures). These optimum patterns or equations are called *models*.

MAJOR ACTIVITIES OF DATA MINING

Major data mining activities include the following general operations (Hand et al., 2001):

1. **Exploratory Data Analysis:** These data exploration activities include interactive and visual techniques that allow you to "view" a data set in terms of summary statistical parameters and graphical display to "get a feel" for any patterns or trends that are in the data set.
2. **Descriptive Modeling:** This activity forms higher-level "views" of a data set, which can include the following:
    a. Determination of overall probability distributions of the data (sometimes called *density estimations*);
    b. Models describing the relationship between variables (sometimes called *dependency modeling*);
    c. Partitioning of the data into groups, either by *cluster analysis* or *segmentation*. Cluster analysis is a little different, as the clustering algorithms try to find "natural groups" either with many "clusters," or in one type of cluster analysis, the user can specify that all the cases "must be" put into x number of clusters (say, for example, three cluster groups). For segmentation, the goal is to find homogeneous groups related to the variable to be modeled (e.g., customer segments like *big-spenders*).
3. **Predictive Modeling: Classification and Regression:** The goal here is to build a model where the value of one variable can be predicted from the values of other variables. Classification is used for "categorical" variables (e.g., Yes/No variables or multiple-choice answers for a variable like 1–5 for "like best" to "like least"). Regression is used for "continuous" variables (e.g., variables where the values can be any number, with decimals, between one number and another; age of a person

would be an example, or blood pressure, or number of cases of a product coming off an assembly line each day).

4. **Discovering Patterns and Rules:** This activity can involve anything from finding the combinations of items that occur frequently together in transaction databases (e.g., products that are usually purchased together, at the same time, by a customer at a convenience store, etc.) or things like finding groupings of stars, maybe new stars, in astronomy, to finding genetic patterns in DNA microarray assays. Analyses like these can be used to generate association rules; e.g., if a person goes to the store to buy milk, he will also buy orange juice. Development of association rules is supported by algorithms in many commercial data mining software products. An advanced association method is Sequence, Association, and Link (SAL) analysis. SAL analysis develops not only the associations, but also the sequences of the associated items. From these sequenced associations, "links" can be calculated, resulting in Web link graphs or rule graphs (see the NTSB Text Mining Tutorial, included with this book, for nice illustrations of both rule graphs and SAL graphs).

5. **Retrieval by Content:** This activity type begins with a known pattern of interest and follows the goal to find similar patterns in the new data set. This approach to pattern recognition is most often used with text material (e.g., written documents, brought into analysis as Word docs, PDFs, or even text content of Web pages) or image data sets.

To those unfamiliar with these data mining activities, their operations might appear magical or invoke images of the wizard.

Contrary to the image of data miners as magicians, their activities are very simple in principle. They perform their activities following a very crude analog to the way the human brain learns. Machine learning algorithms learn case by case, just the way we do. Data input to our senses are stored in our brains not in the form of individual inputs, but in the form of *patterns.* These patterns are composed of a set of neural signal strengths our brains have associated with known inputs in the past. In addition to their abilities to build and store patterns, our brains are very sophisticated pattern recognition engines. We may spend a lifetime building a conceptual pattern of "the good life" event by event and pleasure by pleasure. When we compare our lives with those in other countries, we unconsciously compare what we know about their lives (data inputs) with the patterns of our good lives. Analogously, a machine learning algorithm builds the pattern it "senses" in a data set. The pattern is saved in terms of mathematical weights, constants, or groupings. The mined pattern can be used to compare mathematical patterns in other data sets, to score their quality. Granted, data miners have to perform many detailed numerical operations required by the limitations of our tools. But the principles behind these operations are very similar to the ways our brains work.

Data mining did not arise as a new academic discipline from the studies in universities. Rather, data mining is the logical next step in a series of developments in business to use data and computers to do business better in the future. Table 2.1 shows the historical roots of data mining.

The discussion in Chapter 1 ended with the question of whether the latest data mining algorithms represent the best we can do. The answer was probably no. The human minds

TABLE 2.1   Historical Development of Data Mining

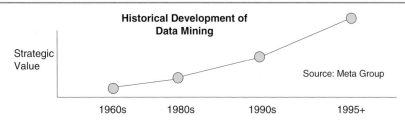

Developmental Step	Data Collection	Data Access	Data Warehousing & Decision Support	Data Mining
Business Question	"What was my total revenue last year?"	"What were sales in Ohio last March?	"What were sales in Ohio last March" – Drill down to Dayton	"What will be the sales in Dayton next month"
Enabling Technology	Computers tapes and disks	Relational databases & SQL	Data warehouses Multidimensional databases	Adv. algorithms multiprocessors massive databases
Characteristics	Delivery of static past data summaries	Delivery of dynamic past data at record level	Delivery of dynamic past data at multiple levels	Prospective Proactive Information delivery

of the data mining algorithm developers will continue to generate novel and increasingly sophisticated methods of emulating the human brain.

MAJOR CHALLENGES OF DATA MINING

Some of the major challenges of data mining projects include

- Use of data in transactional databases for data mining
- Data reduction
- Data transformation
- Data cleaning
- Data sparsity
- Data rarity (rare case pattern recognition and thus "data set balancing")

Each of these challenges will be discussed in the ensuing chapters.

EXAMPLES OF DATA MINING APPLICATIONS

Data mining technology can be applied anywhere a decision is made, based on some body of evidence. The diversity of applications in the past included the following:

- **Sales Forecasting:** One of the earliest applications of data mining technology
- **Shelf Management:** A logical follow-on to sales forecasting
- **Scientific Discovery:** A way to identify which among the half-billion stellar objects are worthy of attention (JPL/Palomar Observatory)
- **Gaming:** A method of predicting which customers have the highest potential for spending
- **Sports:** A method of discovering which players/game situations have the highest potential for high scoring
- **Customer Relationship Management:** Retention, cross-sell/up-sell propensity
- **Customer Acquisition:** A way to identify the prospects most likely to respond to a membership offer

MAJOR ISSUES IN DATA MINING

Some major issues of data mining include the following (adapted from Han and Kamber, 2006):

1. **Mining of different kinds of information in databases:** It is necessary to integrate data from diverse input sources, including data warehouses/data marts, Excel spreadsheets, text documents, and image data. This integration may be quite complex and time consuming.
2. **Interactive mining of knowledge at multiple levels of abstraction:** Account-level data must be combined with individual-level data and coordinated with data with different time-grains (daily, monthly, etc.). This issue requires careful transformation of each type of input data to make them consistent with each other.
3. **Incorporation of background information:** Some of the most powerful predictor variables are those gathered from outside the corporate database. These data can include demographic and firmographic data, historical data, and other third-party data. Integration of this external data with internal data can be very tricky and imprecise. Inexact ("fuzzy") matching is necessary in many cases. This process can be very time consuming also.
4. **Data mining query languages and ad hoc data mining:** Data miners must interface closely with database management systems to access data. Structured Query Language (SQL) is the most common query tool used to extract data from large databases. Sometimes, specialized query languages must be used in the place of SQL. This requirement means that data miners must become proficient (at least to some extent) in

the programming skills with these languages. This is the most important interface between data mining and database management operations.

5. **Presentation and visualization of data mining results:** Presenting highly technical results to nontechnical managers can be very challenging. Graphics and visualizations of the result data can be very valuable to communicate properly with managers who are more graphical rather than numerical in their analytical skills.

6. **Handling "noisy" or incomplete data:** Many items of data ("fields") for a given customer or account (a "record") are often blank. One of the most challenging tasks in data mining is filling those blanks with intuitive values. We will discuss some approaches for filling blank data fields in Chapter 4. In addition to data that is not there, some data present in a record represent randomness and are analogous to noise in a signal transmission. Different data mining algorithms have different sensitivities to missing data and noise. Part of the art of data mining is selecting the algorithm with the right balance of sensitivity to these "distractions" and also to have a relatively high potential to recognize the target pattern.

7. **Pattern evaluation—the "interestingness" problem:** Many patterns may exist in a data set. The challenge for data mining is to distinguish those patterns that are "interesting" and useful to solve the data mining problem at hand. Various measures of interestingness have been proposed for selecting and ranking patterns according to their potential interest to the user. Applying good measures of interestingness can highlight those variables likely to contribute significantly to the model and eliminate unnecessary variables. This activity can save much time and computing "cycles" in the model building process.

8. **Efficiency and scalability of data mining algorithms:** Efficiency of a data mining algorithm can be measured in terms of its predictive power and the time it takes to produce a model. Scalability issues can arise when an algorithm or model built on a relatively small data set is applied to a much larger data set. Good data mining algorithms and models are linearly scalable; that is, time consumed in processing increases geometrically rather than exponentially with the size of the data set.

9. **Parallel, distributed, and incremental mining algorithms:** Large data mining problems can be processed much more efficiently by "dividing and conquering" the problem with multiple processors in parallel computers. Another strategy for processing large data sets is to distribute the processing to multiple computers and compose the results from the combined outputs. Finally, some data mining problems (e.g., power grid controls) must be solved by using incremental algorithms, or those that work on continuous streams of data, rather than large "chunks" of data. A good example of such an algorithm is a Generalized Regression Neural Net (GRNN). Many power grids are controlled by GRNNs.

10. **Handling of relational and complex types of data:** Much input data might come from relational databases (a system of "normalized" tables linked together by common keys). Other input data might come from complex multidimensional databases (elaborations of star-schemas). The data mining process must be flexible enough to encompass both.

11. **Mining information from heterogeneous and global information systems:** Data mining tools must have the ability to process data input from very different database structures. In tools with graphical user interfaces (GUIs), multiple nodes must be configured to input data from very different data structures.

GENERAL REQUIREMENTS FOR SUCCESS IN A DATA MINING PROJECT

Following are general requirements for success of a data mining project:

1. Significant gain is expected. Usually, either
    a. Results will identify "low-hanging fruit," as in a customer acquisition model where analytic techniques haven't been tried before (and anything rational will work better).
    b. Improved results can be highly leveraged; that is, an incremental improvement in a vital process will have a strong bottom-line impact. For instance, reducing "charge-offs" in credit scoring from 10% to 9.8% could make a difference of millions of dollars.
2. A team skilled in each required activity. For other than very small projects, it is unlikely that one person will be sufficiently skilled in all activities. Even if that is so, one person will not have the time to do it all, including data extraction, data integration, analytical modeling, and report generation and presentation. But, more importantly, the analytic and business people must cooperate closely so that analytic expertise can build on the existing domain and process knowledge.
3. Data vigilance: Capture and maintain the accumulating information stream (e.g., model results from a series of marketing campaigns).
4. Time: Learning occurs over multiple cycles. The corporate mantra of Dr. Ferdinand Porsche was "Racing improves the breed." Today, Porsche is the most profitable automobile manufacturer in the world.

Likewise, data mining models must be "raced" against reality. The customer acquisition model used in the first marketing campaign is not very likely to be optimal. Successive iterations breed successive increases in success.

Each of these types of data mining applications followed a common methodology in principle. We will expand on the subject of the data mining process in Chapter 3.

EXAMPLE OF A DATA MINING PROJECT: CLASSIFY A BAT'S SPECIES BY ITS SOUND

Approach:

1. Use time-frequency features of echo-location signals to classify bat species in the field (no capture is necessary).

2. University of Illinois biologists gathered 98 signals from 19 bats representing 6 species.
3. Thirty-five data features were calculated from the signals, such as low frequency at the 3 db level; time position of the signal peak; amplitude ratio of the 1st and 2nd harmonics.
4. Multiple data mining algorithms were employed to relate the features to the species.

Figure 2.3 shows a plot of the 98 bat signals.

The groupings of bat signals in Figure 2.3 are depicted in terms of color and shape of the plotted symbols. From these groupings, we can see that it is likely that a modeling algorithm could distinguish between many of them (as many colored groups cluster), but not all (as there are multiple clusters for most bat types). The first set of models used decision trees and was 46% accurate. A second set used a new tree algorithm that looks two steps ahead (see TX2Step on http://64.78.4.148/PRODUCTS/tabid/56/Default.aspx) and did better at 58%. A third set of models used neural nets with different configurations of inputs. The best neural net solution increased the correct prediction rate to 68%, and it was observed that the simplest neural net architecture (the one with the fewest input variables) did best. The reduced set of inputs for the neural networks had been suggested by the inputs chosen by the two decision tree algorithms. Further models with nearest neighbors, using this same reduced set of inputs, also did as well as the best neural networks. Lastly, an ensemble of the estimates from four different types of models did better than any of the individual models.

The bat signal modeling example illustrates several key points in the process of creating data mining models:

1. Multiple algorithms are better than a single algorithm.
2. Multiple configurations of an algorithm permit identification of the best configuration, which yields the best model.
3. Iteration is important and is the only way to assure that the final model is the right one for a given application.
4. Data mining can speed up the solution cycle, allowing you to concentrate on higher aspects of the problem.
5. Data mining can address information gaps in difficult-to-characterize problems.

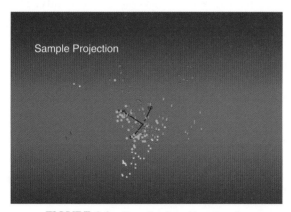

Sample Projection

FIGURE 2.3   Sample plot of bat signals.

THE IMPORTANCE OF DOMAIN KNOWLEDGE

One data mining analyst might build a model with a data set and find very low predictability in the results. Another analyst might start with the same data set but create a model with a much higher predictability. Why the difference? In most cases like this, the difference is in the data preparation, not the modeling algorithm chosen. Granted, some algorithms are clearly superior to others for a particular data set. A model is no better than the predictor variables input to it. The second analyst may know much more about the business domain from which the data came. This intimate knowledge facilitates the derivation of powerful predictor variables from the set of existing variables. In Chapter 16, we will see that the derivation of time-lagged variables (a.k.a. temporal abstractions) permitted the creation of a much more powerful model than without them. There is simply no substitution for domain knowledge. If you don't have it, get it by either learning it before building a model, or bring it into the project team in the form of one who does know it.

POSTSCRIPT

Why Did Data Mining Arise?

Now, we can go on to a final dimension of our subject of analytical theory. Statistical analysis has been around for a long time. Why did data mining development occur when it did? Necessity may indeed be the mother of invention. During the past 50 years, business, industry, and society have accumulated a huge amount of data. It has been estimated that over 90% of the total knowledge we have now has been learned since 1950. Faced with huge data sets, analysts could bring computers to their "knees" with the processing of classical statistical analyses. A new form of learning was needed. A new approach to decision making based on input data had to be created to work in this environment of huge data sets. Scientists in artificial intelligence (AI) disciplines proposed that we use an approach modeled on the human brain rather than on Fisher's Parametric Model. From early AI research, neural nets were developed as crude analogs to the human thought process, and decision trees (hierarchical systems of Yes/No answers to questions) were developed as a systematic approach to discovering "truth" in the world around us.

Data mining approaches were also applied to relatively small data sets, with predictive accuracies equal to or better than statistical techniques. Some medical and pharmaceutical data sets have relatively few cases but many hundreds of thousands of data attributes (fields). One such data set was used in the 2001 KDD Cup competition, which had only about 2,000 cases, but each case had over 139,000 attributes! Such data sets are not very tractable with parametric statistical techniques. But some data mining algorithms (like MARS) can handle data sets like this with relative ease.

Some Caveats with Data Mining Solutions

Hand (2005) summarized some warnings about using data mining tools for pattern discovery:

1. **Data Quality:** Poor data quality may not be explicitly revealed by the data mining methods, and this poor data quality will produce poor models. It is possible that poor data will support the building of a model with relatively high predictability, but the model will be a fantasy.
2. **Opportunity:** Multiple opportunities can transform the seemingly impossible to a very probable event. Hand refers to this as the *problem of multiplicity*, or the law of truly large numbers. For example, the odds of a person winning the lottery in the United States are extremely small, and the odds of that person winning it twice are fantastically so. But the odds of *someone in the United States* winning it twice (in a given year) are actually better than ever. As another example, you can search the digits of pi for "prophetic" strings such as your birthday or significant dates in history and usually find them—given enough digits.
3. **Interventions:** One unintended result of a data mining model is that some changes will be made to invalidate it. For example, developing fraud detection models may lead to some effective short-term preventative measures. But soon thereafter, fraudsters may evolve in their behavior to avoid these interventions in their operations.
4. **Separability:** Often, it is difficult to separate the interesting information from the mundane information in a data set. Many patterns may exist in a data set, but only a few may be of interest to the data miner for solving a given problem. The definition of the target variable is one of the most important factors that determine which pattern the algorithm will find. For one purpose, retention of a customer may be defined very distinctively by using a variable like Close_date to derive the target. In another case, a 70% decline in customer activity over the last two billing periods might be the best way to define the target variable. The pattern found by the data mining algorithm for the first case might be very different from that of the second case.
5. **Obviousness:** Some patterns discovered in a data set might not be useful at all because they are quite obvious, even without data mining analysis. For example, you could find that there is an almost equal number of married men as married women (duh!). Or, you could learn that ovarian cancer occurs primarily in women and that check fraud occurs most often for customers with checking accounts.
6. **Nonstationarity:** Nonstationarity occurs when the process that generates a data set changes of its own accord. For example, a model of deer browsing propensity on leaves of certain species will be quite useless when the deer population declines rapidly. Any historical data on browsing will have little relationship to patterns after the population crash.

Note: Some of the specific details of the "theory" and methods will be discussed in more depth in Part II, especially Chapters 7 and 8, and also "classification" and "numerical prediction" algorithms more thoroughly in Chapters 11 and 12; with an additional smattering of algorithms discussed in Chapters 13, 14, 15, 16, and 17.

References

Accenture Global Services. (2006). Standardized customer application and record for inputting customer data into analytical applications. U.S. Patent #7047251, May 16, 2006.

Fayyad, U., Piatetsky-Shapiro, G., Smyth, P., & Uthurusamy, R. (1996). *Advances in Knowledge Discovery and Data Mining*. Menlo Park, CA: AAAI Press.

Frawley, W., Piatetsky-Shapiro, G., & Matheus, C. (1991). Knowledge discovery in databases—An overview. In *Knowledge Discovery in Databases 1991* (pp. 1–30). Reprinted in *AI Magazine*, Fall 1992.

Han, J., & Kamber, M. (2006). *Data Mining: Concepts and Techniques* (2nd ed.). San Francisco: Morgan Kaufmann.

Hand, D., Mannila, H., & Smyth, P. (2001). *Principles of Data Mining*. Cambridge, MA and London, England: The MIT Press: A Bradford Book.

Hand, D. J. (2005). *What You Get Is What You Want? Some Dangers of Black Box Data Mining*. M2005 Conference Proceedings. Cary, NC: SAS Institute, Inc.

Mannila, H. (2000). Theoretical frameworks for data mining. *SIGKDD Explorations*, *1*(2), 30–32.

The Data Mining Process

PREAMBLE

Data miners are fond of saying that data mining is as much art as it is science. What they mean by this statement is that the data mining process is a scientific endeavor overlain with a significant amount of artistic practice. This chapter will expand on this statement in the context of the many practical challenges of data mining in real-world databases.

THE SCIENCE OF DATA MINING

A very early definition of data mining was "the nontrivial extraction of implicit, previously unknown, and potentially useful information from data" (Frawley et al., 1992). A later definition of data mining expanded on this definition slightly, referring to the application of various algorithms for finding patterns or relationship in a data set (Fayyad et al., 1996).

An attendant term, *knowledge discovery*, referred to the collateral process of data access, data preprocessing, data post-processing, and interpretation of results. The combined process was referred to as the *KDD process*. This is a very useful approach to express all the steps necessary for finding and exploiting relationships in data; however, it was not followed for very long in the development of data mining in business during the 1990s.

The concept of data mining to a business data analyst includes not only the finding of relationships, but also the necessary preprocessing of data, interpretation of results, and provision of the mined information in a form useful in decision making. In other words, a business data analyst includes the classical definitions of data mining and knowledge discovery into one process. While this approach is not very palatable for the academic, it serves the business analyst quite well. We will adopt this approach in this chapter, not because it is best, but because it serves well to communicate both the nature and the scope of the process of leveraging relationship patterns in data to serve business goals.

THE APPROACH TO UNDERSTANDING AND PROBLEM SOLVING

Before an investigation can occur, the basic approach must be defined. There are two basic approaches to discovering truth: the *deductive* approach and the *inductive* approach. The deductive approach starts with a few axioms—simple true statements about how the world works. The understanding of the phenomenon can be deduced from the nature of the axioms. This approach works fine in mathematics, but it does not work very well for describing how the natural world works. During the Renaissance, an alternate approach to truth was formulated, which turned the deductive method upside down. This new method approached truth *inductively* rather than *deductively*. That is, the definition of simple truths describing the phenomenon is the *goal* of the investigation, not the *starting point!* Development of the inductive approach to truth in science led to the formation of the scientific method.

In his classic work on problem solving, George Pólya showed that the mathematical method and the scientific method are similar in their use of an iterative approach but differ in steps followed (Pólya, 1957). Table 3.1 compares these two methods.

TABLE 3.1   Comparison of the Steps in the Mathematical and Scientific Methods

Mathematical Method	Scientific Method
1. Understanding	1. Characterization from experience and observation
2. Analysis	2. Hypothesis: a proposed explanation
3. Synthesis	3. Deduction: prediction from hypothesis
4. Review/Extend	4. Test and experiment

The method followed in the data mining process for business is a blend of the mathematical and scientific methods. The basic data mining process flow follows the mathematical method, but some steps from the scientific method are included (i.e., characterization, and test and experiment). This process has been characterized in numerous but similar formats. The most widespread formats in use are the CRISP-DM format, SEMMA, and DMAIC. In subsequent chapters of this book, we will refer to the data mining process in terms of the CRISP-DM format.

CRISP-DM

The CRISP-DM format for expressing the data mining process is the most complete available. It was created by a consortium of NCR, SPSS, and Daimler-Benz companies. The process defines a hierarchy consisting of major phases, generic tasks, specialized tasks, and process instances. The major phases are related in Figure 3.1 as it is applied to fraud modeling.

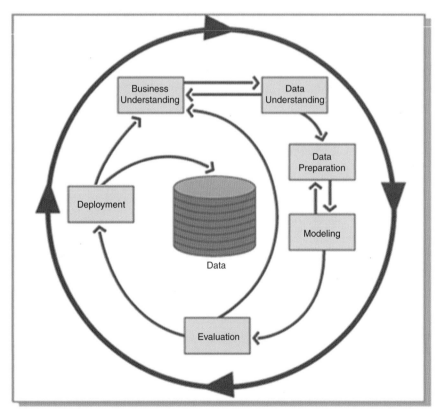

FIGURE 3.1   Phases of the CRISP-DM process (Chapman et al., 2000).

Each phase of the process consists of a number of second-level generic activities, each with several specialized operations. A fourth level (Tasks) could be defined in this process, but these tasks are very domain-specific; that is, they must be defined in terms of the specific business problem to be solved in the context of the specific data used to solve it. This organization can be viewed in terms of the following hierarchy:

Data Mining Phase
  Activities
    Operations
      Tasks

The expanded data flow process hierarchy described in the following sections is based largely on Chapman et al. (2000), but some activities have been added (shown with asterisks). Each phase in the list is annotated with the degree to which it pertains to art or science. The art of data mining will be discussed in further detail in the section titled "The Art of Data Mining."

BUSINESS UNDERSTANDING (MOSTLY ART)

Before you can begin data mining, you must have a clear understanding of what you want to do and what success will look like in terms of business processes that will benefit. Each of the following major tasks that promote understanding of the business problem should be followed in most cases.

Define the Business Objectives of the Data Mining Model

You should understand the background that spawned the business needs that a data mining model might serve. For example, a body of unstructured data might exist in your company, including notes, memos, and reports. Information in these unstructured formats is not present in a database, so it cannot be queried like normal data. The business objective is to find a way to capture relevant information in these unstructured formats into a data format that will support decision making. In this case, a text mining model might be useful to capture relevant information in these documents (see Chapter 9 for a discussion of text mining). An important part of formulating the business objectives is to include individuals from all business units of the company that are affected by the problem and will benefit from its solution. You must compose a set of success criteria from interactions with these "stakeholders." Only in that way will you know from the beginning what a "good" model will look like, in terms of metrics accepted by the stakeholders. In addition to success criteria, all stakeholders should be fully aware of the benefits of success in the project and apprised of its cost in terms of resources (human and financial). This approach is a very important factor in "engineering" success of the data mining project.

Assess the Business Environment for Data Mining

Building a data mining model is a lot like erecting a building. In addition to knowing what the building will look like when it is done, we must plan for its construction. The first thing you must do is to take inventory of your resources. That may mean listing the data integration, data quality, and analytical tools at your disposal. If an important tool is missing, you have to acquire it or figure out how to do specific tasks with the tools you have. A shortfall in tools and materials may increase the risk of schedule slippage (or even failure). Any other significant risks should be identified (e.g., risk of failure of obtaining the necessary approvals from management), and contingency plans should be formed.

In addition to assessing the modeling environment, you should assess one or more deployment environments. Many data mining models have just sat on the shelf because they were impractical or too costly to implement. Restrictions in the deployment environment might dictate the form and power of the model. For example, if the model will be used to guide the Underwriting department to minimize loss-risk, model output might be required in the form of business rules. In that case, a decision tree model might be the best choice, and one from which only a few rules must be induced to guide the underwriters. For deployment of a customer retention model, a prediction produced by a neural net model might be used to drive an interdiction campaign to retain high-value customers with relatively high attrition propensities. Agreement from the Marketing department to conduct such a campaign should be obtained before modeling operations begin.

Finally, results of the business assessment should be fully documented together with sufficient explanatory material and terminology to serve as a standalone document for later reference.

Formulate the Data Mining Goals and Objectives

Formulating the data mining goals and objectives may seem moot (in relation to the business goal), but it is critical to the success of the data mining project. The primary goal of the data mining exercise is *not* to train a good predictive model that per se, but rather to *deploy* a good predictive model to meet the business objective! Often, data miners take this for granted. But, it does not happen automatically. Well-deployed models must be engineered rather than envisioned. Many models that have been envisioned for their usefulness in a company have been relegated to the shelf (as it were) because they could not be implemented efficiently and effectively. In Chapter 21, we will discuss one implementation of a generalized deployment engine (Zementis ADAPA), which employs the Amazon.com Elastic Cloud for deployment of models over the Internet. This product is one of the few serious efforts to shift the emphasis from model building to model deployment.

Data mining goals include may include the following:

- Building (or helping to build) a suitable database, from which modeling data sets can be extracted easily
- Developing and deploying a model, which generates significant business value

- Building a knowledge base of modeling "learnings," which can be leveraged later to do a better job with data mining (easier, faster, cheaper)

Each of these data mining goals is associated with a set of objectives. For example, a good set of objectives for the goal of developing and deploying a model can include the following:

- Acquiring a suitable data set for modeling
- Creating a short-list of predictor variables
- Creating a model of acceptable accuracy
- Deploying the model in production operations
- Monitoring for acceptable performance
- Updating the model with current data
- Providing feedback of intelligence gained by application of the model

Each of these objectives will be implemented by a set of tasks. An example of a task list for the objective of creating a short-list of predictor variables might include these tasks:

- Identification/derivation of the target variable (the variable to be predicted)
- Univariate and bi-variate analysis of candidate predictor variables
- Multivariate analysis of candidate predictor variables
- Correlation analysis of candidate predictor variables
- Preliminary screening of variables with various metrics and screening algorithms (e.g., Gini scoring, or with some measure of "interestingness")
- Preliminary modeling of the target variable (e.g., with a decision tree algorithm) to select variables for the short-list

Each of these tasks will be composed of subtasks or steps followed to accomplish the task. For example, steps followed in the univariate or bi-variate analysis task could include

- Generation of various descriptive statistics (e.g., means, standard deviations, etc.)
- Bi-variate scatterplots
- Association and linkage analysis

By now, maybe you can see where we must go next with this body of methodology. Of course, we must compose all the objectives, tasks, and steps into a project plan and then manage that plan. We must "plan the work" and then "work the plan." This plan should assign start dates and end dates to each objective, task, and step and identify the resources needed to accomplish it (people, money, and equipment).

Microsoft Project is a good project planning package to use for project plan formation and tracking. The embedded goal of good project planning is to finish the plan ahead of schedule and under budget. Much has been written about how to do this, so we won't go into this topic any further here. For more information, buy a good book on project planning. Much good information on project planning is available on the Internet, notably from the Project Management Institute (PMI). PMI certification as a Project Management Professional (PMP) is a good adjunct to successful data mining.

DATA UNDERSTANDING (MOSTLY SCIENCE)

The objectives and tasks in this and subsequent sections will be presented in a rough outline format.

An expanded outline is available on the DVD, which explains in general terms some of the operations usually performed for each objective of a database marketing project. (See database marketing documents on the DVD.) The asterisks next to some of the tasks indicate those added to the standard CRISP-DM methodology document. Where possible, the tutorials will refer to the specific data mining objectives and tasks performed and why they are necessary for the data set and modeling goal in focus.

The CRISP-DM activity for data understanding was specified separately in the diagram, but in this book, we will treat it together with data preparation in Chapter 4. Following are some of the common data understanding activities:

1. Data acquisition
    a. Data access*
    b. Data integration*
    c. Initial data collection report
2. Data description
    a. Variables*
    b. Cases*
    c. Descriptive statistics*
    d. Data description report
3. Data quality assessment
    a. Missing values*
    b. Outliers*
    c. Data quality report

Data Acquisition

Before you can do anything with your data, you have to acquire it. This statement appears on the surface to be self-evident. Determining how to find and extract the right data for modeling, however, is not at all self-evident. First, you have to identify the various data sources available to you. These data may be nicely gathered together in an enterprise data warehouse. While this situation is ideal, most situations will be quite different. Data may be scattered in many business units of the company in various data "silos," spreadsheets, files, and hard-copy (paper) lists. Your next challenge is to put all this information together.

Data Integration

Combining data sets is not an easy thing to do. Usually, data are in different formats or exist in different levels of aggregation or are expressed in different units. A big part of the integration activity is to build a data map, which expresses how each data element in

---

* Tasks included to augment those in Chapman et al. (2000).

each data set must be prepared to express it in a common format and a common record structure. Data in relational databases must be either "flattened" (gathered together into one row, or record), or the data map must be traversed to access the data in the databases directly through in-database access utilities, available in some analytical tools. The advantage of doing in-database mining is that your data do not have to be extracted into an external file for processing; many data mining tools can operate on the data quite nicely right where they are. The disadvantage of in-database mining is that all data must be submitted to the necessary preprocessing activities to make them useful for mining. This preprocessing takes time! So, one of the advantages of extracting data to an external file is that processing and modeling can be performed much faster, particularly if your modeling data set can be contained in main memory.

Data Description

You can make some big mistakes by beginning to prepare data before you adequately describe those data. Data description tasks are not just fodder for a data "readiness" report. These tasks must be properly fit to the characteristics of your data. Before you can do that, you must *know* your data. For each data element available to you, look through as many data records as you can to get a general feeling for each variable. Statistical packages and most data mining tools have some simple descriptive statistical capabilities to help you characterize your data set. See a more detailed discussion of descriptive data operations in Chapter 4.

Data Quality Assessment

I have heard the same refrain many times from data warehousing technicians: "Our data is clean!" But it never is. The reason for this misconception is that analytics require a special kind of "cleanliness." Most analytical algorithms cannot accept blanks in any variable used for prediction. One of the hardest and most important operations in data preparation is filling blank entries in all variables. This process is called *imputation*, and it can be accomplished by some simple operations and by some rather sophisticated operations. Another vexing problem in data sets is outliers, or values beyond the normal range of the response you are modeling. More information about imputing missing values and handling outliers is presented in Chapter 4.

DATA PREPARATION (A MIXTURE OF ART AND SCIENCE)

Basic data preparation operations access, transform, and condition data to create a data set in the proper format suitable for analytical modeling. The major problem with data extracted from databases is that the underlying structure of the data set is not compatible with most statistical and data mining algorithms. Most data in databases are stored at the account level, often in a series of time-stamped activities (e.g., sales). One of the greatest challenges is rearranging these data to express responses on the basis of the entity to be modeled. For example, customer sales records must be gathered together in the same row of a data set for each customer. Additional preparation must be done to condition the data set to fit the input requirements of the modeling algorithm. There are a number of basic issues that must be addressed in this process.

Basic issues that must be resolved in data preparation:

- How do I clean up the data?—Data Cleansing
- How do I express data variables?—Data Transformation
- How do I handle missing values?—Data Imputation
- Are all cases treated the same?—Data Weighting and Balancing
- What do I do about outliers and other unwanted data?—Data Filtering
- How do I handle temporal (time-series) data?—Data Abstraction
- Can I reduce the amount of data to use?—Data Reduction
    - Records?—Data Sampling
    - Variables?—Dimensionality Reduction
    - Values?—Data Discretization
- Can I create some new variables?—Data Derivation

A detailed discussion of the activities and operations of data understanding and data preparation will be presented in Chapter 4.

MODELING (A MIXTURE OF ART AND SCIENCE)

A general discussion of modeling activities is presented in the following sections. You can take a "deep-dive" into some of these activities in ensuing chapters. For example, more detailed presentations of the modeling operations are presented in Chapters 5, 11, 12, and 13.

Steps in the Modeling Phase of CRISP-DM

Note that modeling activities with asterisks have been added to the CRISP-DM list of activities.

1. Select Modeling Techniques.
    a. *Choose modeling algorithms**: How you prepare your data will depend to some degree on what modeling algorithm you choose. If you choose a parametric statistical algorithm (such as multiple linear regression), you may have to transform some variables to account for significant nonlinearity. If you choose a Support Vector Machine, you might have to standardize your data to fit its requirements.
    b. *Choose modeling architecture* (single analysis, ensemble, etc.)*: A simple, straightforward analysis will include submitting your data to the algorithm and evaluating the models created. Sometimes, that is all you have to do, and

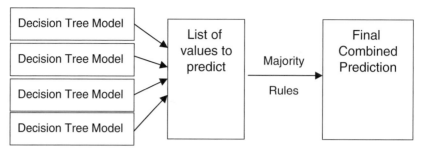

FIGURE 3.2   How a modeling ensemble works.

sometimes it is not. There are many ways to enhance this simplistic approach to refine your models and improve their performance. Some algorithms like neural nets permit you to adjust the algorithm architecture to improve performance (add hidden layers or increase the learning rate). Even these techniques may not be sufficient to optimize performance. You can create a series of models, using different algorithms (ensembles), or you can model on different samples of data and compare or combine results (bootstrap, jackknife resampling, and V-fold cross validation). Finally, you can build some simple feedback processes in your models to iteratively improve your model (boosting). Figure 3.2 shows how an ensemble model works.

c. *Specify modeling assumptions:* Every modeling algorithm makes assumptions. Your challenge is to choose an algorithm whose assumptions fit your data and your modeling goal. For example, you can use a multiple linear regression algorithm safely if your data set does not violate significantly any of the important assumptions of the Parametric Model (the body of assumptions behind parametric statistical theory). You can use a neural net for classification, if your target variable is categorical. Among neural nets, a Radial Basis Function (RBF) neural net can handle outliers better than can an ordinary neural net. Many modeling tools provide both kinds of neural nets. If data outliers are important in your data environment, an RBF neural net would be a good choice.

2. Create an Experimental Design.

Many analytical projects fail because the experimental design was faulty. It is very important to include an analysis of the response under normal conditions to compare results to those under various treatments. For example, an article in the *London Times* on April 22, 2007, described a study of cancer clusters around 7 cell phone towers in different parts of the UK. But, there are about 47,000 cell phone towers in the UK. This means that only about 0.015% of the towers were included in the study. The researchers could have said just as easily that cell phone towers prevent cancer 99.985% of the time! In other words, there was no control. A proper control study would have been to analyze cancer occurrence near a random sample of the 47,000 towers in the UK and compare the results with those of the cancer clusters. Maybe the cell phone towers had nothing whatsoever to do with causing cancer. A more frightening statistic can be gleaned from an article in the *Journal of the National Cancer Institute* (Murray et al., 2008), in which the authors found that in 75 articles on cancer prevention and control published in 41 journals, only about half of the studies had a sufficiently rigorous experimental design to support their conclusions. The next time you read about a report of a cancer study in the newspaper, take it with a grain of salt. You might get just as good an insight from flipping a coin.

(*Note*: In a double-blind experiment, neither the individuals nor the researchers know who belongs to the control group and the experimental group.)

These warnings should not scare you away from doing studies like these, but they may inoculate you against these errors and help you to avoid them. When George Santayana quipped, "Those who cannot remember the past are condemned to repeat it," he was referring to the errors of history. We will not repeat the analytical errors of those in the studies reported by Murray et al. (2008) if we are aware of the dangers of bad experimental design and take steps to correct them before we start the data mining process.

3. Build the Model.

Model building is mostly art, and will be discussed in greater detail in the section "The Art of Data Mining." Here, we can consider the steps to follow in building an analytical model. The general steps in modeling are as follows:

a. Set parameters (if the algorithm is not automatic): Many modeling algorithms (like neural nets) start with various default settings. Study the defaults and the theory behind these settings. The algorithm settings are programmed into the function of the model for a reason. Often, the reason is that different data sets require slightly different model settings. The default settings are a good place to start. Create other models with different settings and see what happens. You may (and often do) find that subsequent models are more powerful predictions compared to the default model.

b. Build various types of models: Using one type of algorithm to model a data set is good, but using multiple algorithms is far better. One algorithm gives one "perspective" on the information pattern in your data set, like looking at the world with one eye. But multiple algorithms will give you multiple perspectives on your information patterns. Let them "vote" on which predicted value or category is right for a case. Then, follow some heuristic (decision rule) to decide which predicted value is to be accepted. You can pick the average of numerical values or let the majority rule in classification. Such a modeling tactic is called *ensemble modeling* (refer to Figure 3.2). Another way to use multiple models is to use resampling methods to build models on different randomly selected subsets of data. We will explore these techniques more fully in Chapter 13.

4. Assess the Model (Mostly Science).

How do you tell how good a model is? The best way is to wait until you can verify the predictions in reality. But you can't wait until then to evaluate various models. You must compare your candidate models several times during the course of analytical modeling. The most common way to evaluate a model is to compare it to what you would expect to happen if you did not use the model. There are some very effective techniques for doing this, using various tables and graphs (coincidence tables, lift charts, ROI curves, normal probability charts). Also, there are some statistical measures of error. These model assessment techniques (and others) will be discussed as a part of the discussion of model enhancement in Chapter 13.

Model assessment is one of the iterative activities performed in modeling. Models should be assessed by one or more of the assessment techniques, which may give some insight on where the model failed. Model parameters can then be adjusted with the help of this insight, and a new model built. This process expresses the CRISP-DM data mining cycle in a more concrete form.

This cycle continues until the assessment techniques converge on an optimum predictive power. Statisticians use the term *convergence* to express the point in the diminishing returns of error minimization that reaches some predetermined expression of the minimum error. For example, a standard error statistic of 0.01 might be set as the stopping point of convergence in a regression algorithm. Sometimes, a minimum rate of convergence is selected as the stopping point. Neural nets use analogous stopping functions to end model training.

5. Evaluate the Model (Mostly Science).

After you create the best model you can under the circumstances of time, budget, and practicality, it is time to evaluate results, review the process as a whole, and determine the next steps to follow with future modeling efforts. A good approach to accomplish these tasks is to include them in a modeling report. This report will help to synthesize conclusions to form general insights about the modeling effort and to point the way to improving the results during the next modeling effort.

Evaluating results of the model may be easy, or it can be rather difficult. An example of an easy evaluation is to observe the number of correct predictions compared to the total number of predictions. If that percentage is relatively high, you might conclude that the model was a success. There is a weakness in this approach, however, because on this basis alone you can't say that this high predictability would not have happened without the model. Ah, we come back to the importance of the experimental design. It becomes clearer now that without a good experimental design, evaluation of the model results is moot. Difficulties of a different kind occur when we try to evaluate what the model did not find but should have (based on actual data). These errors are called *false-negatives* and can occur when your cancer risk model predicts very low cancer risk. It might take years to confirm whether or not that prediction was good, and then it might be too late—the patient is dead. A more insidious error may arise in evaluating results that the model did not find but *might have found* if we had used different data. This type of error points to the very heart of the difference between standard parametric methods and Bayesian methods. Bayesian analysis can incorporate results from past studies, along with data from a present study, to predict more globally what might happen in the future.

Parametric or Bayesian methods, which approach is best? Well, here we have it again: the answer depends on your experimental design. If you want to make a good prediction under all types of cases that might occur, then you must create an experimental design to do that. You may be able to use parametric methods to do it, if you include data likely to cover all case types and you use the right sampling methods to select your data set for analysis. On the other hand, it may be much more efficient to use a Bayesian method (like a Naïve Bayesian belief net) to incorporate results from past studies, which included case types for which you have no available data. Digging for nuggets of information in a large data warehouse is like digging for fossils in the ground. At the very outset, we must say "Caveat fossor" (let the digger, or the miner, beware).

After you evaluate results, you should evaluate the entire process that generated them. This part of the evaluation should consider your modeling goals, your results, and their relationship to the negotiated criteria for success. You should list what went right with the project and what went wrong. One outcome of such an evaluation might be that the results could have been better (that is, you could have built a better model), but the success criteria did meet the goals of the data mining project. Such a project can be deemed a success in the present, but it can point also to ways to accomplish higher goals in future projects.

Finally, evaluation of modeling results should include a list of possible modeling goals for the future and the modeling approaches to accomplish them. The modeling report should discuss briefly the next steps and how to accomplish them. These steps should be expressed in terms of what support must be gained among the stakeholders targeted by these new projects, the processes in the company that must be put in place to accomplish the new projects, and the expected benefit to the company for doing so. In other words, the *business case* for future studies must be built on the merits of the present study. This requirement is one of the major differences between the practice of data mining in business and in academia.

DEPLOYMENT (MOSTLY ART)

1. Plan Model Deployment
   a. Create deployment plan
2. Plan Model Monitoring and Maintenance
   a. Model monitoring plan*
   b. Model maintenance plan*
3. Produce Final Report
   a. Produce final written report
   b. Produce final modeling presentation
4. Review Project

CLOSING THE INFORMATION LOOP* (ART)

As noted previously, dashed arrows could be added in the CRISP-DM diagram in Figure 3.1 to indicate:

1. Feedback of model deployment results to the database*
2. Feedback of model deployment results to the business understanding phase*

Other data mining process flow hierarchies follow the same basic pattern as CRISP-DM. For example,

SEMMA (used by SAS):
Sample
Explore
Manipulate
Model
Assess

DMAIC (a Six Sigma approach designed primarily for industrial applications):
Define
Measure
Analyze
Improve
Control

THE ART OF DATA MINING

Creating a model of relationships in a data set is somewhat like sculpting a statue. The sculptor starts with a block of marble (raw data for the data miner) and a visual concept in his mind (the "true" model in the data for the data miner). After a few chips here and there, the sculptor stands back and looks at his work. The data mining modeler does the same thing after some initial cleaning of the data, imputing missing values, and creating some derived variables. Then the sculptor takes a few more whacks at the block of marble and stands back again to view it. The data miner does the same thing with preliminary modeling using simple algorithms, to identify some of the important variables in the "true" model. Then the data miner makes some adjustments in model parameters or variables (e.g., recoding) and "refines" the model (creates another version). The sculptor continues this iterative process of chipping and viewing, until the finished statue emerges. Likewise, the data miner continues modifying and tuning the model, until there are no further improvements in the predictability of the model. This analogy is rather crude, but it does serve to illustrate the point that a large part of the data mining process is very artistic!

Artistic Steps in Data Mining

Deriving New Variables

Often, several of the strongest predictors in your model will be those you derive yourself. These derived variables can be transforms of existing variables. Common transforms employ

- Log functions
- Power terms (squares, cubes, etc.)
- Trends in existing variables
- Abstractions (collectives, like *Asian*; temporal, like date-offsets; statistical, like means)

This subject will be discussed in greater detail in Chapter 4.

Selecting Predictor Variables

The objective of selecting predictor variables is probably the most artistic among all of the data mining objectives. There are many approaches and methods for selecting the sub-set of variables (the short-list) for use in the model training set. This short-list is an extremely valuable piece of information for the data miner to obtain before beginning to train the model. One reason for this importance is that too few variables will generate a model with a relatively low predictability, and too many variables will just confuse the model "signal." This dynamic follows the principle of Occam's Razor, coined by William of Occam (a fourteenth century clergyman). This principle is often expressed in Latin as the *lex parsimoniae* ("law of parsimony" or "law of succinctness"): *entia non sunt multiplicanda praeter necessitatem*, roughly translated as "entities must not be multiplied beyond necessity". Data mining algorithms follow this principle in their operation. Various methods of developing a short-list are presented in Chapter 5.

POSTSCRIPT

Many business people are very closely attuned to the need for policies and well-defined processes necessary to assure profitability; many analytical people in business are not! Mathematicians and statisticians are very cognizant of the calculation processes and the importance of proper experimental design. But for some reason, it is tempting for data miners to stray from the pathway of the correct process to generate their models. Yielding to this temptation may lead data miners off the "narrow path" into the "Slough of Despond" (as it did to Pilgrim in John Bunyan's *The Pilgrim's Progress*). Maybe this proclivity is due to being too focused on the powerful technology. Or perhaps the problem lies with the academic background of many data miners, which focuses more on theory than practice in the real world. It is *crucial* to the successful completion of a data mining project to follow the well-worn path of accepted process. In Chapter 4, we will see how this process model is followed in preparing data for data mining.

References

Chapman, P., Clinton, J., Kerber, R., Khabaza, T., Reinartz, T., Shearer, C., et al. (2000). *CRISP-DM 1.0*, Chicago, IL: SPSS.

Fayyad, U., Piatetsky-Shapiro, G., Smyth, P., & Uthurusamy, R. (1996). *Advances in Knowledge Discovery and Data Mining*. Cambridge, MA: AAAI/MIT Press.

Foggo, D. (2007, April 22). Cancer clusters at phone masts. *London Times*.

Frawley, W., Piatetsky-Shapiro, G., & Matheus, C. (1992). Knowledge discovery in databases: An overview. *AI Magazine, 13*, 213–228.

Murray, et al. (2008). *Journal of the National Cancer Institute, 100*(7), 483–491.

Pólya, G. (1957). *How to Solve It* (2nd ed.). Princeton, NJ: Princeton University Press.

Santayana, G. (1905–1906). Reason in Common Sense. In *The Life of Reason* (p. 284). New York: Charles Scriber's Sons.

Data Understanding and Preparation

PREAMBLE

Once the data mining process is chosen, the next step is to access, extract, integrate, and prepare the appropriate data set for data mining. Input data must be provided in the amount, structure, and format suited to the modeling algorithm. In this chapter, we will describe the general structure in which we must express our data for modeling and describe the major data cleaning operations that must be performed. In addition, we will describe how to explore your data prior to modeling and how to clean it up. From a database standpoint, a body of data can be regarded as very clean. But from a data mining standpoint, we have to fix many problems like missing data. Having missing data is not a problem for a database manager: what doesn't exist doesn't have to be stored. But for a data miner, what doesn't exist in one field of a customer record might cause the whole record to be omitted from the analysis. The reason is that many data mining algorithms will delete cases that have no data in one or more of the chosen predictor variables.

ACTIVITIES OF DATA UNDERSTANDING AND PREPARATION

Before a useful model can be trained, input data must be provided in the amount, structure, and format suited to the modeling algorithm. The CRISP-DM phases of data understanding and data preparation are discussed together in this chapter because they are related. Often, you must cycle back and forth between data understanding and data preparation activities as you learn more about your data set and perform additional operations on it. We will discuss various data mining activities in both of these phases, together with their component operations necessary to prepare data for both numerical and categorical modeling algorithms. At the end of this chapter, we will organize the activities and operations to form a data description and preparation "cookbook." With this data preparation process in hand, you can begin to prepare your own data sets for modeling in a relatively short period of time. These data preparation steps will be cross-referenced in each tutorial to guide you through the process.

Many books have been written on data analysis and preparation for modeling (some are listed at the end of this chapter). It is not the purpose of this handbook to present a definitive treatise on data preparation (or any of these issues and operations). Rather, we will introduce you to each of them, dive deeper into each operation to address its associated issues, and then direct you to other books to get more detail.

Before we can consider the contribution of each data preparation activity to the modeling effort, we must define some terms. These terms will be used throughout this book to refer to these entities.

Definitions

- **Source data:** Information from any source in any format.
- **Analytical file:** A set of information items from (possibly) multiple sources; that information is composed into one row of information about some entity (e.g., a customer).
- **Record (aka Case):** One row in the analytical file.
- **Attribute:** An item of data that describes the record in some way.
- **Variable:** An attribute installed into a column (field) of the entity record.
- **Target variable:** A variable in the entity record to be predicted by the model.
- **Predictor variable:** A variable in the entity record that is a candidate for inclusion in the model as a predictor of the target variable.
- **Numeric variable:** A variable with only numbers in it; it is treated as a number.
- **Categorical variable:** A variable with any character in it; the character may be a number, but it is treated as text.
- **Dummy variable:** A variable created for each member of the list of possible contents of a categorical variable (e.g., "red," "green," "blue")
- **Surrogate variable:** A variable that has an effect on the target variable very similar to that of another variable in the record.

ISSUES THAT SHOULD BE RESOLVED

The following two lists are a restatement of the basic issues in data understanding and data preparation. These issues will be expanded below to include some approaches resolving them.

Basic Issues That Must Be Resolved in Data Understanding

Following are some of the basic issues you will encounter in the pursuit of an understanding of your data and some activities associated with them:

- How do I find the data I need for modeling?—Data Acquisition
- How do I integrate data I find in multiple disparate data sources?—Data Integration
- What do the data look like?—Data Description
- How clean is the data set?—Data Assessment

Basic Issues That Must Be Resolved in Data Preparation

- How do I clean up the data?—Data Cleansing
- How do I express data variables?—Data Transformation
- How do I handle missing values?—Data Imputation
- Are all cases treated the same?—Data Weighting and Balancing
- What do I do about outliers and other unwanted data?—Data Filtering
- How do I handle temporal (time-series) data?—Data Abstraction
- Can I reduce the amount of data to use?—Data Reduction
    - Records?—Data Sampling
    - Variables?—Dimensionality Reduction
    - Values? —Data Discretization
- Can I create some new variables?—Data Derivation

DATA UNDERSTANDING

Data Acquisition

Gaining access to data you want for modeling is not as easy as it might seem. Many companies have portions of the data you need stored in different data "silos." The separate data stores may exist in different departments, spreadsheets, miscellaneous databases, printed documents, and handwritten notes. The initial challenge is to identify where the data are and how you can get this information. If your data are all in one place (such as in a data warehouse), you still must determine the best way to access that data. If the required data are in one or more database structures, there are three common modes of access to business data:

- Query-based data extracts from the database to flat files
- High-level query languages for direct access to the database
- Low-level connections for direct access to the database

Query-Based Data Extracts

The most common method of extracting data from databases is to use query-based data extracts. The most common tool is SQL-99 (Structured Query Language–1999 version). SQL (originally named SEQUEL) was developed in the 1970s by IBM. Elaborations of a simple SQL Select statement can access data in a database in a large variety of forms. These forms include filtering input and output fields (columns) and records (rows) based on specified levels in a number of variables, aggregations (group by), sorting (order by), and subselects (selects within other selects). This method enables the algorithm to access data the fastest (often in RAM). The problem with this approach is that you have to completely replicate the data and save those data to a file. This requirement may be awkward or impossible to fill with very large data sets.

High-Level Query Languages

Elaborations of SQL optimized for data mining include Modeling Query Language (MQL; Imielinski and Virmani, 1999) and Data Mining Query Language (DMQL; Han et al., 1996). This method is attractive, but the high-level languages are not in standard use. Some day, data mining tools may all support this approach, just as they have come to support XML.

Low-Level and ODBC Database Connections

Many database management systems provide a low-level interface with data stored in data structures. Some data mining tools have incorporated a number of these low-level interfaces to permit access directly to data in the data structures. One of the first systems to do this was NCR Teradata in Warehouse Miner (developed in 1999 as Teraminer Stats). This tool uses the Teradata Call-Level interface to access data directly and create descriptive statistical reports and some analytical modeling operations (e.g., Logistic Regression). Some other data mining tools picked up on that concept to provide in-database access to data via ODBC or other proprietary low-level interfaces (SAS-Enterprise Miner, SPSS Clementine, and *STATISTICA*).

This approach yields several benefits:

- Removes time and space constraints in moving large volumes of data.
- Helps keep management and provisioning of data centralized.
- Reduces unnecessary proliferation of data.
- Facilitates better data governance to satisfy compliance concerns.

In-database mining moves the analytical tasks closer to the data. Closer proximity to the data can significantly increase runtimes and reduce network bottlenecks in the data flow pathway.

Recommendation: If your data are not particularly sensitive and data sets are relatively small, use extracts. If your data are highly confidential or data sets are relatively large, make the extra effort to access your data through the in-database mining capabilities of your tool (if provided).

Data Extraction

Now that you have your data in some form (let's assume it is in flat-file format), how do you put all the pieces together? The challenge before you now is to create a combined data structure suitable for input to the data mining tool. Data mining tools require that all variables be presented as fields in a record. Consider the following data extracts from a Name & Address table and a Product table in the data warehouse:

File #1: Name & Address

Name	Address	City	State	Zipcode
John Brown	1234 E St.	Chicago	IL	60610
Jean Blois	300 Day St.	Houston	TX	77091
Neal Smith	325 Clay St.	Portland	OR	97201

File #2: Product

Name	Address	Product	Sales Date
John Brown	1234 E. St.	Mower	1/3/2007
John Brown	1234 E. St.	Rake	4/16/2006
Neal Smith	325 Clay St.	Shovel	8/23/2005
Jean Blois	300 Day St.	Hoe	9/28/2007

For the data mining tool to analyze names and products in the same analysis, you must organize data from each file to list all relevant items of information (fields) for a given customer on the same line of the output file. Notice that there are two records for John Brown in the Product table, each one with a different sales date. To integrate these records for data mining, you can create separate output records for each product sold to John Brown. This approach will work if you don't need to use Product as a predictor of the buying behavior of John Brown. Usually, however, you do want to include fields like Product as predictors in the model. In this case, you must create separate fields for each record and copy the relevant data into them. The second process is called *flattening* or *denormalizing* the database. The resulting records looks like this:

Name	Address	City	State	Zipcode	Product1	Product2
John Brown	1234 E. St.	Chicago	IL	60610	Mower	Rake
Neal Smith	325 Clay St.	Portland	OR	97201	Shovel	
Jean Blois	300 Day St.	Houston	TX	77091	Hoe	

In this case, the sales date was not extracted. All relevant data for each customer are listed in the same record. Sometimes, this is called the Customer Analytical Record, or CAR. In other industries (or modeling entities other than customers), this data structure may be referred to as just the Analytical Record. We will use the term *Analytical Record* from here on.

To create the Analytical Record, you might have to combine data in different fields in several different data extract files to form one field in the output file. You might have to combine data from several different fields into one field. Most data mining tools provide some data integration capabilities (e.g., merging, lookups, record concatenation, etc.).

A second activity in data integration is the transformation of existing data to meet your modeling needs. For example, you might want to recode variables or transform them mathematically to fit a different data distribution (for working with specific statistical algorithms). Most data mining tools have some capability to do this to prepare data sets for modeling.

Both of these activities are similar to processes followed in building data warehouses. But there is a third activity followed in building data warehouses, which is missing in data preparation for data mining: loading the data warehouse. The extract, transform, and load activities in data warehousing are referred to by the acronym *ETL*. If your data integration needs are rather complex, you may decide to use one of the many ETL tools designed for data warehousing (e.g., Informatica, Abinitio, DataFlux). These tools can process complex data preparation tasks with relative ease.

Recommendation: ETL tools are very expensive. Unless you plan to use such tools a lot, they will not be a cost-effective choice. Most data mining tools have some extraction and transform functions (data integration), but not load functions. It is probable that you can do most of your data integration with the data mining tool, or by using Excel, and a good text editor (MS Notepad will work).

Data Description

This activity is composed of the analysis of descriptive statistical metrics for individual variables (univariate analysis), assessment of relationships between pairs of variables (bivariate analysis), and visual/graphical techniques for viewing more complex relationships between variables. Many books have been written on each of these subjects; they are a better source of detailed descriptions and examples of these techniques. In this handbook, we will provide an overview of these techniques sufficient to permit you to get started with the process of data preparation for data mining. Basic descriptive statistics include:

- **Mean:** Shows the average value; shows the central tendency of a data set
- **Standard deviation:** Shows the distribution of data around the mean
- **Minimum:** Shows the lowest value
- **Maximum:** Shows the highest value
- **Standard deviation:** Shows the distribution of data around the mean
- **Frequency tables:** Show the frequency distribution of values in variables
- **Histograms:** Provide graphical technique to show frequency values in a variable

Analysis and evaluation of these descriptive statistical metrics permit you to determine how to prepare your data for modeling. For example, a variable with relatively low mean and a relatively high standard deviation has a relatively low potential for predicting the target. Analysis of the minimum, maximum, and mean may alert you to the fact that you have some significant outlier values in the data set. For example, you might include the following data in a descriptive statistical report.

N	Mean	Min	Max	StDev
10000	9.769700	0.00	454.0000	15.10153

The maximum is 454, but the mean is only about 9.7, and the standard deviation is only about 15. This is a very suspicious situation. The next step is to look at a frequency table or histogram to see how the data values are distributed between 0 and 454. Figure 4.1 shows a histogram of this variable.

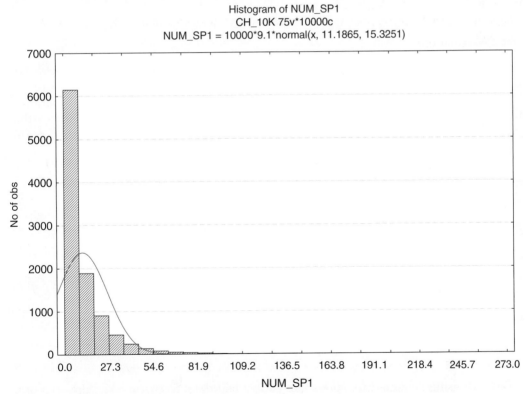

FIGURE 4.1   Histogram of the distribution of NUM_SP1 variable, compared to a normal distribution.

The maximum value of 454 is certainly an outlier, even in this long-tailed distribution. You might be justified in deleting it from the data set, if your interest is to use this variable as a predictor. But this is not all you can learn from this graphic. This distribution is skewed significantly to the right and forms a negative exponential curve. Distributions like this cannot be analyzed adequately by standard statistical modeling algorithms (like ordinary linear regression), which assume a normal distribution (bell-shaped) shown in red on the histogram. This distribution certainly is *not* normal. The distribution can be modeled adequately, however, with logistic regression, which assumes a distribution like this. Thus, you can see that descriptive statistical analysis can provide valuable information to help you choose your data values and your modeling algorithm.

Recommendation: Acquire a good statistical package that can calculate the statistical metrics listed in this section. Most data mining packages will have some descriptive statistical capabilities, but you may want to go beyond those capabilities. For simple descriptive metrics, the Microsoft Excel Analysis Tool Pack add-on will be sufficient. To add the Analysis Tool Pack, click on Tools and select Data Analysis. An option for Data Analysis will be added to the Tools menu. Open your data set in Excel; click on the Data Analysis option in the Tools menu; and fill in the boxes for input range, output range, and descriptive statistics. For more robust descriptive tools, look at *STATISTICA* and SPSS.

Data Assessment

Before you can fix any problems in the data set, you must find them and decide how to handle them. Some problems will become evident during data description operations. Data auditing is similar to auditing in accounting, and includes two operations: data profiling and the analysis of the impact of poor-quality data. Based on these two operations, the data miner can decide what the problems are and if they require fixing.

Data Profiling

You should look at the data distributions of each variable, and note the following:

- The central tendency of data in the variable
- Any potential outliers
- The number of and distribution of blanks across all the cases
- Any suspicious data, like miscodes, training data, system test data, or just plain garbage

Your findings should be presented in the form of a report and listed as a milestone in the project plan.

Data Cleansing

Data cleansing includes operations that correct bad data, filter some bad data out of the data set, and filter out data that are too detailed for use in your model.

Validating Codes Against Lists of Acceptable Values

Human input of data is subject to errors. Also, some codes are not in current use. In either event, you must check the contents of each variable in all records to make sure that all of the contents are valid entries for each variable. Many data mining tools provide some sort of data profiling capabilities. For example, SPSS Clementine provides the Distribution Node, which outputs a list of possible data values for categorical variables, together with the percentage occurrences. *STATISTICA* Data Miner provides the Variable Specs option in the data spreadsheet, which provides a list of unique values in the variable across all cases. If you find codes that are wrong or out of date, you can filter the cases with either of these tools to display those cases with the invalid codes. Most data mining tools offer some sort of expression language in the tool interface that you can use to search and replace invalid codes in the data set.

Deleting Particularly "Dirty" Records

Not uncommonly, many variables have values (or blanks) that are inappropriate for the data set. You should delete these records. Their inclusion in the modeling data set will only confuse the model "signal" and decrease the predictive power of the model.

Witten and Frank (2005) discuss some automatic data cleansing techniques. Some data mining tools (like KXEN) have automated routines that clean input data without operator intervention. Another powerful data cleaning technique is to reduce the variability of time-series data by applying filters similar to those used in signal processing (see "Time-Series Filtering" in the section on Aggregation or Selection to Set the Time-Grain of the Analysis in this chapter).

Data Transformation

Numerical Variables

Many parametric statistical routines (such as Ordinary Least Squares, or OLS, regression) assume that effects of each variable on the target are linear. This means that as variable-X increases by an amount-a, then the target variable increases by some constant multiple of the amount-a. This pattern of increase forms a geometric progression. But, when the multiple is not constant, the pattern of increase forms an exponential pattern. If you want to use parametric statistical modeling algorithm, you should transform any variables forming exponential (nonlinear) curves. Otherwise, estimation errors caused by the violation of the assumption of linearity could invalidate predictions made by the model.

Some statistical and machine learning algorithms operate best if the numerical values are standardized. In statistical parlance, *standardization* means to transform all numerical values to a common range. One common strategy is to use z-values, based on the mean and standard deviation. Each value in a variable is replaced by its z-value (or normal deviate), expressed by

$$z = (\text{value} - \text{mean})/\text{standard deviation}$$

The z-value scale ranges from $-$infinity to $+$infinity, but in a normal distribution, 99.75% of the values will lie between $z = -3$ and $z = +3$. Most statistical and data mining packages have utilities to standardize values. Other standardizing algorithms constrain the transformed values to between -1.0 and $+1.0$. Many Support Vector Machines (SVMs) require that data be submitted to them in this format.

Categorical Variables

Categorical variables have their own problems. Some categorical variables having values consisting of the integers 1–9 will be assumed to be continuous numbers by the parametric statistical modeling algorithm. Such variables can be used safely, even though values between the integers (e.g., 1.56) are not defined in the data set. But other variables may contain textual categories rather than numeric values. For example, entries consisting of the colors red, blue, yellow, and green might require the definition of "dummy" variables. A dummy variable is a binary variable (coded as 1 or 0) to reflect the presence or absence of a particular categorical code in a given variable. For example, a variable like *color* may have a number of possible entries: red, blue, yellow, or green. For this variable, four dummy variables would be created (Color-Red, Color-Blue, Color-Yellow, and Color-Green), and all cases in the data set would be coded as 1 or 0 for the presence or absence of this color.

Coding of Dummy Variables for the Variable Color

Case	Color	Color-Red	Color-Blue	Color-Yellow	Color-Green
1	Red	1	0	0	0
2	Blue	0	1	0	0
3	Yellow	0	0	1	0
4	Green	0	0	0	1
5	Blue	0	1	0	0

Algorithms that depend on calculations of covariance (e.g., regression) or that require other numerical operations (e.g., most neural nets) must operate on numbers. Dummy variables transform categorical (discrete) data into numerical data. Adding dummy variables to the analysis will help to create a better fit of the model, but you pay a price for doing so. Each raw variable that you represent by using a group of dummies causes you to lose 1 degree of freedom in the analysis. The number of degrees of freedom represents the number of independent items of information available to estimate another item of information (the target variable). Therefore, the more tightly you fit your model (the more *precise* your model is), the more degrees of freedom you lose. Consequently, you have less information to work with, and you are left with less ability to apply the model successfully on other data sets, which may have a slightly different target pattern than the one you fit tightly with the model. This situation is called *reducing the generality* of the model. Generality is just as important as (maybe even more so than) the accuracy of the model.

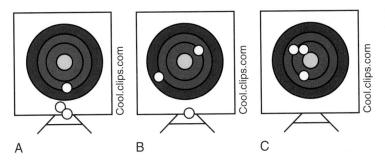

FIGURE 4.2 Relationship between the terms "accuracy" and "precision."

The target hits in Figure 4.2 show three different patterns of accuracy and precision: (1) diagram A shows a pattern that is very precise but not very accurate; (2) diagram B shows a pattern that is not very precise but is relatively accurate (one hit is near the bull's-eye); and (3) diagram C shows the ideal pattern that is very accurate and very precise. Overfitting a model is likely to form a pattern of accuracy on another data set like diagram A. A poorly fitted model might exhibit a pattern of accuracy in predicting the target variable like diagram B. Yes, some of the predictions would be accurate, but most would not. We will talk more about this concept in Chapter 13.

This problem with dummy variables occurs primarily in parametric statistical algorithms, but it is reflected also in machine learning algorithms (e.g., decision trees and neural nets) by the greater tendency of dummy variables to cause *overfitting*. Overfitting is the result of so tightly fitting the model solution to the training data set that it does not perform well on other data sets. It is optimized to a specific set of data. It is possible to train an almost perfect model with a machine learning algorithm, but it will not be very useful for predicting the outcome (the target variable) for other data sets.

Recommendation: The contents of some variables are expressed in terms of a series of categories, but they represent an underlying numerical progression. For example, in an analysis of increasing stress in the workplace, you might include a variable for day of the week. This variable might be coded Monday, Tuesday, Wednesday, Thursday, or Friday. At first glance, these entries appear to be categories. But when you relate them to stress building up during the work week, you can treat them as a numeric variable. Some statistical packages recode categorical data with a set of sequential numbers automatically and treat them numerically. Alternatively, you can recode them yourself to facilitate interpretation of the results of the analysis. Other categorical variables really are categories and do not reflect any consistent numerical basis. For these categorical variables, you should create dummy variables for each category and enter them into the model separately.

Data Imputation

When data are missing in a variable of a particular case, it is very important to fill this variable with some intuitive data, if possible. Adding a reasonable estimate of a suitable data value for this variable is better than leaving it blank. The operation of deciding what data to use to fill these blanks is called *data imputation*. This term means that you assign data to the blank based on some reasonable heuristic (a rule or set of rules). In deciding what values to use to fill blanks in the record, you should follow the cardinal rule of data imputation, "Do the least harm" (Allison, 2002).

Selection of the proper technique for handling missing values depends on making the right assumption about the pattern of "missingness" in the data set. If there is a strong pattern among the missing values of a variable (e.g., caused by a broken sensor), the variable should be eliminated from the model.

Assumption of Missing Completely at Random (MCAR)

The assumption of Missing Completely at Random (MCAR) is satisfied when the probability of missing values in one variable is unrelated to the value of the variable itself, or to values of any other variable. If this assumption is satisfied, then values of each variable can be considered to be a random sample of all values of this variable in the underlying population from which this data set was drawn. This assumption may be unreasonable (may be violated) when older people refuse to list their ages more often than younger people. On the other hand, this assumption may be reasonable when some variable is very expensive to measure and is measured for only a subset of the data set.

Assumption of Missing at Random (MAR)

The assumption of Missing at Random (MAR) is satisfied when the probability of a value's being missing in one variable is unrelated to the probability of missing data in another variable, but may be related to the value of the variable itself. Allison (2002) considers this to be a weaker assumption than MCAR. For example, MAR would be satisfied if the probability of missing income was related to marital status, but unrelated to income within a marital status (Allison, 2002). If MAR is satisfied, the mechanism causing the missing data may be considered to be "ignorable." That is, it doesn't matter *why* MAR occurred, only *that* it occurred.

Techniques for Imputing Data

Most statistical and data mining packages have some facility for handling missing values. Often, this facility is limited to simple recoding (replacing the missing value with some value). Some statistical tool packages (like SPSS) have more complete missing value modules that provide some multivariate tools for filling missing values. We will describe briefly a few of the techniques. Following the discussion of methods, some guidelines will be presented to help you choose which method to use.

List-wise (or case-wise) deletion. This means that the entire record is deleted from the analysis. This technique is usually the default method used by many statistical and machine learning algorithms. This technique has a number of advantages:

- It can be used for any kind of data mining analysis.
- No special statistical methods are needed to accomplish it.
- This is the safest method when data are MCAR.
- It is good for data with variables that are completely independent (the effect of each variable on the dependent variable is *not* affected by the effect of any other variable).
- Usually, it is applicable to data sets suitable for linear regression and is even more appropriate for use with logistic and Poisson regression.

Regardless of its safety and ease of use, there are some disadvantages to its use:

- You lose the nonmissing information in the record, and the total information content of your data set will be reduced.
- If data are MAR, list-wise deletion can produce biased estimates (if salary level depends positively on education level—that is, salary level rises as education level rises—then list-wise deletion of cases with missing salary level data will bias the analysis toward lower education levels).

Pair-wise deletion. This means that all the cases with values for a variable will be used to calculate the covariance of that variable. The advantage of this approach is that a linear regression can be estimated from only sample means and a covariance matrix (listing covariances for each variable). Regression algorithms in some statistical packages use this method to preserve the inclusion of all cases (e.g., PROC CORR in SAS and napredict in R). The major advantage of pair-wise deletion is that it generates internally consistent metrics (e.g., correlation matrices). This approach is justified only if the data are MCAR. If data are only MAR, this approach can lead to significant bias in the estimators.

Reasonable value imputation. Imputation of missing values with the mean of the nonmissing cases is referred to often as *mean substitution*. If you can safely apply some decision rule to supply a specific value to the missing value, it may be closer to the true value than even the mean substitution would be. For example, it is more reasonable to replace a missing value for number of children with zero rather than replace it with the mean or the median number of children based on all the other records (many couples are childless). For some variables, filling blanks with means might make sense; in other cases, use of the median might more appropriate. So, you may have a variety of missing value situations, and you must have some way to decide which values to use for imputation. In SAS, this approach is facilitated by the availability of 28 missing value codes, which can be dedicated to different reasons for the missing value. Cases (rows) for each of these codes can be imputed with a different reasonable value.

Maximum Likelihood Imputation

The technique of maximum likelihood imputation assumes that the predictor variables are independent. It uses a function that describes the probability density map (analogous to a topographic map) to calculate the likelihood of a given missing value, using cases where the value is not missing. A second routine maximizes this likelihood, analogous to finding the highest point on the topographic map.

Multiple Imputation

Rather than just pick a value (like the mean) to fill blanks, a more robust approach is to let the data decide what value to use. This multiple imputation approach uses multiple variables to predict what values for missing data are most likely or probable.

Simple Random Imputation. This technique calculates a regression on all the nonmissing values in all of the variables to estimate the value that is missing. This approach tends to underestimate standard error estimates. A better approach is to do this multiple times.

Multiple Random Imputation. In these techniques, a simple random imputation is repeated multiple times (*n*-times). This method is more realistic because it treats regression parameters (e.g., means) as sample values within a data distribution. An elaboration of this approach is to perform multiple random imputation *m*-times with different data samples. The global mean imputed value is calculated across multiple samples and multiple imputations.

SAS includes a procedure MI to perform multiple imputations. The following is a simple SAS program for multiple imputation, using PROC MI:

PROC MI data = <your data set>

         out = <your output file>

         var <your variable list>

Output to the <your output file> are five data sets collated into one data set, each characterized by its own value for a new variable *.imputation*.

Another SAS PROC MIANALYZE can be used to analyze the output of PROC MI. It provides an output column titled "Fraction Missing Information," which shows an estimate of how much information was lost by imputing the missing values. Allison (2002) provides much greater detail and gives very practical instructions for how to use these two SAS PROCs for multiple imputation of missing values.

See Table 4.1 for guidelines to follow for choosing the right imputation techniques for a given variable.

Recommendations

1. If you have a lot of cases, delete records with missing values.
2. If you are using linear regression as a modeling algorithm and have few missing values, use pair-wise deletion.
3. If you are using SAS, use PROC MI.
4. If you have any insight as to what the value ought to be, fill missing values with reasonable values.
5. Otherwise, use mean imputation.

Most data mining and statistical packages provide a facility for imputation with mean values.

Data Weighting and Balancing

Parametric statistical algorithms measure how far various derived metrics (e.g., means, standard deviations, etc.) are from critical values defined by the characteristics of the data distribution. For example, if a value in a data set is beyond 1.96 standard deviation units from the mean (= the *z*-value), it is beyond the value where it could be a part of the other data 95% of the time. This limit is called the 95% Confidence Level (or 95% CL). Parametric statistical algorithms like OLS learn things about the data by using all cases to calculate the metrics (e.g., mean and standard deviation), and compare all data values in relation to those metrics and standard tables of probability to decide if a relationship exists between two

TABLE 4.1 Guidelines for Choosing the Right Data Imputation Technique

Case-Wise Deletion	Pair-Wise Deletion	Substitution	ML Imputation	Expectation Maximization	Simple Random Imputation	Multiple Random Imputation
Simplest and easiest	Preserves cases	Good when a decision rule is known	Relatively unbiased with large samples	An iterative process	Tends to overestimate correlations	Best for nonlinear algorithms
Sacrifices cases			Consistently estimates under a wide range of conditions	Uses all other variables to predict missing values		Not good for determining interaction effects
Acceptable if the number of cases is large and the event to be modeled is not rare			Data should be MAR	Data should be MAR		Must be matched to the analysis model
Most valid statistically			Best when data is monotonic	Assumption of a normal distribution		Appropriate if data deleted by case-wise deletion is intolerable
Safe for any kind of data mining analysis			Appropriate if number of cases deleted by case-wise deletion is intolerable			
Good for data sets where the variables are completely independent						

variables. Sometimes, you may want to weight each data point with data in another variable to calculate relationships consistent with reality. For example, a data value input from sensor-A may be twice as accurate as data from sensor-B. In this case, it would be wise to apply a weight of 2 for all values input by sensor-A and a weight of 1 for values input from sensor-B.

Machine learning (ML) algorithms learn in a very different way. Instead of going through all of the cases to calculate summary metrics, machine learning algorithms learn case by case. For example, neural nets assign random weights to each variable on the first pass through the data. On subsequent passes through the data (usually about 100 or more), the weights are adjusted in some process like back-propagation, according to the effects of variables in each case. Without case weighting, variables in all cases have the same potential effect on adjusting the weights. Think of the weight applied to a rare event case as if it were a frequency applied for calculating a weighted mean. If the rare event is present in 5% of the cases, you could weight the effect of the rare event cases by a factor = 0.95 and weight all other cases with 0.05 (the reciprocal of the frequency). Then the back-propagation algorithm would be affected by the patterns in the rare cases equally as by the common cases. That is the best way for the neural net to distinguish the rare pattern in the data. This is just what is done by the Balance node in SPSS Clementine. The Balance node puts out a report showing this weighting factor for each target state. The Balance node can be generated by the Distribution node, which creates a frequency table on the target variable. This frequency factor becomes the weighting factor in the Balance node.

This dynamic of ML tools like neural nets is very different from the operation of parametric statistical analysis (sometimes called *frequentist* methods). Frequentist methods look at the data only once and make a judgment based on a number of metrics calculated on the data. ML tools learn incrementally, like we do. Variable weights in the neural net are like signal strengths between neurons in the human brain. We learn by iteratively strengthening or weakening signal transmission and timing between neurons in our brains. Case weights could be picked up by the neural net algorithm to modify the effect of variables of the rare cases on the solution. *STATISTICA* Data Miner provides an option to weight input data cases.

Data Filtering and Smoothing

Data filtering refers to eliminating rows (cases) to remove unnecessary information. This is done to clarify the signal of the variables you are trying to model. Removing unnecessary information reduces the noise below the level that can confuse the analysis. Expressed this way, it sounds as though we are doing signal processing—and that is exactly what we *are* doing. However, the signal here is not a radar signal, but a data signal! Both kinds of signals are just expressions of an underlying informational domain. A radar signal is an expression of the underlying domain of distance and location. A satellite image signal is an expression of a visual domain. Analogously, a customer attrition signal in a corporate database is an expression of a customer retention domain in a company.

Removal of Outliers

The simplest kind of filtering is removal of outliers. Sometimes, you want to keep the outliers (abnormal values). In fact, some outliers are of primary interest to the modeling of credit risk, fraud, and other rare events. For models of normal responses, it might be a good idea to remove the extreme outliers. The reason for this is that you want to model the data that help to define the normal response. If you leave the outliers in the data set, they will just inject noise, which will reduce the predictability of the model. But you might reason that you should keep all values in the data because you have to score values like this in production operations of the model. Well, yes, you have to score outliers, but you can afford to be wrong in your predictions 5% of the time (for example), for the sake of being very predictive on the other 95% of the data.

Hawkins (1980) defines an outlier as "... an observation that deviates so much from other observations as to arouse suspicion that it was generated by a different mechanism." Hawkins discusses four kinds of outlier detection algorithms:

- Those based on critical distance measures
- Those based on density measures
- Those based on projection characteristics
- Those based on data distribution characteristics

Some data mining packages have special routines for identification of outliers. For example, *STATISTICA* Data Miner provides a Recipe module, which permits automatic checking and removal of outliers beyond a given range of value or proportion of the frequency distribution (in this case, 95%). Invoking the distributional outlier option in this tool will trim off cases in the tails beyond the 95% confidence interval for distance from the mean value.

More complicated filtering may be necessary when analyzing time-series data. And it is in the analysis of time-series data that we see the closest analogies to signal processing of transmission and image signals.

Aggregation or Selection to Set the Time-Grain of the Analysis

Time-Series Filtering. One of the best treatments of filtering of time-series data is provided by Masters (1995). Masters provides intuitive explanations of what filtering is and how it can be used effectively to help model time-series data. Signal filters remove high-frequency signal fluctuations ("jitter") either at the top of the range, the bottom of the range, or both.

Low-Pass Filter. A low-pass filter passes data below a specified highest level of acceptability. See Masters (1995) for a good example of this filter type.

High-Pass Filter. A high-pass filter does the opposite of a low-pass filter; it removes data above a specified lower level of acceptability. See Masters (1995) for a good example of this filter type.

At this point, you might be wondering why we emphasize time-series data processing, when most data sets we must prepare for modeling are *not* time-series! The reason is that signal processing techniques can be applied very effectively to data sets other than time-series data. When you view a predictor variable in the context of its contribution in predicting a target variable, you can think of that contribution as a signal of the state or level of the target variable. Some variables might provide a stronger signal (be more predictive), and other variables may be less so. There will be a certain amount of noise (confusion in the target signal) in the values of the predictor variable. You can selectively remove some of this noise in ways analogous to time-series signal filtering. The basis for these data filtering operations is well grounded in engineering theory.

A low-pass data filter can be implemented simply by eliminating cases with values below a certain threshold value. The effect of this operation will be to remove trivial inputs to the modeling algorithm. On the other hand, a high-pass filter can be used to remove cases above a threshold (outliers). One of the arts practiced in data mining is picking the right thresholds for these operations. To do this right, you must know the data domain very well. For example, data for telephone minutes of use (MOU) each month show that average call durations are about 3 minutes in length. But the curve tails off to the right for a long time! Back in the days of analog modems, connect times could be days long, presumably because modems were left on and forgotten. Retention data sets using MOU as a predictor variable for attrition ("churn") should be filtered with a high-pass filter to remove these outliers. The resulting model will be a much better predictor, just like a radio signal passed through a digital high-pass filter in a radio set will be much clearer to the listener.

Data Abstraction

Data preparation for data mining may include some very complex rearrangements of your data set. For example, you extract into an intermediate data set a year of call records by month for a group of customers. This data set will have up to 12 records for each customer, which is really a time-series data set. Analyzing time-series data directly is complex. But there is an indirect way to do it by performing a *reverse pivot* on your intermediate data set.

Consider the telephone usage data in Table 4.2. These records are similar to those that could be extracted from a telephone company billing system. These records constitute a time-series of up to 12 monthly billings for each customer. You will notice the first customer was active for the entire 12 months (hence 12 records). But the second customer has only 9 records because this person left the company. In many industries (including telecommunications), this loss of business is called attrition or *churn*. One of the authors of this book led the team that developed one of the first applications of data mining technology to the telecommunications industry, churn modeling in 1998. This model was based on Analytical Records created by reverse pivoting the time-series data set extracted from the billing systems of telephone companies.

This data set could be analyzed by standard statistical or machine learning time-series tools. Alternatively, you can create Analytical Records from this data set by doing a reverse pivot. This operation copies data from rows for a given customer and installs them in

TABLE 4.2 Telephone Company Billing System Data Extract. (*Note:* The "churn month" for customer #2 is 9, or September.)

Cust_ID	Month	MOU	Due	$Paid	$Balance
1	1	26	$21.19	$21.19	0
1	2	91	$74.17	$37.08	$37.08
1	3	43	$35.05	$17.52	$17.52
1	4	74	$60.31	$60.31	$0.00
1	5	87	$70.91	$35.45	$35.45
1	6	99	$80.69	$40.34	$40.34
1	7	60	$48.90	$24.45	$24.45
1	8	99	$80.69	$40.34	$40.34
1	9	68	$55.42	$55.42	$0.00
1	10	50	$40.75	$20.38	$20.38
1	11	38	$30.97	$15.49	$15.49
1	12	92	$74.98	$37.49	$37.49
2	1	20	$16.30	$8.15	$8.15
2	2	26	$21.19	$21.19	0
2	3	38	$30.97	$15.49	$15.49
2	4	61	$49.72	$24.86	$24.86
2	5	84	$68.46	$68.46	$0.00
2	6	84	$68.46	$34.23	$34.23
2	7	35	$28.53	$14.26	$14.26
2	8	31	$25.27	$12.63	$12.63
2	9	26	$21.19	$10.60	$10.60

columns of the output record. A normal pivot does just the opposite by copying column data to separate rows. This output record for the reverse pivot would appear as the following record shows for the first 9 months:

Cust_ID	MOU1	MOU2	MOU3	MOU4	MOU5	MOU6	MOU7	MOU8	MOU9
1	26	91	43	74	87	99	60	99	68

This record is suitable for analysis by all statistical and machine learning tools. Now that we have the time-dimensional data flattened out into an Analytical Record, we can re-express the elements of this record in a manner that will show churn patterns in the data. But in its present format, various customers could churn in any month. If we submit the current form of the Analytical Record to the modeling algorithm, there may not be enough signal to relate churn to MOU and other variables in a specific month. How can we rearrange our data to intensify the signal of the churn patterns? We can take a page out of the playbook for analyzing radio signals by performing an operation analogous to signal amplification. We do this with our data set by deriving a set of *temporal abstractions,* in which the values in each variable are related to the churn month. Instead of analyzing the relationship of churn (in whatever month it occurred)

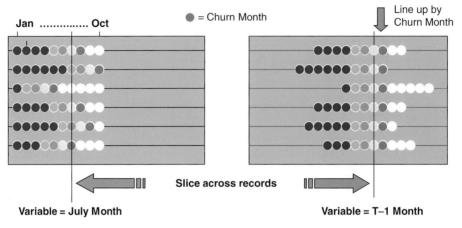

FIGURE 4.3   Demonstration of the power of temporal abstractions for "amplifying" the churn signal to the analysis system (our eyes and brains).

to specific monthly values (e.g., January MOU), we relate the churn month to the MOU value for the previous month, and the month before the previous month, and so forth. Figure 4.3 illustrates the power of these temporal abstractions for clarifying the churn signal to our visual senses, and it will appear more clearly to the data mining algorithm also.

Other Data Abstractions

In the experience of many data miners, some of the most predictive variables are those you derive yourself. The temporal abstractions discussed in the previous sections are an example of these derived variables. Other data abstractions can be created also. These abstractions can be classified into four groups (Lavrac et al., 2000):

- **Qualitative abstraction:** A numeric expression is mapped to a qualitative expression. For example, in an analysis of teenage customer demand, compared to that of others, customers with ages between 13 and 19 could be abstracted as a value of 1 to a variable "teenager," while others are abstracted to a value of 0.
- **Generalization abstraction:** An instance of an occurrence is mapped to its class. For example, in an analysis of Asian preferences, compared to non-Asian, listings of "Chinese," "Japanese," and "Korean" in the Race variable could be abstracted to 1 in the Asian variable, while others are abstracted to a value of 0.
- **Definitional abstraction:** An instance in which one data element from one conceptual category is mapped to its counterpart in another conceptual category. For example, when combining data sets from different sources for an analysis of customer demand among African-Americans, you might want to map "Caucasian" in a demographic data set and "White Anglo-Saxon Protestant" in a sociological data set to a separate variable of "Non-black."
- **Temporal abstraction:** See the preceding discussion.

Data miners usually refer to the first three types of data abstractions as forms of recoding. The fourth type, temporal abstraction, is not commonly used. However, the methodologies of several data mining tool vendors do have forms of temporal abstractions integrated into their design.

Recommendation: If you want to model time-series data but don't want to use time-series algorithms (e.g., ARIMA), you can create temporal abstractions and model on them. Some tools (e.g., KXEN) provide a facility for creating these variables. Sometimes, they are called *lag* variables because the response modeled lags behind the causes of the response. The DVD contains a Perl program you can modify; it will permit you to create temporal abstractions from your time-series data.

Data Reduction

Data reduction includes three general processes:

- Reduction of dimensionality (number of variables)
- Reduction of cases (records)—Data Sampling
- Discretization of values

Data Sampling

In data mining, data sampling serves four purposes:

1. *It can reduce the number of data cases submitted to the modeling algorithm.* In many cases, you can build a relatively predictive model on 10–20% of the data cases. After that, the addition of more cases has sharply diminishing returns. In some cases, like retail market-basket analysis, you need all the cases (purchases) available. But usually, only a relatively small sample of data is necessary.

   This kind of sampling is called *simple random sampling*. The theory underlying this method is that each sample case selected has an equal chance of being selected as does any other case.

2. *It can help you select only those cases in which the response patterns are relatively homogeneous.* If you want to model telephone calling behavior patterns, for example, you might judge that calling behaviors are distinctly different in urban and rural areas. Dividing your data set into urban and rural segments is called *partitioning* the database. It is a good idea to build separate models on each partition.

   When this partitioning is done, you should randomly select cases within each defined partition. Such a sampling is called *stratified random sampling*. The partitions are the "strata" that are sampled separately.

3. *It can help you balance the occurrence of rare events for analysis by machine learning tools.*

   As mentioned earlier in this chapter, machine learning tools like neural nets and decision trees are very sensitive to unbalanced data sets. An unbalanced data set is one in which one category of the target variable is relatively rare compared to the other ones. Balancing the data set involves sampling the rare categories more than average (oversampling) or sampling the common categories less often (undersampling).

4. *Finally, simple random sampling can be used to divide the data set into three data sets for analysis*:

    a. Training set: These cases are randomly selected for use in training the model.

    b. Testing set: These cases are used to assess the predictability of the model, before refining the model or adding model enhancements.

    c. Validation set: These cases are used to test the final performance of the model after all modeling is done.

Some data miners define the second set as the validation set and the third set as the testing set. Whatever you prefer in nomenclature, this testing process should proceed in the manner described. The need for the second testing set is that it is not appropriate to report model performance on the basis of the second data set, which was used in the process of creating the model. That situation would create a logical tautology, or using a thing to describe itself. The validation data set should be kept completely separate from the iterative modeling process. We describe this iterative process in greater detail in Chapter 6.

Reduction of Dimensionality

Now that we have our Analytical Record (amplified, if necessary, with temporal abstractions), we can proceed to weed out unnecessary variables. But how do you determine which variables are unnecessary to the model before you train the model? This is almost a Catch-22 situation. But fortunately, there are several techniques you can perform to give you some insights into identifying the proper variables to submit to the algorithm and which ones to delete from your Analytical Record.

Correlation Coefficients

One of the simplest ways to assess variable relationships is to calculate the simple correlation coefficients between variables. Table 4.3 is a correlation matrix, showing pair-wise correlation coefficients. These data have been used in many texts and papers as examples of predictor variables used to predict the target variable, crime rate.

From the correlation matrix in Table 4.3, we can learn two very useful things about our data set. First, we can see that the correlations of most of the variables with crime rate are

TABLE 4.3   Correlation Coefficients for Some Variables in the Boston Housing Data Set

Correlations of Some Variables in the Boston Housing Data Set. Correlations in Red Are Significant at the 95% Confidence Level				
	Crime Rate	**Nonretail Bus Acres**	**Charles River**	**Dist to Empl Centers**	**Property Tax Rate**
Crime Rate	1.000000	0.406583	−0.055892	−0.379670	0.582764
Nonretail Bus Acres	0.406583	1.000000	0.062938	−0.708027	0.720760
Charles River	−0.055892	0.062938	1.000000	−0.099176	−0.035587
Dist to Empl Centers	−0.379670	−0.708027	−0.099176	1.000000	−0.534432
Property Tax Rate	0.582764	0.720760	−0.035587	−0.534432	1.000000

relatively high and significant, but that for Charles River proximity is relatively low and insignificant. This means that Charles River may not be a good variable to include in the model. The other thing we can learn is that none of the correlations of the predictor variables is greater than 0.90. If a correlation between two variables exceeded 0.90 (a common rule-of-thumb threshold), their effects would be too *collinear* to include in the model. Collinearity occurs when plots of two variables against the target lie on almost the same line. Too much collinearity among variables in a model (multicollinearity) will render the solution ill-behaved, which means that there is no unique optimum solution. Rather, there will be too much overlap in the effects of the collinear variables, making interpretation of the results problematic.

Chi-Square Automatic Interaction Detection (CHAID)

The CHAID algorithm is used occasionally as the final modeling algorithm, but it has a number of disadvantages that limit its effectiveness as a multivariate predictor. It is used more commonly for variable screening to reduce dimensionality. But even here, there is a problem of bias toward variables with more levels for splits, which can skew the interpretation of the relative importance of the predictors in explaining responses on the dependent variable (Breiman et al., 1984).

Despite the possible bias in variable selection, it is used commonly as a variable screen method in several data mining tools (e.g., *STATISTICA* Data Miner).

Principal Components Analysis (PCA)

Often, PCA is used to identify some of the strong predictor variables in a data set. PCA is a technique for revealing the relationships between variables in a data set by identifying and quantifying a group of *principal components*. These principal components are composed of transformations of specific combinations of input variables that relate to a given output (or target) variable (Jolliffe, 2002). Each principal component accounts for a decreasing amount of the variations in the raw data set. Consequently, the first few principal components express most of the underlying structure in the data set. Principal components have been used frequently in studies as a means to reduce the number of raw variables in a data set (Fodor, 2002; Hall and Holmes, 2003). When this is done, the original variables are replaced by the first several principal components. This approach to the analysis of variable relationships does not specifically relate the input variables to any target variable. Consequently, the principal components may hide class differences in relation to the target variable (Hand et al., 2001). In one study of the application of PCA to face recognition, the principal components tended to express the wrong characteristics suitable for face recognition (Belhumeur et al., 1997). Therefore, you may see varying success with the uses of PCA for dimensionality reduction to create the proper set of variables to submit to a modeling algorithm.

Gini Index

The Gini Index was developed by the Italian statistician Corrado Gini in 1912, for the purpose of rating countries by income distribution. The maximum Gini Index = 1 would mean that all the income belongs to one country. The minimum Gini Index = 0 would mean that the income is evenly distributed among all countries. This index measures the degree of unevenness in the spread of values in the range of a variable. The theory is that variables with a relatively large amount of unevenness in the frequency distribution of values in its range (a high Gini Index value) have a higher probability to serve as a predictor variable for another related variable. We will revisit the Gini Index in Chapter 5.

Graphical Methods

You can look at correlation coefficients to gain some insight into relationships between numeric variables, but what about categorical variables? All is not lost. Some data mining tools have specialized graphics for helping you to determine strength of relationships between these categorical variables. For example, SPSS Clementine provides the Web node that draws lines between categorical variables positioned on the periphery of a circle (Figure 4.4). The width of the connecting lines represents the strength of the relationship between the two variables. The following Web diagram shows a strong relationship between preferences of "No" for Diet Pepsi (located at about 4 p.m. on the periphery of the diagram) and No for Diet 7UP (located at about 2 p.m. in the diagram). There are no links between "Yes" for Diet Pepsi and Diet 7UP (5:30 p.m. on the diagram and 2:30 p.m., respectively). Therefore, you might expect that there might be a relatively strong relationship between the "No" preferences of these beverages.

Other common techniques used for reduction of dimensionality are

- Multidimensional scaling
- Factor analysis
- Singular value decomposition
- Employing the "Kernel Trick" to map data into higher-dimensional spaces. This approach is used in Support Vector Machines and other Kernel Learning Machines, like KXEN. (See Aizerman et al., 1964.)

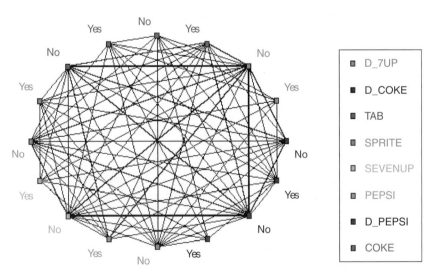

FIGURE 4.4 Web diagram from SPSS Clementine.

Recommendation. You can gain a lot of insight on relationships from simple operations, like calculating the means, standard deviations, minimum, maximum, and simple correlation coefficients. You can employ the Gini score easily by running your data through one of the Gini programs on the DVD. Finally, you can get a quick multivariate view of data relationships by training a simple decision tree and inferring business rules. Then you will be ready to proceed in preparing your data for modeling.

These techniques are described and included in many statistical and data mining packages.

Data Discretization

Some machine learning techniques can work with only categorical predictor variables, not continuous numeric variables. You can convert a continuous numeric variable into a series of categories by assigned subranges of the value range to a group of new variables. For example, a variable ranging from 1–100 could be discretized (converted into discrete values) by dividing the range in four subranges: 0–25, 26–50, 51–75, and 76–100. Another name for these subranges is *bins*. In the binning process, each value in the range of a variable is replaced by a bin number. Many data mining packages have binning facilities to create these subranges automatically. One of the attributes of the binning process is that it reduces noise in the data. To that extent, binning is a form of data smoothing. Credit scores are created using bins, in which bin boundaries are tuned and engineered to maximize the predictive power of the credit scoring model. The Scorecard Module in the Fair Isaac Model Builder tool is used to produce the FICO credit score. It uses a range engineering approach in the process of interactive binning to maximize the Information Content (IV) and Weight of Evidence (WOE) associated with a specific binning design. The IV provides a measure of the loss of information when bins are combined. The WOE relates the proportion of good credit scores with bad credit scores in each bin for that variable in the training data set. This approach to prediction engineers the *data* to maximize the predictability of a very simple linear programming modeling algorithm. This focus on data engineering is very different from the model engineering approach presented in Chapter 6. But both approaches can yield very predictive models. Even though you may choose the model engineering approach, you can leverage data engineering concepts to some degree in the data preparation process by

- Recoding data
- Transforming data
- Binning data
- Smoothing data
- Clustering data

Data Derivation

Assignment or Derivation of the Target Variable

The operation involving assignment or derivation of the target variable defines the modeling goal in terms of available input variables. The modeling goal is to "hit" the target variable with the prediction of the model. Often, the target variable can be selected from among the existing variables in the data set. For example, the target variable for a model of equipment failure could be the presence or absence of a failure date in the data record. In other cases, the target variable may be defined in a more complex manner. The target variable for customer attrition in a model created by one of the authors was defined as the month in which customer phone usage declined at least 70% over the previous two billing periods. This variable was derived by comparing the usage of all customers for each month in the time-series data set with the usage two billing periods in the past. The billing period of this cellular phone company was every two months, so the usage four months previous to each month was used as the value of comparison. Most often, the target variable must be derived following some heuristic (logical rule). The simplest version of an attrition target variable in that cellular phone company would have been to identify the month in which the service was discontinued. Insurance companies define attrition in that manner also.

Derivation of New Predictor Variables

New variables can be created from a combination of existing variables. For example, if you have access to latitude and longitude values for each case (for example, a customer list), you might create a new variable, Distance to Store, by employing one of the simple equations for calculating distance on the surface of the earth between two pairs of latitude-longitude coordinates. One common formula for calculating this distance is based on the Law of Cosines and expressed here in the form of an Excel cell formula:

$$= \text{ACOS}(\text{SIN}(\text{Lat1}) * \text{SIN}(\text{Lat2}) + \text{COS}(\text{Lat1}) * \text{COS}(\text{Lat2}) *$$
$$\text{COS}(\text{Lon2} - \text{Lon1})) * 3934.31$$

The value 3934.31 is the average radius of the earth in miles, and output is the distance in miles between the two points. The latitude and longitude values must be expressed in radians so that the trig functions work properly.

Other transformations might include calculation of rates. For example, you could divide one variable (number of purchases) by another variable (time-interval) to yield the purchase rate over time. The raw values of the number of purchases may show little relationship to attrition (customers leaving the company), while decline in purchase rates might be very predictive.

Attribute-Oriented Induction of Generalization Variables

Han and Kamber (2006) define the technique of attribute-oriented induction as generalizing from a list of detailed categories in a variable to form a higher-level (more general) expression of a variable. For example, you might lack information about a customer's occupation. You could form a concept generalization, White_Collar_Worker, based on specific levels in a number of other variables (e.g., Yearly_Salary, Homeowner, and Number_of_Cars). That induced variable might be very predictive of your target variable. See Han and Kamber (2006) for more details on concept generalization and attribute-oriented induction of variables.

You can also induce segmentation variables using this technique. For example, you might query the database of banking customer prospects against the customer database to find indirect relationships with these prospects, considering matching addresses, phone numbers, or secondary signer information on customer accounts. All matches could be coded as *Y* in a new variable, Indirect_Relationship, and all others are coded as *N*. All prospects with *Y* in the Indirect_Relationship variable could be used as targets for a specific marketing campaign to sell them direct banking services.

POSTSCRIPT

After you are done with data preparation, you are ready to choose your list of variables to submit to the modeling algorithm. If this list of variables (or features) is determined manually, it may take a long time to complete. But there is help in the wings of most data mining tools. Most tools have some form of Feature Selection facility to help you select which features you want to use. Chapter 5 will present some feature selection techniques available for you to use, depending on your tool of choice.

References

Aizerman, M., Braverman, E., & Rozonoer, L. (1964). Theoretical foundations of the potential function method in pattern recognition learning. *Automation and Remote Control, 25,* 821–837.

Allison, P. D. (2002). *Missing Data.* Thousand Oaks, CA: Sage Publishing.

Belhumeur, P. N., Hespanha, J. P., & Kriegman, D. J. (1997). Eigenfaces vs. fisherfaces: Recognition using class specific linear projection. *IEEE Transactions on Pattern Analysis and Machine Intelligence, 19*(7), 711–720.

Breiman, L., Friedman, J. H., Olshen, R. A., & Stone, C. J. (1984). *Classification and Regression Trees.* Boca Raton, FL: Chapman & Hall.

Fodor, I. K. (2002). *A survey of dimension reduction techniques. LLNL Technical Report,* UCRL-ID-148494.

Hall, M., & Holmes, G. (2003, November/December). Benchmarking attribute selection techniques for discrete class data mining. *IEEE Transactions on Knowledge and Data Engineering, 15*(3), 1437–1447.

Han, J., & Kamber, M. (2006). *Data Mining: Concepts and Techniques.* San Francisco: Morgan Kaufmann.

Han, J., Fu, Y., Wang, W., Koperski, K., & Zaiane, O. (1996). DMQL: A data mining query language for relational databases. In *1st ACM SIGMOD Workshop on Research Issues in Data Mining and Knowledge Discovery* (pp. 27–33). June.

Hand, D., Mannila, H., & Smyth, P. (2001). *Principals of Data Mining.* Cambridge, MA: MIT Press.

Hawkins, D. (1980). *Identification of Outliers.* London: Chapman and Hall.

Imielinski, T., & Virmani, A. (1999). MSQL: A query language for database mining. *Data Mining and Knowledge Discovery: An International Journal, 3,* 373–408.

Jolliffe, I. T. (2002). *Principal Component Analysis* (2nd ed.). New York, NY: Springer.

Lavrac, N., Keravnou, E., & Zupan, B. (2000). *Intelligent data analysis in medicine.* White Paper. Slovenia: Faculty of Computer and Information Sciences, University of Ljubljana.

Masters, T. (1995). *Neural, Novel & Hybrid Algorithms for Time-Series Predictions.* New York: J.Wiley & Sons.

Ramesh, G., Maniatty, W.A., & Zaki, M.J. (2001). *Indexing and data access methods for database mining.* Technical Report UMI order number SUNYA-CS-01-01: 1–8. Albany, NY, USA: Dept. of Computer Science, University at Albany.

Witten, I. H., & Frank, E. (2005). *Data Mining: Practical Machine Learning Tools and Techniques.* San Francisco: Morgak Kaufmann Publishers.

Feature Selection

PREAMBLE

After your analytical data set is prepared for modeling, you must select those variables (or features) to use as predictors. This process of feature selection is a very important strategy to follow in preparing data for data mining. A major problem in data mining in large data sets with many potential predictor variables is *the curse of dimensionality*. This expression was coined by Richard Bellman (1961) to describe the problem that increases as more variables are added to a model. As additional variables are added to a model, it may be able to predict a number better in regression models or discriminate better between classes in a classification model. The problem is that *convergence* on those solutions during either the error minimization process or the iterative learning process gets increasingly slow as additional variables are added to the analysis. Feature selection aims to reduce the number of variables in the model, so it lessens the effect of the curse by removing irrelevant or redundant variables, or noisy data. It has the following immediate positive effects for the analysis:

- Speeds up processing of the algorithm
- Enhances data quality
- Increases the predictive power of the algorithm
- Makes the results more understandable

Therefore, one of the first jobs of the data miner is to develop a *short-list* of variables. This abbreviated list will include (one hopes) only those variables that significantly increase the predictive power and the generalizing ability of the model.

VARIABLES AS FEATURES

Variables are also known as *attributes*, *predictors*, or *features*. But in some areas of machine learning, a distinction is made between a variable and a feature. Kernel learning machines (including Support Vector Machines) transform variables with some mathematical function to relate them to higher-order theoretical spaces. Humans can understand a third-order space (with three dimensions) and even a fourth-order space, when you consider objects occupying the same three-dimensional space at different times. Mathematics can define theoretical spaces with N-dimensions (up to infinity), in which dimensions are defined with a mathematical function. When this is done, the transformed variables are called *features*, not variables. The calculation process of converting these variables to features is called *mapping*. Each of the variables is mapped into the higher-dimensional space called a *hyperspace*. This space can be mathematically defined so that it is possible to separate clusters of mapped data points with a plane defined by the dimensions. This *hyperplane* can be configured to maximally separate clusters of data in a classification problem. This is how a Support Vector Machine functions. Even though variables will be mapped into hyperspaces by some modeling algorithms, we will use the terms *features* and *variables* interchangeably in this book.

TYPES OF FEATURE SELECTIONS

There are two types of feature selection strategies: feature ranking methods and best subset selection.

FEATURE RANKING METHODS

Simple feature ranking methods include the use of statistical metrics, like the correlation coefficient (described in Chapter 4). A more complex feature ranking method is the Gini Index (introduced in Chapter 4).

Gini Index

The Gini Index can be used to quantify the unevenness in variable distributions, as well as income distributions among countries. The theory behind the Gini Index relies on the difference between a theoretical equality of some quantity and its actual value over the

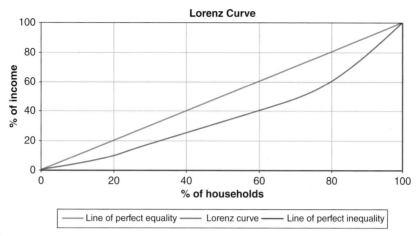

FIGURE 5.1   The Lorenz curve relating the distribution of income among households in a population.

range of a related variable. This concept was introduced by Max O. Lorenz in 1905 to represent the unequal distribution of income among countries, and can be illustrated by Figure 5.1.

The theoretical even distribution of income among households is symbolized by the straight line through the center of the figure. The inequality in incomes among households is shown by the red line below the line of perfect equality. If the red line remained near the bottom of the figure until the 80th percentile, for example, it would represent a population with a few very rich people and a lot of very poor people.

Cerrado Gini incorporated the Lorenz concept in 1912 to quantify the change in relative frequency of income values along the range of a population of countries. For example, if you divide the % Households into deciles (every 10%), you can count the number of households in each decile and express the quantity as a relative frequency. This binning approach allows you to use a frequency-based calculation method instead of an integration method to find the area under the Lorenz curve at each point along the % Households axis (the x-axis). With this approach, you can calculate the *relative mean difference* (RMD) of all the binned values (frequency of a bin/mean frequency across all bins), and divide it by $2 \times$ the mean frequency value, expressed for a population of size n, with a sequence of values y_i, $i = 1$ to n:

$$RMD = \frac{MD}{\text{arithmetic mean}}$$

where

$$MD = \frac{1}{n^2} \sum_{i=1}^{n} \sum_{j=1}^{n} |y_i - y_j|$$

You can use this method as a guide in selecting a short-list of variables to submit to the modeling algorithm. For example, you might select all variables with a Gini score greater than 0.6 for entry into the model. The disadvantage of using this method is that it combines effects of data in a given range of one variable that may not reflect the combined effects of all variables interacting with it. But that is the problem with most feature ranking methods.

A slightly more integrative approach is to use bi-variate methods like the scatterplots and web diagrams described in Chapter 4.

Bi-variate Methods

Other bi-variate methods like mutual information calculate the distance between the actual joint distribution of features X and Y and what the joint distribution would be if X and Y were independent. The joint distribution is the probability distribution of cases where both events X and Y occur together. Formally, the mutual information of two discrete random variables X and Y can be defined as

$$I(X;Y) = \sum_{y \in Y} \sum_{x \in X} p(x,y) \, \log \left(\frac{p(x,y)}{p_1(x)p_2(y)} \right),$$

where $p(x,y)$ is the joint probability distribution function, and $p_1(x)$ and $p_2(y)$ are the independent probability (or marginal probability) density functions of X and Y, respectively. If you are a statistician, this likely all makes sense to you, and you can derive this metric easily. Otherwise, we suggest that you look for some approach that makes more sense to you intuitively. If this is the case, you might be more comfortable with one of the multivariate methods implemented in many statistical packages. Two of those methods are stepwise regression and partial least squares regression.

Multivariate Methods

Stepwise Linear Regression

A slightly more sophisticated method is the one used in stepwise regression. This classical statistics method calculates the F-value for the incremental inclusion of each variable in the regression. The F-value is equivalent to the square root of the student's t-value, expressing how different two samples are from each other, where one sample includes the variable and the other sample does not. The t-value is calculated by

$t =$ difference in the sample means/standard deviation of differences

and so

$$F = \sqrt{t\text{-value}}$$

The F-value is sensitive to the number of variables used to calculate the numerator of this ratio and for the denominator. Stepwise regression calculates the F-value both with and without using a particular variable and compares it with a critical F-value to either include the variable (forward stepwise selection) or to eliminate the variable from the regression (backward stepwise selection). In this way, the algorithm can select the set of variables that meets the F-value criterion. It is assumed that these variables account for a sufficient amount of the total variance in the target variable to predict it at a given level of confidence specified for the F-value (usually 95%).

If your variables are numeric (or can be converted to numbers), you can use stepwise regression to select the variables you use for other data mining algorithms. But there is a fly in this ointment. Stepwise regression is a *parametric* procedure and is based on the same assumptions characterizing other classical statistical methods. Even so, stepwise regression can be used to give you one perspective on the short-list of variables. You should use other methods and compare lists. Don't trust necessarily the list of variables included in the regression solution because their inclusion assumes linear relationships of variables with the target variable, which in reality may be quite nonlinear in nature.

Partial Least Squares Regression

A slightly more complex variant of multiple stepwise regression keeps track of the partial sums of squares in the regression calculation. These partial values can be related to the contribution of each variable to the regression model. *STATISTICA* provides an output report from partial least squares regression, which can give another perspective on which to base feature selection. Table 5.1 shows an example of this output report for an analysis of manufacturing failures.

It is obvious that variables 1 and 2 (and marginally, variable 3) provide significant contributions to the predictive power of the model (total $R^2 = 0.934$). On the basis of this analysis, we might consider eliminating variables 4 through 6 from our variable short-list.

Sensitivity Analysis

Some machine learning algorithms (like neural nets) provide an output report that evaluates the final weights assigned to each variable to calculate how *sensitive* the solution is to the inclusion of that variable. These sensitivity values are analogous to the F-values calculated for inclusion of each variable in stepwise regression. Both SPSS Clementine and *STATISTICA* Data Miner provide sensitivity reports for their automated neural nets. These sensitivity values can be used as another reflection of the best set of variables to include in a model. One strategy that can be followed is to train a neural net with default

TABLE 5.1 Marginal Contributions of Six Predictor Variables to the Target Variable (Total Defects)

Summary of PLS (fail_tsf.STA) Responses: TOT_DEFS Options: NO-INTERCEPT AUTOSCALE	
	Increase - R^2 of Y
Variable 1	0.799304
Variable 2	0.094925
Variable 3	0.014726
Variable 4	0.000161
Variable 5	0.000011
Variable 6	0.000000

characteristics and include in your short-list all variables with greater than a threshold level of sensitivity. Granted, this approach is less precise than the linear stepwise regression, but the neural net set of variables may be much more generalizable, by virtue of their ability to capture nonlinear relationships effectively.

Complex Methods

A piecewise linear network uses a distance measure to assign incoming cases to an appropriate cluster. The clusters can be defined by any appropriate clustering method. A separate function called a *basis function* is defined for each cluster of cases. A pruning algorithm can be applied to eliminate the least important clusters, one at a time, leading to a more compact network. This approach can be viewed as a nonlinear form of stepwise linear regression.

Multivariate Adaptive Regression Splines (MARS)

The MARS algorithm was popularized by Friedman (1991) to solve regression and classification problems with multiple outcomes (target variables). This approach can be viewed as a form of piecewise linear regression, which adapts a solution to local data regions of similar linear response. Each of the local regions is expressed by a different basis function. MARS algorithms can also be viewed as a form of regression tree, in which the hard splits into separate branches of the tree are replaced by the smooth basis functions. In *STATISTICA* Data Miner (for example), the MARSplines algorithm includes a pruning routine, which provides a very powerful tool for feature selection. The MARSplines algorithm will pick up only those basis functions (and those predictor variables) that provide a sizeable contribution to the prediction. The output of the MARSplines module will retain only those variables associated with basis functions that were retained for the final solution of the model and rank them according to the number of times they are used in different parts of the model.

You can run your data through a procedure like the *STATISTICA* MARSplines module to gain some insights for building your variable short-list. Refer to Hastie et al. (2001) for additional details.

SUBSET SELECTION METHODS

The subset selection approach to feature selection evaluates a subset of features that have significant effect as a group for predicting a target variable. The most common subset selection approaches are wrapper-based. Wrappers use a search algorithm to search through the space of possible features and evaluate each subset by running a model on the subset. Some wrapper methods perform this evaluation with different randomly selected subsets, using a cross-validation method. Cross-validation divides the data set into a number of subsets for each group of features and evaluates a model trained on all but one subset. The subset not used for the model is used to validate the model for that iteration. During the next iteration, a different random subset is used for validation.

The freely available Weka software has wrapper ability. But another way to use wrapper-based feature selection methods cheaply is to use RapidMiner, a GNU open-source data mining package. RapidMiner provides four feature selection methods:

- Backward feature selection, using multiple subsets
- Feature weighting using nearest-neighbor
- Wrapper-based feature selection
- Automatic feature selection

RapidMiner provides a wizard to help you create a new analysis process (Figure 5.2). The wizard guides you through creating a new process. You start by selecting a template process from a list. This template serves as a kind of skeleton for your process.

You can process your variable list through RapidMiner and submit the variable short-list to your favorite modeling algorithm or ensemble.

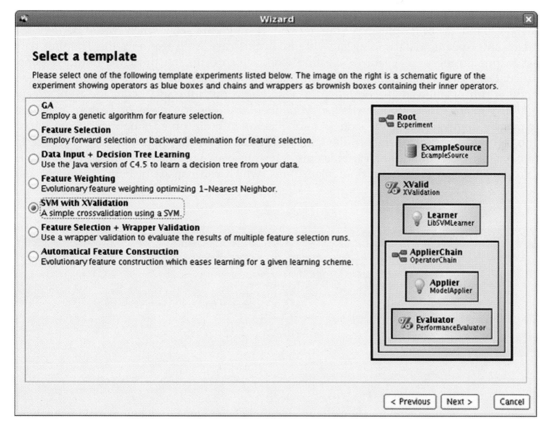

FIGURE 5.2 The RapidMiner Process Creation Wizard is a simple way to create basic process definitions.

STATISTICA's Feature Selection tool is very easy to use, especially in the Data Miner Workspace, and also automatically, behind the scenes without your having to do anything in the Data Miner Recipe format. *STATISTICA* has three formats for doing data mining:

1. Interactive module, where Feature Selection is available;
2. Data Miner Workspace, where you have the most control over Feature Selection; and
3. Data Miner Recipe, where Feature Selection is basically automatic.

The Data Miner Workspace (analogous to SPSS Clementine and SAS-EM Workspace) provides one of the easiest ways for you to manipulate the parameters of feature selection, and after the process is completed, to easily copy and paste the features (= variables of importance) as the selected variables into any of the other three formats—e.g., (1) data mining interactive module, (2) Data Miner Workspace for competitive evaluation of several algorithms, and (3) DMRecipe for allowing you to control the variables selected, instead of using the default automatic selection that is available in this format.

Use of the Feature Selection node in the *STATISTICA* Data Miner Workspace is illustrated in the following sequence of figures.

In the Data Miner Workspace (Figure 5.3), the data, a credit scoring set, are embedded as an icon in the left panel. In the next panel, called Data Preparation, Cleaning, Transformation, the data set is split into Training and Testing data sets, which would be used for taking into specific data mining algorithms. In this case we'll use only the Training data set to put through the Feature Selection icon, which is in the third red-outlined panel called Data Analysis, Modeling, Classification, Forecasting. When the Feature Selection node is run or allowed to compute, the results are put into an icon in the right green-outlined panel called Reports. When you open this Reports node of Feature Selection (either by double-clicking the icon or right-clicking to bring up a flying menu from which you can select View Document), the variables selected as being important are revealed. In most cases, these important variables are the only ones you need to put as the variable selections that go into the various algorithms, which can consist of both traditional and

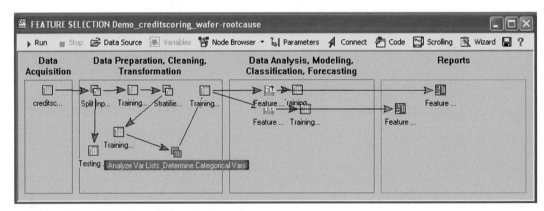

FIGURE 5.3   Statistical Data Miner Workspace.

statistical learning theory algorithms. The reason is that the variables *not* selected are either redundant or not correlated (e.g., predictive) of the dependent (or sometimes called target) variable.

Let's look at some of the specific dialogs and results workbooks that come out of this Feature Selection node. If we right-click this Training Data icon and select the View Data option, the lists of variables shown in Figure 5.4 appears.

Next, we'll look at the Feature Selection icons in Figure 5.5 to see what parameters have been set for this example.

Again, by right-clicking on the Feature Selection and Root Cause Analysis node, and then selecting Edit Parameters, we bring up the dialog shown in Figure 5.6.

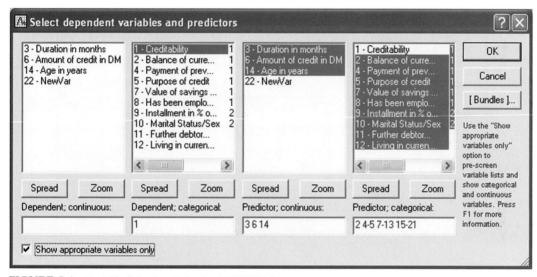

FIGURE 5.4   Variable Selection dialog in *STATISTICA*.

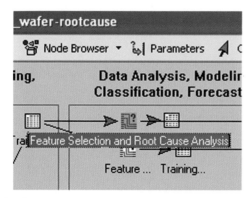

FIGURE 5.5   Feature Selection icon.

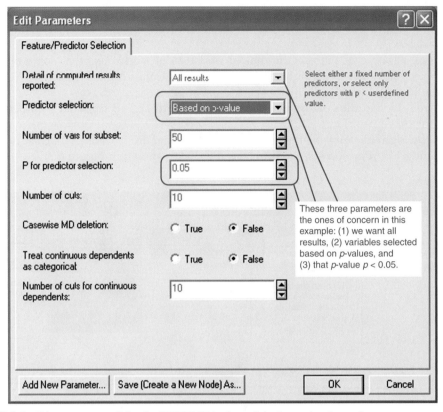

FIGURE 5.6   Edit parameters dialog in *STATISTICA* where defaults can be changed.

Next, by double-clicking (or right-clicking and selecting View Document), we can see what variables have been selected that have a *p*-value of less than 0.95, based on the iterative method used for this feature selection (Figure 5.7).

In Figure 5.7, we can see that only nine variables had a *p*-value of <0.05, and of those, Balance of Current Account was the most important or most predictive variable to determine whether a client was worthy of being given credit.

We can also look at the table, giving these variables and *p*-values, by clicking on the Best Predictor icon in the tree hierarchy on the left panel to get the view shown in Figure 5.8.

We can also click on the Best Predictor List, the bottom item in the tree hierarchy of this results workbook, to get a listing of the variable numbers that are important (Figure 5.9).

The list in Figure 5.9 is very useful. For selecting the variables that will be used for the analyses in the data mining algorithms, the easiest way is to just highlight the longer list of categorical predictors and copy and paste into the subsequent Variables Selection dialog; in this case, you can probably remember the two continuous predictors, e.g., the 3 and 6, so that only one copy/paste operation needs to be done.

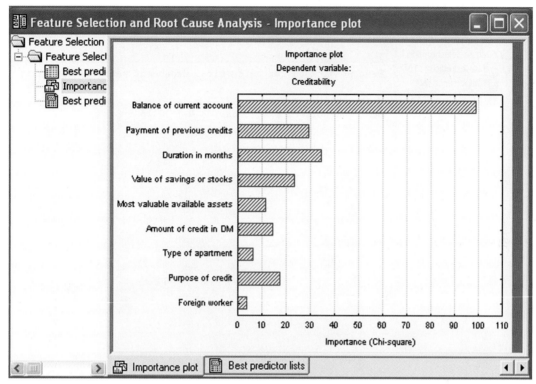

FIGURE 5.7 Importance plot result from Feature Selection Analysis.

Feature Selection and Root Cause Analysis - Best predictors for categorical dependent var: Cr...

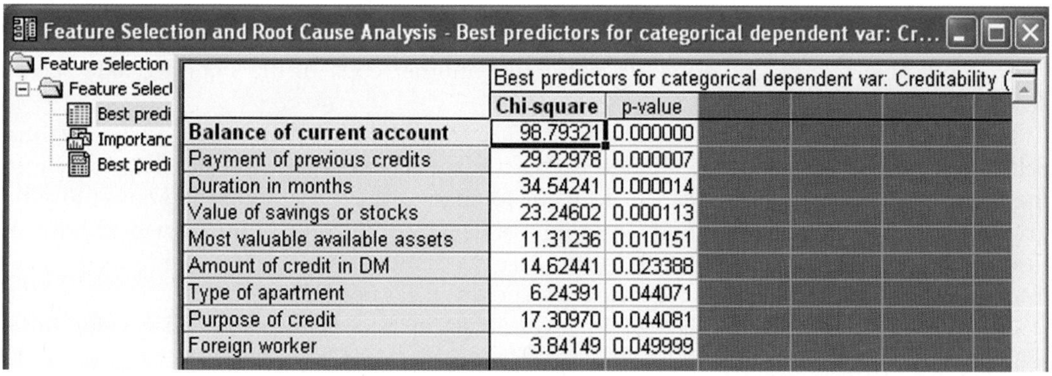

	Best predictors for categorical dependent var: Creditability (	
	Chi-square	p-value
Balance of current account	98.79321	0.000000
Payment of previous credits	29.22978	0.000007
Duration in months	34.54241	0.000014
Value of savings or stocks	23.24602	0.000113
Most valuable available assets	11.31236	0.010151
Amount of credit in DM	14.62441	0.023388
Type of apartment	6.24391	0.044071
Purpose of credit	17.30970	0.044081
Foreign worker	3.84149	0.049999

FIGURE 5.8 Best predictor variables ordered top to bottom on basis of highest chi-square to lowest.

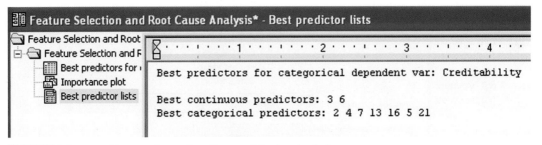

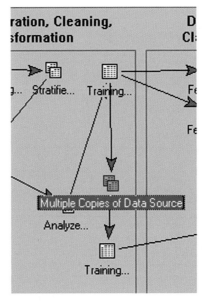

FIGURE 5.9   List of best predictors from Feature Selection Analysis.

FIGURE 5.10   Multiple copies of data source.

To show how this can be done, we can make another copy of the Training data set, as shown in Figure 5.10.

Multiple Copies of Data Source copies the Training data set that was used for Feature Selection (as we don't want to change the entire set of variables selected that went into the feature selection process, *but* we do want to change the variable list to show only the important variables in this second Training data set, so that only those variables are run through the Data Miner algorithms). In this case, to demonstrate, a Classification Tree data mining algorithm is represented by the icon shown in Figure 5.11.

Next, we'll highlight the categorical variables in the Feature Selection workbook (Figure 5.12).

Now we'll copy these variables, remembering the 3 and 6 for the continuous predictors. Then, opening this copy of the Training data set node, we'll make the important variable selections, as shown in Figure 5.13.

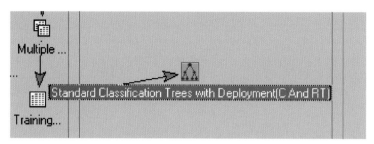

FIGURE 5.11   C and RT (classification tree) icon.

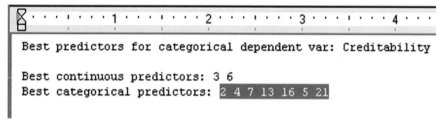

FIGURE 5.12   List of best predictors from Feature Selection.

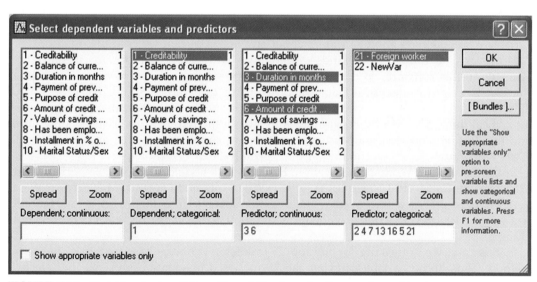

FIGURE 5.13   Variable selection dialog with variables from the "best predictors" list (as seen in Figure 5.12) selected.

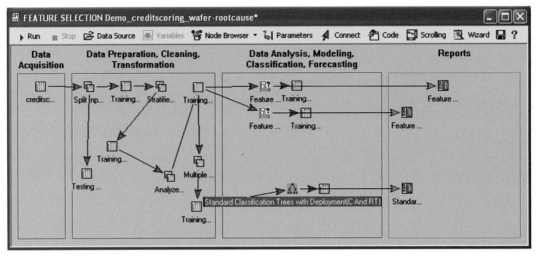

FIGURE 5.14   Addition of duplicate training data icon and standard Classification tree icon to this Data Miner project.

The Data Miner Workspace that has been added to this duplicate Training data set and the classification tree algorithm now looks like Figure 5.14 after running this algorithm.

And the Classification tree results, obtained from this Results icon in the green panel on the right, and illustrated in Figure 5.15, shows in this simple example that only one variable could be used to determine if a particular credit transaction was advisable. (Of course, it is more complicated than this, and on the DVD accompanying this book is a complete Credit Scoring Tutorial to work through, to see further details, and discover how accurate this one variable is in making a good judgment about any one applicant for credit.)

However, you can make all kinds of variations with the flexibility and customization available in this software, so one will be demonstrated in an example using a computer chip/wafer chip manufacturing data set consisting of 2858 variables and 2062 cases. Naturally, 2858 variables are too many to keep track of for quality control on an assembly line, so the critical thing needed in this example is to reduce the number of variables being input into the Quality Control Data Mining algorithms to as few as possible and yet maintain 95–99% quality wafer chips coming off the assembly line.

In the Data Miner Workspace shown in Figure 5.16, a node called Analyze Variable Lists to Determine Categorical Variables is added. When this node is run, it makes two Wafer Yields data sets. They will be used to make Scatterplots by Time (Figure 5.17); note that this is a special Feature Selection icon, with a different name than the one in the previous example.

In this case, we are asking for the top 25 predictor variables based on the Chi-square method. Thus, those variables having the 25 highest Chi-square scores will be selected; the p-values are also given, but the variables are ordered in decreasing value of the Chi-square, as seen in the results table in Figure 5.18.

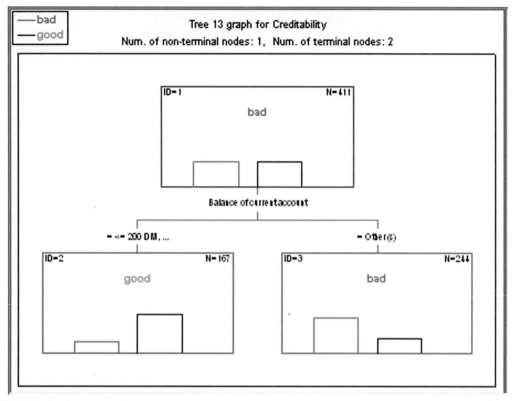

FIGURE 5.15 Classification tree results; balance of current account acts as a single variable to determine a "good" or "bad" credit risk.

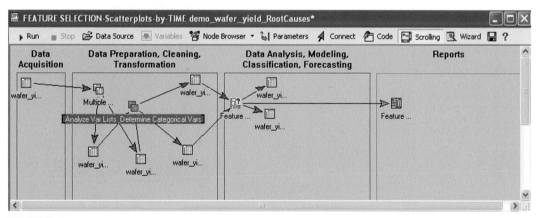

FIGURE 5.16 Analyze variable lists to determine categorical variables node are highlighted.

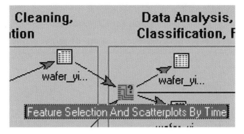

FIGURE 5.17   A specially constructed icon called "Feature Selection and Scatterplots by Time".

The importance plot is shown in Figure 5.19.

From the plot in Figure 5.19, if we wanted to select the fewest variables for subsequent predictive modeling, we could either take the top 3 or maybe the top 4. Or we can look for other inflection points in the curve and maybe select the top 9 or maximally maybe the top 12, because after the top 12 variables, the remainder level off in a plateau effect.

	Best predictors for categorical dependent var: LowHiYield (wafer_yield)	
	Chi-square	p-value
Equipment NW00/5722	137.8842	0.000000
Equipment NW00/5822	137.8248	0.000000
Equipment NW00/5922	137.5644	0.000000
Equipment NW60/8751	84.1724	0.000000
Equipment NW30/9002	61.1891	0.000000
Equipment NW30/5922	58.6283	0.000100
Equipment NW30/5777	57.5475	0.000086
Equipment NW30/5722	57.4447	0.000089
Equipment NW20/6772	51.8290	0.000040
Equipment NW00/5922	46.7004	0.005318
Equipment NW00/5777	46.2775	0.005958
Equipment NW00/5722	46.2775	0.005958
Equipment NW00/5922	41.8120	0.000132
Equipment NW80/8799	41.2171	0.000005
Equipment NW20/6772	41.0879	0.000539
Equipment NW00/5822	40.6859	0.000107
Equipment NW00/5722	40.6078	0.000110
Equipment NW80/11223	39.6548	0.000001
Equipment NW00/5922	39.0139	0.000053
Equipment NW20/8900	38.9455	0.000107
Equipment NW00/5722	38.9216	0.000055
Equipment NW00/5822	38.2135	0.000072
Equipment NW00/5922	37.9806	0.002473
Equipment NW51/11223	37.8189	0.000001
Equipment NW00/5722	35.4158	0.003485

FIGURE 5.18   Top 25 predictor variables ordered by decreasing value of chi-square.

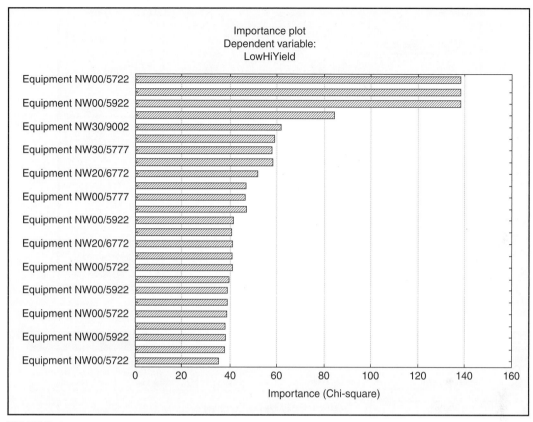

FIGURE 5.19   Importance plot generated by Feature Selection.

This special Feature Selection with Scatterplots by Time node produced a large number of plots, as you can see from the tree hierarchy of the workbook, plots taken at various time points in the production line/assembly line process of manufacturing wafer chips.

The Other Two Ways of Using Feature Selection in *STATISTICA*: Interactive Workspace

With the interactive Feature Selection workspace shown in Figure 5.20, you can use the more conventional point-and-click procedures to produce a list of variables selected to your specifications.

STATISTICA DMRecipe Method

The data redundancy methods shown in Figures 5.21 and 5.22 can be applied in the DMRecipe format to reduce dimensionality, e.g., reduce the variables going into the subsequent

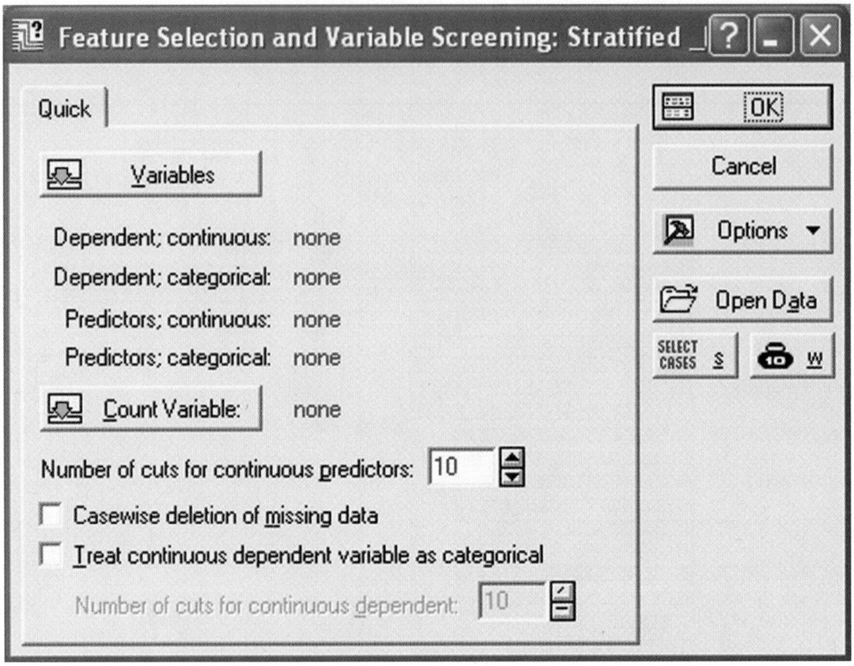

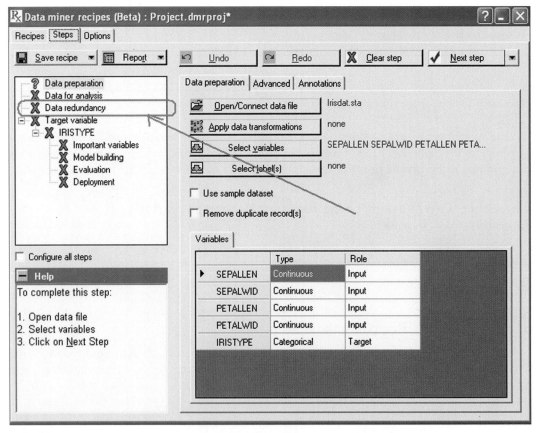

FIGURE 5.20 Interactive module dialog for Feature Selection.

FIGURE 5.21 Data redundancy selection in DMRecipe format.

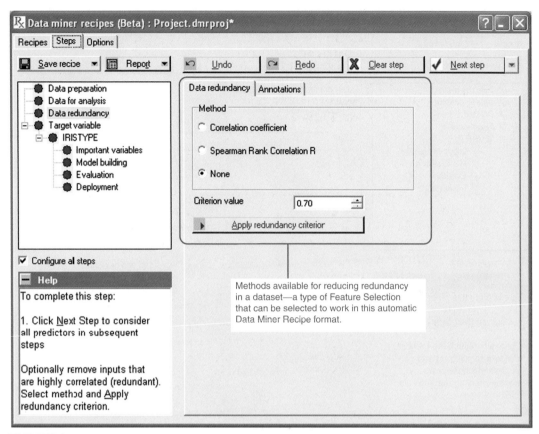

FIGURE 5.22   DMRecipe data redundancy methods.

data mining statistical learning theory algorithms. These DMRecipe methods of data redundancy use computations differently than those used in the Feature Selection methods of both the interactive Feature Selection module and the Data Miner Workspace, but they do a very good job of getting rid of extraneous variables in a model. The DMRecipe fast approach might be considered a quick-and-dirty method by pure academics, but for business purposes the results obtained are completely satisfactory for making important bottom-line business decisions. And one of the coauthors of this book has found that the accuracy scores obtained from the DMRecipe are essentially as good as the interactive algorithm, usually within a few percentage points. However, this same coauthor prefers to use Feature Selection in the Data Miner Workspace to find variables of interest and then select them by hand using the Select Variables button in the DMRecipe workspace.

But the DMRecipe does have a Feature Selection option, although the computations don't work identically to the Feature Selection in the workspace format of *STATISTICA*. Figure 5.23 shows where they can be selected, if you wish, rather than just taking the simple default of hitting Run.

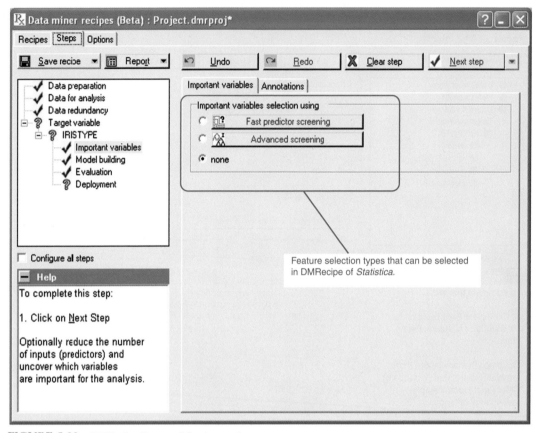

FIGURE 5.23   DMRecipe Feature Selection options.

POSTSCRIPT

We hope that the preceding information will provide new users with an understanding about how to use feature selection to reduce dimensionality in your data, which usually will provide much more accurate models/predictions. If you are used to using traditional statistics, especially factor analysis, you may see an analogy in this feature selection process; the difference, however, is that in factor analysis you can reduce dimensionality, but you have a much more difficult time defining what these new factors (e.g., the reduced dimensions) really represent; whereas with feature selection, you retain the originally recorded variable, reducing redundancy (e.g., variables basically recording the same concept) in a more understandable manner.

References

Bellman, R. (1961). *Adaptive Control Processes*. New Jersey: Princeton University Press.

Friedman, J. H. (1991). Multivariate Adaptive Regression Splines (with discussion). *Annals of Statistics, 19*, 1.

Hastie, T., Tibshirani, R., & Friedman, J. H. (2001). *Elements of Statistical Learning*. New York, Berlin and Heidelberg: Springer.

6

Accessory Tools for Doing Data Mining

PREAMBLE

Before moving into a discussion of the proper algorithms to use for a data mining project, we must take a side trip to help you understand that modeling algorithms are just one set of data mining tools you will use to complete a data mining project. The practice of data mining includes the use of a number of techniques that have been developed to serve as a set of tools in the data miner's toolbox. In the early days of data mining, many of these tools had to be built (usually in SQL or Perl) and used in an ad hoc fashion for every job. Many of these functions have been included as separate objects in data mining packages or "productized" separately. Most jobs will require the data miner to become proficient in even those tools that are not included in a given data mining package. The following tools can help the data miner:

- **Data access tools:** SQL and other database query languages.
- **Data integration tools:** Extract-transform-load (ETL) tools to access, modify, and load data from different structures and formats into a common output format (e.g., database, flat file).

- **Data exploration tools:** Basic descriptive statistics, particularly frequency tables; slicing, dicing, and drill-downs.
- **Model management tools:** Data mining workspace libraries, templates, and projects.
- **Modeling analysis tools:** Feature selection; model evaluation tools. (*Note:* This topic will be expanded in Chapter 13.)
- **Miscellaneous tools:** In-place data processing (IDP) tools, rapid deployment tools, model monitoring tools.

Being able to use these tools properly can be very helpful in the identification of significant variables, facilitating rapid decision-making necessary to compete successfully in the global marketplace.

DATA ACCESS TOOLS

Structured Query Language (SQL) Tools

Many SQL tools are available to extract data from databases, including MS SQL Server, Linux SQL tools, MySQL, Embarcadero, and others. These tools can be used to explore the nature of data in databases, prior to extraction. They can be used to extract data also, but other tools (such as ETL tools described later) may serve better. Some data access and data integration tools (e.g., Business Objects, DataFlux, DataStage) can serve as SQL generators to access and process data. Some data mining tools offer SQL query capabilities. For example, *STATISTICA* Data Miner provides a query generator for extraction of data from database tables (Figure 6.1).

Most extraction, transformation, and loading of data can be performed in native SQL programs, but it is most often the case that specialized ETL tools can perform these tasks more efficiently.

Extract, Transform, and Load (ETL) Capabilities

Most data mining packages provide at least some ETL functions. For the sake of example, we will show how one data mining tool package, *STATISTICA* Data Miner, can be used to perform ETL tasks.

Extracting Data: Connections can be made with various types of databases, including process databases (e.g., via the specialized *STATISTICA* OSI PI Connector). *STATISTICA* stores the metadata describing the nature of the tables that are queried, such as control limits, specification limits, valid data ranges, etc.

Transforming Data: STATISTICA Data Transformation nodes include standard operations for transposing, sorting, and ranking of data, in addition to standardizing, transforming, and stacking variables. Data can be aggregated and/or smoothed so that meaningful subsequent process monitoring methods (e.g., for change-point or trend detection) can be applied to robust or smoothed estimates of process averages within aggregated time intervals.

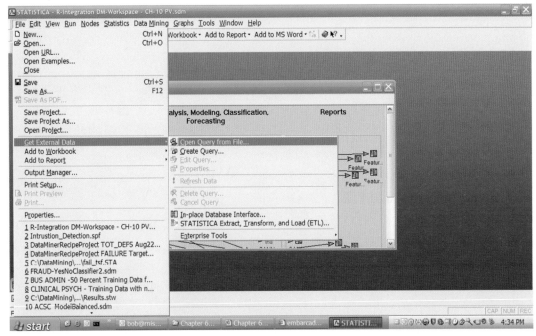

FIGURE 6.1   *STATISTICA* Data Miner's SQL generator.

These capabilities in *STATISTICA* are accessible in two places in the interface, as shown in Figures 6.2A and 6.2B.

Loading Data: Data loading tools in *STATISTICA* can automate the process of validating and aligning multiple diverse data sources into a single source suitable for ad hoc or automated analyses. In the Enterprise version of *STATISTICA*, data can be written back to database tables or to *STATISTICA* spreadsheet data sets. This write-back capability provides analysts and process engineers a convenient access to real-time performance data, without the need to perform tedious data preprocessing or cleaning before any actionable information can be extracted.

DATA EXPLORATION TOOLS

Basic Descriptive Statistics

Measures of Location

- **Mean:** The average for all observations in the range of a variable.
- **Median:** The middle observation in a sorted list of values in the range for a given variable.
- **Mode:** The most frequently occurring value.

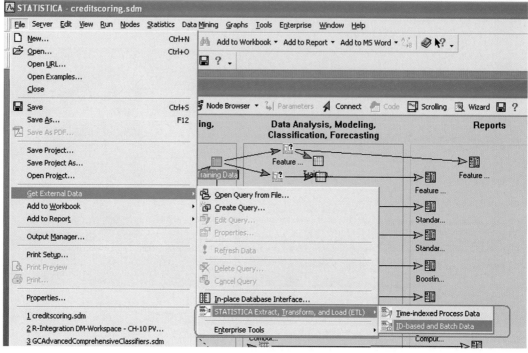

FIGURE 6.2A   ETL functions available in the File menu in *STATISTICA* Data Miner.

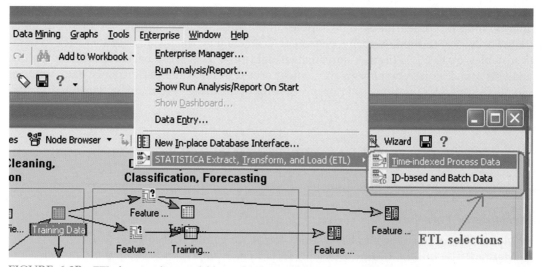

FIGURE 6.2B   ETL functionality available in the Enterprise menu in *STATISTICA* Data Miner, both on a time-based and ID-based format.

Measures of Dispersion

- **Variance:** A measure of the variability of squared values around the mean.
- **Standard Deviation:** The square root of the variance.

If the data are tightly clustered around the mean, the variance and standard deviation are relatively low.

If the data are widely scattered around the mean, the variance and standard deviation are relatively high.

Range

- **Maximum:** The highest value in the range of a variable.
- **Minimum:** The lowest value in the range of a variable.

Together with the mean and standard deviation, the maximum and minimum values can be useful in identifying *outliers* (values so much higher or so much lower than the vast majority of values that they appear to be the result of another process). Outliers may be mistaken readings, garbage data, or they may be very rare but valid measurements. Sometimes apparent outliers are the very values that may contain a disproportionately large amount of the signal of the target variable. The data miner is justified in deleting mistaken readings and garbage data. Under certain conditions, you might be justified in deleting even the very rare by valid measurements, because doing so will reduce the variance in the range of a variable, making it a stronger predictor of the target. In any event, the data miner should decide how to handle outliers in the context of the problem and his or her domain knowledge.

Measures of Position

- **Quantiles:** A portion of the total number of observations. Quantiles are usually names according to the number of portions into which the range is divided.
  - **Quartiles:** 4 portions
  - **Quintiles:** 5 portions
  - **Deciles:** 10 portions
  - **Percentiles:** 100 portions

There are many types of percentiles, including:

- **The PTH percentile:** Value where at least p percent of the items are less than or equal to this value, and $(100 - p)\%$ of the items are greater than or equal to this value.
- **Median Percentile:** 50th percentile
- **Q1:** 1st quartile = 25th percentile
- **Q3:** 3rd quartile = 75th percentile

Measures of Shape

- **Skewness:** The degree to which the distribution of data for a variable is largely to one side of the mean.
- **Kurtosis:** The degree to which distribution of the data for a variable is closely arranged around the mean.

Robust Measures of Location

- **Trimmed mean** is calculated by removing a percentage of values from both ends of the data set. A trimmed mean, therefore, is the arithmetic average after x-percentage of values has been removed from the highest and lowest ends of the data set.
- **Winsorized mean** is the mean computed after the x-percentage highest and lowest values are replaced by the next adjacent value in the distribution. For example, consider an ordered data set with 100 observations: $x1, x2, x3, \ldots, x98, x99, x100$. If you request a Winsorized mean with 5%, then the bottom 5% of values ($x1, x2, x3, x4$, and $x5$) will be replaced with the next adjacent value in the distribution ($x6$). Likewise, the top 5% ($x96, x97, x98, x99, x100$) will be replaced with $x95$.

Frequency Tables

In practically every research project, an initial examination of the data set usually includes frequency tables. In survey research, for example, frequency tables can show the number of males and females who participated in the survey, the number of respondents from particular ethnic and racial backgrounds, and so on. Responses on some labeled attitude measurement scales (e.g., interest in watching football) can also be nicely summarized via the frequency table. In medical research, you may tabulate the number of patients displaying specific symptoms; in industrial research, you may tabulate the frequency of different causes leading to catastrophic failure of products during stress tests (e.g., which parts are actually responsible for the complete malfunction of television sets under extreme temperatures). Customarily, if a data set includes any categorical data, then one of the first steps in the data analysis is to compute a frequency table for those categorical variables.

Frequency or *one-way tables* represent the simplest method for analyzing categorical (nominal) data. They are used often to review how different categories of values are distributed in the sample. For example, in a survey of spectator interest in different sports, we could summarize the respondents' interest in watching football in a *frequency table*, as shown in Table 6.1.

Table 6.1 shows the number, proportion, and cumulative proportion of respondents who characterized their interest in watching football as (1) Always interested, (2) Usually interested, (3) Sometimes interested, or (4) Never interested.

TABLE 6.1   Frequency of Respondents' Interest in Watching Football Games

Category	Frequency Table: Football: "Watching Football"			
	Count	Cumulative Count	Percent	Cumulative Percent
Always: Always interested	39	39	39.00000	39.0000
Usually: Usually interested	16	55	16.00000	55.0000
Sometimes: Sometimes interested	26	81	26.00000	81.0000
Never: Never interested	19	100	19.00000	100.0000
Missing	0	100	0.00000	100.0000

Frequency tables can also be tabulated for continuous data. In *STATISTICA* Data Miner, the Frequency Table function generates frequency tables and histograms for both *continuous* and *categorical* variables. Users can specify the number of intervals for continuous variables. *STATISTICA* will automatically categorize categorical variables by codes if they are specified; otherwise, all distinct values in the categorical variables will be identified. Users have control over two additional aspects of frequency tables: (1) *Type of categorization,* where users specify the method of categorization for continuous variables (for categorical variables, either specific codes are used or all integer values are identified); and (2) *Number of intervals,* where you can change the number of significant digits that are used when labeling the category levels in the graph by specifying the desired number of intervals.

Combining Groups (Classes) for Predictive Data Mining

Many data mining programs have tools for combining groups or classes. Sometimes, this capability is combined with binning tools. Figure 6.3 shows where to find this tool in

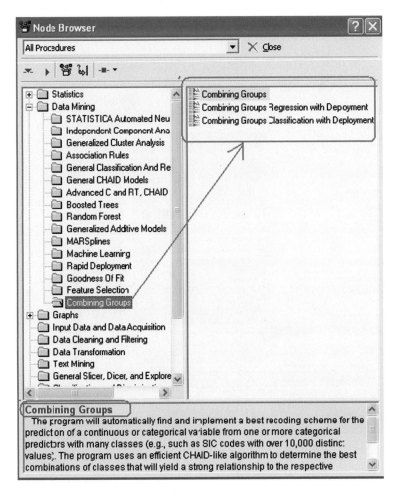

FIGURE 6.3   Location of the Combining Groups tool in *STATISTICA* Data Miner.

STATISTICA Data Miner. The program will automatically find and implement a best recoding scheme for the prediction of a continuous or categorical variable from one or more categorical predictors with many classes (e.g., such as SIC codes with over 10,000 distinct values). The program uses an efficient CHAID-like algorithm to determine the best combinations of classes that will yield a strong relationship to the respective outcome variable of interest. The recoded (aggregated) class variables (now with fewer distinct values) can then be submitted to subsequent analyses with the various tools for predictive data mining.

Slicing/Dicing and Drilling Down into Data Sets/Results Spreadsheets

Using the *STATISTICA* Data Miner tool (provided on the DVD), we can show how to use this capability to take a "deep dive" into details and aspects of a data set (Figure 6.4).

If you select Interactive Drill Down, you can use the interactive interface to access the General Slicer/Dicer. If you select Build Your Own Project, you will get the Node Browser, which enables you to put these "data preparation/drill down" nodes into the Data Miner Workspace. Figure 6.5 shows how to access the General Slicer/Dicer in the Node Browser.

If you select the Interactive Drill Down option, an interactive dialog box will appear (Figure 6.6), allowing you to specify which variables to analyze.

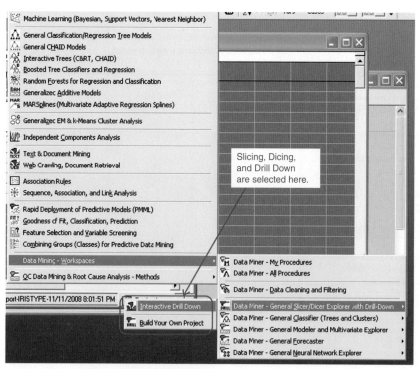

FIGURE 6.4   The menu pathway in *STATISTICA* for accessing the Interactive Drill Down tool.

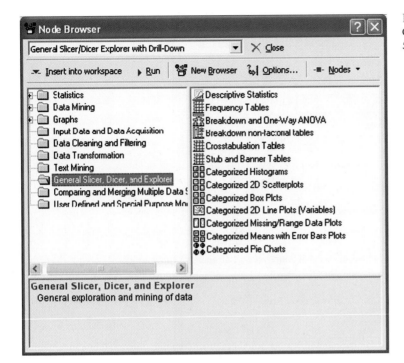

FIGURE 6.5 Location of the General Slicer/Dicer in the *STATISTICA* Node Browser.

MODELING MANAGEMENT TOOLS

Data Miner Workspace Templates

As you get used to making data mining projects, you may want to start from a blank Data Miner Workspace, adding each thing needed as you create the project. But a good way to start is to use predefined templates. These templates already have DM Nodes placed in the workspace; thus, you only have to input the data set and any other nodes to use these templates as a fast method for initial exploration of a data set.

Figures 6.7 and 6.8 show how to access these templates.

MODELING ANALYSIS TOOLS

Feature Selection

Most data mining software packages have some form of tool to help you select the best features to use in the model. Feature Selection can save a lot of time by reducing the number of variables (the "dimensionality") in the data set, which in turn increases the

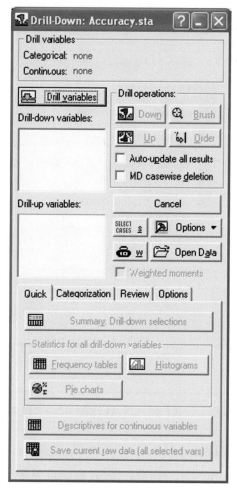

FIGURE 6.6   The dialog box for the Interactive Drill Down option.

probability that the model will be more robust (do well against new data sets). This topic was described in detail in Chapter 5.

Importance Plots of Variables

Importance values were introduced in Chapter 5. We include a more extensive presentation in this chapter, which will show you how to use feature importance values properly. We will use the Credit Scoring data set, similar to those used by bankers and credit card companies to determine whether to give credit to an applicant. We can look at the importance plots from Feature Selection in two ways: by selecting a maximum number of variables (15 variables here) and also by looking at only those importance values

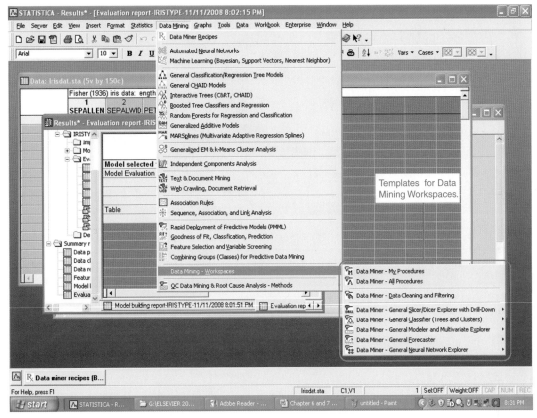

FIGURE 6.7   List of template categories for *STATISTICA* Data Miner Workspaces.

that are significant according to their *p*-values. Figure 6.9 shows the importance values of the top 15 variables. Table 6.2 shows the significance table associated with Figure 6.9.

Importance values are shown in Figure 6.10 for those variables with a *p*-value ≥ 0.05; the associated significance table is shown in Table 6.3.

Importance plots are also generated for some of the data mining algorithms, such as the importance plot generated with the Classification Trees algorithm (Figure 6.11).

Remember that a data mining algorithm provides only one perspective of patterns in a data set. Different algorithms may generate different importance values in different orders of magnitude, depending on how each algorithm "views" data. We can see some differences in importance values in Figure 6.12, generated by the Boosted Trees algorithm.

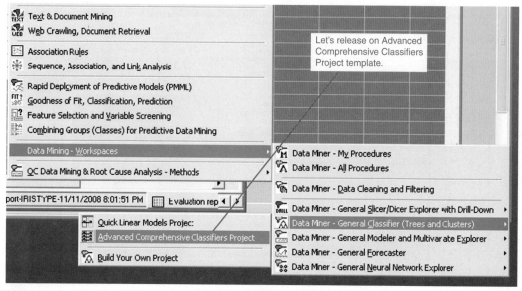

FIGURE 6.8 List of available templates in the General Classifier category.

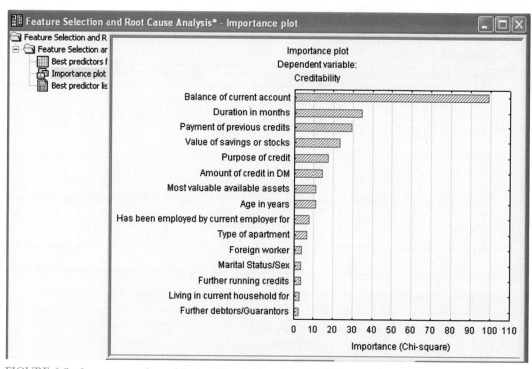

FIGURE 6.9 Importance values of the top 15 variables of the Credit Scoring data set.

TABLE 6.2   Importance Values of the Top 15 Variables, Using Chi-Square

	Best Predictors for Category	
	Chi-Square	p-Value
Balance of current account	98.79321	0.000000
Duration in months	34.54241	0.000014
Payment of previous credits	29.22978	0.000007
Value of savings or stocks	23.24602	0.000113
Purpose of credit	17.30970	0.044081
Amount of credit in DM	14.62441	0.023388
Most valuable available assets	11.31236	0.010151
Age in years	11.15297	0.132084
Has been employed by current employer for	7.59635	0.107535
Type of apartment	6.24391	0.044071
Foreign worker	3.84149	0.049999
Marital status/sex	3.45523	0.326616
Further running credits	3.23490	0.198404
Living in current household for	2.49301	0.476556
Further debtors/guarantors	2.18696	0.335049

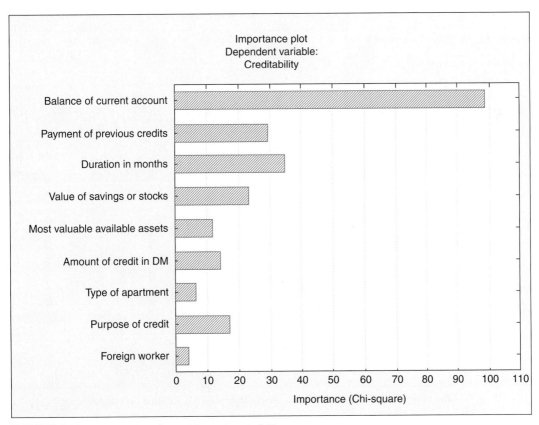

FIGURE 6.10   Importance values with p-values ≥ 0.05.

TABLE 6.3   Chi-Square and p-Values for Variables with p-Values ≤ 0.05

	Best Predictors for Category	
	Chi-Square	p-Value
Balance of current account	98.79321	0.000000
Payment of previous credits	29.22978	0.000007
Duration in months	34.54241	0.000014
Value of savings or stocks	23.24602	0.000113
Most valuable available assets	11.31236	0.010151
Amount of credit in DM	14.62441	0.023388
Type of apartment	6.24391	0.044071
Purpose of credit	17.30970	0.044081
Foreign worker	3.84149	0.049999

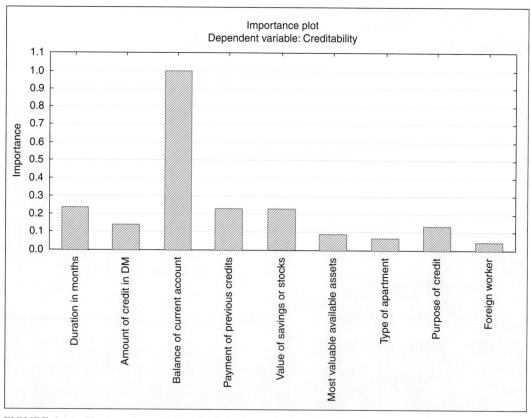

FIGURE 6.11   The importance plot available in the results from the Classification Tree algorithm.

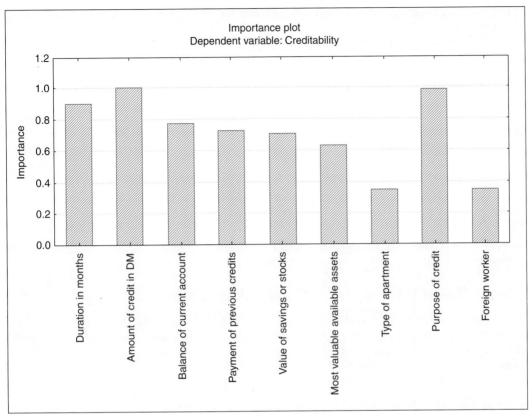

FIGURE 6.12   Variable importance values generated by the Boosted Trees algorithm for the Credit data set.

Although both of these data mining algorithms, Trees and Boosted Trees, used the same nine variables, the relative importance of specific variables was different for the two modeling algorithms. Different algorithms can give different "opinions" for variable importance, and they may generate different predictions of the target variable. You can combine these opinions with the Ensemble modeling approach, described in greater detail in Chapters 13 and 18.

IN-PLACE DATA PROCESSING (IDP)

The conventional way to access data in database tables is to extract that information using an Open Database Connectivity (ODBC) driver. Major problems with this approach include

- The space required to hold the extracted data in the form of flat files;
- The need to duplicate data on an analytical computing system;

- The need to integrate multiple extracts to form the analytic record for data mining processing;
- The time required for download, scheduling of downloads;
- The difficulty in working with very large data sets; and
- The need for the ODBC driver software to be available and properly configured for the two systems participating in the download operation.

Fortunately, there are several approaches available to permit analytical processing of data without extraction to external flat files. Several data mining tool packages provide a facility for accessing data directly in tables in a database (SAS-Enterprise Miner, *STATISTICA* Data Miner). SPSS Clementine provides links to data mining tools for various database management system vendors, which enable Clementine to work in tandem with the embedded vendor mining tools (e.g., Oracle Data Mining). Some data mining tools operate completely within the database management system itself (Teradata Warehouse Miner and Oracle Data Miner). In-place data processing allows direct access to data in tables in multiple databases of differing formats, with subsequent processing and return of results to the database requiring only one pass through the data.

Example: The IDP Facility of *STATISTICA* Data Miner

To access the IDP facility, you click on the File menu, choose Get External Data, and then choose In Place Database Interface. Choose the option to place the analysis in a standalone window and click OK. You can enter an SQL query string in the Query options or edit an existing query string saved to disk.

How to Use the SQL

As an example, *STATISTICA* provides access to most databases (including many large system databases, such as Oracle, Sybase, etc.) via an automatic query system where all you have to do is select the database(s), select the variables, and connect or "join" the different spreadsheets via a common variable, and then select SQL Statement to have the SQL query statement displayed on the screen, as shown in Figure 6.13.

You can view the SQL generated by the configuration in Figure 6.13 by clicking on the SQL Statement tab shown at the bottom of the screen. This SQL Statement can be edited or added to as desired. When you run the SQL Statement, the variables of interest will be pulled into a new data sheet on your computer, and then you can use this new data set for further analysis of these variables.

RAPID DEPLOYMENT OF PREDICTIVE MODELS

In *STATISTICA* Data Miner, for example, new cases can be scored rapidly with models saved in Predictive Modeling Markup Language (PMML) format. You can score new data in the interactive dialog shown in Figure 6.14.

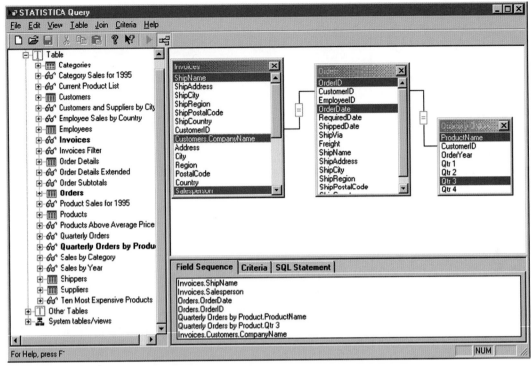

FIGURE 6.13 Table linkages in the SQL Query Builder.

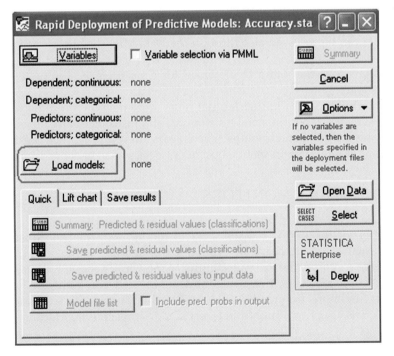

FIGURE 6.14 The interactive dialog box for rapid deployment of models.

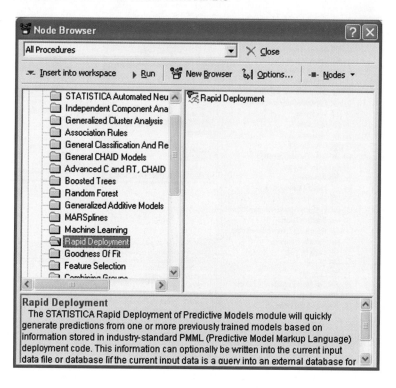

FIGURE 6.15 Accessing the Rapid Deployment tool from the Data Miner Workspace.

Click on the Load Models button to access models saved in PMML format. You can select the variables you want to work with manually (by clicking on the Variables button) or let the PMML file specify the variable list.

The Rapid Deployment tool can be accessed also through the Data Miner Workspace, as shown in Figure 6.15.

Figure 6.16 shows how to integrate the Rapid Deployment tool into the process flow of an existing Data Miner Workspace.

Rapid Deployment is an essential tool for industries that must score new data routinely, such as credit card companies, banks, other financial institutions, and insurance companies, among others. Some companies will spend months, even years, developing their best models and then deploy them on new data in many subsequent cycles.

MODEL MONITORS

Some data mining tools permit the periodic assessment of model performance. Most models will degrade in performance, due to changing economic conditions, business conditions, or cultural conditions. For example, the insurance industry may only need to reassess its models yearly or every couple of years, but banks and credit card companies may need to do this twice a year, or more frequently, as we write this during the week of the "stock market financial crisis," just one month before the 2008 presidential elections in the United States.

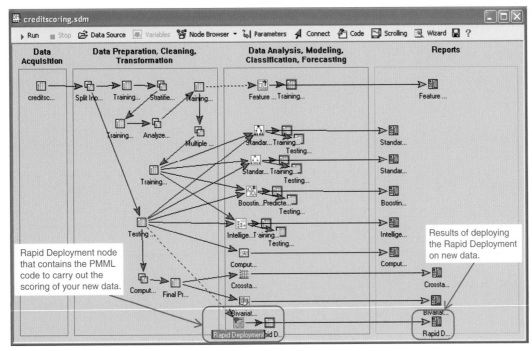

FIGURE 6.16   Integration of the Rapid Deployment tool in a workspace.

POSTSCRIPT

In Chapters 7 and 8, we will move into a subject that terrifies many people new to data mining: mathematics! But this presentation will be very different from most discussions on algorithms in data mining books. We will *not* present a lot of equations to express the nature of these algorithms. Rather, we will provide intuitive explanations of their nature and operation, which will be tied whenever possible to common things in the world.

Bibliography

SAS-EM 5.3 Getting Started Guide. (2008). *SAS-EM 5.3*. Cary, NC: SAS.

Makridakis, S.G., Wheelwright, S.C., & McGee, V.E. (1983). *Forecasting: Methods and Applications* (2nd ed.). New York: Wiley.

On-Line Help from STATISTICA: StatSoft, Inc. (2008). STATISTICA (data analysis software system), version 8.0. www.statsoft.com.

SAS-EM 5.3 Getting Started Guide. (2008). SA+S-EM 5.3. Cary, NC: SAS.

Witten, I.H., & Frank, E. (2000). *Data Mining: Practical Machine Learning Tools and Techniques*. New York: Morgan Kaufmann.

THE ALGORITHMS IN DATA MINING AND TEXT MINING, THE ORGANIZATION OF THE THREE MOST COMMON DATA MINING TOOLS, AND SELECTED SPECIALIZED AREAS USING DATA MINING

This part of the book provides a general introduction to some basic and advanced algorithms, some general solution methods for discrete and continuous target variables, and several common business areas where data mining is used. The list of algorithms is not intended to be comprehensive. The distinction between basic and advanced algorithms is related more to the historical sequence of development than to the complexity of algorithms. The business areas chosen for treatment in this part of the book cover two from science (medical informatics and bioinformatics) and two that relate to important business challenges today (customer relationship management and fraud detection). As with the presentation of the algorithms, this treatment of business areas is representative only, not comprehensive. Your business area may not be included, but that does not mean that we think it is not important! The choice of these four application areas is one of choice by the authors, based primarily on their experience.

As you read these chapters, the general design motif of this book will become clearer. We have avoided formal mathematical descriptions of these tools, methods, and applications in favor of providing intuitive explanations that beginning business data miners can understand easily. Our purpose is to get you up and running to create good data mining models in as short a time as possible. All of the authors are (or have been) deeply involved for many years in data mining educational activities, in colleges, universities, medical schools, and in private instruction seminars. Our choices of topics, tools, methods, applications, and presentation style are conditioned as much by our background in educational instruction as they are formed by our technical experiences and understanding. The "sweet-meat" of this book, though, is the many tutorials included in Part III and on the accompanying DVD, which will help you learn by example, not just by the precepts in the printed pages. These tutorials cover a much broader range of data mining applications than are introduced in Part II of this book. We hope that you will profit by this didactic approach and use this book as a springboard to launch you into the exciting practice of data mining in the twenty-first century.

Basic Algorithms for Data Mining: A Brief Overview

PREAMBLE

Armed with your prepared data set and the list of predictors, you are ready to make one of the most important decisions in the practice of data mining: selecting the right modeling algorithm to start with. In Chapters 13 and 18, we will make the case that groups of algorithms working in ensembles can create better predictions than one algorithm alone. But for now, you must make a selection of the algorithm to start with. This chapter will present the basic algorithms used in data mining and help you to select the right one to use in the beginning.

Before we get into a discussion of specific algorithms, let's look at the names of all of the algorithms we will be discussing in this chapter and also in Chapter 8.

Data Mining Algorithms and Procedural Analyses

Basic Data Mining Algorithms:

- Association Rules
- Automated Neural Networks
- Generalized Additive Models (e.g., Regression Models)
- General Classification/Regression Tree Models
- General CHAID Models
- Generalized EM and k-Means Cluster Analysis

Advanced Data Mining Algorithms (see Chapter 8 for detailed discussions):

- Interactive Trees (CART or C&RT, CHAID)
- Boosted Tree Classifiers and Regression
- MARSplines (Multivariate Adaptive Regression Splines)
- Random Forests for Regression and Classification
- Machine Learning (Bayesian, Support Vectors, Nearest Neighbor)
- Sequence, Association, and Link Analysis
- Independent Components Analysis

Special-Purpose Algorithms

Text and Document Mining, Web Crawling:

- File, Document, and Web (URL) Retrieval
- Text Mining and Document Retrieval

Quality Control Data Mining and Root Cause Analysis:

- Quality Control Charts
- Quality Control Charts for Variable Lists
- Predictive Quality Control
- Root Cause Analysis
- Response Optimization for Data Mining Models

Before describing individual algorithms you can use in most data mining packages, we will present two semi-automated approaches to performing all the necessary operations from accessing data to producing model results. The first example will show how *STATISTICA* Data Miner Recipe Interface packages all basic steps of a data mining project into an easy-to-use interface. The second example is KXEN (Knowledge Extraction Engine). Both tools select the modeling algorithms, permit you to enter a few settings (the default selections work very well in both tools), and automatically generate model results. Use of either tool might be the best way for beginning data miners to build their first model.

STATISTICA Data Miner Recipe (DMRecipe)

After you create your first model, you might feel a new sense of empowerment. It is exciting to see patterns in your data that you couldn't see before! The process can involve just a few mouse clicks and is so easy that you could even write these few steps on a sheet of paper, leave it on your assistant's desk, asking him to run this analysis the next day, while you are away on a business trip, or a meeting across town. The DMRecipe process will be described here briefly. A detailed description of this modeling option is presented in the DMRecipe tutorial.

The DMRecipe Interface consists of several interactive steps that guide you through the process to create very powerful models the first time you try it.

1. To select the DMRecipe Interface from the *STATISTICA* Data Miner toolbar, click on Data Mining and then Data Miner Recipe.
2. Select New on the Recipe screen.
3. Click on Open/Connect Data File to select the input data set.
4. Click on Apply Data Transformations, if needed.
5. Click on Select Variables to select initial input variables.
6. Click on the downward-pointing triangle (▼) symbol (in the upper right of the screen) to run the recipe.

When model training is complete, the results screen shown in Figure 7.1 will display.

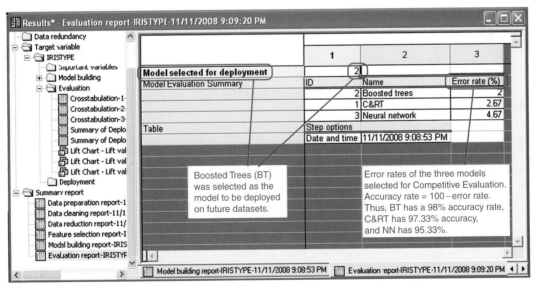

FIGURE 7.1   DMRecipe results screen showing that the Boosted Trees model had the lowest prediction error.

The DMRecipe Interface provides an almost automatic method for building data mining models. The results screen (Figure 7.1) provides several reports (e.g., lift charts) to help the modeler evaluate the predictor power of the models. The results displayed in Figure 7.1 show that the Boosted Trees algorithm had the lowest error rate (*Note:* accuracy = 100 − error rate). The DMRecipe Interface selects several modeling algorithms by default (C&RT, Boosted Trees, Neural Network). As you can see from the results in this figure, Boosted Trees had the lowest error rate among the algorithms trained, with an accuracy rate of 98%. Since Boosted Trees had the highest accuracy rate, it was selected as the model to be used for deployment to score future data sets.

You may find that some data sets do not generate acceptable prediction accuracies with any model. In these cases, hybrid models can be made, usually called consensus models. This type of model will be discussed later, primarily in Chapter 13, and will be used in some of the tutorials.

STATISTICA Data Miner Recipe (DMRecipe) is a semi-automatic method for building relatively complex analytical models for classification (with categorical target variables) or numerical prediction (with continuous target variables). The DMRecipe Interface provides a step-by-step approach to data preparation, variable selection, and dimensionality reduction, resulting in models trained with different algorithms.

Data Preparation. The first major activity in the data mining process is to prepare the data set for modeling. Common data cleaning and transformation operations can be performed to provide data in the format suitable for the modeling algorithms. Also, you can create a "blind-holdout-sample" for use later in the validation models.

Data Analysis. After the data set is properly prepared, you can conduct descriptive statistical analysis of the variables. You can evaluate each variable on the basis of mean, standard deviation, skewness, kurtosis, and observed maximum and minimum.

Data Redundancy. Some variables may carry information very similar to that of other variables, making them redundant. The DMRecipe tool provides measures of this redundancy for continuous variables. You should let DMRecipe eliminate all but one from a group of redundant variables. The resulting variable set will generate a much better model.

Dimensionality Reduction. In addition to eliminating redundant variables, you can reduce the number of variables (dimensionality) even further by eliminating variables highly correlated with the target variable. This operation will reduce the multicollinearity of the data set and increase the likelihood of generating an optimum model. Review Chapter 4 for more details on the problem of multicollinearity.

Model Building. In this step, multiple models are trained automatically. A large number of graphic displays are available to help you evaluate results from each model.

Model Deployment. After building your data mining models, you can use your models to score new data sets. A good model will perform with acceptable accuracy on data that was not used for training.

KXEN

An analogous series of steps are followed in the semi-automatic Modeling Assistant interface of KXEN for processing all steps of a simple data mining operation from model selection to model results.

1. The introductory screen permits you to select from among a number of data mining operations:
   a. Classification or regression (numerical prediction)
   b. Clustering
   c. Time-series analysis
   d. Data exploration
   e. Data manipulation
2. The second screen prompts you for the input data file.
3. The third screen allows you to view the data set or continue to data analysis.
4. The fourth screen provides a metadata report for each variable.
5. The fifth screen provides a facility for variable selection. In KXEN, the variable selection screen prompts you for a list of variables to *exclude* rather than *include.* The counterintuitive response is puzzling at first, but it makes good sense when viewed in context with the strengths of KXEN. This tool can accept variables that are redundant or collinear, and it automatically analyzes them and excludes inappropriate variables from further analysis. Variable exclusion is performed to permit you to avoid submitting totally inappropriate variables to the modeling engine (e.g., customer names, sequential numbering codes, etc.).
6. The final screen provides a model results report (Figure 7.2).

A number of model reports are available in KXEN similar to those in the DMRecipe tool (e.g., lift charts and coincidence matrices).

Like DMRecipe, KXEN does a lot of things behind the scenes. One of the most powerful of those hidden features is the operation of the Consistent Coder. This facility does a lot of data preparation automatically, like binning and recoding. Also, KXEN derives

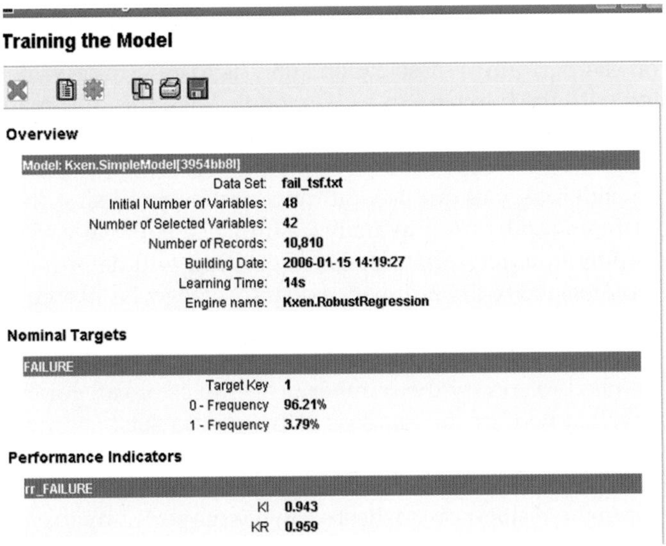

Training the Model

Overview

Model: Kxen.SimpleModel[3954bb8l]

Data Set:	fail_tsf.txt
Initial Number of Variables:	48
Number of Selected Variables:	42
Number of Records:	10,810
Building Date:	2006-01-15 14:19:27
Learning Time:	14s
Engine name:	Kxen.RobustRegression

Nominal Targets

FAILURE

Target Key	1
0 - Frequency	96.21%
1 - Frequency	3.79%

Performance Indicators

rr_FAILURE

KI	0.943
KR	0.959

FIGURE 7.2   Results screen of KXEN.

automatically a number of new variables as combinations and transformations of existing variables. This automated reduction of unnecessary dimensionality and addition of potentially useful dimensionality (new features) can produce very powerful models without significant danger of multicollinearity.

KXEN is very efficient for modeling problems in which many (even thousands) of models must be developed, such as in Retail domains. There, separate sales forecasting models must be created for each category of sales merchandise in large retail operations (such as retail supermarkets). KXEN can create huge numbers of models very quickly.

The (almost) automated functions of DMRecipe and KXEN Modeling Assistant provide a glimpse of one direction in which data mining is developing. These tools provide a close analogy to the ideal described in the Preface of this book, in which data mining is as easy to use as the automobile interface.

BASIC DATA MINING ALGORITHMS

Association Rules

The goal of *association rules* is to detect relationships or associations between specific values of categorical variables in large data sets. This technique allows analysts and researchers to uncover hidden patterns in large data sets. The classic example of an early association analysis found that beer tended to be sold with diapers, pointing to the co-occurrence of watching Monday night football and caring for family concerns at the same time. Variants like the *a priori* algorithm use predefined threshold values for detection of associations (see Agrawal et al., 1993; Agrawal and Srikant, 1994; Han et al., 2001; see also Witten and Frank, 2000). This algorithm is provided by SAS Enterprise Miner, SPSS Clementine, and *STATISTICA* Data Miner.

How Association Rules Work. Assuming you have a record of each customer transaction at a large bookstore, you can perform an association analysis to determine which other book purchases are associated with the purchase of a given book. With this information in hand at the time of purchase, you could recommend to the customer a list of other books the customer may wish to purchase. Such an application of association analysis is called a *recommender engine*. Such recommender engines are used at many online retail sites (like Amazon.com).

Association algorithms can be used to analyze simple categorical variables, dichotomous variables, and/or multiple target variables. The algorithm will determine association rules without requiring you to specify the number of distinct categories present in the data or any prior knowledge regarding the maximum factorial degree or complexity of the important associations (except in the *a priori* variant). A form of cross-tabulation table can be constructed without the need to specify the number of variables or categories. Hence, this technique is especially well suited for the analysis of huge data sets.

Table 7.1 shows an example of a tabular representation of results from an association rules algorithm.

Note that the rules in the results spreadsheet shown were sorted by the Correlation column.

Graphical representations of association rules are shown in Figures 7.3 and 7.4.

TABLE 7.1 Word Correlations, with Their Support and Confidence Values. Support Is Expressed by the Joint Probability of Word 1 and Word 2 Occurring Together; Confidence Is the Conditional Probability of Word 1 Given Word 2.

Summary of association rules (Scene 1.sta)
Min. support = 5.0%, Min. confidence = 5.0%, Min. correlation = 5.0%
Max. size of body = 10, Max. size of head = 10

	Body	==>	Head	Support(%)	Confidence(%)	Correlation(%)
154	and, that	==>	like	6.94444	83.3333	91.28709
126	like	==>	and, that	6.94444	100.0000	91.28709
163	and, PAROLLES	==>	will	5.55556	80.0000	73.02967
148	will	==>	and, PAROLLES	5.55556	66.6667	73.02967
155	and, you	==>	your	5.55556	80.0000	67.61234
122	your	==>	and, virginity	5.55556	57.1429	67.61234
164	and, virginity	==>	your	5.55556	80.0000	67.61234
121	your	==>	and, you	5.55556	57.1429	67.61234
73	that	==>	like	6.94444	41.6667	64.54972
75	that	==>	and, like	6.94444	41.6667	64.54972
161	and, like	==>	that	6.94444	100.0000	64.54972

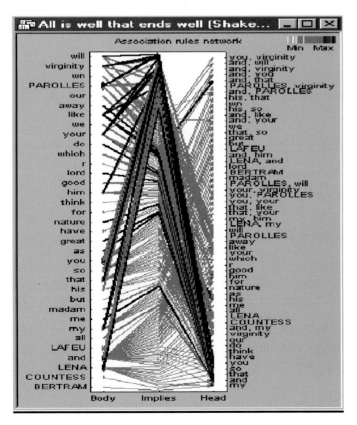

FIGURE 7.3 Link graph for words spoken in *All's Well That Ends Well.* The thickness of the line linking words is a measure of the strength of the association. (*Source:* StatSoft Inc.)

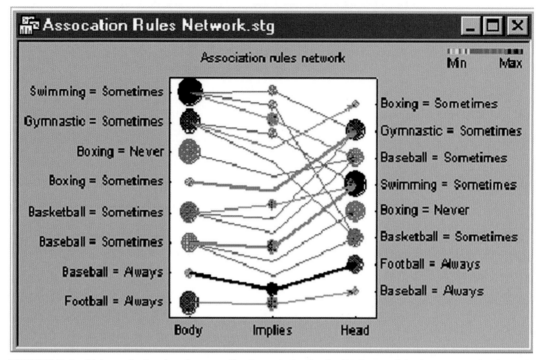

FIGURE 7.4   Link graph showing the strength of association by the thickness of the line connecting the *Body* and *Head* words of some association rules. (*Source:* StatSoft Inc.)

In Figure 7.4, the support values for the *Body* and *Head* portions of each association rule are indicated by the sizes and colors of each. The thickness of each line indicates the confidence value (conditional probability of *Head* given *Body*) for the respective association rule; the sizes and colors of the circles in the center, above the *Implies* label, indicate the joint support (for the co-occurrences) of the respective *Body* and *Head* components of the respective association rules.

Neural Networks

Neural networks used for computation were based on early understandings of the structure and function of the human brain. They were proposed as a means for mathematical computation by McCulloch and Pitts (1943). But it was not until the 1980s that the concept was developed for use with digital computers. The underlying assertion of neural nets is that all of the operations of a digital computer can be performed with a set of interconnected McCulloch-Pitts "neurons" (Abu-Mostafa, 1986).

Figure 7.5 shows how the human neurons are structured.

Neuron cells receive electrical impulses from neighboring cells and accumulate them until a threshold value is exceeded. Then they "fire" an impulse to an adjacent cell. The

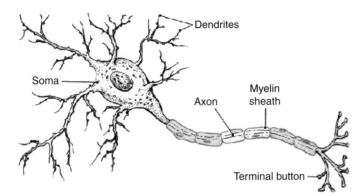

FIGURE 7.5 Structure of the human neuron [*Source:* Carlson, Neil R. (1992). *Foundations of Physiological Psychology.* Needham Heights, Massachusetts: Simon & Schuster. pp. 36].

capacity of the cell to store electrical impulses and the threshold are controlled by biochemical processes, which change over time. This change is under the control of the autonomic nervous system and is the primary means by which we "learn" to think or activate our bodies.

Artificial neurons in networks (Figure 7.6) incorporate these two processes and vary them numerically rather than biochemically. The aggregation process accepts data inputs by summing them (usually). The activation process is represented by some mathematical function, usually linear or logistic. Linear activation functions work best for numerical estimation problems (i.e., regression), and the logistic activation function works best for classification problems. A sharp threshold, as is used in decision trees, is shown in Figure 7.6.

The symbol X_i represents input variables, representing the number of other neurons connected to a human neuron. W_i represents the numerical weights associated with each linkage, and they are analogous to the strength of the interconnections. This strength of connection represents the proximity of connections between two neurons in the region called the *synapse* (Bishop, 1995).

Artificial neurons are connected together into an *architecture* or processing structure. This architecture forms a network in which each input variable (called an input *node*) is connected to one or more output nodes. This network is called an artificial neural network (neural net for short). When the input nodes with summation aggregation function and a logistic activation function are directly connected to an output node, the mathematical processing is analogous to a logistic regression with a binary output. This configuration

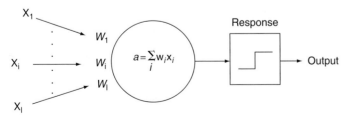

FIGURE 7.6 Architecture of a neuron, with a number of inputs (X_i).

FIGURE 7.7   A plot of the logistic
function.

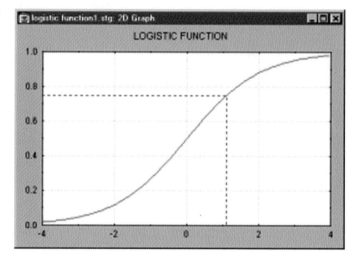

of a neural net is a powerful classifier. It has the ability to handle nonlinear relationships between the output and the input variables, by virtue of the logistic function shown in Figure 7.7.

The logistic function fits many binary classification problems and can express much of the nonlinear effects of the predictors.

The most interesting property of a neural net comes into view when you intercalate a middle layer of neurons (nodes) between the input and output node, as shown in Figure 7.8.

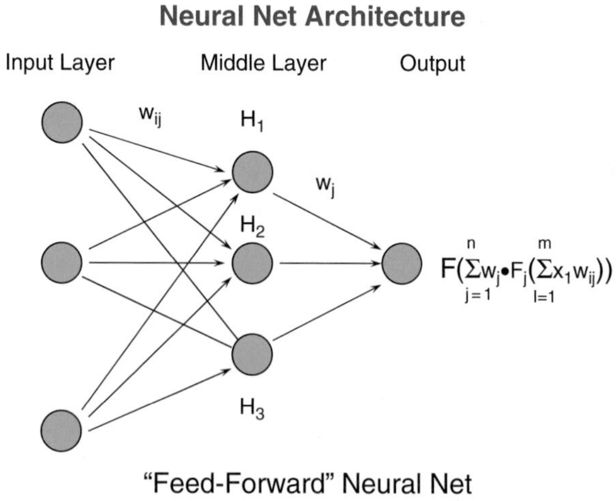

FIGURE 7.8   Architecture of a neural net.

Neural Net Architecture

Input Layer   Middle Layer   Output

w_{ij} H_1

w_j

H_2

$$F\left(\sum_{j=1}^{n} w_j \cdot F_j\left(\sum_{l=1}^{m} x_1 w_{ij}\right)\right)$$

H_3

"Feed-Forward" Neural Net

Weights (W_{ij}) are assigned to each connection between the input nodes and middle layer nodes, and between the middle layer nodes and the output node(s). These weights have the capacity to model nonlinear relationships between the input nodes and output node(s). Herein lies the great value of a neural net for solving data mining problems. The nodes in the middle layer provide the capacity to model nonlinear relationships between the input nodes and the output node (the decision). The greater the number of nodes in the middle layer, the greater the capacity of the neural net to recognize non-linear patterns in the data set. But, as the number of nodes increases in the middle layer, the training time increases exponentially, and it increases the probability of overtraining the model. An overtrained model may fit the training set very well but not perform very well on another data set. Unfortunately, there are no great rules of thumb to define the number of middle layer nodes to use. The only guideline is to use more nodes when you have a lot of training cases and use fewer nodes with fewer cases. If your classification problem is complex, use more nodes in the middle layer; if it is simple, use fewer nodes.

The neural net architecture can be constructed to contain only one output node and be configured to function as a regression (for numerical outputs) or binary classification (yes/no or 1/0). Alternatively, the net architecture can be constructed to contain multiple output nodes for estimation or classification, or it can even function as a clustering algorithm.

The learning process of the human neuron is reflected (crudely) by performing one of a number of weight adjustment processes, the most common of which is called *backpropagation*, shown in the diagram of Figure 7.9.

The backpropagation operation adjusts weights of misclassified cases based on the magnitude of the prediction error. This adaptive process iteratively retrains the model and improves its fit and predictive power.

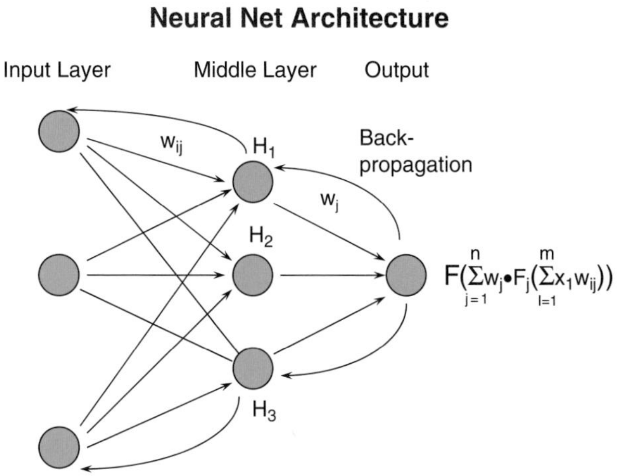

Neural Net Architecture

FIGURE 7.9 A feed-forward neural net with backpropagation.

How Does Backpropagation Work?

The processing steps for backpropagation are as follows:

1. Randomly assign weights to each connection.
2. Read the first record and calculate the values at each node as the sum of the inputs times their weights.
3. Specify a threshold value above which the output is evaluated to 1 and below which it is evaluated to 0. In the following example, the threshold value is set to 0.01.
4. Calculate the prediction error as

$$\text{Error} = \text{Expected prediction} - \text{Actual prediction}.$$

5. Adjust the weights as

$$\text{Adjustment} = \text{Error} \times \text{output weight}.$$

6. Calculate the new weight as

$$\text{Old input weight} + \text{Adjustment (assume as 0.1).}$$

7. Do the same for all inputs.
8. Repeat a number of iterations through the data (often 100–200).

Figure 7.10 shows the evaluation of all weights after the first record is processed.
For example, the new weight of input variable X_1 connected to the middle layer (or hidden layer) $H_1 = -4.9 + (-1 \times 2.2) = -7.1$, as shown in Figure 7.11.

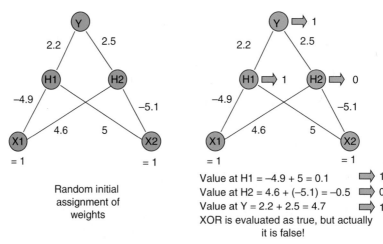

How Does Backpropagation Work?
Example: Solve the XOR Case
(Assume a threshold value of 0.01)

Random initial assignment of weights

Value at H1 = −4.9 + 5 = 0.1 ⟹ 1
Value at H2 = 4.6 + (−5.1) = −0.5 ⟹ 0
Value at Y = 2.2 + 2.5 = 4.7 ⟹ 1
XOR is evaluated as true, but actually it is false!

FIGURE 7.10   How backpropagation solves the XOR case. (Note that these NN flow from the bottom to the top; previous models flowed from left to right.)

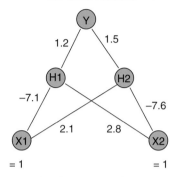

FIGURE 7.11   Weights after backpropagation for one record.

Advantages of neural nets include the facts that they

- Are general classifiers. They can handle problems with very many parameters, and they are able to classify objects well even when the distribution of objects in the N-dimensional parameter space is very complex.
- Can handle a large amount of nonlinearity in the predictor variables.
- Can be used for numerical prediction problems (like regression).
- Require no underlying assumption about the distribution of data.
- Are very good at finding nonlinear relationships. The hidden layer(s) of the neural net architecture provides this ability to model highly nonlinear functions efficiently.

Disadvantages of neural nets include

- They may be relatively slow, especially in the training phase but also in the application phase.
- It is difficult to determine *how* the net is making its decision. It is for this reason that neural nets have the reputation of being a "black box."
- No hypotheses are tested and no *p*-values are available in the output for comparing variables.

Modern implementations of neural nets in many data mining tools open up the "black box" to a significant extent by showing the *effect* of what it does related to the contribution of each variable. As a result, many modern neural net implementations are referred to as "gray boxes." These effects of the trained neural net are often displayed in the form of a *sensitivity analysis*. In this context, the term *sensitivity* has a slightly different meaning than it does in classical statistical analysis. Classical statistical sensitivity is determined either by calculation of the statistical model with all but one of the variables, and leaving out a different variable in each of a number of iterations (equal to #variables), or by keeping track of the partial least squares values (as is done in partial least squares regression). In neural net analysis, sensitivities are calculated from the normalized weights associated with each variable in the model.

Training a Neural Net

Training a neural net is analogous to a ball rolling over a series of hills and valleys. The size (actually, the mass) of the ball represents its momentum, and the learning rate is analogous to the slope of the error pathway over which the ball rolls (Figure 7.12).

The low momentum of the search path means that the solution (depicted by the ball) may get stuck in the local minimum error region of the surface (A) rather than finding the global minimum (B). This happens when the search algorithm does not have enough tendency to continue searching (momentum) to climb the hill over to the global minimum. This problem is compounded when the learning rate is relatively high, analogous to a steep slope of the decision surface (Figure 7.13).

The configuration of the neural net search path depicted in Figure 7.12 is much more likely to permit the error minimization routine to find the global minimum. This increased likelihood is due to the relatively large momentum (shown by the large ball), which may carry the ball long enough to find the global minimum. In manually configured neural nets, the best learning rate is often 0.9, and the momentum is often set to 0.1 to 0.3.

Another neural net setting that must be optimized is the learning decay rate. Most implementations of a neural net start at the preset learning rate and then reduce it incrementally during subsequent runs. This decay process has the effect of progressively flattening the search surface. The run with the lowest error rate is selected for the training run.

Modeling with a manual neural net is very much an art. The modeling process usually includes a number of training runs with different combinations of

- Learning rate
- Learning rate decay
- Momentum
- Number of nodes in the middle layer
- Number of middle layers to add

FIGURE 7.12 The topology of a learning surface associated with a low momentum and a high learning rate.

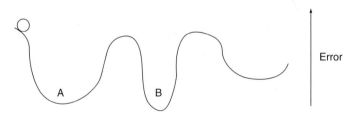

Error

FIGURE 7.13 The topology of a learning surface associated with a high momentum and a low learning rate.

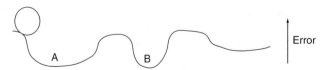

Error

Because of the artistic nature of neural net modeling, it is difficult for novice data miners to use many implementations of neural nets successfully. But there are some implementations that are highly automated, permitting even novice data miners to use them effectively.

The type of neural net described here is sometimes called a Multilayer Percep-tron (MLP).

Additional Types of Neural Networks

Following are some additional types of neural nets:

- **Linear Networks:** These networks have two layers: input and output layers. They do not handle complexities well but can be considered as a "baseline model."
- **Bayesian Networks:** Networks that employ Bayesian probability theory which can be used to control model complexity, and can be used to optimize weight decay rates, and to automatically find the most important input variables.
- **Probabilistic Networks:** These networks consist of three to four layers.
- **Generalized Regression:** These networks train quickly but execute slowly. Probabilistic (PNN) and Generalized Regression (GRNN) neural networks operate in a manner similar to that of Nearest-Neighbor algorithms (see Chapter 12), except the PNN operates only with categorical target variables and the GRNN operates only with numerical target variables. PNN and GRNN networks have advantages and disadvantages compared to MLP networks (adapted from http://www.dtreg.com/pnn.htm):
  - It is usually much faster to train a PNN/GRNN network than an MLP network.
  - PNN/GRNN networks often are more accurate than MLP networks.
  - PNN/GRNN networks are relatively insensitive to outliers (wild points).
  - PNN networks generate accurate predicted target probability scores.
  - PNN networks approach Bayes optimal classification.
  - PNN/GRNN networks are slower than MLP networks at classifying new cases.
  - PNN/GRNN networks require more memory space to store the model.
- **Kohonen:** This type of neural network is used for classification. It is sometimes called a "self-organizing" neural net. It iteratively classifies inputs, until the combined difference between classes is maximized. This algorithm can be used as a simple way to cluster data, if the number of cases or categories is not particularly large. For data sets with a large number of categories, training the network can take a very long time.

MLPs can be used to solve most logical problems, but only those in which the classes are *linearly separable*. Figure 7.14 shows a classification problem in which it is possible to sepa-rate the classes with a straight line in the space defined by their dimensions.

Figure 7.15 shows two classes that cannot be separated with a straight line (i.e., are not linearly separable).

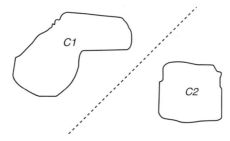

(*Patterns are linearly separable*)

FIGURE 7.14   Two pattern classes that are linearly separable.

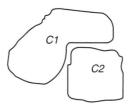

(Patterns are **NOT** linearly separable)

FIGURE 7.15   Nonseparable classes.

Radial Basis Function (RBF) Networks

RBFs are similar to MLPs with three layers (input, middle or "hidden" layer, and output). Also like MLPs, RBFs can model any nonlinear function easily. The major difference between the two networks is that an RBF does not use raw input data, but rather passes a *distance measure* from the inputs to the hidden layer. This distance is measured from some center value in the range of the variable (sometimes the mean) to a given input value in terms of a Gaussian function (Figure 7.16). These distances are transformed into similarities that become the data features in a succeeding regression step. This nonlinear function can permit the mapping operation to capture many nonlinear patterns in the input data.

The processing of RBFs (like any neural network) is iterative. The weights associated with the hidden nodes are adjusted following some strategy (like backpropagation). If a large enough RBF is run through enough iterations, it can approximate almost any function almost perfectly; that is, it is theoretically a *universal approximator*. The problem with RBF processing (like with the MLP) is the tendency to overtrain the model.

Advantages of RBFs

RBFs can model any nonlinear function using a single hidden layer, which removes some design decisions about numbers of layers to use for the networks like the MLP. The simple linear transformation in the output layer can be optimized fully using traditional linear modeling techniques; these techniques are fast and do not suffer from problems such as local minima that plague MLP training techniques. RBF networks can therefore be trained extremely quickly (i.e., orders of magnitude faster than MLPs).

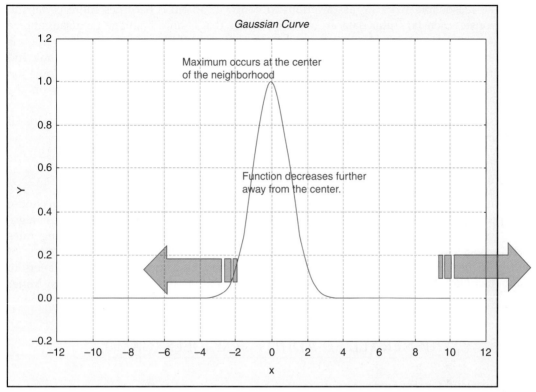

FIGURE 7.16   Plot of a Gaussian function, similar to the aggregation function in an RBF.

Disadvantages of RBFs

On the other hand, before linear optimization can be applied to the output layer of an RBF network, the number of radial units must be decided, and their centers and deviations must be set. Although faster than MLP training, the algorithms to do this are equally prone to discovering suboptimal combinations. (In compensation, the *STATISTICA* Neural Networks Intelligent Problem Solver can perform the inevitable experimental stage for you.)

RBF's more eccentric response surface requires a lot more units to adequately model most functions. Of course, it is always possible to draw shapes that are most easily represented one way or the other, but the balance in practice does not seem to favor RBFs. Consequently, an RBF solution will tend to be slower to execute and more space consuming than the corresponding MLP (although faster to train, which is sometimes more of a constraint).

RBFs are not good for extrapolating beyond known data; the response drops off rapidly towards zero if data points far from the training data are used (due to the Gaussian basis function). Often the RBF output layer optimization will have set a bias level, more or less equal to the mean output level, so in fact the extrapolated output is the observed mean—a reasonable working assumption. In contrast, an MLP becomes more certain in its response when far-flung data are used. Whether this is an advantage or disadvantage depends largely on the application, but on the whole the MLP's uncritical extrapolation is regarded as a bad point; extrapolation far from training data is usually dangerous and unjustified. However, both methods, like logistic regression, are far better at extrapolation than methods like regression or polynomial networks that have no constraints on the output estimate.

RBFs are also more sensitive to the curse of dimensionality and have greater difficulties if the number of input units is large.

Automated Neural Nets

Several data mining tools offer neural nets that have "smart" search algorithms to choose the appropriate starting points for their parameters. But the biggest benefit of these algorithms is that they search over the decision surface with different initial learning rates (which also decay between iterations), different momentums, and different number of nodes in the middle layer. Usually, you have to choose the number of middle layers to use before the algorithm takes over. Both SPSS Clementine and *STATISTICA* Data Miner have very powerful automated neural nets.

GENERALIZED ADDITIVE MODELS (GAMs)

As theory of Generalized Linear Models (GLMs) developed in the 1980s, the need for an increasing number of predictor variables was recognized as a key issue. The problem with increasing the number of predictor variables is that the variance increases also. The higher the variance, the harder it is for a prediction algorithm to perform well (perform acceptably on new data). This is one aspect of the "curse of dimensionality." To bypass this problem, Stone (1986) proposed modification of the GLM by replacing the definition of each predictor variable with an additive approximation term. This approximation is performed with a linear univariate smoothing function. This approach avoided the curse of dimensionality by performing a simple fitting of each predictor variable to the dependent variable. The new approach also expressed the definition of each predictor variable such that it was possible to relate *how* the variable affected the dependent variable. Remember, in the standard Multiple Linear Regression (MLR) equation, the estimated coefficients represent effects of differing scale, as well as differing relationships to the dependent variable. Consequently, you can't analyze the MLR coefficients directly to determine relationships. But with the enhancement by Stone, you can see these relationships directly. Still, the cost of that enhancement was a decrease in generalization (the ability to perform acceptably on new data).

Hastie and Tibshirani (1990) incorporated Stone's idea into a formal definition of Generalized Additive Models (GAMs). A GAM uses a nonlinear link function to map input data into a solution space, similar to a GLM. This flexible approach to mapping of inputs can fit the response probability distribution of any member of the exponential family of data distributions ($Y = X^n$). Choice of the appropriate link function depends on the distribution of the data set. For Normal, Poisson, and Gamma distributions, appropriate link functions include

Identity link [$Y = f(x)$]
Log link [$Y = \log(x)$]
Inverse link [$Y = 1/x$]

For binomial distributions, the Logit link is used [$Y = \log(x/(1 - x))$]

Outputs of GAMs

Typical outputs of GAMs include

- Iteration history of model fitting
- Summary statistics, including R^2
- Residual tables and plots
- Scatterplots of observed versus predicted values
- Normal probability plots

Interpreting Results of GAMs

Model interpretation is a vital step after model fitting. For example, analysis of residual values helps to identify outliers; analysis of normal probability plots shows how "normal" the predictions were across the range of values for the dependent variable. For example, Figure 7.17 shows a *STATISTICA* plot of partial residuals (residuals after effects of other variables have been removed).

This plot allows you to evaluate the nature of the relationship between the predictor with the residualized (adjusted) dependent variable values. You can see that most of the adjusted residuals are within the 95% confidence limits of normally expected values, but some points on the upper left and lower right of the plot appear to be outliers. Subsequent analysis of these data points might yield some valuable insights for improving the fit of the model.

CLASSIFICATION AND REGRESSION TREES (CART)

 Decision Trees

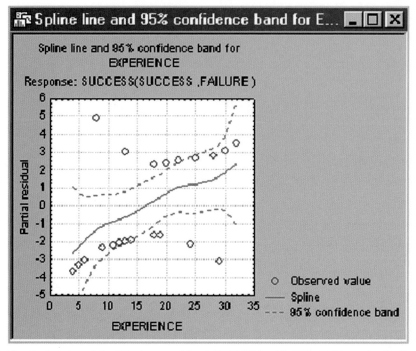

FIGURE 7.17   A plot of partial residuals created by *STATISTICA*, with 95% confidence levels (dashed lines).

Classification and Regression Tree (CART or C&RT) methodology was introduced in 1984 by UC Berkley and Stanford researchers Leo Breiman, Jerome Friedman, Richard Olshen, and Charles Stone. CART processing is structured as a sequence of simple questions. The answers to these questions determine what next question, if any, is posed. The result is a network of questions that forms a tree-like structure. The "ends" of the tree are terminal "leaf" nodes, beyond which there are no more questions.

The two most popular algorithms are

1. **CART:** Classification and Regression Tree (with generic versions often denoted C&RT)
2. **CHAID:** Chi-Square Automatic Interaction Detection (Kass, 1980)

Key elements defining a decision tree algorithm are as follows:

- Rules at a node for splitting the data according to its value on one variable
- A stopping rule for deciding when a subtree is complete
- Assigning each terminal leaf node to a class outcome (prediction)

Trees recursively partition the data, creating at each step more homogenous groups. The resulting rules are the paths it takes to get from the root node to each leaf.

Consider the tree shown in Figure 7.18, created to classify the Iris data set (a standard data set used in data mining discussions and benchmarks). Only two variables are needed to classify Iris type: petal length and petal width.

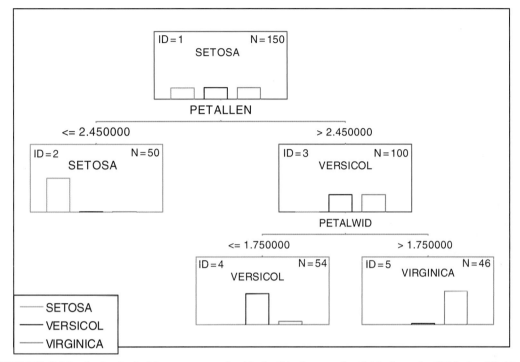

Tree 3 graph for IRISTYPE
Num. of non-terminal nodes: 2, Num. of terminal nodes: 3

FIGURE 7.18 A simple decision tree created with the Iris data set (available from the UC-Irvine Machine Learning Repository: http://archive.ics.uci.edu/ml/).

When you consider two of the predictor variables, petal length and width, you'll see how the CART algorithm processes the first question at node 1. Figure 7.19 shows a categorized scatterplot of the results of this simple decision tree model.

You can see that the algorithm found a splitting point that perfectly distinguished the species *Setosa* from other two species (*Versicolor* and *Virginica*). The rule is: When petal length is less than 2.45 mm, then *Setosa* is characterized. Subsequent questions will distinguish *Versicolor* from *Virginica*.

The second split is on petal width. Figure 7.20 shows the tree with the decision rules for each node.

From Figure 7.20, we can see that *Versicolor* and *Virginica* can be distinguished adequately (but not perfectly) by asking the second question, "Is the petal width greater than or equal to or less than 1.75 mm?"

The final categorized scatterplot is shown in Figure 7.21.

The final tree is shown in Figure 7.22.

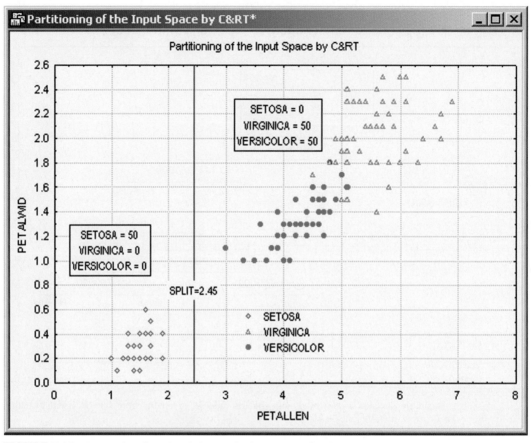

FIGURE 7.19　Scatterplot of results of the simple decision tree for the Iris data set.

FIGURE 7.20　The second split in building the decision tree for the Iris data set.

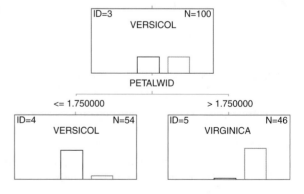

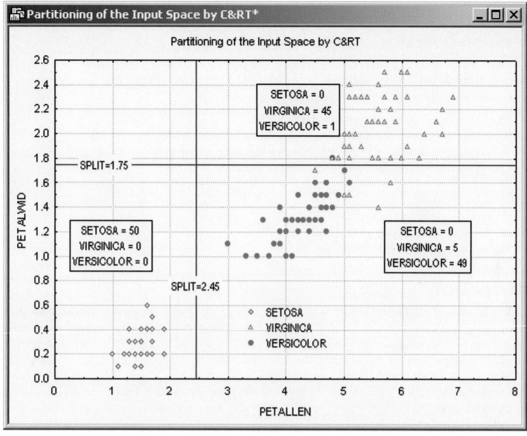

FIGURE 7.21   The categorized scatterplot after two splits of the Iris data set.

Tree 3 graph for IRISTYPE
Num. of non-terminal nodes: 2, Num. of terminal nodes: 3

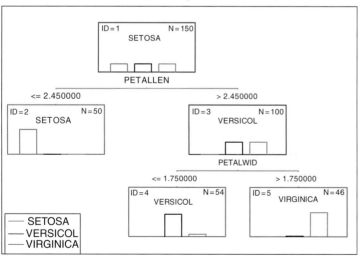

FIGURE 7.22   The final tree to classify three Iris species.

Recursive Partitioning

Once a best split is found, CART repeats the search process for each node below (child nodes), until either further splitting is stopped by a criterion, or splitting is impossible.

Common stopping conditions include

- Minimum number of cases has been reached.
- A certain fraction of the total number of cases is in the node.
- A maximum number of levels of splitting has been achieved.
- The maximum number of nodes has been reached.

Conditions under which further splitting is impossible include when

- Only one case is left in a node.
- All cases are duplicates of each other.
- The node is pure (all target values agree).

Pruning Trees

Rather than focusing on when to stop pruning, CART trees are grown larger than they need to be and then pruned back to find the best tree. CART determines the best tree by using the testing data set or by using the process of V-fold cross-validation. The testing validation is performed by scoring the tree with the data set not used for training the model. Cross-validation is a form of *resampling*, which draws a number of samples from the entire distribution and trains models on all samples. The V-fold cross-validation is performed by

1. Partitioning the entire data set into a number (V) of parts (folds);
2. Training V models on different combinations of $V - 1$ folds, with the error estimated each time using the Vth fold;
3. Using the mean (and sigma) of the V error measurements to estimate tree accuracy on new data;
4. Choosing the design parameters (e.g., complexity penalty) that minimize the error in step 3;
5. Refitting the tree, using all the data, using the parameters of step 4.

Figure 7.23 shows a 3-fold cross-validation operation.

The cross-validation process provides a number of independent estimates of the error associated with the algorithm itself rather than due to the randomness in the data. A model created with a CART algorithm (or any other algorithm, for that matter) should not be accepted until the prediction error is partitioned in this manner.

General Comments about CART for Statisticians

1. CART is nonparametric and does not require specification of a data distribution.
2. The final modeling variables are not selected beforehand but selected automatically by the algorithm.

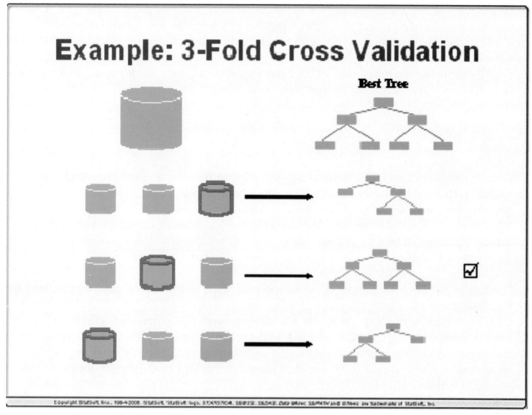

FIGURE 7.23   How a 3-fold cross-validation design works.

3. There is no need to transform data to be consistent with a given mathematical function. Monotonic transformations will have no effect.
4. Very complex interaction patterns can be analyzed.
5. CART is not significantly affected by outliers in the input space.
6. CART is affected, but only locally, by outliers in the output variable.
7. CART can accept any combination of categorical and continuous variables.
8. CART can adjust for samples stratified on a categorical dependent variable.
9. CART can process cases with missing values; the cases are not deleted.

Advantages of CART over Other Decision Trees

1. You can relax the stopping rules to "overgrow" decision trees and then prune back the tree to the optimal size. This approach minimizes the probability that important structure in the data set will be overlooked by stopping too soon.
2. CART incorporates both testing with a test data set and cross-validation to assess the goodness of fit more accurately.

3. CART can use the same variables more than once in different parts of the tree. This capability can uncover complex interdependencies between sets of variables.
4. CART can be used in conjunction with other prediction methods to select the input set of variables.
5. CART can be incorporated into hybrid models, where CART feeds inputs to a neural network model (which itself cannot select variables).

Uses of CART

1. CART is *simple!*
2. *Data preparation.* Classical statistical models require that the analyst has a clear understanding of the nature of the function inherent in the data to be modeled. CART requires very little input for the beginning data miner.
3. *Variable selection.* CART can be used to create the short-list of predictor variables to submit to the modeling algorithm. There is no guarantee that the variables most useful for a tree will also prove most useful for a neural network or other function, but in practice this is a useful technique.
4. The use of predictors multiple times in the tree helps to detect complex interactions in the data.
5. CART can handle missing values by identifying surrogate (alternate) splitting rules. During training, after the best split is found for a node, new splits using *other* variables are scored according to their similarity in distributing the data to the left and right child nodes. The best five or so are then stored as backup or surrogate questions to ask should the main variable not be available.

GENERAL CHAID MODELS

CHAID is an acronym for Chi-Square Automatic Interaction Detector. CHAID differs from CART in that it allows multiple splits on a variable. For classification problems, it relies on the Chi-square test to determine the best split at each step. For regression problems (with a continuous target variable), it uses the F-test.

Key elements of the CHAID process are

1. *Preparing the predictor variables:* Continuous variables are "binned" to create a set of categories, where each category is a subrange along the entire range of the variable. This binning operation permits CHAID to accept both categorical and continuous inputs, although it internally works only with categorical variables.
2. *Merging categories:* The categories of each variable are analyzed to determine which ones can be merged safely to reduce the number of categories.
3. *Selecting the best split:* The algorithm searches for the split point with the smallest adjusted *p*-value (probability value, which can be related to significance).

Advantages of CHAID

1. It is fast!
2. CHAID builds "wider" decision trees because it is not constrained (like CART) to make binary splits, making it very popular in market research.
3. CHAID may yield many terminal nodes connected to a single branch, which can be conveniently summarized in a simple two-way contingency table, with multiple categories for each variable.

Disadvantages of CHAID

1. Since multiple splits fragment the variable's range into smaller subranges, the algorithm requires larger quantities of data to get dependable results.
2. The CHAID tree may be unrealistically short and uninteresting because the multiple splits are hard to relate to real business conditions.
3. Real variables are forced into categorical bins before analysis, which may not be helpful, particularly if the order in the values should be preserved. (Categories are inherently unordered; it is possible for CHAID to group "low" and "high" versus "middle," which may not be desired.)

GENERALIZED EM AND *k*-MEANS CLUSTER ANALYSIS—AN OVERVIEW

The purpose of clustering techniques is to detect similar subgroups among a large collection of cases and to assign those observations to the clusters as illustrated in Figure 7.24. The clusters are assigned a sequential number to identify them in results reports. A good clustering algorithm will find the number of clusters as well as the members of each. Cases within a group should be much more similar to each other than to cases in other clusters.

A typical sample application of cluster analysis is a marketing research study in which a number of variables related to consumer behavior are measured for a large sample of respondents. The purpose of the study is to detect "market segments," i.e., groups of respondents that are somehow more similar to each other than they are to respondents in other groups (clusters). Just as important as identifying such clusters is the need to determine how those clusters are different.

k-Means Clustering

The classic *k*-means algorithm was introduced by Hartigan (1975; see also Hartigan and Wong, 1978). Its basic operation is simple: given a fixed number (k) of clusters, assign observations to those clusters so that the means across clusters (for all variables) are as different from each other as possible. The difference between observations is measured in terms of one of several distance measures, which commonly include Euclidean, Squared Euclidean, City-Block, and Chebychev.

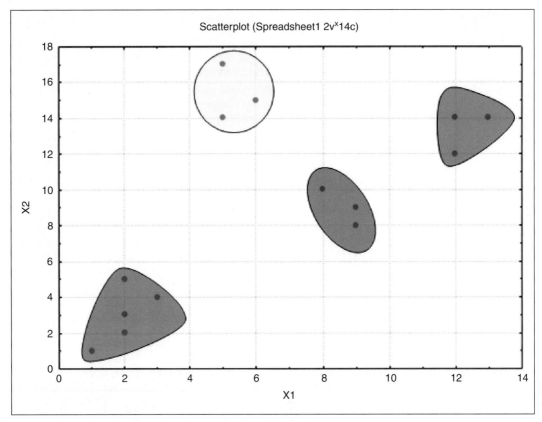

FIGURE 7.24   Data clusters in a clustering problem.

For categorical variables, all distances are binary (0 or 1). The variable is assigned as 0 when the category of an observation is the same as the one with the highest frequency in a cluster; otherwise, it is assigned a value of 1. So, with the exception of the Chebychev distance, for categorical variables, the different distance measures will yield identical results.

EM Cluster Analysis

The goal of the EM clustering method is to find the most likely set of clusters for the observations (together with prior expectations). The basis for this technique is a body of statistical theory called *finite mixtures*. A *mixture* is a set of probability distributions, representing k clusters, which govern the attribute values of that cluster. This means that each of the distributions gives the probability that a particular observation would have one of a certain set of attribute values, if it were truly a part of that cluster. An observation belongs to only one cluster, but which one is not known at the start of analysis.

Consider two clusters A and B, where each has a normal distribution characterized by means, standard deviations, and a prior probability (P) of belonging to clusters A and B adds to 1, such that P(A), the probability of belonging to cluster A $= 1 - P(B)$. The prior probability represents the *expectation*, and the calculation of the distribution parameters is the process of *maximization*.

Processing Steps of the EM Algorithm

The steps for processing the EM algorithm are as follows:

1. Start with initial guesses of the distributional parameters for each observation.
2. Use the initial guesses to calculate the cluster probabilities for each observation.
3. Use the calculated probabilities to re-estimate the parameters.
4. Go back to step 2 and do it again (until your time budget runs out).

The algorithm converges toward a fixed point but never gets there. But we can calculate the likelihood that the observation came from the data set, given the values for the parameters. The overall likelihood across all observations is the "goodness" of the clustering solution, and it increases during each iteration through the process. This likelihood may be only a "local" maximum (greater than all values near it), and there may be another maximum in another part of the probability landscape that is higher. The highest maximum across the entire probability landscape is the "global" maximum.

V-fold Cross-Validation as Applied to Clustering

The general idea of *V*-fold cross-validation as it is applied to clustering is to divide the overall sample into *V* folds, or randomly drawn (disjoint) subsamples. The same type of analysis is then successively applied to the observations belonging to the $V - 1$ folds (training sample), and the results of the analyses are applied to sample *V* (the sample or fold that was not used to estimate the parameters, build the tree, determine the clusters, etc.; i.e., this is the testing sample) to compute some index of predictive validity. The results for the *V* replications are aggregated (averaged) to yield a single measure of the stability of the respective model, i.e., the validity of the model for predicting new observations.

Cluster analysis is an unsupervised learning technique, and we cannot observe the (real) number of clusters in the data. However, it is reasonable to replace the usual notion (applicable to supervised learning) of "accuracy" with that of "distance": In general, we can apply the *V*-fold cross-validation method to a range of numbers of clusters and observe the resulting average distance of the observations (in the cross-validation or testing samples) from their cluster centers (for *k*-means clustering); for EM clustering, an appropriate equivalent measure would be the average negative log-likelihood computed for the observations in the testing samples.

Note: The preceding discussion on *k*-means and EM clustering is based on Witten and Frank (2005).

POSTSCRIPT

These basic algorithms will work relatively well for most data sets. But there are some advanced algorithms available that may do even better. In Chapter 8, we will describe some advanced algorithms that also incorporate a higher degree of automation than the basic algorithms implemented in most data mining tools.

References

Abu-Mostafa, Y. (1986). In J. Denker (Ed.), *Neural Networks for Computing. 1986 American Institute of Physics Conference* (pp. 1–5).

Agrawal, R., Imielinski, T., & Swami, A. (1993). Mining association rules between sets of items in large databases. In *Proceedings of the 1993 ACM SIGMOD Conference* (pp. 207–216). Washington, DC: (SIGMOD 93), May 1993.

Agrawal, R., & Srikant, R. (1994). Fast algorithms for mining association rules. In *Proceedings of the 20th VLDB Conference* (pp. 487–499). Santiago, Chile: (VLDB'94).

Bishop, C. (1995). *Neural Networks for Pattern Recognition*. Oxford, UK: Oxford University Press.

Han, J., Lakshmanan, L. V. S., & Pei, J. (2001). Scalable frequent-pattern mining methods: An overview. In T. Fawcett (Ed.), *KDD 2001: Tutorial Notes of the Seventh ACM SIGKDD International Conference on Knowledge Discovery and Data Mining* New York: The Association for Computing Machinery.

Hartigan, J. A. (1975). *Clustering Algorithms*. New York: Wiley.

Hartigan, J. A., & Wong, M. A. (1978). Algorithm 136. A *k*-means clustering algorithm. *Applied Statistics, 28*, 100.

Hastie, T. J., & Tibshirani, R. J. (1990). *Generalized Additive Models*. New York: Chapman & Hall.

Kass, G. V. (1980). An exploratory technique for investigating large quantities of categorical data. *Applied Statistics, 29*, 119–127.

McColloch, W., & Pitts, W. (1943). A logical calculus of the ideas immanent in nervous activity. *Bulletin of Mathematical Biophysics, 7*, 115–133.

Schimek, M. G. (2000). *Smoothing and Regression: Approaches, Computations, and Application*. New York: Wiley.

Witten, I. H., & Frank, E. (2000). *Data Mining: Practical Machine Learning Tools and Techniques*. New York: Morgan Kaufmann.

Bibliography

Carling, A. (1992). *Introducing Neural Networks*. Wilmslow, UK: Sigma Press.

Carlson, N. A. (1992). *Foundations of Physiological Psychology*. Needham Heights, Massachusetts: Simon & Schuster.

Fausett, L. (1994). *Fundamentals of Neural Networks*. New York: Prentice Hall.

Haykin, S. (1994). *Neural Networks: A Comprehensive Foundation*. New York: Macmillan Publishing.

Kohonen, T. (1982). Self-organized formation of topologically correct feature maps. *Biological Cybernetics, 43*, 59–69.

Patterson, D. (1996). *Artificial Neural Networks*. Singapore: Prentice Hall.

Ripley, B. D. (1996). *Pattern Recognition and Neural Networks*. Cambridge, UK: Cambridge University Press.

Rumelhart, D. E., & McClelland, J. L. (1986). *Parallel Distributed Processing* (Vol. 1). Cambridge, MA: The MIT Press Foundations.

Stone, C. (1986). The dimensionality reduction principle for Generalized Additive Models. *Ann. Statist., 14*(2), 590–606.

Tryon, R. C. (1939). *Cluster Analysis*. New York: McGraw-Hill.

Advanced Algorithms for Data Mining

PREAMBLE

You can perform most general data mining tasks with the basic algorithms presented in Chapter 7. But eventually, you may need to perform some specialized data mining tasks. This chapter describes some advanced algorithms that can "supercharge" your data mining jobs. They include the following:

1. Advanced General-Purpose Machine Learning Algorithms
    *   Interactive Trees (C&RT or CART, CHAID)
    *   Boosted Tree Classifiers and Regression
    *   MARSplines (Multivariate Adaptive Regression Splines)
    *   Random Forests for Regression and Classification (discussed in Chapter 11)
    *   Machine Learning—Naïve Bayesian Classifier and Nearest Neighbor (discussed in Chapter 11)
    *   Statistical Learning Theory—Support Vector Machines
    *   Sequence, Association, and Link Analysis

- Independent Components Analysis
- Kohonen Clustering

2. Text Mining Algorithms (discussed in Chapter 9—Text Mining and Natural Language Processing)
3. Quality Control Data Mining and Root Cause Analysis
   - Quality Control Charts
   - Quality Control Charts for Variable Lists
   - Predictive Quality Control
   - Root Cause Analysis
   - Response Optimization for Data Mining Models
4. Image and Object Data Mining: Visualization and 3D-Medical and Other Scanning Imaging

You may wonder why there are so many algorithms available. Research during the past 30 years has generated many kinds and variants of data mining algorithms that are suited to particular areas in the solution landscape. Figure 8.1 illustrates where specific

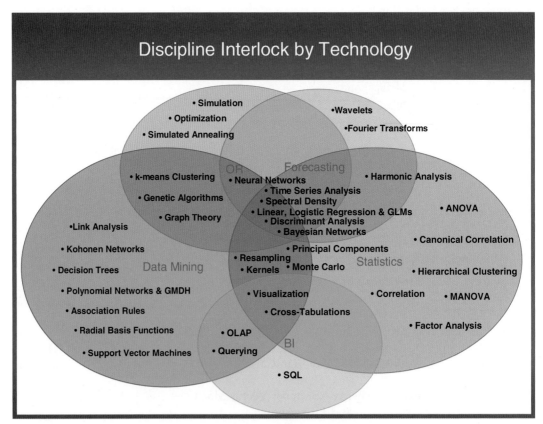

FIGURE 8.1 The relationship between specific algorithms and business analytical problem areas (and the overlap between them).

data mining algorithms fit into the solution landscape of various business analytical problem areas: operations research—OR, forecasting, data mining, statistics, and business intelligence—BI.

Figure 8.1 came from studies by Dustin Hux and John Elder (both of Elder Research, Inc.) on algorithms used in journal articles in different domains. From Figure 8.1, you can see which field uses what technique and also what techniques are suited to overlaps between areas. For example, visualization and cross-tabulations are used in business intelligence, data mining, and statistics.

Data miners use many analysis techniques from statistics, but often ignore some techniques like factor analysis (not always wisely). In addition, data mining includes a lot of techniques that are not considered typically in the world of statistics (such as radial basis function networks, genetic algorithms). Operations research (OR) uses clustering, graph theory, neural networks, and time series, but also depends very heavily on simulation and optimization. Forecasting overlaps data mining, statistics, and OR, and adds a few algorithms like Fourier transforms and wavelets.

In addition to the overlap of algorithms in different areas, some of them are known by different names. For example, Principal Components Analysis (PCA) is known in electrical engineering as the Karhounen-Loeve transform, and in statistics as the eigenvalue-eigenvector decomposition.

In our early college years, we take courses in many different disciplines, and it looks as though techniques are developed in them independently. One of the important byproducts of higher education (especially graduate school) is that we begin to see the interconnections between these ideas in different disciplines. The Ph.D. degree is short for Doctor of Philosophy. Doctoral degrees are handed out in many very technical disciplines, and it might seem strange that "philosophy" is still in the name. What does philosophy have to do with Recombinant DNA Genetics? The answer is "Everything." One of the jokes often heard in graduate schools is "You learn more and more about less and less, until you know everything about nothing." Well, a very highly constrained subject matter discipline is the end point (not quite "nothing"), and through the process of getting there, you can see the connections with a great many other disciplines. And this connected view of a broad subject area (e.g., genetics) provides the necessary philosophical framework for the study of your specific area. You are not educated properly in a discipline until you can view it in the context of its relationship with many other disciplines. So it is with the study of analytical algorithms. This book will take you far along that path (books like the one by Hastie et al., 2001, do it better), but this introduction will provide enough background to help you navigate through the plethora of data mining and statistical analysis algorithms available in most data mining tool packages.

Now, we will turn to the main job at hand in this chapter and look at each of the advanced algorithms individually. Because these algorithms are implemented in slightly different ways in each data mining or statistical package, we will cast the explanations in terms of how they are implemented in *STATISTICA* Data Miner (for which a free 90-day copy is available on the enclosed DVD). In addition to the free software, we have provided numerous tutorials (many of them use *STATISTICA* Data Miner; several with other tools, particularly SPSS Clementine and SAS Enterprise Miner). Some of the following text was adapted from the *STATISTICA* software online help: StatSoft, Inc. (2008). *STATISTICA* (data analysis software system), version 8.0. www.statsoft.com. You can experiment with these algorithms during the 90-day license period, even if you normally use a different package in your organization.

ADVANCED DATA MINING ALGORITHMS

Interactive Trees

The *STATISTICA Interactive Trees* (I-Trees) module builds trees for predicting dependent variables that are continuous (estimation) or categorical (classification). The program supports the classic Classification and Regression Tree (CART) algorithm (Breiman et al.; see also Ripley, 1996) as well as the CHAID algorithm (Chi-Square Automatic Interaction Detector; Kass, 1980). The module can use algorithms, user-defined rules, criteria specified via an interactive graphical user interface (brushing tools), or a combination of those methods. This enables users to try various predictors and splitting criteria in combination with almost all the functions of automatic tree building.

Figure 8.2 displays the tree results layout in the I-Trees module. The main results screen is shown in Figure 8.3.

Manually Building the Tree

The I-Trees module doesn't build trees by default, so when you first display the Trees Results dialog, no trees have been built. (If you click the Tree Graph button at this point, a single box will be displayed with a single root node, as in Figure 8.4.)

The Tree Browser

The final tree results are displayed in the workbook tree browser, which clearly identifies the number of splitting nodes and terminal nodes of the tree (Figure 8.5).

To review the statistics and other information (e.g., splitting rule) associated with each node, simply highlight it and review the summary graph in the right pane. The split nodes can be collapsed or expanded in the manner that most users are accustomed to from standard MS Windows-style tree browser controls. Another useful feature of the workbook tree browser is the ability to quickly review the effect of consecutive splits on the resulting child nodes in an animation-like manner.

Advantages of I-Trees

- I-Trees is particularly optimized for very large data sets, and in many cases the raw data do not have to be stored locally for the analyses.
- It is more flexible in the handling of missing data. Because the Interactive Trees module does not support ANCOVA-like design matrices, it is more flexible in the handling of missing data; for example, in CHAID analyses, the program will handle predictors one at a time to determine a best (next) split; in the General CHAID (GCHAID) Models module, observations with missing data for any categorical predictor are eliminated from the analysis.
- You can perform "what-if" analyses to gain better insights into your data by interactively deleting individual branches, growing other branches, and observing various results statistics for the different trees (tree models).

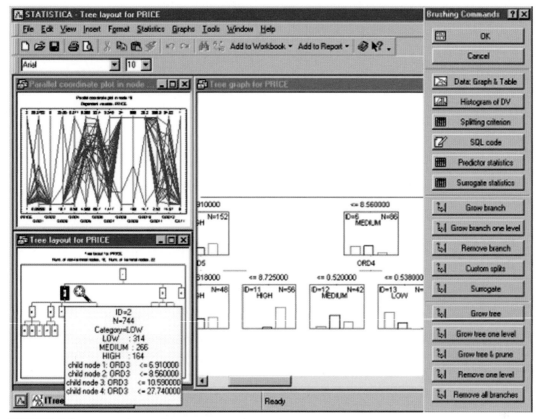

FIGURE 8.2   Layout of part of the I-Trees interface in *STATISTICA* Data Miner.

FIGURE 8.3   The results screen of the I-Trees module.

II. THE ALGORITHMS IN DATA MINING AND TEXT MINING

FIGURE 8.4 Initial tree graph showing only one node (no splitting yet).

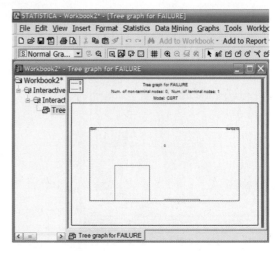

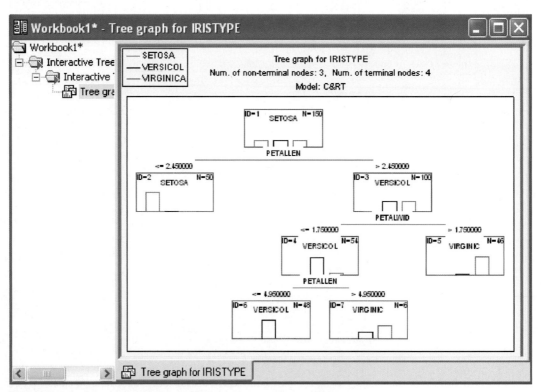

FIGURE 8.5 Final tree results from I-Trees, using example of Iris data set.

- You can automatically grow some parts of the tree but manually specify splits for other branches or nodes.
- You can define specific splits and select alternative important predictors other than those chosen automatically by the program.
- You can quickly copy trees into new projects to explore alternative splits and methods for growing branches.
- You can save entire trees (projects) for later use.

Building Trees Interactively

Building trees interactively has proven popular in applied research, and data exploration is based on experts' knowledge about the domain or area under investigation, and relies on interactive choices (for how to grow the tree) by such experts to arrive at "good" (valid) models for prediction or predictive classification. In other words, instead of building trees automatically, using sophisticated algorithms for choosing good predictors and splits (for growing the branches of the tree), a user may want to determine manually which variables to include in the tree and how to split those variables to create the branches of the tree. This enables the user to experiment with different variables and scenarios and ideally to derive a better understanding of the phenomenon under investigation by combining her or his expertise with the analytic capabilities and options for building the tree (see also the next section).

Combining Techniques

In practice, it may often be most useful to combine the automatic methods for building trees with educated guesses and domain-specific expertise. You may want to grow some portions of the tree using automatic methods and refine and modify the choices made by the program (for how to grow the branches of the tree) based on your expertise. Another common situation in which this type of combination is called for is when some variables that are chosen automatically for some splits are not easily observable because they cannot be measured reliably or economically (i.e., obtaining such measurements would be too expensive). For example, suppose the automatic analysis at some point selects a variable *Income* as a good predictor for the next split; however, you may not be able to obtain reliable data on income from the new sample to which you want to apply the results of the current analysis (e.g., for predicting some behavior of interest, such as whether or not the person will purchase something from your catalog). In this case, you may want to select a surrogate variable, i.e., a variable that you can observe easily and that is likely related or similar to variable *Income* (with respect to its predictive power; for example, a variable *Number of years of education* may be related to *Income* and have similar predictive power; while most people are reluctant to reveal their level of income, they are more likely to report their level of education, and hence, this latter variable is more easily measured).

The I-Trees module provides a large number of options to enable users to interactively determine all aspects of the tree-building process. You can select the variables to use for each split (branch) from a list of suggested variables, determine how and where to split a variable, interactively grow the tree branch by branch or level by level, grow the entire tree automatically, delete ("prune back") individual branches of trees, and more. All of these options are provided in an efficient graphical user interface in which you can "brush" the current tree, i.e., select a specific node to grow a branch, delete a branch, etc. As in all modules for predictive data mining, the decision rules contained in the final tree built for regression or classification prediction can optionally be saved in a variety of ways for deployment in data mining projects, including C/C++, *STATISTICA* Visual Basic, or Predictive Model Markup Language (PMML). Hence, final trees computed via this module can quickly and efficiently be turned into solutions for predicting or classifying new observations.

Multivariate Adaptive Regression Splines (MARSplines)

We'll use the *STATISTICA* Data Miner software tool to describe the MARSplines algorithm, but the ideas described can be applied to whatever software package you use.

Note: Many of the paragraphs in this section are adapted from the *STATISTICA* online help, StatSoft, Inc. (2008). *STATISTICA* (data analysis software system), version 8.0. www.statsoft.com.

The *STATISTICA* Multivariate Adaptive Regression Splines (MARSplines) module is a generalization of techniques (called MARS) popularized by Friedman (1991) for solving regression- and classification-type problems, with the goal to predict the value of a set of dependent or outcome variables from a set of independent or predictor variables. MARSplines can handle both categorical and continuous variables (whether response or predictors). With categorical responses, MARSplines will treat the problem as a classification problem; with continuous dependent variables, as a regression problem. MARSplines will automatically determine that for you.

MARSplines is a nonparametric procedure that makes no assumption about the underlying functional relationship between the dependent and independent variables. Instead, it constructs the model from a set of coefficients and basis functions that are entirely "driven" from the data. In a sense, the method follows decision trees in being based on the "divide and conquer" strategy, which partitions the input space into regions, each with its own regression or classification equation. This makes MARSplines particularly suitable for problems with higher input dimensions (i.e., with more than two variables), where the *curse of dimensionality* would likely create problems for other techniques.

The MARSplines technique has become particularly popular in data mining because it does not assume or impose any particular type or class of relationship (e.g., linear, logistic, etc.) between the predictor variables and the dependent (outcome) variable of interest. Instead, useful models (i.e., models that yield accurate predictions) can be derived even in situations in which the relationships between the predictors and the dependent variables are non-monotone and difficult to approximate with parametric models. For more information about this technique and how it compares to other methods for nonlinear regression (or regression trees), see Hastie et al. (2001).

In linear regression, the response variable is hypothesized to depend linearly on the predictor variables. It's a parametric method, which assumes that the nature of the relationships (but not the specific parameters) between the dependent and independent variables is known *a priori* (e.g., is linear). By contrast, nonparametric methods do not make any such assumption as to how the dependent variables are related to the predictors. Instead, it allows the model function to be "driven" directly from data.

Multivariate Adaptive Regression Splines (MARSplines) constructs a model from a set of coefficients and features or "basis functions" that are determined from the data. You can think of the general "mechanism" by which the MARSplines algorithm operates as multiple piecewise linear regression where each breakpoint (estimated from the data) defines the "region of application" for a particular (very simple) linear equation.

Basis Functions

Specifically, MARSplines uses two-sided truncated functions of the form (as shown in Figure 8.6) as basis functions for linear or nonlinear expansion, which approximates the relationships between the response and predictor variables.

Figure 8.6 shows a simple example of two basis functions $(t - x)+$ and $(x - t)+$ (adapted from Hastie et al., 2001, Figure 9.9). Parameter t is the knot of the basis functions (defining the "pieces" of the piecewise linear regression); these knots (t parameters) are also determined from the data. The plus (+) signs next to the terms $(t - x)$ and $(x - t)$ simply denote that only positive results of the respective equations are considered; otherwise, the respective functions evaluate to zero. This can also be seen in the illustration.

The MARSplines Model

The basis functions together with the model parameters (estimated via least squares estimation) are combined to produce the predictions given the inputs. The general *MARSplines* model equation (see Hastie et al., 2001, equation 9.19) is given as

$$y = f(X) = \beta_o + \sum_{m=1}^{M} \beta_m h_m(X)$$

where the summation is over the M nonconstant terms in the model. To summarize, y is predicted as a function of the predictor variables X (and their interactions); this function

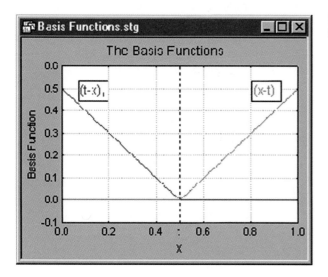

FIGURE 8.6 MARSplines basis functions for linear and nonlinear analysis.

consists of an intercept parameter (β_o) and the weighted (by β_m) sum of one or more basis functions $h_m(X)$, of the kind illustrated earlier. You can also think of this model as "selecting" a weighted sum of basis functions from the set of (a large number of) basis functions that span all values of each predictor (i.e., that set would consist of one basis function and parameter t, for each distinct value for each predictor variable). The MARSplines algorithm then searches over the space of all inputs and predictor values (knot locations t) as well as interactions between variables. During this search, an increasingly larger number of basis functions is added to the model (selected from the set of possible basis functions), to maximize an overall least squares goodness-of-fit criterion. As a result of these operations, MARSplines automatically determines the most important independent variables as well as the most significant interactions among them. The details of this algorithm are further described in Hastie et al. (2001).

Categorical Predictors

MARSplines is well suited for tasks involving categorical predictors variables. Different basis functions are computed for each distinct value for each predictor, and the usual techniques for handling categorical variables are applied. Therefore, categorical variables (with class codes rather than continuous or ordered data values) can be accommodated by this algorithm without requiring any further modifications.

Multiple Dependent (Outcome) Variables

The MARSplines algorithm can be applied to multiple dependent (outcome) variables, whether continuous or categorical. When the dependent variables are continuous, the algorithm will treat the task as regression; otherwise, as a classification problem. When the outputs are multiple, the algorithm will determine a common set of basis functions in the predictors but estimate different coefficients for each dependent variable. This method of treating multiple outcome variables is not unlike some neural network architectures, where multiple outcome variables can be predicted from common neurons and hidden layers; in the case of MARSplines, multiple outcome variables are predicted from common basis functions, with different coefficients.

MARSplines and Classification Problems

Because MARSplines can handle multiple dependent variables, it is easy to apply the algorithm to classification problems as well. First, it will code the classes in the categorical response variable into multiple indicator variables (e.g., 1 = observation belongs to class k, 0 = observation does not belong to class k); then MARSplines will fit a model and compute predicted (continuous) values or scores; and finally, for prediction, it will assign each case to the class for which the highest score is predicted (see also Hastie et al., 2001, for a description of this procedure).

Model Selection and Pruning

In general, nonparametric models are adaptive and can exhibit a high degree of flexibility that may ultimately result in overfitting if no measures are taken to counteract it. Although overfit models can achieve zero error on training data (provided they have a sufficiently large number of parameters), they will almost certainly perform poorly when presented with new observations or instances (i.e., they do not generalize well to the prediction of "new" cases). MARSplines tends to overfit the data as well. To combat this problem, it uses a pruning technique (similar to that in classification trees) to limit the complexity of the model by reducing the number of its basis functions.

MARSplines as a Predictor (Feature) Selection Method

The selection of and pruning of basis functions in MARSplines makes this method a very powerful tool for predictor selection. The MARSplines algorithm will pick up only those basis functions (and those predictor variables) that make a "sizeable" contribution to the prediction. The Results dialog of the Multivariate Adaptive Regression Splines (MARSplines) module will clearly identify (highlight) only those variables associated with basis functions that were retained for the final solution (model).

Applications

MARSplines has become very popular recently for finding predictive models for "difficult" data mining problems, i.e., when the predictor variables do not exhibit simple and/or monotone relationships to the dependent variable of interest. Because of the specific manner in which MARSplines selects predictors (basis functions) for the model, it generally does well in situations in which regression-tree models are also appropriate, i.e., where hierarchically organized successive splits on the predictor variables yield accurate predictions. In fact, this technique is as much a generalization of regression trees as it is of multiple regression. The "hard" binary splits are replaced by "smooth" basis functions.

A large number of graphs can be computed to evaluate the quality of the fit and to aid with the interpretation of results. Various code generator options are available for saving estimated (fully parameterized) models for deployment in C/C++/C#, Visual Basic, or PMML.

The MARSplines Algorithm

Implementing MARSplines involves a two-step procedure that is applied successively until a desired model is found. In the first step, we build the model (increase its complexity) by repeatedly adding basis functions until a user-defined maximum level of complexity is reached. (We start with the simplest—the constant; then we iteratively add the next term, of all possible, that most reduces training error.) Once we have built a very complex model, we begin a backward procedure to iteratively remove the least significant basis functions from the model, i.e., those whose removal leads to the least reduction in the (least-squares) goodness of fit.

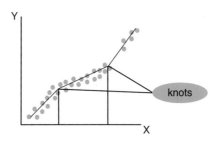

FIGURE 8.7   Piecewise regression plot showing locations of the knots.

MARSplines is a *local nonparametric method* that builds a piecewise linear regression model; it uses separate regression slopes in distinct intervals of the predictor variable space (Figure 8.7).

The slope of the piecewise regression line is allowed to change from one interval to the other as the two "knots" points are crossed; knots mark the end of one region and beginning of another. Like CART, its structure is found first by overfitting and then pruning back.

The major advantage of MARSplines is that it automates all those aspects of regression modeling that are difficult and time consuming to conduct by hand:

- Selecting which predictors to use for building models;
- Transforming variables to account for nonlinear relationships;
- Detecting interactions that are important;
- Self-testing to ensure that the model will work on future data.

The result is a more accurate and more complete model that could be handcrafted—especially by inexperienced modelers.

Statistical Learning Theory: Support Vector Machines

Support Vector Machines are based on the Statistical Learning Theory concept of decision planes that define decision boundaries. A decision plane ideally separates objects having different class memberships, as shown in Figure 8.8. There, the separating line defines a boundary on the right side of which all objects are *GREEN* and to the left of which all objects are *RED*. Any new object falling to the right is classified as *GREEN* (or as *RED* should it fall to the left of the separating line).

Most classification tasks, however, are not that simple, and often more complex structures are needed to make an optimal separation, i.e., correctly classify new objects (test cases) on the basis of the examples that are available (train cases). In Figure 8.9, it is clear that a full separation of the *GREEN* and *RED* objects would require a curve, which is more complex than a line. Classification tasks based on drawing separating lines to distinguish between objects of different class memberships are known as *hyperplane classifiers*. Support Vector Machines are particularly suited to handle such tasks.

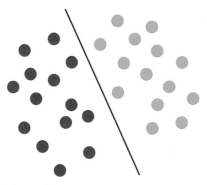

FIGURE 8.8   Linear separation in input data space.

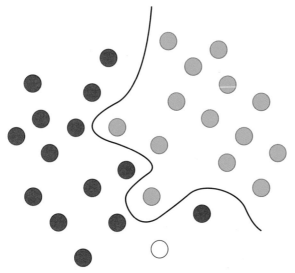

FIGURE 8.9   Nonlinear separation in input data space.

Figure 8.10 shows the basic idea behind Support Vector Machines. Here, we see the original objects (left side of the schematic) mapped, i.e., rearranged, using a set of mathematical functions known as kernels. The process of rearranging the objects is known as *mapping* (transformation) to a new space with different dimensions called *feature space*. Note that in this new space, the mapped objects (right side of the schematic) are linearly separable and, thus, instead of constructing the complex curve (left schematic), all we have to do is find an optimal line that can separate the *GREEN* and *RED* objects.

STATISTICA Support Vector Machine (SVM) is a classifier method that performs classification tasks by constructing hyperplanes in a multidimensional space that separates cases of different class labels. It supports both regression and classification tasks and can handle multiple continuous and categorical variables. For categorical variables, a dummy variable

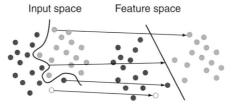

Input space   Feature space

FIGURE 8.10   Mapping of input data points to feature space, where linear separation is possible.

is created with case values of either *0* or *1*. So a categorical dependent variable consisting of three levels—*A, B, C*—is represented by a set of three dummy variables:

$$A: \{1\ 0\ 0\}, \quad B: \{0\ 1\ 0\}, \quad C: \{0\ 0\ 1\}$$

To construct an optimal hyperplane, SVM employs an iterative training algorithm, minimizing an error function. According to the form of this error function, SVM models can be classified into four distinct groups:

- Classification SVM Type 1 (also known as C-SVM classification);
- Classification SVM Type 2 (also known as nu-SVM classification);
- Regression SVM Type 1 (also known as epsilon-SVM regression);
- Regression SVM Type 2 (also known as nu-SVM regression).

(See "Support Vector Machines Introduction" in *STATISTICA Online* help for a complete description of Type 1 and Type 2 SVM: Install the *STATISTICA* program on the DVD bound with this book to access this online help.)

Kernel Functions

Support Vector Machines use kernels that can be linear, polynomial, Radial Basis Function (RBF), or sigmoid. The RBF is by far the most popular choice of kernel types used, mainly because of their localized and finite responses across the entire range of the real *x*-axis.

Sequence, Association, and Link Analyses

Sequence, Association, and Link Analyses compose a group of related techniques for extracting rules from data sets that can be generally characterized as "marketbaskets"—a metaphor for a group of items purchased by the customer, either in a single transaction, or over time in a sequence of transactions. Such products can be goods displayed in a supermarket, spanning a wide range of items from groceries to electrical appliances, or they can be insurance packages that customers might be willing to purchase, etc. Customers fill their basket with only a fraction of what is on display or on offer.

Association Rules

The first step in marketbasket analysis is to infer association rules, which express which products are frequently purchased together. For example, you might find that purchases of flashlights also typically coincide with purchases of batteries in the same basket.

Sequence Analysis

Sequence analysis is concerned with the order in which a group of items was purchased. For instance, buying an extended warranty is more likely to follow (in that specific sequential order) the purchase of a TV or other expensive electric appliance. Useful sequence rules, however, are not always obvious, and sequence analysis helps you to extract such rules no matter how hidden they may be in your transactional data. There is a wide range of applications for sequence analysis in many areas of industry, including customer shopping patterns, phone call patterns, insider trading evidence in the stock market, DNA sequencing, and Web log streams.

Link Analysis

Link analysis provides information on the strength of the association rules or sequence rules. Once extracted, rules about associations or the sequences of items as they occur in a transaction database can be extremely useful for numerous applications. Obviously, in retailing or marketing, knowledge of purchase "patterns" can help with the direct marketing of special offers to the "right" or "ready" customers (i.e., those who, according to the rules, are most likely to purchase specific items given their observed past consumption patterns). However, transaction databases occur in many areas of business, such as banking. In fact, the term *link analysis* is often used when these techniques for extracting sequential or nonsequential association rules are applied to organize complex "evidence." It is easy to see how the "transactions" or "shopping basket" metaphor can be applied to situations in which individuals engage in certain actions, open accounts, contact other specific individuals, and so on. Applying the technologies described here to such databases may quickly extract patterns and associations between individuals and actions and, for example, reveal the patterns and structure of some clandestine illegal network.

Association Rule Details

An association is an expression of the form

Body --> Head (Support, Confidence)

Following this form, an example of an association rule is

If a customer buys a flashlight, he/she will buy batteries (250, 81%).

More than one dimension can be used to define the Body portion of the association rule. For example, the rule might be expanded as

If a customer is a plumber and buys a flashlight, he/she will buy batteries (150, 89%).

Figure 8.11 illustrates an example of an association problem.

Support value is computed as the joint probability (relative frequency of co-occurrences) of the Body and the Head of each association rule. This is expressed by the quantity

$$Support = \frac{\#purchases\_of\_A}{Total\_Purchases}$$

FIGURE 8.11 An association example.

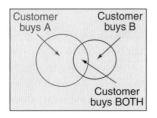

Transaction ID	Items Bought
2000	A, B, C
1000	A, C
4000	A, D
5000	B, E, F

Confidence value denotes the conditional probability of the Head of the association rule, given the Body of the association rule, expressed as

$$Confidence = \frac{\#purchases\_of\_A\_then\_B}{Support\_for\_A}$$

Lift value measures the confidence of a rule and the expected confidence that the second product will be purchased depending on the purchase of the first product, expressed as

$$Lift = \frac{\#Confidence\_of\_A\_then\_B}{Support\_for\_C}$$

The association example shown in Figure 8.11 can be evaluated for each item and reported in Figure 8.12.

For rule $A \Rightarrow C$:

Support = support ($\{A \Rightarrow C\}$) = 50%
Confidence = support ($\{A \Rightarrow C\}$)/support ($\{A\}$) = 66.6%
Lift = confidence ($\{A \Rightarrow C\}$)/support ($\{C\}$) = 1.33

This rule has 66.6% confidence (strength) meaning 66.6% of customers who bought A also bought C. The support value of 50% means that this combination covers 50% of transactions in the database. The Lift value of 1.33 gives us information about the increase in probability of the "then" condition, given the "if" condition. It is 33% more likely than independence suggests, that is, than assuming the purchases are unrelated.

FIGURE 8.12 Association results for an example.

Transaction ID	Items Bought
2000	A, B, C
1000	A, C
4000	A, D
5000	B, E, F

Min. Support 50%
Min. Confidence 50%

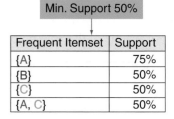

Min. Support 50%

Frequent Itemset	Support
{A}	75%
{B}	50%
{C}	50%
{A, C}	50%

Sequence Analysis Applications

Temporal order is important in many situations; for instance:

- Time-series databases and sequence databases
- Frequent patterns ⇒ (frequent) sequential patterns
- Applications of sequential pattern mining
- Customer shopping sequences
- Medical treatment
- Natural disasters (e.g., earthquakes)
- Science and engineering processes
- Stocks and markets
- Telephone calling patterns
- Weblog click streams
- DNA sequences
- Gene structures, and many more

Link Analysis—Employing Visualization

In Figure 8.13, the support values for the body and head portions of each association rule are indicated by the size of each circle. The thickness of each line indicates the relative joint support of two items, and its color indicates their relative lift. A minimum of two items in the Item name list view must be selected to produce a web graph.

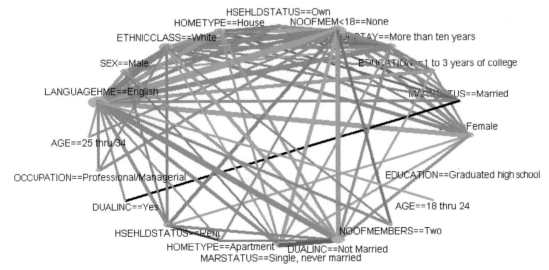

FIGURE 8.13 Web graph, showing support and lift.

Independent Components Analysis (ICA)

ICA is designed for signal separation in the process Statistical Signal Processing, which has a wide range of applications in many areas of technology, ranging from audio and image processing to biomedical signal processing, telecommunications, and econometrics. Imagine being in a room with a crowd of people and two speakers giving presentations at the same time. The crowd is making comments and noises in the background. We are interested in what the speakers say and not the comments emanating from the crowd. There are two microphones at different locations, recording the speakers' voices as well as the noise coming from the crowd. Our task is to separate the voice of each speaker while ignoring the background noise (Figure 8.14).

ICA can be used as a method of blind source separation, meaning that it can separate independent signals from linear mixtures with virtually no prior knowledge of the signals. An example is decomposition of electro or magneto-encephalographic signals. In computational neuroscience, ICA has been used for feature extraction, where it seems to mimic the basic cortical processing of visual and auditory information. New application areas are being discovered at an increasing pace.

STATISTICA *Fast Independent Component Analysis (FICA)*

FICA uses state-of-the-art methods for applying the Independent Component Analysis algorithm to virtually any practical problem requiring separation of mixed signals into their original components. These methods include Simultaneous Extraction and Deflation techniques. Other features supported in the program include data preprocessing and case selection. The program also supports the implementation of the ICA methods to either new analyses (i.e., model creation) or the deployment of existing models that have been previously prepared and saved. Thus, while you can use the ICA module for creating new models, you can also rerun existing models for deployment and further analysis.

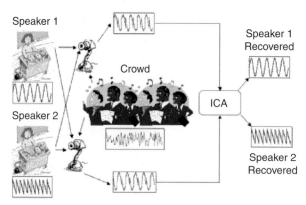

FIGURE 8.14   How Independent Component Analysis is used.

A large number of graphs and spreadsheets can be computed to evaluate the quality of the FICA models to help interpret results and conclusions. Various code generator options are available for saving estimated (fully parameterized) models for deployment in C/C++/C#, Visual Basic, or PMML.

Note: Many of the preceding paragraphs on advanced algorithms were adapted from the online help of *STATISTICA*; StatSoft, Inc. (2008). *STATISTICA* (data analysis software system), version 8.0. www.statsoft.com.

Kohonen Networks

A form of neural network in which there are no known dependent variables was proposed by Kohonen (1982) for use in unsupervised clustering. The network is trained by assigning cluster centers to a radial layer by iteratively submitting training patterns to the network and adjusting the winning (nearest) radial unit center and its neighbors toward the training pattern (Kohonen, 1982; Fausett, 1994; Haykin, 1994; Patterson, 1996). The resulting operation causes data points to "self-organize" into clusters. A short-hand acronym for Kohonen networks is a self-organizing feature map (SOFM).

Characteristics of a Kohonen Network

A Kohonen network has the following characteristics:

- *Competition.* For each input pattern, the neurons compete with one another.
- *Cooperation.* The winning neuron determines the spatial location of a topological neighborhood of excited neurons, thereby providing the basis for cooperation among the neurons.
- *Synaptic Adaptation.* The excited neurons adjust their synaptic weights to enhance their responsiveness to a similar input pattern.

In *STATISTICA*, a Kohonen network—self-organizing feature maps—can be obtained by selecting the Graphs (Kohonen) tab of the SANN–Results dialog. (Note that for all graph types, you can include cases in the Train, Test, and/or Validation subsets by selecting the appropriate check boxes in the Sample group box. For example, to view a histogram of the outputs for the Validation subset, select Output in the X-axis box, select the Validation check box in the Sample group, and click the Histograms of X button. Only the predictions for cases in the validation sample will be plotted.)

Quality Control Data Mining and Root Cause Analysis

Quality control algorithms modified for use in Data Miner Workspaces that are not available as standalone statistical modules are available in some of the data mining software available commercially. The Quality Control module of *STATISTICA* takes full advantage of the *STATISTICA* dynamic data transfer/update technology, and this module is preconfigured to optimally support applications in which the output (charts, tables) needs to dynamically reflect changes in data streams of practically arbitrary volume. It is also designed to work as part of the client/server and distributed processing architectures in the most demanding manufacturing environments where data streams from multiple channels need to be processed in real time. It is optimized for online, real-time quality control charting and processing (e.g., online automated alarms), and other user-defined decision support (automatic or interactive) operations with dynamic "live" data streams.

IMAGE AND OBJECT DATA MINING: VISUALIZATION AND 3D-MEDICAL AND OTHER SCANNING IMAGING

Image and object data mining is an area of active current research, involving development of new and modified algorithms that can better deal with the complexities of three-dimensional object identification. Most searching for a desired image is currently based on text metadata, such as filename or associated text. But this does not adequately allow identification of objects at accuracy rates of 90% or better, which is what is really needed in this field. Content-based image retrieval emerged during the 1990s, focusing on color and texture cues, and although this was easier than text data, it turned out to be less useful. Visual object analysis offered a better solution, as the last century drew to a close, because

- Machine learning methods offered greater accuracy in object identification; and
- Large amounts of training data were now available (computer storage space now not a problem).

We will not discuss this topic thoroughly in this chapter, since this is a developing area, but we want to make you aware of visualization and the role in our lives it will provide in the coming decades by providing a list of terms and concepts and also returning to this topic in Chapter 21, where we list it as a "prospect for future development."

Following are terms, concepts, places, and names of data mining algorithms associated with visual object identification:

- CUBIC project at IBM
- Visual imaging projects at UC Santa Barbara, and UC Berkeley
- PROBLEM: "Object Category Recognition"—humans can recognize these, but computers cannot easily
- Using newer "modified algorithms," researchers in this field have gone from 16% recognition in 2004 work to 90% correct "object category recognition" in 2008, with a selected number of objects, using the following algorithms:
  - Nonlinear kernelized SVM (slower but more accurate)
  - Boosted trees (works slowly)
  - Linear SVM (fast but not accurate)
  - Intersection kernels in SVM (gone to 90% correct "object category recognition in 2008")

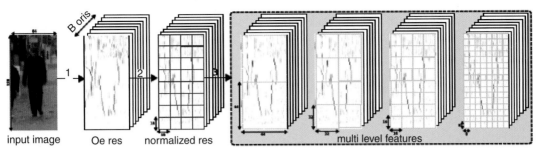

FIGURE 8.15   The three-stage pipeline of the feature computation process. (1) The input grayscale image of size 64 × 128 is convolved with oriented filters (! = 1) in 8 directions to obtain oriented energy responses. (2) The responses are then L1 normalized over all directions in each nonoverlapping 16 × 16 block independently to obtain normalized responses. (3) Multilevel features are then extracted by constructing histograms of oriented gradients by summing up the normalized response in each cell. The diagram depicts progressively smaller cell sizes from 64 × 64 to 8 × 8. [Taken from http://www.eecs.berkeley.edu/Research/Projects/CS/vision/shape/papers/mbm08cvpr_tag.pdf. Maji, S., Berg, A. C., Malik, J. (2008) Classification using intersection kernel support vector machines is efficient, by in *IEEE Computer Vision and Pattern Recognition 2008, Anchorage.* Computer Science Division, EECS University of California at Berkeley, CA 94720, USA]

Today, 100 objects can be categorized with the methods listed here, but we need to go to several levels of magnitude larger for these methods to be fully successful in accurately recognizing images for such things as national security surveillance and precise medical identification of conditions when using three-dimensional imaging procedures. But we believe that shape-based object recognition is the key for the future in this area.

Figure 8.15 offers a schematic of this emerging process.

POSTSCRIPT

By now, you may be very tired of theory and be itching to get your hands on the tools. In fact, you might do just that by going through one of the tutorials. But you should come back to this section of the book later to learn more about some common application areas of data mining. The shape recognition research mentioned earlier is one direction of development in the general area of pattern recognition. But a more commonly applied form of pattern recognition is text mining, which we will explore in Chapter 9.

References

Breiman, L., Friedman, J. H., Olshen, R. A., & Stone, C. J. (1984). *Classification and regression trees*. Monterey, CA: Wadsworth & Brooks/Cole Advanced Books & Software.

Fausett, L. (1994). *Fundamentals of Neural Networks*. New York: Prentice Hall.

Friedman, J. (1991). Multivariate adaptive regression splines (with discussion). *Annals of Statistics, 19*, 1–141.

Hastie, T., Tibshirani, R., & Friedman, J. H. (2001). *The Elements of Statistical Learning: Data Mining, Inference, and Prediction*. New York: Springer Verlag.

Haykin, S. (1994). *Neural Networks: A Comprehensive Foundation.* New York: Macmillan Publishing.

Kass, G. V. (1980). An exploratory technique for investigating large quantities of categorical data. *Applied Statistics, 29,* 119–127.

Kohonen, T. (1982). Self-organized formation of topologically correct feature maps. *Biological Cybernetics, 43,* 59–69.

Malik, J. (2008). *The future of image research.* KDD'08, August 24–27, 2008, Las Vegas, Nevada, USA; ACM 978-1-60558-193-4/08/08; PLENARY INVITED TALK.

Patterson, D. (1996). *Artificial Neural Networks.* Singapore: Prentice Hall.

Ripley, B. D. (1996). *Pattern Recognition and Neural Networks.* Cambridge and New York: Cambridge University Press.

Text Mining and Natural Language Processing

PREAMBLE

Pattern recognition is the most basic description of what is done in the process of data mining. Usually, these patterns are stored in structured databases and organized into *records*, which are composed of *rows* and *columns* of data. The columns are attributes (numbers or text strings) associated with a *table* (entity), accessed by links between attributes among the tables (relations). This entity-relational structure of data is called a *relational database*. Large relational databases store huge quantities of data in *data warehouses* in large companies. Despite the rather large amount of business information that exists in data warehouses, the vast majority of business data is stored in documents that are virtually unstructured. According to a study by Merrill Lynch and Gartner, 85–90% of all corporate data are stored in some sort of unstructured form (i.e., as text) (McKnight, 2005). This is where text mining fits into the picture: it is the process of discovering new, previously unknown, potentially useful information from a variety of *unstructured* data sources including business documents, customer comments, web pages, and XML files.

THE DEVELOPMENT OF TEXT MINING

Text mining (also known as *text data mining* and *knowledge discovery* in textual databases) is the process of deriving novel information from a collection of texts (also known as a *corpus*). By "novel information," we mean associations, hypotheses, or trends that are not explicitly present in the text sources being analyzed. Though rightly considered a part of the general field of data mining, text mining differs significantly in many details due to the patterns being extracted from natural language text rather than from structured databases of facts (Yang and Lee, 2005). Databases are designed for programs to process automatically; text is written for people to read. We do not have programs that can "read" and "understand" text (at least not in the manner human beings do). Furthermore, despite the phenomenal advances achieved in the field of natural language processing (Manning and Schutze, 1999), we will not have such programs for the foreseeable future. Many researchers think it will require a full simulation of how the mind works before we can write programs that read and understand the way people do (Hearst, 2003).

A concise summary of what text mining does was provided by Delen and Crossland (2008):

> So what does text mining do? On the most basic level, it numericalizes an unstructured text document and then, using data mining tools and techniques, extracts patterns from them. (p. 1707)

Thus, text mining can be applied to many applications in a variety of fields, including

1. Marketing
2. National security and corporate security applications
3. Medical and biomedical
4. Legal—attorneys—law cases
5. Corporate finance—for business intelligence
6. Patent analysis—for the U.S. Patent & Trademark Office
7. Public relations—comparing web pages of comparable businesses, colleges, or organizations

Current sources of text mining software—both commercial and open source freeware include

1. SAS-Text Mining
2. SPSS-Text Mining and Text Analysis for Surveys
3. *STATISTICA* Text Miner
4. GATE—Natural Language Open Source
5. RapidMiner—with its Word Vector Tool plug-in
6. R-Language programming text mining—Open Source
7. Practical text mining with Perl—Open Source
8. ODM—Oracle Data Mining
9. Megaputer's "TextAnalyst"

A comprehensive discussion of text mining with Perl just came out in mid-2008 in the form of a professional book that discusses text mining from a practical, nonmathematical standpoint. It also includes many kinds of sample texts and programs that are in the public domain; thus, you can download and use them free (Bilisoly, 2008). The author, Bilisoly, is a professor at Central Connecticut State University, where he developed and teaches a graduate-level course in text mining for that school's new data mining program. He presents text mining in such an easy-to-understand format that we will paraphrase and emphasize some of the introductory comments in the book. The only thing that Bilisoly requires of his students is to be willing to learn to write some very simple programs using Perl, which is a programming language specifically designed to work with text. As Bilisoly points out, there are minimally three broad themes in text mining:

- "Text mining is built upon counting and text pattern matching;
- Language, although complex, can be studied by considering its simpler properties; and
- Combining computer and human strengths is a powerful way to study language" (pp. xviii–xix of Preface, Bilisoly, 2008).

Text pattern matching can mean identifying a pattern of letters in a document, or it may refer to the rejection of words that contain a string of letters of interest but are not related. The process of counting the number of matches to a text pattern occurs repeatedly in text mining, such that you can compare two different documents by counting how many times different words occur in each document. Additionally, even the language (regardless of its complexity) can be checked against large language collections, called *corpora*. After the computer has found the examples of usage, analysts can attribute meaning to these discovered patterns. This blend of human and computer discovery constitutes the third theme of text mining referred to in the preceding list. Specifically, analysts choose the best way to analyze the text further, by either combining groups of words that appear to mean the same thing or directing the computer to do it automatically in a second iteration of the process, and then analyze the results. As Bilisoly points out, "this back and forth process is repeated as many times as necessary." This is very similar to the Design of Experiment (DOE) process followed for industrial process control and quality control. Following a given DOE, you may proceed through several iterations, harvesting preliminary insights from previous iterations of the experiment, then adjusting input parameters and running the experiment again, until the results show an F-value significant at the 95% level of confidence. The process followed in text mining is very similar to this sequence of analyses.

A PRACTICAL EXAMPLE: NTSB

Next, we will consider an example concerning the National Transportation Safety Board, or NTSB, a U.S. governmental agency, as a lead-in to one of the tutorials in the accompanying DVD of this book, called "Tutorial—NTSB Text Mining."

We will use the software package *STATISTICA* Text Miner for this practical example so that we have something tangible to "sink our learning teeth" into and thus learn the concepts as rapidly as possible. Many of the graphic illustrations presented here and some of the associated text discussions were originally written by guest author Kiron Mathew, Key-Bank, Cleveland, Ohio, formerly of StatSoft Data Mining Consulting Group.

Basic *STATISTICA* Text Miner features include

- Design as a general open-architecture tool for mining unstructured information;
- Options for accessing documents in different formats in many different languages;
- Support for full web-crawling capabilities to extract documents from the Internet at various link levels defined by the user;
- Ability to process actual text stored in documents, as well as access text though links to URLs;
- Numerous statistical and data mining analyses that can be applied to the processed textual information, leading to discovery of new insights, hypotheses, and information.

The options mentioned here are illustrated in Figure 9.1.

The Quick tab shown in Figure 9.1 permits you to load text from a file in a folder saved on your computer, such as a Word .doc or .pdf file, or from a variable column (where text has been cut and pasted from the Clipboard). Additionally, the Load Documents button allows you to load in documents from other sources.

The Advanced tab, shown in Figure 9.2, permits you to set the language and specify the number of words to include in the string viewed in the text document.

In addition to the list of available languages shown in the figure, there are several more below Norwegian at the bottom of the list, when you pull down the slider bar on the right. The Filters tab, shown in Figure 9.3, allows you to make various selections on length of words, e.g., by minimum and maximum size of word and other parameters.

When you use Text Miner for the first time, it might be wise to allow the default settings shown in Figure 9.3. After you get some initial results, you can experiment with changing these parameters to reduce the number of words, for example, in a particular size category that you do not deem important to your project.

FIGURE 9.1　The entry screen in the *STATISTICA* Text Miner interface.

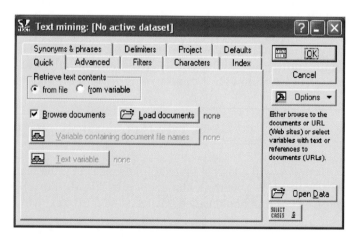

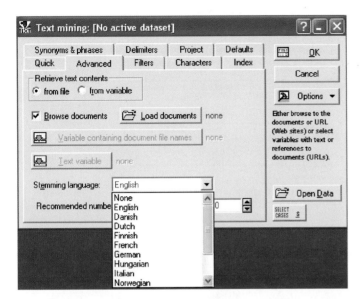

FIGURE 9.2 The Advanced tab of the entry screen.

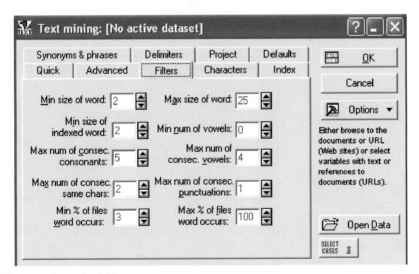

FIGURE 9.3 The Filter tab of the entry screen.

The Characters tab, shown in Figure 9.4, allows you to set the characters that can (1) be in a word, (2) begin a word, or (3) end a word.

You can edit the characters on this tab to meet your needs, if necessary. When you examine these characters, you will notice that the default set shown in Figure 9.4 includes most if not all of the alphanumeric characters used in English-speaking countries.

The Index tab, shown in Figure 9.5, permits you to make an "inclusion file of words" that you are interested in and use only those for text mining analysis.

FIGURE 9.4   The Characters tab of the entry screen.

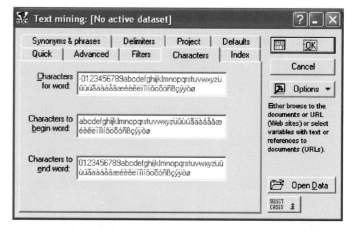

FIGURE 9.5   The Index tab of the entry screen.

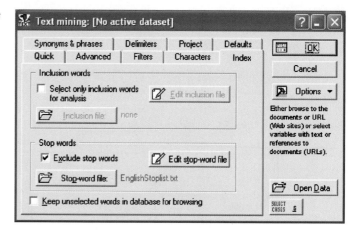

Additionally, you can make a "stop-word file" and set it to exclude these words. Any words that would not have significant meaning for your task can be added to this list. This capability can reduce the number of words that the text mining process will look at, making the results of a lesser number of different words easier to work with in the data mining analysis and other analyses, like graphical analysis, that follow.

The Synonyms and Phrases tab, shown in Figure 9.6, permits you to combine synonyms and also to make phrases of "selected multiwords" that are of special interest in your current project.

The Delimiters tab, shown in Figure 9.7, allows you to have the Text Miner look only at words that are between "starting" and "ending" phrases, if this is important to your project.

The Project tab, shown in Figure 9.8, allows you to do the following:

1. Create a new project.
2. Use an existing project.

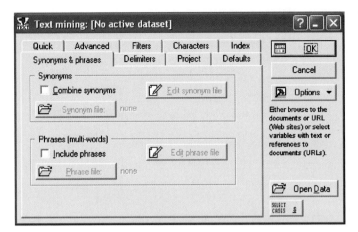

FIGURE 9.6   Synonyms and Phrases tab of the entry screen.

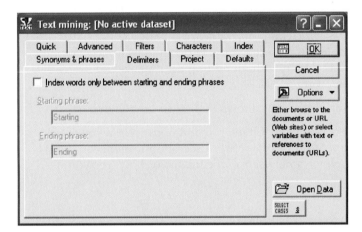

FIGURE 9.7   The Delimiters tab of the entry screen.

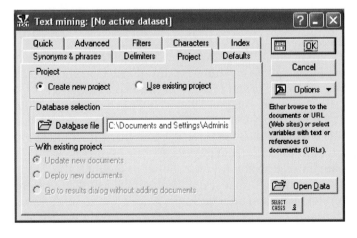

FIGURE 9.8   The Project tab of the entry screen.

FIGURE 9.9   The Defaults tab of the entry screen.

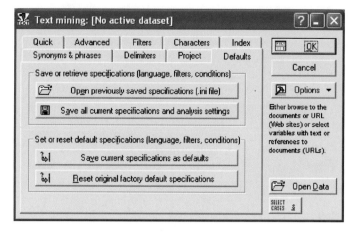

3. Select your database file from a specific computer folder/file pathway.
4. With an existing project, (a) update the project with new documents, (b) deploy new documents, or (c) go to the Results dialog without adding documents and select the kinds of results that you want to examine.

Finally, the Defaults tab, shown in Figure 9.9, allows you to do the following:

1. Save or retrieve specifications like language, filters, or conditions from a previously saved file called an .ini file (where a set of previously used specifications is saved) as a "template" that you can apply to a new set of documents or however you need this information.
2. Set or reset default specifications.

If you don't have the text documents needed for your project already saved as Word.doc files or other files, an alternative method is provided in *STATISTICA* Text Miner to get text information and even "crawl" the Web to whatever level of links necessary within web pages to gather the information needed. This is done through the Web Crawling, Document Retrieval dialog shown in Figure 9.10.

The File Filter pull-down menu in Figure 9.10 shows Word Document highlighted by default. Other choices include Web Pages (e.g., .html or .htm); All Document Files (.txt, .doc, .rtf, .pdf, etc.); Text (.txt); Rich Text Format (.rtf); Mail files; and PDF.

Now, let's look at other information you can get from the Web Crawling dialog. Note that the level of depth in Figure 9.11 (in the upper-left corner) has been set to 3. This means that when you run this dialog, it will "crawl the web" to three levels of links; it will follow a path directed by a link to a link to a link (three linkage levels).

Additional web page addresses are entered in the list and shown in Figure 9.12.

In the dialog shown in Figure 9.12, you can see that seven different web site URLs were typed into the Target (URL or Folder) text box, and after each, the Add to Crawl button was clicked to add the URL to the list.

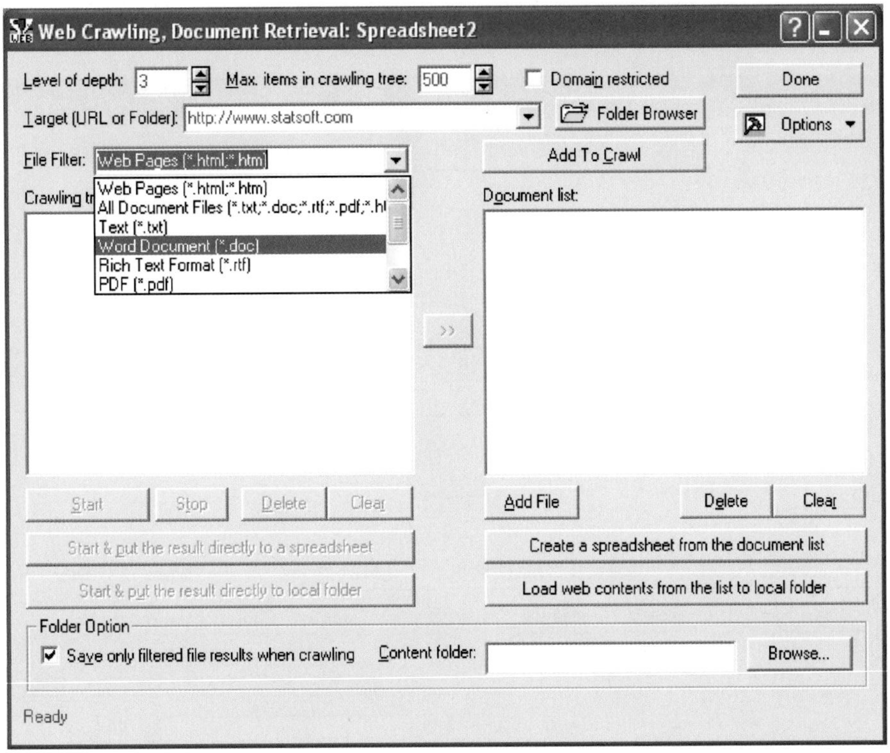

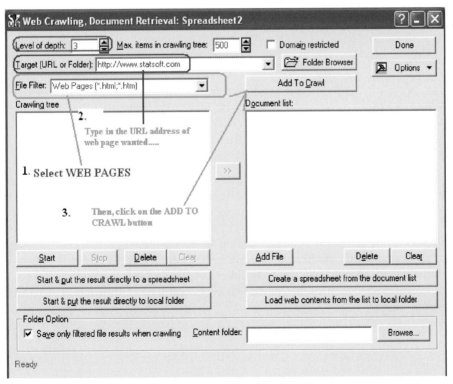

FIGURE 9.10   The Web Crawling, Document Retrieval dialog.

FIGURE 9.11   Additional information available in the Web Crawling, Document Retrieval dialog.

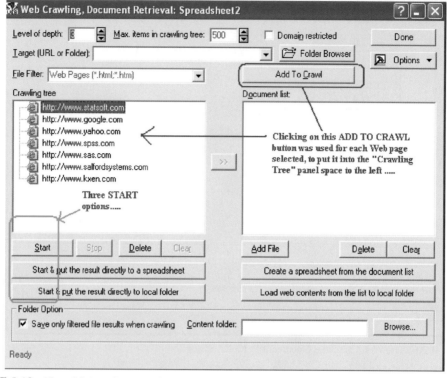

FIGURE 9.12   The addition of more web page addresses to the dialog.

The next option to set is File Filter, shown in Figure 9.13 as set to PDF files only. Now you have just one more thing to do, in case you want to select something other than HTML pages in your search.

The next step is to click on one of the three start buttons, according to your choice of where to put the results. You can then start the crawling process in several ways:

- Click the Start button to transfer the selected files and subdirectories into the left pane, which provides some visual feedback as the files are identified, allowing you to stop this process any time. Next, select the (references to) files or URLs you want to retain, transfer them (by clicking the >> button) to the right pane, and then click the Create a Spreadsheet from the Document List button to create the spreadsheet.
- Click the Start and Put the Result Directly to a Spreadsheet button; this option is recommended to retrieve very large numbers of file or URL references (the scrollable pane user interface becomes inefficient when several tens of thousands of references are retrieved!).

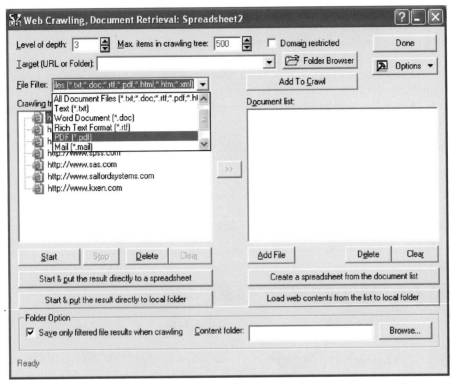

FIGURE 9.13   Setting File Filter in the Web Crawling, Document Retrieval dialog.

• Click the Start and Put the Result Directly to Local Folder button to transfer the actual documents or web pages (.htm or .html documents) as well as the respective subdirectory structure to the location specified in the Folder Option box. This method is recommended to create a permanent repository of the actual documents, for example, for later repeated processing with *STATISTICA* Text Mining and Document Retrieval.

Documents will show in the left panel as the crawl process proceeds; you can select the ones you want.

Selected folders and documents are displayed as a directory tree in the left pane of the Web Crawling, Document Retrieval dialog. You can select and transfer them to the Document list. You can also add the local file to this list by clicking the Add File button. From the links and file references in the Document list, you can make a spreadsheet or load their contents into a local directory structure (as specified in the Content Folder field).

You can transfer items (file references or Web URLs) in the left pane of this dialog by selecting (highlighting) the desired items and then clicking the >> button to transfer them to the right pane (the Document list).

(*Note*: The preceding information was adapted from the *STATISTICA* Version 8 On-Line Help documentation.)

To see a more complete description of all the parameter settings that were used for the NTSB-Aircraft Accident Example, go to the Tutorial titled "NTSB Aircraft Accident" that is on the DVD that accompanies this book. We have left out many details of this tutorial in this chapter, but we need to go on to see the overall concept at this point.

Figure 9.14 shows the overall process of text mining with *STATISTICA* Text Miner.

Goals of Text Mining of NTSB Accident Reports

The goals of text mining include

- Identification of sets of related words in documents (e.g., accident reports of the NTSB);
- Identification of clusters of similar reports;
- Identification of clusters associated with fatal accidents;

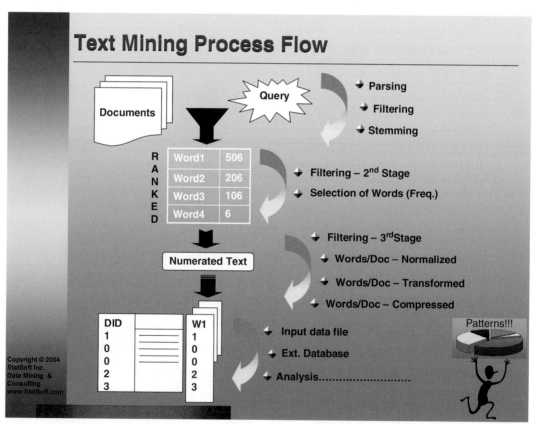

FIGURE 9.14   The text mining process flow.

- Exploratory analysis using structured (fields in a record) and unstructured data (textual information) to discover hidden patterns that could provide some useful insights related to causes of fatal accidents;
- Identification of frequent item sets (clusters of words) related to key words.

Figure 9.15 shows how you can use Principal Component Analysis (PCA) to identify significant groups of words associated with fatal accidents.

In Figure 9.15, you can see that PCA creates major axes of variation (components), in which the first component accounts for the largest amount of variation in the data. Subsequent components account for decreasing amounts of the variation in the data. This approach identifies three groups of words (circled in Figure 9.15) that are distinguished from the large majority of words plotted on the left. You can drill down into the words of these groupings to gain insights about which factors are associated with fatal accidents.

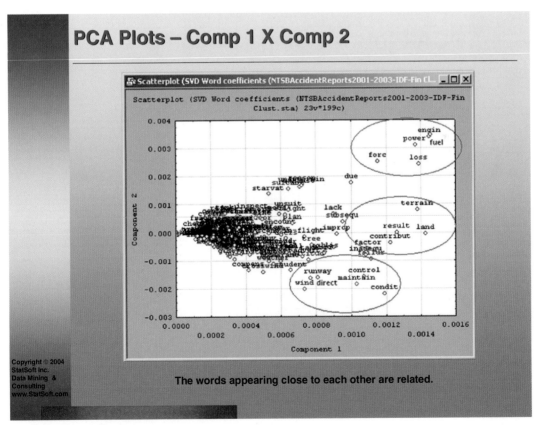

FIGURE 9.15   Use of PCA to show relationships between words.

In the top group, it appears that the words are all related to similar things that may be important, indicating things like "loss of engine power," "loss of forward force," maybe even "loss of fuel." In the middle group, the words *terrain, result, land, contribute, factor, inadequate,* and *failure* (plus others) seem to be related to concepts such as "terrain approaching," "need to land," "factors on inadequate terrain contributing to failure to make a forced landing," and other similar concepts. In the bottom group, words like *runway, control, maintain, wind, conditions,* and others appear to be associated with concepts like "need to maintain control," "runway conditions," and "wind conditions."

Figure 9.16 illustrates how you can gain some insight into relationships between clusters by plotting component #1 against component #3.

From the expanded scatterplot shown in Figure 9.17, you can gain additional insight into aircraft accidents.

A group of words (from the black box recorders and recordings between the pilots of the aircraft and the control tower) include *weather* (conditions), *low, altitude, clearance, night,*

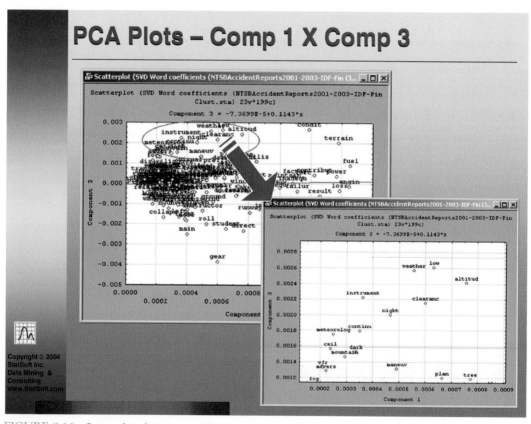

FIGURE 9.16　Scatterplot of component #1 against component #3 for a group of words (circled).

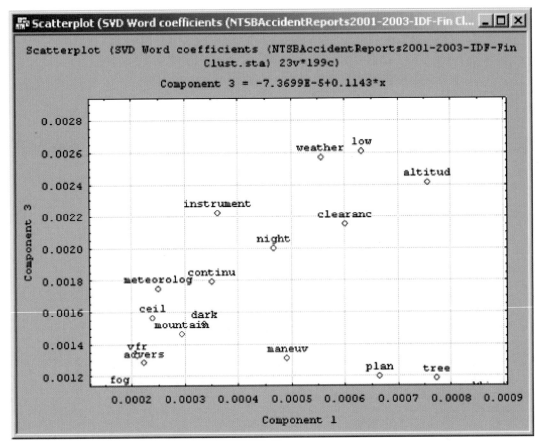

FIGURE 9.17   Expanded scatterplot of a group of words in the PCA plot.

instrument, meteorological (conditions), *dark, ceiling, mountain, adverse* (adverse something), *fog, maneuvering, plan* (perhaps ability to "follow through on original flight plan"), and *tree*.

The situation implied by these words sounds ominous. Concepts related to these words might include fog, bad weather, low altitude, statements of "don't have clearance to land" or "don't have clearance" to get over a mountain, etc., and maybe even "tree(s) looming ahead."

Another way to look at the word data is to compare frequencies of words as new variables to analyze. You can create a cross-tabulation of word clusters with the level of injury in the accidents, as shown in Figure 9.18.

You can use these clustered words to drill down into the messages in which they were imbedded (see Figure 9.19).

You then can conclude from cluster 3 that some fatal accidents were related to pilot failure to maintain airspeed, contributing to a stall. You can do the same thing for cluster #14, as shown in Figure 9.20.

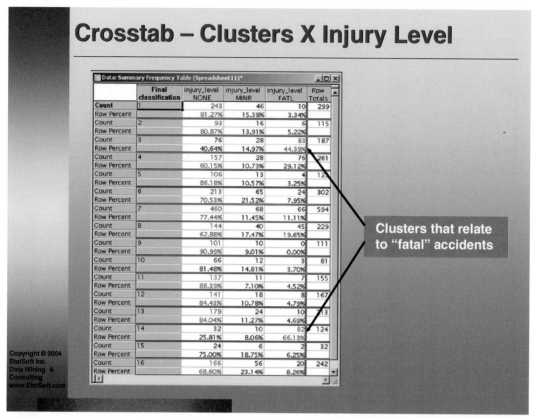

FIGURE 9.18   Cross-tabulation of word clusters with injury level, showing clusters related to fatal accidents.

You can conclude from cluster #14 that some fatal accidents were related to "pilot – visual flight rules (VFR) – instrument – weather conditions – low ceiling."

You can drill down even further by analyzing a given word. Figure 9.21 shows some additional insights gained by analyzing the word *instrument*.

You can expand this analysis by drawing on the text in the NTSB tutorial by guest author Kiron Mathew.

Drilling into Words of Interest

You have seen that the word *instrument* was one of the important words that appeared or was repeated in most of the accident reports related to cluster #14. By reading the prototype descriptions (see earlier), and by reviewing the scatterplots of terms in the semantic space shown earlier, you can clearly see that *instrument* here is used to describe "instrument meteorological conditions," or, in laymen's terms, "bad weather" where pilots had to fly in

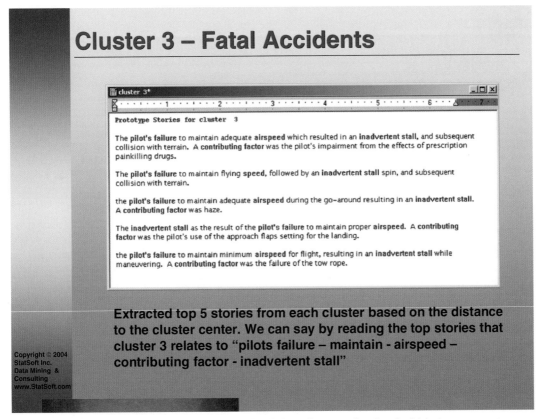

FIGURE 9.19 Phrases in cluster #3 in which embedded words were related to fatal injuries.

clouds, with low cloud ceilings, poor general visibility, and so on. Flying in those conditions is a generally more demanding task, requiring special pilot training and certification (the so-called Instrument Rating).

As a next logical step, you can try to drill into such words of interest to further analyze and extract useful information.

Means with Error Plots

To analyze the words further, you can use the Means with Error Plots option (available on the Graphs pull-down menu, found along the top toolbar in *STATISTICA*) to further substantiate the fact that the occurrence of word *instrument* related to injury level (see Figure 9.22).

In Figure 9.23, you can see another aspect in the relationship between the word *instrument* and the month of occurrence.

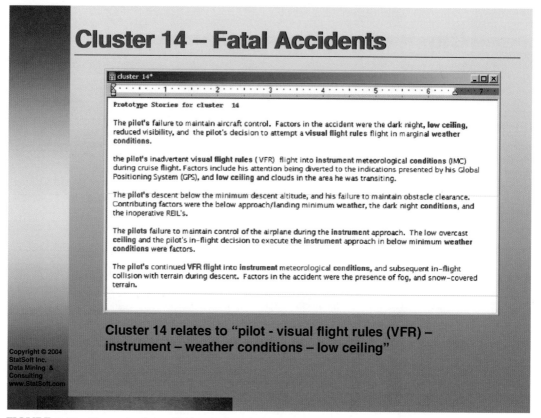

FIGURE 9.20   Phrases in cluster #14 in which embedded words were related to fatal injuries.

Feature Selection Tool

The Feature Selection and Variable Screening tool is extremely useful for reducing the dimensionality of analytic problems; in this case, it was used to select the few specific words (out of the remaining 198 words) that are likely candidates related to the term *instrument*. This tool is located in the Data Miner pull-down menu in the top toolbar or in the Data Mining Workspace. In the workspace, you can connect a data input node with the Feature Selection and Variable Screening (FS) node. You should set the node to output all available reports by right-clicking on the FS node and selecting All Results.

After the node is run, right-click on the Results icon and select Importance Plot. This plot will appear as shown in Figure 9.24.

Figure 9.24 displays the top 15 important words that are related to the word *instrument* ranked according to their importance. From this histogram, you can tell that when the word *instrument* was mentioned, reviewers also used words like *meteorological, condition, in-flight, continue, Visual Flight Rules (VFR), land, collision*, etc. Now you can quickly focus on the few

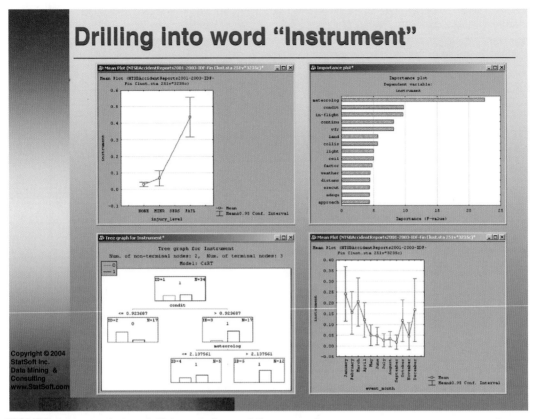

FIGURE 9.21   Analysis to drill down to the word "instrument."

specific words that are likely to co-occur with the word *instrument*, which helps you to further understand the probable causes of instrument-related accidents.

You can gain additional insight into other words related to *instrument* by right-clicking on the Results icon and selecting Tree Graph for Instrument (shown in Figure 9.25).

The decision tree in Figure 9.25 displays some rules followed in deciding where to split each variable's range to make the tree. You see that weather conditions and meteorological conditions were of great importance in aircraft accidents, at least in the data set used for this example.

A Conclusion: Losing Control of the Aircraft in Bad Weather Is Often Fatal

The analyses described up to this point clearly identified one of the main causes of fatal accidents among general aviation (GA) aircraft: loss of control in bad weather. Interestingly, the same conclusion is regularly summarized in the annual NALL reports published by the AOPA Air Safety Foundation (see http://www.aopa.org/asf/publications/).

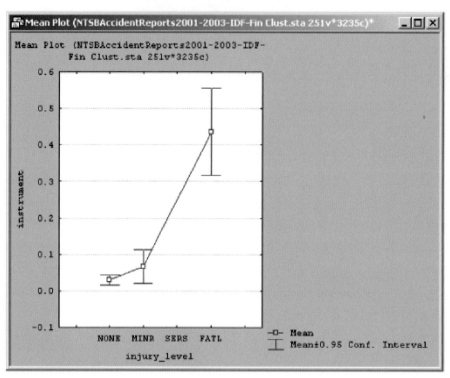

FIGURE 9.22   Mean with error bars for the word *instrument* related to injury level.

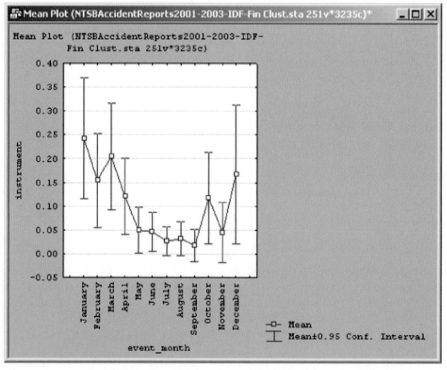

FIGURE 9.23   Relationship between the mean occurrence of the word *instrument* and the month of occurrence.

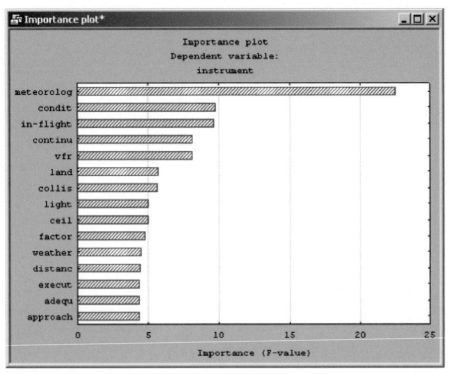

FIGURE 9.24 The relative importance of significant predictors output by the Feature Selection and Variable Screening node.

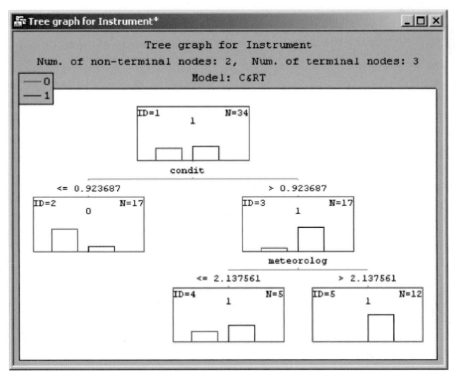

FIGURE 9.25 A simple decision tree for the most important variables.

However, in the case of the analyses presented here, we arrived at this conclusion without the benefit of much domain knowledge, utilizing only the analytic tools available in the software. Hence, this example provides an excellent "proxy" for situations in which a corpus of text with unknown structure needs to be analyzed to derive meaningful relations to the dimensions of interest in the respective domain.

Summary

The preceding description is not a comprehensive overview of text mining, but is a fast overview, showing straightforward ways to make serious progress on an unstructured problem.

TEXT MINING CONCEPTS USED IN CONDUCTING TEXT MINING STUDIES

The following concepts in text mining are used in conducting text mining studies, and some are very useful in analyzing the results obtained from text mining software:

- Preprocessing of text material
- Categorization of text material
- Clustering in text analysis
- Information extraction from text
- Probabilistic models like Hidden Markov
- Hybrid methods for text analysis
- Bootstrapping
- Link analysis—visualization of text material
- Rule structure
- Rule constraints
- Pattern matching
- Concept guards
- Centrality

The preceding concepts are not discussed in this chapter, but you can go to the "Text Mining Terminology" section of the Glossary to get definitions, and you will also find many of these terms used in the tutorials, which are located in three places: (1) in Part III of this book, (2) on the DVD that accompanies this book, and (3) in the new tutorials and "updated discussions," which will be periodically put up on this book's web page at Elsevier.

POSTSCRIPT

In Chapters 1–9, we introduced you to the history and basic theory underlying data mining and exposed you to the spectrum of methods and techniques used to practice it. In Chapter 10, we will introduce you to the interfaces of three of the most popular data mining tool packages in use today: SAS-Enterprise Miner, SPSS Clementine, and *STATISTICA* Data Miner. In Chapters 11–17, we will introduce you to some of the common application types of data mining and some of the principal areas in which they are practiced. In these chapters, we will use one or another of the three data mining packages introduced in Chapter 10 to illustrate the concepts and methods of analysis used commonly in these application areas.

References

Bilisoly, R. (2008). *Practical Text Mining with Perl*. New York: John Wiley & Sons.

Delen, D., & Crossland, M. D. (2008). Seeding the survey and analysis of research literature with text mining. In *Expert Systems with Applications* (Vol. 34, Issue 3, pp. 1707–1720). Elsevier. available online at www.sciencedirect.com.

Hearst, M. (2003). *What Is Text Mining?*. UC Berkeley: SIMS.

Manning, C. D., & Schütze, H. (1999) Foundations of statistical natural language processing. Cambridge, MA: MIT Press.

McKnight, W. (2005). Building business intelligence: Text data mining in business intelligence. *DM Review*, 21–22.

Yang, H. C., & Lee, C. H. (2005). A text mining approach for automatic construction of hypertexts. In *Expert Systems with Applications* (Vol. 29, Issue 4, pp. 723–734). Elsevier. (SCI, EI).

The Three Most Common
Data Mining Software Tools

PREAMBLE

In this chapter, we will introduce you to the interfaces of three of the common data mining tools on the market today: SPSS Clementine, SAS-Enterprise Miner, and *STATISTICA* Data Miner. The first tool we'll describe is arguably the most popular data mining tool package on the market today (and the oldest): SPSS Clementine.

SPSS CLEMENTINE OVERVIEW

SPSS Clementine is the most mature among the major data mining packages on the market today. Since 1993, many thousands of data miners have used Clementine to create very powerful models for business. It was the first data mining package to use the graphical programming user interface. It enables you to quickly develop predictive models and deploy them in business processes to improve decision making.

197

Overall Organization of Clementine Components

After purchasing Clementine from Integral Solutions in 1999, SPSS teamed up with Daimler-Benz and NCR Corporation to produce the Cross Industry Standard Process for Data Mining (CRISP-DM). One of the authors of this book contributed to that project. The Clementine package was integrated with CRISP-DM to help guide the modeling process flow. Today, Clementine can be purchased in standalone client mode or with a number of modules, including a client/server version. The modules of version 11.1 include the following components:

The Basic Module

- Classification and Regression Tree node (CART or C&RT)
- Quest node for binary classification
- CHAID node for creating decision trees based on the Chi-square statistic
- K-Means node for clustering
- Generalized Rule Induction node for creating rule sets
- Factor/PCA node for data reduction
- Linear Regression node for ordinary linear regression

Classification Module

- Binary Classifier node
- Neural Net node for classification
- C5.0 node for classification
- Decision List node for creating rule sets of a group of segments
- Time-Series node to create Autoregressive Integrated Moving Average (ARIMA) models
- Feature Selection node for ranking input variables by importance in predicting the target variable
- Logistic Regression node for creating logit models
- Discriminant Analysis node for creating parametric versions of logistic regression models
- Generalized Linear Model (GLM) for GLM modeling
- Self-Learning Response Model node for retraining a model on as few as one new case

Segmentation Module

- Kohonen net for unsupervised clustering using a form of a neural net
- TwoStep node for clustering in two steps in tandem to reduce the number of cases to subclusters and then clustering the subclusters

Association Module

- Apriori node optimized for categorical rule induction
- CARMA node for unsupervised association of cases
- Sequence node to discover association rules in sequential or time-series data

The following are optional add-on modules for Clementine:

- Solution Publisher option for export of models for scoring outside the Clementine environment
- Text Mining option for mining unstructured data
- Web Mining option
- Database Modeling and Optimization option for integration with other data mining packages

For complete information, see the Clementine home page (http://www.spss.com/clementine/).

Organization of the Clementine Interface

The Clementine interface employs an intuitive visual programming interface to permit you to draw logical data flows the way you think of them. This approach to visual programming was pioneered by the I-Think analytical tool in the 1980s, which allowed scientists to build a data flow as if on a blackboard. The Clementine data flow is called a *stream*.

Clementine Interface Overview

Figure 10.1 shows the Clementine stream used in the CRM tutorial in this book.

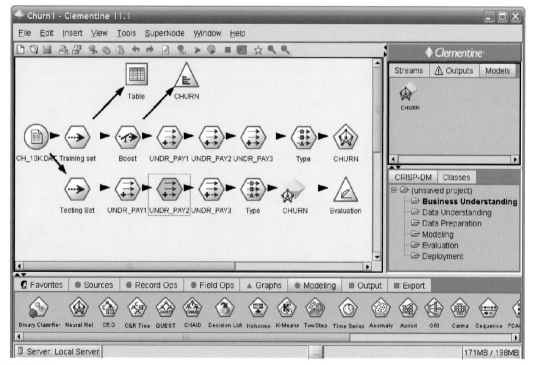

FIGURE 10.1   The Clementine user interface.

Stream Canvas

The stream canvas is the largest pane in the interface. You can work on multiple streams in the same canvas (as shown in Figure 10.1), or multiple stream files, listed in the Streams tab in the upper-right Managers window.

Palettes

The palettes are groups of processing nodes listed across the bottom of the screen. The palettes include Favorites, Sources, Record Ops (operations), Field Ops, Graphs, Modeling, Output, and Export. When you click on one of the palette names, the list of nodes in that palette is displayed below.

Managers

The upper right window holds three managers: Streams, Outputs, and Models. You can switch the display of the manager window to show any of these items. The Streams tab will display the active streams, which you can choose to work on in a given session. The Output tab will show all graphs and tables output during a session. The Models tab will show all the trained models you have created in the session.

Toolbars

At the top of the Clementine window, you can see a list of icons, which are shortcuts to common operations.

Mouse Operations

The Clementine user interface is designed for use with a three-button mouse. If you don't have one, you can simulate the third button by holding down the Alt key when clicking the mouse and dragging items around the canvas. You use the mouse as follows:

- Single-click to show context-sensitive menus for a node.
- Double-click to add nodes to the canvas or edit existing nodes.
- Middle-click to connect nodes with arrows to show the pathway of data flow.

There are two other mouse operations you can perform, which will save you some time in modifying the stream architecture. The first operation is the delete function. If you want to delete an arrow connector between two nodes, place the mouse pointer in the line and right-click. A small box labeled Delete Connection will pop up. Left-click on the pop-up box, and the arrow will disappear.

The second additional mouse operation adds a node between two other nodes. To add nodes this way, make sure that the space between the nodes is large enough to fit another node. (*Note*: You can just drag one node or the other to increase the space between nodes, and the arrow will stretch or shrink.) To see how this works, paste a type node onto the modeling canvas of Figure 10.1 just above the middle of the arrow from the top SuperNode. Then place your mouse pointer on the middle of the arrow line, and press and hold the middle mouse button to drag the arrow line to connect the new node. After it is connected,

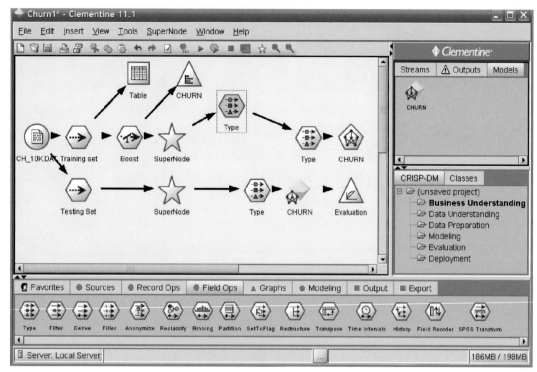

FIGURE 10.2   Modeling canvas after adding a type node after the top SuperNode icon.

you can left-click on the new node and drag it and the connector arrow back into a straight configuration. Figure 10.2 shows the modeling canvas after the connection of the new node but before straightening out the arrow line.

Setting the Default Directory

The Clementine system looks for files in the default directory (in the installation folder). You can save a lot of time if you change the default directory to the one you are working in. To do this, first click on File and then Set Directory, browse to the appropriate directory, and then click Set. From then on, Clementine will assume that files you load or save to are in the directory you chose.

SuperNodes

Clementine includes the option to create SuperNodes, which are groups of nodes indicated by a SuperNode icon (a large blue star). We could have combined the three derive nodes to create a SuperNode in the training stream and then used that SuperNode in the testing stream, as shown in Figure 10.3.

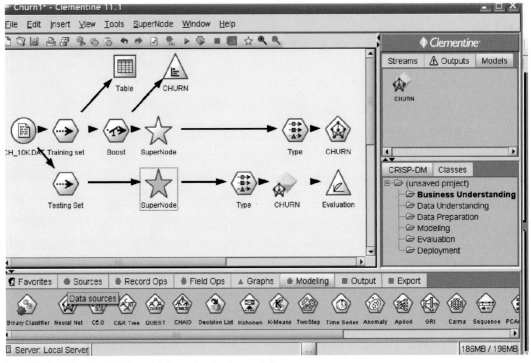

FIGURE 10.3   Use the SuperNode option in the modeling canvas.

Compare Figure 10.1 with Figure 10.3. You can see that the three derived nodes in Figure 10.1 are replaced by the SuperNode icon in Figure 10.2. In the form of the stream shown in Figure 10.3, the three derive nodes function exactly as they did in the stream shown in Figure 10.1.

Execution of Streams

In Clementine, you have a number of options for executing streams:

- **Execute the entire stream:** Right-click on a source node and select the option Execute from Here.
- **Execute any part of the stream that begins with a source node and ends with an output node:** Right-click on the output node and select Execute.
- **Execute from any point in the stream:** Right-click and select Execute. This operation executes all operations from that point to the output nodes.
- **Execute a stream from the toolbar:** Click a node in a stream and then click on the small triangle on the toolbar at the top of the modeling canvas.
- **Execute a stream from the Tools menu:** Click on a node in a stream and choose Execute from the Tools menu.

For additional help on using the interface or any of the nodes, click on the Help menu.

SAS-ENTERPRISE MINER (SAS-EM) OVERVIEW

Note that much of the following text is abstracted from the SAS-EM version 5.3, "Introduction to SAS Enterprise Miner Software," Copyright by SAS, Cary, North Carolina, USA.

Overall Organization of SAS-EM Version 5.3 Components

As in SPSS Clementine, the SAS-EM data mining process consists of a process flow diagram, which is a form of a graphical user interface, where you can add nodes, modify nodes, connect nodes with arrows for the direction of flow of the computations, modify nodes, and save the entire workspace as a data mining project. Like in SPSS Clementine, this workspace is designed for use by business analysts (in business, industry, governmental agencies, etc.) with little statistical expertise but who can navigate through the data mining methodology fairly easily. However, you probably can't give it to your assistant or any person randomly picked off the street and expect that person to make an "intelligent" model or even to get the model to run. At the same time, the quantitative statistical or engineering expert can go "behind the nodes" to customize the analytical processes.

The analytical tools include

- Clustering
- Association and sequence discovery
- Marketbasket analysis
- Path analysis
- Self-organizing maps/Kohonen
- Variable selection (analogous to *Feature Selection* as termed in *STATISTICA* Data Miner)
- Decision trees and gradient boosting
- Linear and logistic regression
- Two-stage modeling
- Partial least squares
- Support Vector Machines
- Neural networking

   Data preparation tools include

- Outlier detection
- Variable transformations
- Variable clustering
- Interactive binning
- Principal components
- Rule building and induction
- Data imputation
- Random sampling
- Partitioning of data sets (into train, test, and validate data sets)

Advanced visualization tools can be used to create multidimensional histograms and graphically compare different algorithm models (via gains or lift charts).

Layout of the SAS-Enterprise Miner Window

The Enterprise Miner window, as shown in Figure 10.4, contains the following interface components:

- **Toolbar and toolbar shortcut buttons:** The Enterprise Miner Toolbar is a graphic set of node icons that are organized by SEMMA categories. Above the toolbar is a collection of toolbar shortcut buttons that are commonly used to build process flow diagrams in the diagram workspace. Move the mouse pointer over any node or shortcut button to see the text name. Drag a node into the diagram workspace to use it. The toolbar icon remains in place and the node in the diagram workspace is ready to be connected and configured for use in your process flow diagram. Click on a shortcut button to use it.

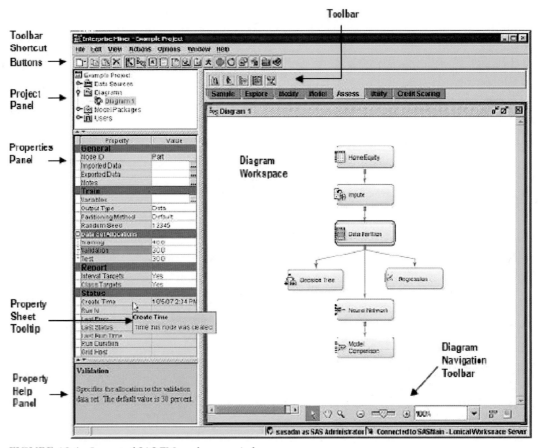

FIGURE 10.4   Layout of SAS-EM workspace window.

- **Project panel:** Use the Project panel to manage and view data sources, diagrams, model packages, and project users.
- **Properties panel:** Use the Properties panel to view and edit the settings of data sources, diagrams, nodes, and model packages.
- **Diagram workspace:** Use the diagram workspace to build, edit, run, and save process flow diagrams. This is the place where you graphically build, order, sequence, and connect the nodes that you use to mine your data and generate reports.
- **Property Help panel:** The Property Help panel displays a short description of the property that you select in the Properties panel. Extended help can be found in the Help Topics selection from the Help main menu or from the Help button on many windows.
- **Status bar:** The Status bar is a single pane at the bottom of the window that indicates the execution status of an SAS-Enterprise Miner task.

Various SAS-EM Menus, Dialogs, and Windows Useful During the Data Mining Process

Clicking Preferences opens the Preferences window, as shown in Figure 10.5. In this window, you can use the following options to change the user interface (this process is similar to setting parameters by right-clicking on Nodes in the *STATISTICA* Data Miner Workspace):

- **Look and Feel:** Allows you to select Cross Platform, which uses a standard appearance scheme that is the same on all platforms, or System, which uses the appearance scheme that you have chosen for your platform.

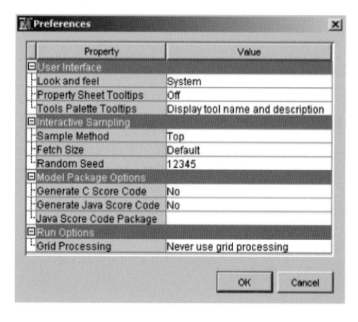

FIGURE 10.5 The Preferences window, where you can set parameters for SAS-EM.

- **Property Sheet Tooltips:** Controls whether tooltips are displayed on various property sheets appearing throughout the user interface.
- **Tools Palette Tooltips:** Controls how much tooltip information you want displayed for the tool icons in the toolbar.
- **Sample Method:** Generates a sample that will be used for graphical displays. You can specify either Top or Random.
- **Fetch Size:** Specifies the number of observations to download for graphical displays. You can choose either Default or Max.
- **Random Seed:** Specifies the value you want to use to randomly sample observations from your input data.
- **Generate C Score Code:** Creates C score code when you create a report. The default is No.
- **Generate Java Score Code:** Creates Java score code when you create a report. The default is No. If you select Yes, you must enter a filename for the score code package in the Java Score Code Package box.
- **Java Score Code Package:** Identifies the filename of the Java Score Code Package.
- **Grid Processing:** Enables you to use grid processing when you are running data mining flows on grid-enabled servers.

Figure 10.6 shows another parameters dialog in which you can specify a data source. Figure 10.7 provides a closer look at the Properties dialog.

As you can see in the dialog in Figure 10.7, you can set parameters, such as the number of bins, shown as 2 in the figure, but you also can change this to any number desired. Some of the other parameters are categorical, so you either select Yes or No.

After setting the parameters, you then need to build the "Data Miner workflow/Workspace," selecting the "nodes" needed from the following lists. Figure 10.8 shows a sample flow diagram in SAS-EM.

After a model is created, then it is run, following which, results can be opened. Various kinds of output are illustrated next. A basic stats example is shown in Figure 10.9.

A decision tree analysis example is shown in Figure 10.10.

A lift chart is shown in Figure 10.11.

Figure 10.12 shows how decision tree nodes are expressed.

Figure 10.13 shows one type of profit chart.

A comparison of several DM algorithms is shown in Figure 10.14.

A model comparison of Train and Validate data sets, using several algorithms, is shown in Figure 10.15.

Software Requirements to Run SAS-EM 5.3 Software

To re-create this example, you must have access to SAS Enterprise Miner 5.3 software, either as a client/server application, or as a complete client on your local machine.

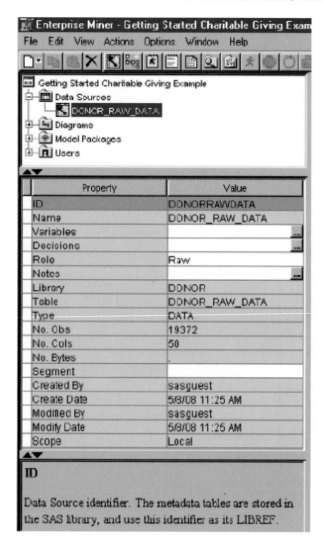

FIGURE 10.6   Another parameters dialog in SAS-EM where properties and their values can be user-specified.

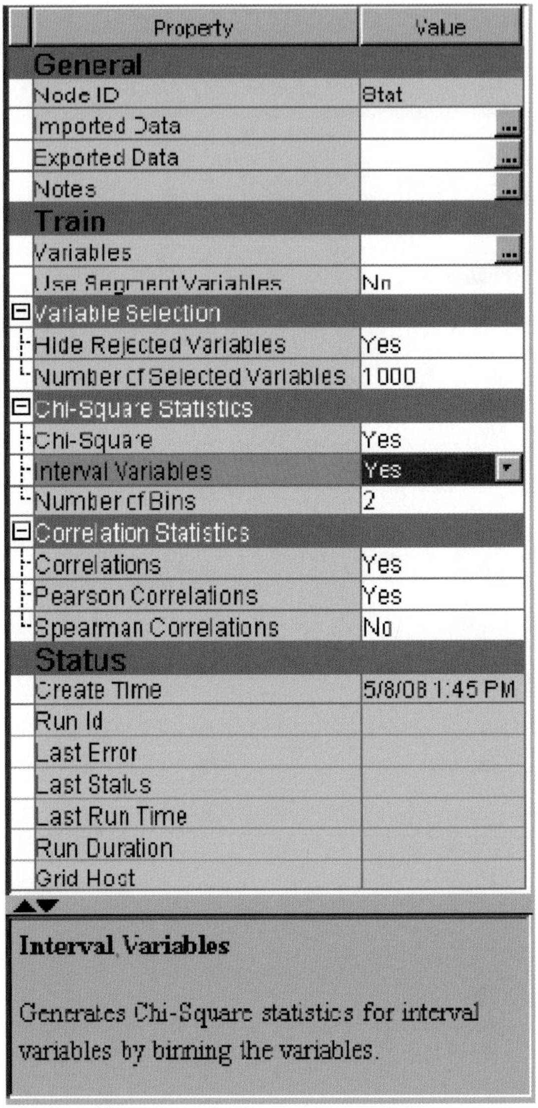

FIGURE 10.7 A closer look at the Properties dialog of SAS-EM where the parameters of various statistical functions can be user-specified.

Property	Value
General	
Node ID	Stat
Imported Data	...
Exported Data	...
Notes	...
Train	
Variables	...
Use Segment Variables	No
⊟ Variable Selection	
Hide Rejected Variables	Yes
Number of Selected Variables	1000
⊟ Chi-Square Statistics	
Chi-Square	Yes
Interval Variables	Yes
Number of Bins	2
⊟ Correlation Statistics	
Correlations	Yes
Pearson Correlations	Yes
Spearman Correlations	No
Status	
Create Time	5/8/08 1:45 PM
Run Id	
Last Error	
Last Status	
Last Run Time	
Run Duration	
Grid Host	

Interval Variables

Generates Chi-Square statistics for interval variables by binning the variables.

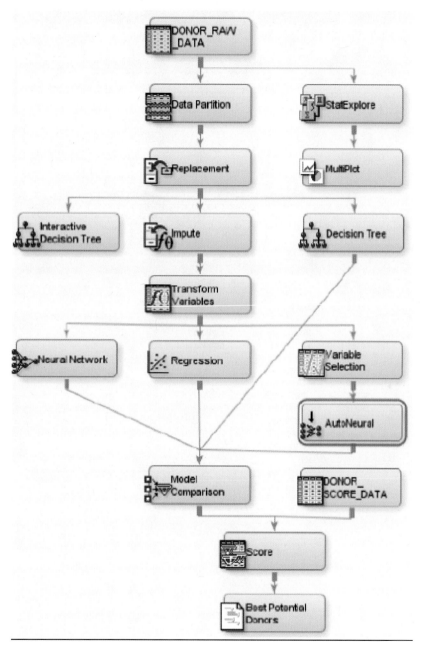

FIGURE 10.8   Flow of an SAS-EM workspace.

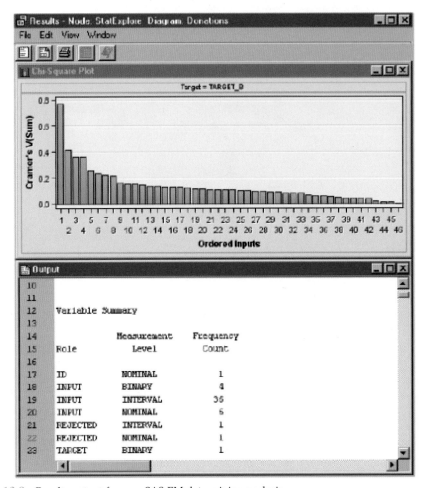

FIGURE 10.9   Results output from an SAS-EM data mining analysis.

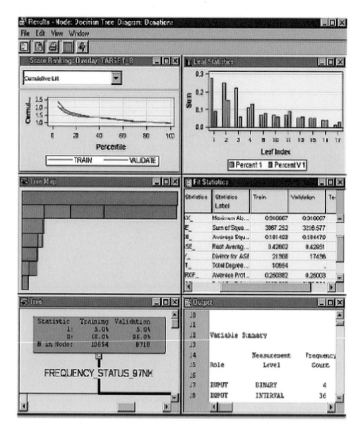

FIGURE 10.10 Another results example from an SAS-EM data mining analysis.

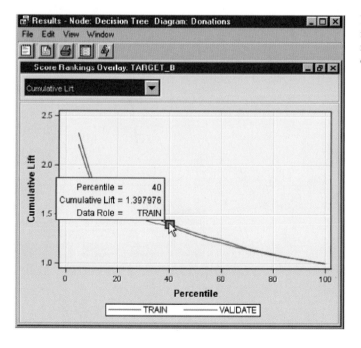

FIGURE 10.11 Graph example of results output from SAS-EM data mining analysis; this is a cumulative lift chart.

FIGURE 10.12 Decision tree output from SAS-EM data mining analysis.

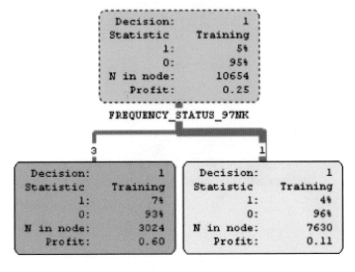

FIGURE 10.13 Profit type chart from the SAS-EM data mining results analysis.

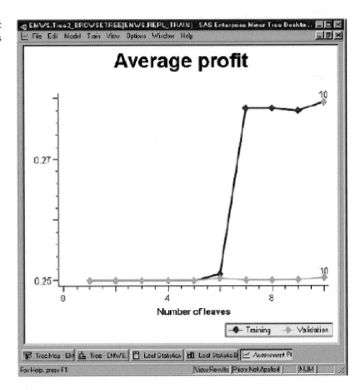

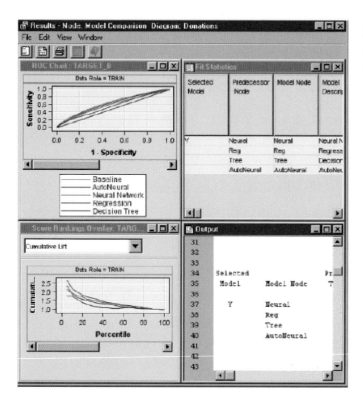

FIGURE 10.14 SAS-EM data mining analysis results windows, showing several types of graphs and tabulated data.

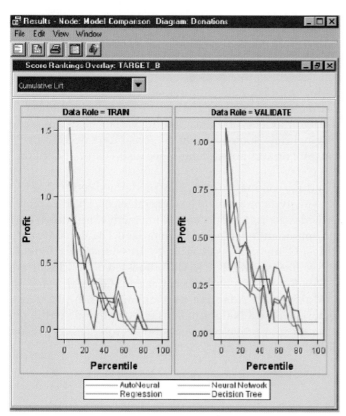

FIGURE 10.15 Graphical results from SAS-EM data mining analysis.

STATISTICA DATA MINER, QC-MINER, AND TEXT MINER OVERVIEW

Much of the following text is taken directly from *STATISTICA* Data Miner On-Line Help; StatSoft, Inc. (2008). *STATISTICA* (data analysis software system), version 8.0. www.statsoft.com.

Overall Organization and Use of *STATISTICA* Data Miner

To use the *STATISTICA* data mining tools, follow the simple steps outlined in the following sections.

1. Select the desired option from the Data Miner menu.

The *STATISTICA* Data Miner menu contains commands to create a Data Miner Workspace where you can build and maintain complex models, commands to select predefined templates of Data Miner Workspaces for simple and complex tasks, and commands to select *STATISTICA* analysis modules for particular specialized analyses (see Figure 10.16).

You can choose from the following:

- **Data Miner – My Procedures/Data Miner – All Procedures.** Select either of these commands to create a new data mining workspace.
- **Data Miner – Data Cleaning and Filtering.** Select this command to choose from a large number of nodes for "cleaning" the data, i.e., for filtering out invalid data values, missing data replacement, user-defined transformations, ranking, standardization, etc. The very powerful Feature Selection and Variable Screening node enables you to quickly process very large lists of continuous and categorical predictors for regression and classification problems, and to select a subset that is most strongly related to the dependent (outcome) variables of interest. The algorithm for selecting those variables is not biased in favor of a single method for subsequent analyses (e.g., pick the highest correlations for later analyses via linear models), and the resulting variable lists are made available as mappings into the original data source so that no actual data need to be copied (e.g., from a remote database). For additional details, see also Feature Selection and Variable Screening.
- **Data Miner – General.** Select any of these commands to display predefined sets of data mining templates for typical types of analysis problems. The General Slicer/Dicer Explorer with Drill-Down command also provides access to a specialized interactive drill-down tool.
- **Neural Networks, Independent Components Analysis, Generalized EM & *k*-Means Cluster Analysis, Association Rules, General Classification/Regression Tree Models, General CHAID Models, Interactive Trees (C&RT, CHAID), Boosted Tree Classifiers and Regression, Random Forests for Regression and Classification, Generalized Additive Models, MARSplines (Multivariate Adaptive Regression Splines), Machine Learning (Bayesian, Support Vectors, Nearest Neighbors).** These commands will display the modules for performing the respective types of analyses interactively, using the standard *STATISTICA* user interface.

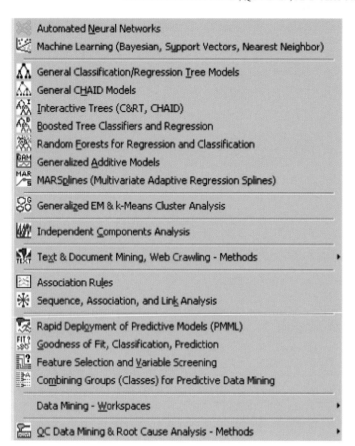

FIGURE 10.16 *STATISTICA* Data Miner menu, from the Data Miner pull-down menu.

- **Rapid Deployment of Predictive Models (PMML); Goodness of Fit, Classification, Prediction; Feature Selection and Variable Filtering; Combining Groups (Classes) for Predictive Data-Mining.** These commands will display the respective specialized modules; Rapid Deployment of Predictive Models will quickly generate predictions from one or more previously trained models based on information stored in industry-standard Predictive Model Markup Language (PMML) deployment code. Goodness of Fit will compute various goodness-of-fit statistics and graphs for regression and classification problems. Feature Selection and Variable Filtering is used to select variables (columns) from very large data sets or external databases, e.g., to select subsets of predictors from hundreds of thousands of predictors, or even more than one million predictors. Combining Groups (Classes) for Predictive Data-Mining is used to automatically find and implement a best recoding scheme for the prediction of a continuous or categorical variable from one or more categorical predictors with many classes (e.g., such as SIC codes with more than 10,000 distinct values).

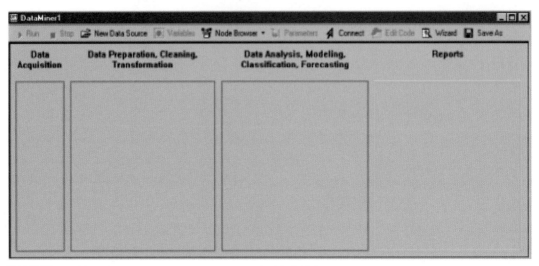

FIGURE 10.17   A blank Data Miner Workspace in *STATISTICA*.

Example: Select Data Miner - All Procedures from the Statistics - Data Mining submenu to display a standard Data Miner Workspace, as shown in Figure 10.17.

2. Select a new data source.

Next, specify the input data for the data mining project. Click the New Data Source button on the Data Miner Workspace to display a standard data file selection dialog where you can select either a *STATISTICA* data file (*STATISTICA* spreadsheet designated for input) or a database connection for in-place processing of data in remote databases by In-Place Database methods.

Example: Select the Boston2.sta data file from the sample data files of *STATISTICA*, as shown in Figure 10.18.

FIGURE 10.18   The Select Spreadsheet dialog allows you to select a data set.

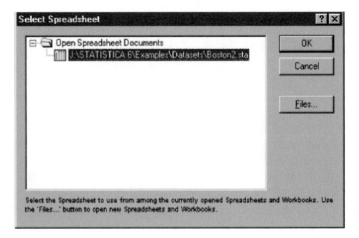

This sample data file contains the data from the Boston housing study (Harrison & Rubin-feld, 1978; reported in Lim et al., 1997). Click OK to select the variables for the analyses.

3. Select the variables for the analyses.

Next, select the variables for the analyses. *STATISTICA* Data Miner distinguishes between categorical and continuous variables, and dependent and predictor (independent variables). Categorical variables are those that contain information about some discrete quantity or characteristic describing the observations in the data file (e.g., Gender: Male or Female); continuous variables are measured on some continuous scale (e.g., Height, Weight, Cost). Dependent variables are the ones you want to predict; they are also some-times called *outcome variables*; predictor (independent) variables are those that you want to use for the prediction or classification (of categorical outcomes).

You don't have to select variables into each list; in fact, some types of analyses expect only a single list of variables (e.g., cluster analysis). You can also make additional selections, such as specify certain codes for categorical variables, case selection conditions, or case weights; or you can specify censoring, a learning/testing variable, etc.

Example: Select the variable Price as the categorical dependent variable that is to be pre-dicted, select variable Cat1 as a categorical predictor, and select variables ORD1 through ORD12 as continuous predictors, as shown in Figure 10.19.

Now click OK to add this data source to the data mining workspace.

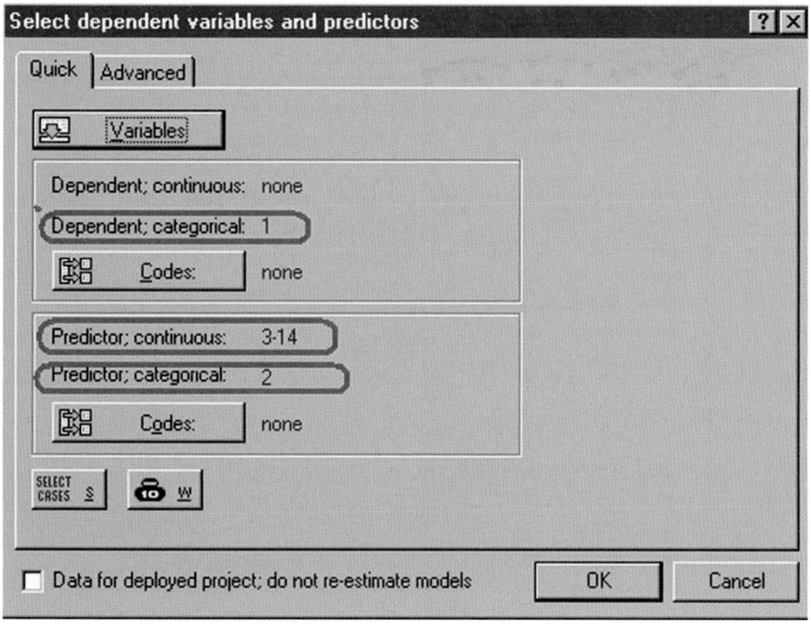

FIGURE 10.19   Variable selection dialog in *STATISTICA* Data Miner.

4. Display the Node Browser and select the desired analyses or data management operation.

Next, click the Node Browser button on the Data Miner Workspace or display the Node Browser by selecting that command from the Nodes menu; you can also press Ctrl+B on your keyboard to display the Node Browser (see Figure 10.20).

The Node Browser contains all the procedures available for data mining in the Data Miner Workspace; you can choose from more than 260 procedures for data filtering and cleaning and for data analysis. By default, all procedures are organized in folders along with the types of analyses that they perform. However, the Node Browser is fully configurable. You can specify multiple Node Browser configurations, and these customizations will automatically be saved along with the Data Miner Workspace. Thus, you can greatly simplify routine analyses by fully customizing the default Node Browser configuration for your work.

To select analyses (analysis nodes), highlight them in the right pane and click the Insert into Workspace button on the Node Browser toolbar; you can also simply double-click on the analysis node of interest to insert it into the workspace. The lower pane of the Node Browser contains a description of the currently highlighted selection.

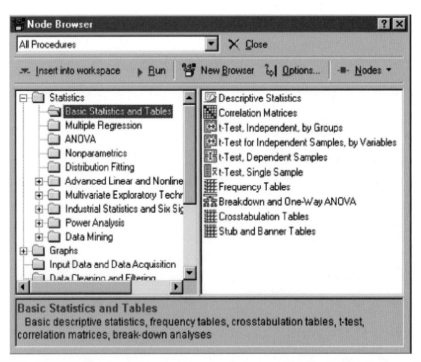

FIGURE 10.20   The Node Browser dialog allows you to select any or all of the DM Workspace nodes available for use in the Data Mining Workspace of *STATISTICA*.

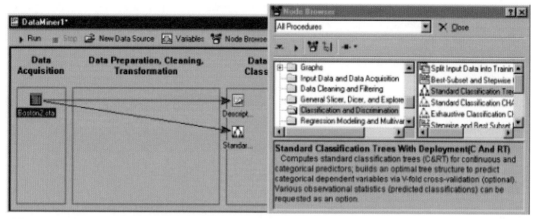

FIGURE 10.21   Node Browser and Data Miner Workspace in *STATISTICA* showing how nodes are either double-clicked on in the Node Browser or "dragged" into the workspace.

Example: Select the nodes for Descriptive Statistics. Then scroll down in the left pane of the Node Browser and select the folder labeled Classification and Discrimination, and in the right pane, select Standard Classification Trees with Deployment (see Figure 10.21).

If a data source in the workspace is currently highlighted, it will be connected automatically to the nodes as they are selected (inserted) into the workspace. You can also use the Connect toolbar button to connect data sources to nodes. To delete an arrow, click on it and select Delete from the shortcut menu (displayed by right-clicking your mouse), or press the Del key on your keyboard. You can temporarily disable an arrow by selecting Disable from the shortcut menu. Arrows that are disabled will not be updated or recomputed.

5. Run (update) the Data Miner project.

Next, run the Data Miner project. All nodes connected to data sources via (nondisabled) arrows will be updated, and the respective analyses will be produced.

Let us share a note on data cleaning, filtering, and Exploratory Data Analysis (EDA). The *STATISTICA* Data Miner project workspace is fully integrated into the *STATISTICA* data analysis environment. At any point, you can click on a data source or results workbook (spreadsheet, report), either in the Data Acquisition area or in any other area (i.e., data sources created by analyses), to review the respective document. Also, you can use any of the interactive analyses available in *STATISTICA* to further explore those documents—for example, to run simple descriptive statistics or create descriptive graphs to explore the respective results. These types of EDA techniques are indispensable for data cleaning and verification. For example, it is useful to always run simple descriptive statistics, computing the minima and maxima for variables in the analyses, to ensure that data errors (impossible values) are corrected before they lead to erroneous conclusions. Also, the various options on the Data menu of the data spreadsheet toolbar are very useful for cleaning and verifying the data in interactive analyses before submitting them to further analyses.

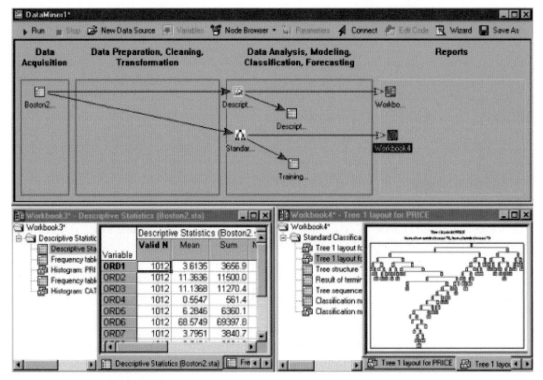

FIGURE 10.22   Example of a simple Data Miner Workspace "project" and the results that are obtained from each of the Results icons, located on the right; when these icons are double-clicked, the workbooks show at the bottom pop-up on your computer screen.

Example: Click the Run button, select Run All Nodes from the Run menu, or press F5.

Detailed results are created by default for each type of analysis in *STATISTICA* workbooks; double-click on a workbook to review its contents (see Figure 10.22). You can also connect all the green arrows to the workbooks into a single workbook to direct all results to a single container. The complete functionality of *STATISTICA* workbooks is available for these results, so you can save these results; drag, drop, and edit individual results and graphs; mark certain spreadsheets as input data for subsequent analyses; and so on.

6. Customize analyses, edit results, and save results.

The next step is to review the results, edit the analyses, and save the results, as follows:

- In general, click on any icon and then use the shortcut menu to review the various options available for the object (analysis, data source, output document, result, etc.).
- To review results, double-click on the workbooks or other documents created by the analyses. Use the options on the Data Miner tab of the Options dialog (accessible from the Tools menu) to configure *STATISTICA* Data Miner, for example, to direct output to reports instead of workbooks.

- To edit analyses (change the parameters for the analyses), double-click on the respective analysis icons; this will display the Edit Parameters dialog, which contains parameters and settings specific to the respective node.
- To edit documents created by analyses for downstream analyses, click on the item and select View Document from the shortcut menu.
- You can delete nodes by highlighting them and pressing the Del key, or selecting Delete from the shortcut menu, or using the standard Undo methods (pressing Ctrl+Z or clicking the Undo button on the toolbar) to undo your changes to the Data Miner Workspace.
- To save the workspace, select Save from the File menu; the default filename extension for the Data Miner Workspace is .sdm. By default, the program will save all input data sources embedded into the data mining project; this default can be changed by clearing the Embed Input Files in Data Miner Project Files When Saving check box on the Data Miner tab of the Options dialog (accessible from the Tools menu).

Example: To compute various graphical summaries, double-click on the Descriptive Statistics node and set the Detail of Reported Results parameter to All results; then click OK. Next, double-click on the Standard Classification Trees with Deployment node, select the V-Fold Cross-Validation tab (see Figure 10.23), and request V-fold cross-validation; this is a very important safeguard against overlearning.

Next, click on the General tab and set the Minimum n Per Node (of the Final Tree) to 50; this will cause the tree-growing procedure to terminate when the node size falls below that number and hence to create less complex trees. Then click OK.

You will see that the two analysis nodes as well as the workbook nodes are now displayed with a red frame around them (see Figure 10.24); this denotes that these nodes are not up to date, or they are dirty. Let us finally move the arrow from Standard Classification Trees with Deployment to point to the same (first) workbook where the descriptive statistics are displayed. To do this, click on the head of the arrow and drag it over to the first

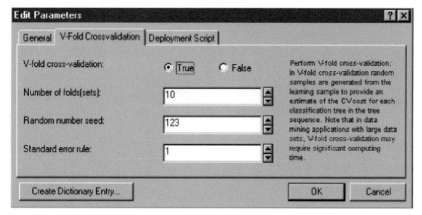

FIGURE 10.23   The Edit Parameters dialog, in this case allowing the selection of V-fold cross-validation parameters.

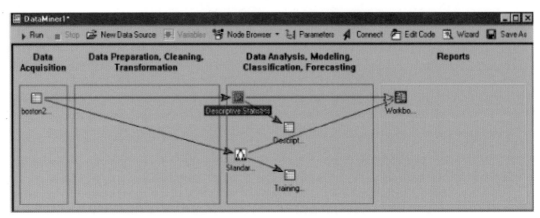

FIGURE 10.24   A simple Data Miner Workspace project, with two analysis nodes, with results being sent to one "results workbook" node.

workbook (release the mouse button as your cursor is hovering over the workbook node); also delete the now "disconnected" node (unless you would like to keep it for reference).

Next click the Update button or press F5 to see the result shown in Figure 10.25.

The "results workbook" is illustrated, which is embedded in the Workbook 1 result node that you see in the upper right within the Data Miner Workspace.

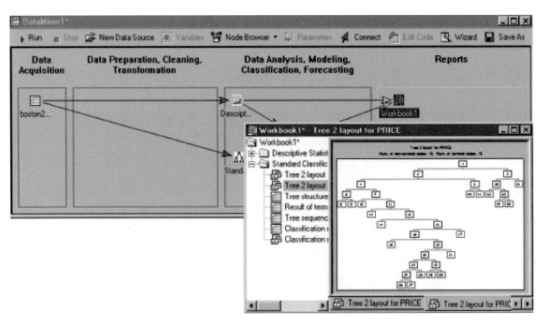

FIGURE 10.25   Example of Data Miner Workspace (top illustration) and Tree 2 layout Standard Classification Tree algorithm model.

Note that *V*-fold cross-validation is a time-consuming procedure that will validate each tree in the tree sequence several times; however, a significant benefit is that the program will now pick for you the best tree, i.e., the one with the best cross-validation cost and node complexity trade-off.

After all nodes are updated, double-click on the (now single) results node to review all results in the workbook. Note that both the Descriptive Statistics as well as the results of the Standard Classification Trees analysis are displayed in the same workbook.

7. Deploy solution (models) for new data.

STATISTICA Data Miner includes a complete deployment engine for Data Miner solutions that comprises various tools. For example,

- You can create Visual Basic or C/C++/C# program code in most interactive analysis modules that will compute predictions, predicted classifications, and clusters assignments (such as General Regression Models, Generalized Linear Models, General Discriminant Function Analysis, General Classification and Regression Trees (GC&RT), Generalized EM & *k*-Means Cluster Analysis, etc.).
- You can create XML-syntax-based PMML files with deployment information in most interactive modules that will compute predictions, predicted classifications, or cluster assignments (i.e., the same modules mentioned in the preceding paragraph). One or more PMML files with deployment information based on trained models can be loaded by the Rapid Deployment of Predictive Models modules to compute predictions or predicted classifications (and related summary statistics) in a single pass through the data; hence, this method is extremely fast and efficient for scoring (predicting or classifying) large numbers of new observations.
- General Classification and Regression Trees and General CHAID modules can be used to create SQL query code to retrieve observations classified to particular nodes or to assign observations to a node (i.e., to write the node assignments back into the database).
- Complex neural networks and neural network ensembles (sets of different neural network architectures producing an average or weighted predicted response or classification) can also be saved in binary form and later applied to new data.

In addition, *STATISTICA* Data Miner contains various designated procedures in the (Node Browser) folders titled Classification and Discrimination, Regression Modeling and Multivariate Exploration, and General Forecaster and Time Series, to perform complex analyses with automatic deployment and cooperative and competitive evaluation of models (see Figure 10.26).

For example, the Classification and Discrimination folder contains nodes for stepwise and best-subset linear discriminant function analysis, various tree classification methods, generalized linear models procedures, and different neural network architectures (see Figure 10.27).

The analysis nodes with automatic deployment are generally named *TypeOfAnalysis* with Deployment. Simply connect these nodes to an input data source, update (train) the project, and you are ready for deployment. To the node, connect a data source marked for deployment (i.e., select the Data for Deployed Project check box in the dialog specifying the variables for

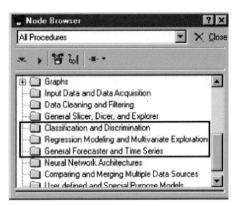

FIGURE 10.26   Node Browser showing All Procedures and focusing on three folders: (1) Classification and Discrimination, (2) Regression Modeling and Multivariate Exploration, and (3) General Forecaster and Time Series.

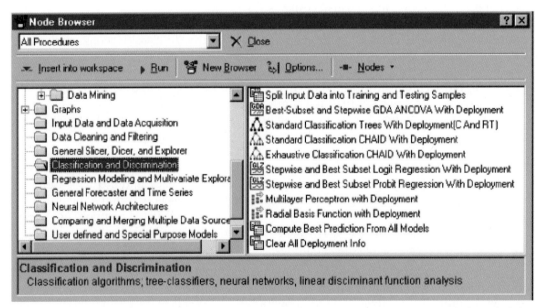

FIGURE 10.27   Here, the Node Browser focuses on the Classification and Discrimination folder, which has been selected, thus showing the specific nodes available in the right panel.

the analysis), and the program will automatically apply the most current model (e.g., tree classifier, neural network architecture) to compute predictions or predicted classifications.

Example: Start a new Data Miner project by selecting a predefined project for classification. From the Data Mining–Workspaces–Data Miner–General Classifier (Trees and Clusters) submenu, select Advanced Comprehensive Classifiers Project, as shown in Figure 10.28.

Then click the New Data Source button and select the Boston2.sta data file again. Specify Price as the categorical dependent variable, select variable Cat1 as a categorical predictor,

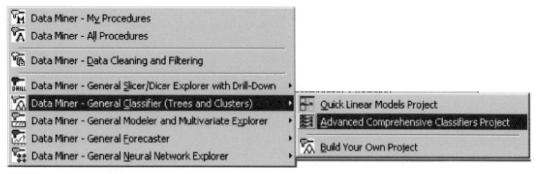

FIGURE 10.28   The lower part of the Data Miner menu, showing some of the accessory tools available, such as ready-made templates that can be obtained from the Advanced Comprehensive Classifiers Project, if selected.

and select variables ORD1 through ORD12 as continuous predictors. Click the Connect button and connect the data icon to the Split Input node, which is the main connection point for the Advanced Comprehensive Classifiers Project (see Figure 10.29).

   Now click the Run button. A number of very advanced and somewhat time-consuming analyses are then performed:

- The Split Input node in the Data Preparation, Cleaning, Transformation area will randomly select two samples from the input data: one for training the various models for classification and the other to evaluate the models; i.e., for the observations in the Testing sample, the program will automatically compute predicted classification and misclassification rates so that the Compute Best Prediction from all Models node (the one

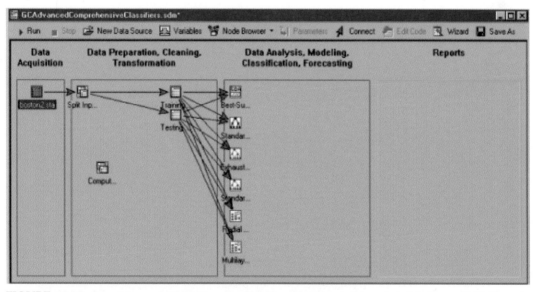

FIGURE 10.29   The Advanced Comprehensive Classifiers Project template, selected in Figure 10.28. The only thing that has been added here is the data set icon, shown in the left most panel.

that initially is not connected to anything in the Data Preparation, Cleaning, Transformation area) can automatically pick the best classifier or compute a voted best classification (i.e., apply a meta-learner).

- The program will automatically apply to the Training sample the following classification methods: linear discriminant analysis, standard classification trees (C&RT) analysis, CHAID, Exhaustive CHAID, a Radial Basis Function neural network analysis, and a multilayer perceptron.
- Next, the program will automatically apply the trained models to the "new" data, i.e., the testing sample; the observations in that sample have not been used for any computations so far (estimation of the models), so they provide a good basis for evaluating the accuracy of the predicted classifications for each model.

A large amount of output will be created, as shown in Figure 10.30.

You can review the results for each model in the respective results nuggets in the Reports areas; during the initial research stage of your data mining project, you probably would want to review carefully the models and how well they predict the response of interest. You can also double-click on each of the analysis nodes to select different types of parameters for the respective analyses; in that case you can use the Run to Node option (on the shortcut menu or the Run menu) to update only the selected node.

You can also now connect new data, marked for deployment, to the Compute Best Prediction from All Models node (the one that is not connected to anything at this point in the Data Preparation, Cleaning, Transformation area). For this example, simply connect the Testing data (which was created as a random sample from the original input data source); then use the option Run to Node to compute the predicted classifications for each model (see Figure 10.31).

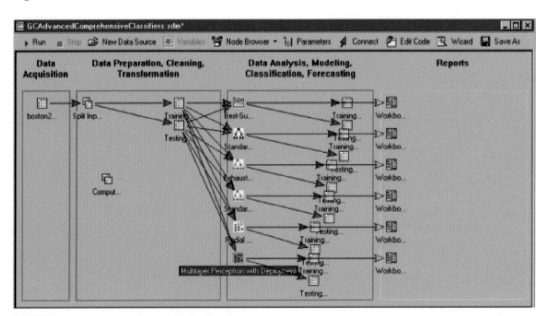

FIGURE 10.30   A completed Data Miner project.

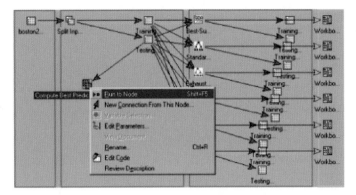

FIGURE 10.31 By using the short-cut menu on a node (accessible by right-clicking on the node) and then selecting Run to Node to make the node work, you can produce the results, embedded in a results icon, shown at the right with the green arrows.

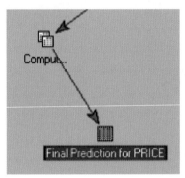

FIGURE 10.32 A final predictions spreadsheet that was created in the second panel of the Data Miner Workspace, e.g., the Data Preparation, Cleaning, Transformation area.

After a few seconds, the results spreadsheet with predictions will be created as another node in the Data Preparation, Cleaning, Transformation area (see Figure 10.32).

You can review the final predictions by selecting View Document from the shortcut menu, after clicking on the Final Prediction for PRICE icon (which contains the predicted classifications for the variable PRICE from all models). For example, you can compute a multiple histogram for the accuracy for each classifier (also reported in the Final Prediction for PRICE spreadsheet), as shown in Figure 10.33.

You could also look at the accuracy of classification, broken down by each category, and so on. In this case, it appears that all algorithms were reasonably accurate. By default, the Final Prediction spreadsheet will also contain a column with a voted classification from all classifiers. Experience has shown that predicted classification is often most accurate when it is based on multiple classification techniques, which are combined by voting (the predicted class that receives the most votes from participating models is the best predic-tion). Figure 10.34 shows the categorized histogram of the voted classifications by the observed classifications.

Clearly, the voted classification produces excellent accuracy in the test sample. Remember that the test sample was randomly selected from the original data and was not used to estimate the models (train the networks, etc.).

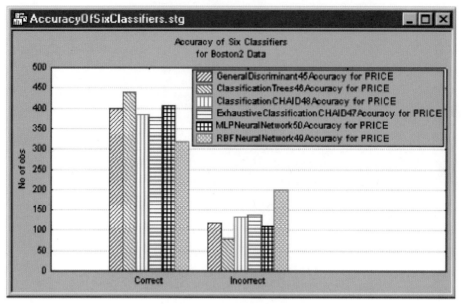

FIGURE 10.33   Bar graph results comparing the group of different data mining algorithms that were competitively compared in this data mining project.

FIGURE 10.34   A 3D graphical representation of the classifications from the data mining model, using the Voted Predictions (e.g., a hybrid model by voting together the individual algorithms to make a consensus model).

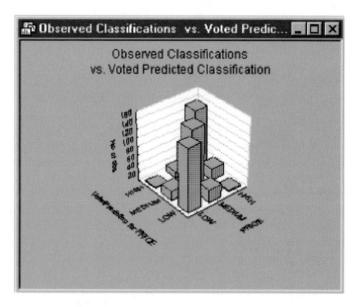

8. Prepare project for final customer deployment (in the field).

Once deployment information is stored after training nodes marked with deployment, for classification or prediction (regression problems), you can save the entire project and later retrieve that file to compute predicted values for new observations. For example, a loan officer may want to predict credit risk based on the information provided on a loan application. The loan officer (end user or customer) will not have to retrain the models in the current project again; instead, he or she can simply connect new data to the prediction node (usually labeled Compute Best Prediction from All Models) and proceed to process the new data. In fact, the data analyst who created the prediction model from training (learning) data can delete all computational (analysis) nodes from the project and leave only a single node for computing predicted responses. In a sense, such projects are "locked"; i.e., there is no risk of losing the deployment information due to accidentally starting a retraining of the models.

Let's consider advanced methods for deployment in the field. If you are familiar with *STATISTICA* Visual Basic (SVB), you might also consider writing a custom program that would further customize the user interface for the end user of the deployed solution. If you review the function available for the *STATISTICA* Data Miner library in the SVB Object Browser, you can see that practically all aspects of the *STATISTICA* Data Miner user interface can be customized programmatically; for example, you could "attach" the automatic application of a deployed solution to new data to a toolbar button so that a loan officer would only have to fill out a form with an applicant's data, click a button, and retrieve scores for credit risk and fraud probability.

Example: Suppose you want to send a fully trained and deployed solution to a customer or client, based on the analyses briefly discussed in step 7. You can simply delete all nodes other than the one labeled Compute Best Prediction from All Models and save the project as MyDeployedProject.sdm. You may also want to rename the lengthy Compute Best Prediction from All Models to the simple "instruction" Connect New Data Here, as shown in Figure 10.35.

All deployment information will automatically be saved along with the project file.

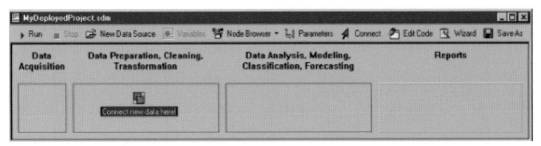

FIGURE 10.35   Connect New Data Here, renamed from Compute Best Prediction from All Models, has embedded in it all the training information and thus is a deployable model that you could send to a customer to use on that customer's new data.

Three Formats for Doing Data Mining in *STATISTICA*

The previous description of how *STATISTICA* Data Miner works represented only one of the three methods/formats of doing data mining in *STATISTICA*: the Data Miner Workspace format. *STATISTICA* has these three formats:

- Data Miner Workspace;
- Interactive data mining modules; and
- "Automated data mining" via Data Miner Recipe (DMRecipe).

Examples of each of these two additional formats are illustrated next.

Interactive Data Mining Using the Decision Trees—C&RT Module

Classification trees can be created in an interactive way by selecting the dialog in Figure 10.36 from the Data Mining pull-down menu and selecting Classification Trees (C&RT). From this point, you can use this dialog in a point-and-click format, selecting variables, changing any of the parameters from its default settings, if desired, and clicking OK to run the computations, with a results dialog popping up when all computations are complete. At this point, all you have to do is click on the buttons of the results dialog, which represent various outputs, all of which will be put into a workbook. This workbook can be saved for later reference.

Data Miner Recipe (DMRecipe)

Using DMRecipe is the new "automatic" way of doing data mining, where as few as four to seven clicks are required. You just click Run, go away to have lunch, and return to a

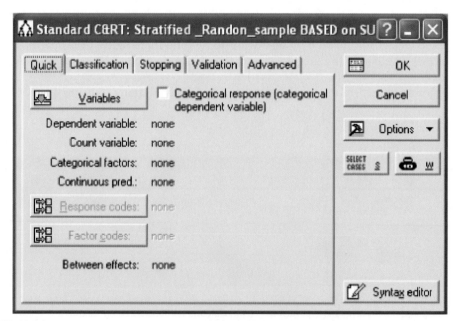

FIGURE 10.36   Interactive decision trees (e.g., Standard C&RT) dialog in *STATISTICA*.

completed program and workbook, with the topmost screen being a comparative evaluation of all algorithms, giving the accuracy scores for each algorithm. If you find one algorithm to be of particular interest, then you can go to the Interactive format of that algorithm and try to tweak it further, if desired, but usually the DMRecipe output will be sufficient for most needs. Figure 10.37 shows the DMRecipe dialog, and Figure 10.38 shows a workbook of results from running DMRecipe.

As you can see in Figure 10.38, the boosted trees algorithm had the lowest error rate, only 0.67%; in other terms, the accuracy rate of boosted trees is 99.33%.

STATISTICA, in addition to a data mining format, also has two additional data mining tools that integrate fully with the basic data mining algorithms; they are

1. QC-Mining algorithms, for quality control, industrial applications (see Figure 10.39)
   As you can see in Figure 10.39, there are five separate selections on the QC-Mining group, the last one listed above being "Response Optimization for Data Mining Models"
2. Text mining (see Figures 10.40 and 10.41)

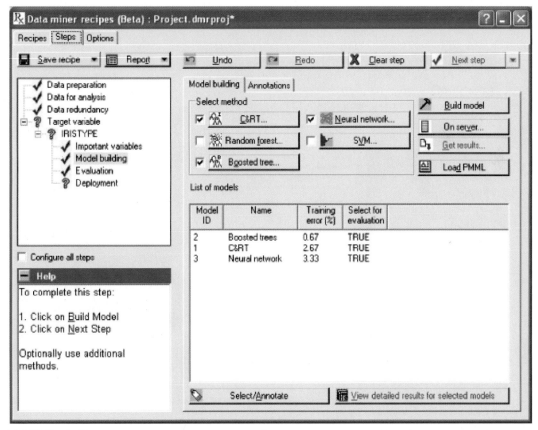

FIGURE 10.37   The automatic Data Miner Recipe format in *STATISTICA*, where data mining can be accomplished with as few as four to seven clicks of the mouse.

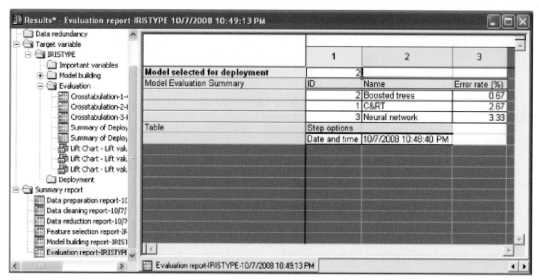

FIGURE 10.38 The results of the Data Miner Recipe automatically pop up on the screen at the completion of the computations.

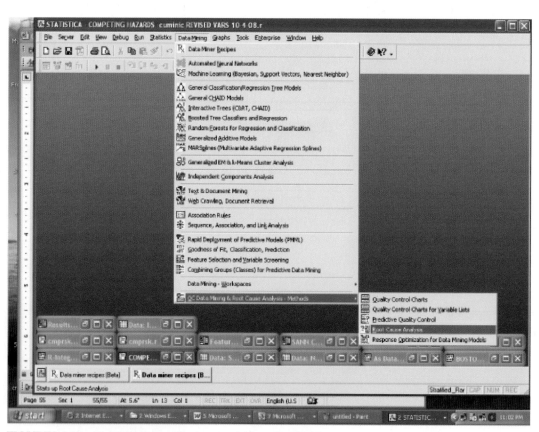

FIGURE 10.39 The QC-Mining selections available in *STATISTICA*.

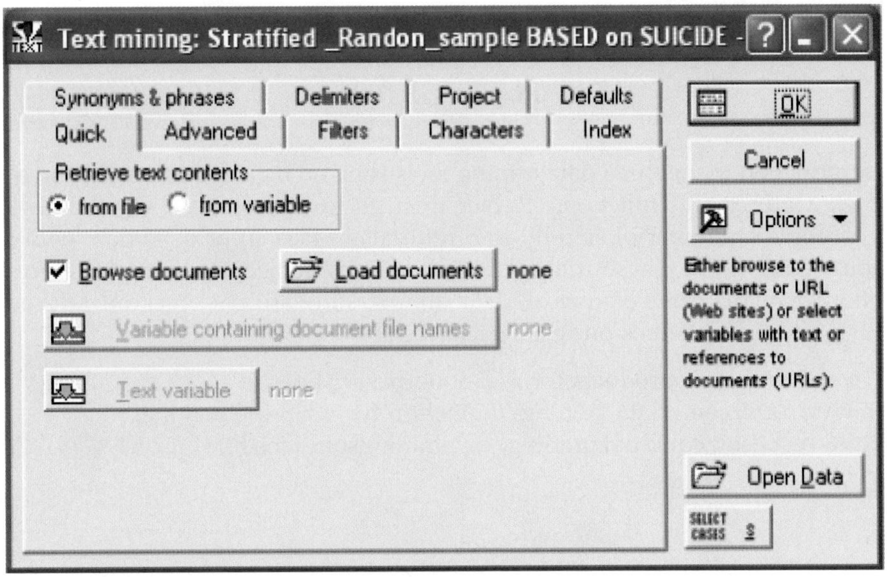

FIGURE 10.40   The main interactive Text Mining dialog in *STATISTICA*.

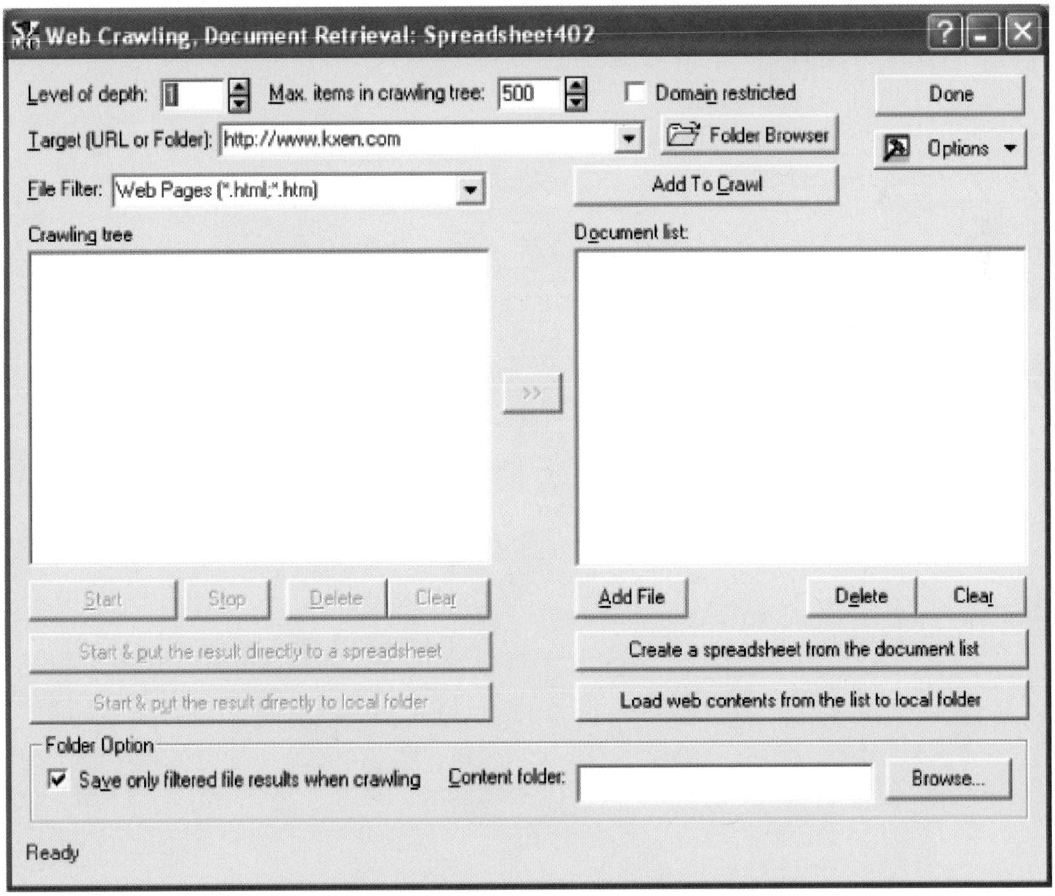

FIGURE 10.41   The Document Retrieval/Web Crawling dialog in *STATISTICA* Text Miner.

POSTSCRIPT

This introduction to the three data mining tools is by no means a comprehensive description of their features and functions. Rather than go further into the capabilities of these tools, we provide a number of step-by-step tutorials in Part III of this book, with more on the accompanying DVD. These tutorials cover various subject domains of interest, using one or another of the three statistical/data mining software packages introduced in this chapter. For additional details on these packages see

- http://www.SAS.com/products for SAS-Enterprise Miner;
- http://www.SPSS.com/data_mining/index.htm for SPSS Clementine;
- http://www.statsoft.com/datamining/dataminingsolutions.html for *STATISTICA* Data Miner.

References

Harrison, D., & Rubinfield, D. L. (1978). Hedonic prices and the demand for clean air. *Journal of Environmental Economics and Management, 5*, 81–102.

Lim, T.-S., Loh, W.-Y., & Shih, Y.-S. (1997). An emprical comparison of decision trees and other classification methods. *Technical Report 979*. Madison: Department of Statistics, University of Wisconsin.

Classification

PREAMBLE

The first general set of data mining applications predicts to which category of the target variable each case belongs. This grouping activity is called *classification*.

WHAT IS CLASSIFICATION?

Classification is the operation of separating various entities into several classes. These classes can be defined by business rules, class boundaries, or some mathematical function. The classification operation may be based on a relationship between a known class assignment and characteristics of the entity to be classified. This type of classification is called *supervised*. If no known examples of a class are available, the classification is *unsupervised*. The most common unsupervised classification approach is *clustering*. The most common applications of clustering technology are in retail product affinity analysis (including marketbasket analysis) and fraud detection. In this chapter, we will confine the discussion to supervised classification methods. Unsupervised classification methods will be discussed in Chapter 17, in relation to the detection and modeling of fraud.

There are two general kinds of supervised classification problems in data mining: (1) binary classification—only one target variable and (2) multiple classification—more than one target variable. An example of analyses with only one target variable are models to identify high-probability responders to direct mail campaigns. An example of analyses with multiple target variables is a diagnostic model that may have several possible outcomes (influenza, strep throat, etc.).

INITIAL OPERATIONS IN CLASSIFICATION

Before classification can begin, there are some initial tasks you must perform:

1. Determine what kind of classification problem you have. This means that you have to determine how many target classes you have and define them, at least in general terms.
2. Define the boundaries of each class in terms of the input variables.
3. Construct a set of decision rules from class boundaries to define each class.
4. Determine the *prior-probability* of each class, based on the frequency of occurrence of a class in the entire data set.
5. If appropriate, you should determine the cost of making the wrong choice in assigning cases to a given class. This task is extremely important in some classification situations (e.g., medical diagnosis).

MAJOR ISSUES WITH CLASSIFICATION

There are a number of issues that you must face before proceeding with the classification project. It is important to consider each of these issues, and either resolve the issues before modeling or set some expectations surrounding them.

What Is the Nature of the Data Set to Be Classified?

The purpose of the classification should be specified, and it should be related to the expected interpretation of the results. For example, a classification of breast cancer propensity should be accompanied with costs of a wrong diagnosis. You probably would not worry much about misclassifying response rate in a mailing campaign, but failing to diagnose a breast tumor could be fatal.

How Accurate Does the Classification Have to Be?

If you are in a time-crunch, a model with 80% sensitivity to prediction accuracy built in 2 days may be good enough to serve the model's purpose. In business, time is money!

How Understandable Do the Classes Have to Be?

One of the strengths of a decision tree model is that it produces results that are easy to understand in terms of the predictor variables and target variables. Results from a neural net model may be more predictive, but the understandability of the results in terms of the predictor variables is less clear. Often, there is a trade-off between accuracy and understandability of the results. This trade-off may be related to the choice of modeling algorithm. Some algorithms do better for some data sets than others. The *STATISTICA* Data Miner Recipe Interface uses several modeling techniques in the form of a recipe that provides a basis for choosing the right trade-off combination of accuracy and understandability of the model results.

ASSUMPTIONS OF CLASSIFICATION PROCEDURES

Classification in general requires that you accept a number of assumptions. The fidelity of your classes and their predictive ability will depend on how close your data set fits these assumptions. In Chapter 4, we stressed the importance of describing your data set in terms of the nature of its variables, their possible interactions with the target variable and with each other, and their underlying distributional pattern. In classification, you should try to satisfy these assumptions as much as possible.

Numerical Variables Operate Best

Categorical variables can be used, but they should be decomposed into dummy variables, if possible (cf. Chapter 4 for an introduction to dummy variables).

No Missing Values

By default, most data mining algorithms (including those for classification) will eliminate cases with missing values in predictor variables. Imputation of missing values is one way to fix this problem (see Chapter 4). Another way that some classification algorithms (e.g., C&RT) may fill missing values is to use surrogate variables. A surrogate variable has a similar splitting behavior to the variable with the missing value, and its value in this case can be used to replace a missing value in predictor variable.

Variables Are Linear and Independent in Their Effects on the Target Variable

When we say that variables are linear and independent in their effects on the target variable, we mean that there is a straight-line (linear) change in the target variable as each variable is varied over its range and that the effect of one variable is not related to (is independent of) effects of any other variable. There is not much you can do about the independency assumption, if it is false. But you can increase the linearity of the predictor variables by transforming their scales with common nonlinear functions, like $\log_{10}$, $\log_e$, squares, cubes, and higher-level functions.

According to probability theory, target variables must be independent also. Classification targets selected to define categories must be mutually exclusive and categorically exhaustive (MECE). Categorically exhaustive means that the outcome is at least one category. For example, a data set used to classify shades of red balls cannot contain any blue balls. Mutually exclusive means one and only one target can be assigned to each case. If one candidate target variable is "Residential dwelling," another target variable cannot be "Single-family dwellling." In this situation, single-family dwellings are a subset of residential dwellings, and both categories may have an equal probability in the classification operation. If MECE is not satisfied, assignment of some cases into categories may be arbitrary (to some extent), and not related exclusively to the predictor variables.

METHODS FOR CLASSIFICATION

There are many techniques used for classification in statistical analysis and data mining. For the sake of parsimony, we will limit this discussion to machine learning techniques available in many data mining packages. The following classification algorithms were introduced and described in Chapters 7 and 8. Some additional information will be presented here to help you apply them to classification problems.

1. Decision trees
2. CHAID
3. Random forest and boosted trees
4. Logistic regression
5. Neural nets

Other useful classification algorithms include

1. *K*-nearest neighbor
2. Naïve Bayesian classifier

Any classification method uses a set of *features* to characterize each object, where these features should be relevant to the task at hand. We consider here methods for *supervised* classification, meaning that a human expert both has determined into what classes an object may be categorized and also has provided a set of sample objects with known classes. This set of known objects is called the *training set* because it is used by the classification programs to learn how to classify objects. There are two phases to constructing a classifier. In the training phase, the training set is used to decide how the features ought to be weighted and combined in order to separate the various classes of objects. In the application phase, the weights determined in the training set are applied to a set of objects that do *not* have known classes in order to determine what their classes are likely to be.

If a problem has only a few (two or three) important features, then classification is usually an easy problem. For example, with two parameters you can often simply make a scatterplot of the feature values and can determine graphically how to divide the plane into homogeneous regions where the objects are of the same classes. The classification problem becomes very hard, though, when there are many parameters to consider. Not only is the resulting high-dimensional space difficult to visualize, but there are so many different combinations of parameters that techniques based on exhaustive searches of the parameter space rapidly become computationally infeasible. Practical methods for classification always involve a heuristic approach intended to find a "good-enough" solution to the optimization problem.

Nearest-Neighbor Classifiers

A very simple classifier can be based on a nearest-neighbor approach. In this method, you simply find in the N-dimensional feature space the closest object from the training set to an object being classified. Since the neighbor is nearby, it is likely to be similar to the object being classified and so is likely to be the same class as that object.

Classification by a nearest-neighbor algorithm searches for the closest value. Several issues, though, are associated with the use of the algorithm: (1) the inclusion of irrelevant variables lowers the classification accuracy; (2) the algorithm works primarily on numerical variables; categorical variables can be handled but must be specially treated by the algorithm; and (3) if the scales of variables are not in proportion to their importance, classification accuracy will be degraded.

The k-nearest-neighbor algorithm looks for the closest data point in the data set. The k-parameter specifies how many nearest neighbors to consider (an odd number is usually chosen to prevent ties). The "closeness" is defined by the difference ("distance") along the scale of each variable, which is converted to a similarity measure. This distance is defined as the Euclidian distance. Alternatively, the Manhattan Distance can be used, which is defined for a plane with a data point p_1 at coordinates (x_1, y_1) and its nearest neighbor p_2 at coordinates (x_2, y_2) as

$$\text{Manhattan Distance} = |x_1 - x_2| + |y_1 - y_2|. \qquad \text{(Eq. 1)}$$

An analogous relationship can be defined in a higher-dimensional space.

Nearest-neighbor methods have the advantage that they are easy to implement. They can also give quite good results if the features are chosen carefully (and if they are weighted carefully in the computation of the distance). There are several serious disadvantages of the nearest-neighbor methods. First, they (like the neural networks) do not simplify the distribution of objects in parameter space to a comprehensible set of features. Instead, the training set is retained in its entirety as a description of the object distribution. (There are some thinning methods that can be used on the training set, but the result still does not usually constitute a compact description of the object distribution.) The method is also rather slow if the training set has many examples. The most serious shortcoming of nearest-neighbor methods is that they are very sensitive to the presence of irrelevant parameters. Adding a single parameter that has a random value for all objects (so that it does not separate the classes) can cause these methods to fail miserably.

The best choice of k depends largely on the data. In general, larger values of k tend to create larger classes in terms of the range of values included in them. This effect reduces the noise in the classification but makes the classification more choppy (relatively large distinctions between classes). A suitable k for a given data set can be estimated by a decision rule or by using a resampling method (like cross-validation) to assign the mean value among samples. The accuracy of the k-NN algorithm can be severely degraded by the presence of noisy or irrelevant features, or if the feature scales are not consistent with their importance.

Analyzing Imbalanced Data Sets with Machine Learning Programs

Imbalanced Data Sets

Many problems in data mining involve the analysis of rare patterns of occurrence. For example, responses from a sales campaign can be very rare (typically, about 1%). You can just adjust the classification threshold to account for imbalanced data sets. Models built with many neural net and decision tree algorithms are very sensitive to imbalanced data sets. This imbalance between the rare category (customer response) and the common category (no response) can cause significant bias toward the common category in resulting models.

A neural net learns one case at a time. The error minimization routine (e.g., backpropagation, described in Chapter 7) adjusts the weights one case at a time. This adjustment process will be dominated by the most frequent class. If the most frequent class is "0" 99% of the time, the learning process will be 99% biased toward recognition of any data pattern as a "0." Balancing data sets is necessary to balance the bias in the learning process. Clementine (for example) provides the ability to change the classification threshold in the Expert Options, and it provides the ability to balance the learning bias in the Balance node. If you have Clementine, run the same data set through the net with the threshold set appropriately. Then run the data through a Balance node (easily generated by a Distribution node), using the default threshold. Compare the results.

Another way to accomplish this balance is to weight the input cases appropriately. If the neural net can accept weights and use them appropriately in the error minimization process, the results can be comparable to using balanced data sets. *STATISTICA* does this. A final way to balance the learning bias is to adjust the prior probabilities of the "0" and "1" categories. SAS-Enterprise Miner and *STATISTICA* Data Miner use this approach.

If your data mining tool doesn't have one of the preceding methods for removing the learning bias, you may have to physically balance the data by resampling. But resampling can be done two ways: by increasing the sample rate of the rare category (oversampling) or reducing the sample rate of the common category (undersampling). Undersampling the common category eliminates some of the common signal pattern. If the data set is not large, it is better to oversample the rare category. That approach retains all of the signal pattern of the common category and just duplicates the signal pattern of the rare category.

Decision Trees

Interactive trees and boosted trees were introduced in Chapter 7. For purposes of classification, some general information on decision trees is presented here to help you apply these algorithms successfully to classification problems.

A *decision tree* is a hierarchical group of relationships organized into a tree-like structure, starting with one variable (like the trunk of an oak tree) called the *root node*. This root node is split into two to many branches, representing separate classes of the root node (if it is categorical) or specific ranges along the scale of the node (if it is continuous). At each split, a question is "asked," which has an answer in terms of the classes or range of the variable being split. One question might be, "Is this case a male or a female?" Questions like this would be used to build a decision tree with *binary* splits. Decision trees can also be built with multiple splits. Questions asked at each split are defined in terms of some *impurity measure,* reflecting how uniform the resulting cases must be in the splits. Each branch is split further using the classes or ranges of other variables. At each split, the node that is split is referred to as the *parent node,* and the nodes into which it is split are called the *child nodes.* This process continues until some stopping rule is satisfied, such as the minimum number of cases resides in the final node (*terminal leaf node*) along one splitting pathway. This process is called *recursive partitioning.* Figure 11.1 shows an example of a decision tree for classifying colored balls.

In the trees structure shown in the figure, the root node is the first parent. Child nodes 1a and 1b are the children of the first split, and they become the parents of children formed by the second split (e.g., Child 1a-1). Child nodes 1a-1, 1a-2, 1b-1, and 1b-2 are terminal leaf nodes. For simplicity, the terminal nodes of the other child nodes are not shown. To understand how this tree was built, we have to consider the questions asked and define the impurity measure and the stopping rule used.

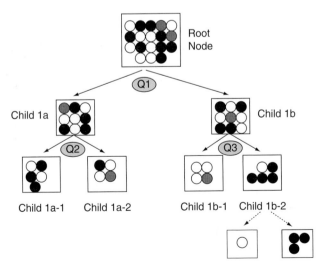

FIGURE 11.1 A simple binary decision tree structure.

The questions here might be

Q1 = What 2 groups would have at least 4 white balls and 4 black balls?
Q2 = What 2 groups would have at least 4 balls, among which 2 are white?
Q3 = What 2 groups would have balls of only two colors?

The impurity measure for this example might be the rule that no more than one ball may be of a different color. The stopping rule might be that there be a minimum of four balls in a group. The potential split at Child node 1b-2 would not be permitted due to the stopping rule. You might be thinking that it is possible to create multiple splits in this example, and you are right! But the structure of this decision tree is constrained to binary splits.

Decision trees are well suited for classification (particularly binary classification), but are also useful in difficult estimation problems, where their simple piecewise-constant response surface and lack of smoothness constraints make them very tolerant of outliers. Also, trees are also probably the easiest model form to interpret (so long as they are small). The primary problem with decision trees is that they require a data volume that increases exponentially as the depth of the tree increases. Therefore, large data sets are required to fit complex patterns in the data. The other major problem involves using multiple splits, namely that the decision of where to split a continuous variable range is very problematic. Much research has been done to propose various methods for making multiple splits.

There are many forms of decision tree algorithms included in data mining tool packages, the most common of which are CHAID and Classification and Regression Trees (C&RT or CART). These basic algorithms are described in Chapter 7. Newer tree-based algorithms composed of groups of trees (such as random forests and boosted trees) are described in greater detail in Chapter 8.

Classification and Regression Trees (C&RT)

For purposes of classification, we should review some of the basic information about C&RT to understand how to use this algorithm effectively. The C&RT algorithm, popularized by Brieman et al. (1984), grew out of the development of a method for identifying high-risk patients at the University of California, San Diego Medical Center. For that purpose, the basic design of C&RT was that of a binary decision tree. The C&RT algorithm is a form of decision tree that can be used for either classification or estimation problems (like regression). Predictor variables can be nominal (text strings), ordinal (ordered strings, like first, second), or continuous (like real numbers).

Some initial settings in most C&RT algorithms are common to most classification procedures. The first setting is the prior probability of the target variables (frequency of the classes). Often, this is done behind the scenes in a C&RT implementation. Another important setting (usually done manually) is to select the measure of the "impurity" to use in evaluating candidate split points. By default, this setting is the Gini score, but other methods such as twoing are available in many data mining packages. The Gini score is based on the relative frequency of subranges in the predictor variables. Twoing divides the cases into the best two subclasses and calculates the variance of each subclass used to define the split point. Sometimes, options for case weights and frequency weights are provided by the algorithm. Missing values in one variable are handled commonly by substituting the value of a *surrogate variable,* one with similar splitting characteristics as the variable with the missing value. The algorithm continues making splits along the ranges of the predictor variables, until some stopping function is satisfied, like the variance of a node declining below a threshold level. When the splitting stops along a given branch, the unsplit node is termed the *terminal node.* Terminal nodes are assigned the most frequent class among the cases in that node.

C&RT trees are pruned most commonly by cross-validation. In this case, cross-validation selects either the smallest subtree whose error rate is within 1 standard error unit of the tree with the lowest error rate, or the subtree with that lowest error rate. The quality of the final tree is determined either as the one with the lowest error rate, when compared to test data, or the one with a stable error rate for new data sets.

Example

The Adult data set from the UCI data mining data set archive was used to create a C&RT decision tree for classification (http://archive.ics.uci.edu/ml/). The data set contains the following:

Target:

- Income <= $50,000/yr.; > $50,000/yr. (categorical)

Predictors:

- Age: - continuous
- Workclass: Private, Self-emp-not-inc . . . - categorical
- Final weight: Weights assigned within a State to correct for differences in demographic structure by age, race, and sex
- Education: Bachelors, Some-college . . . - continuous
- Education-years: - continuous
- Marital-status: Married, Single, Divorced . . . - categorical
- Occupation: Teller, Mechanic. . . - categorical
- Relationship: Wife, Husband . . . - categorical
- Race: White, Black, Other . . . - categorical
- Sex: Female, Male - categorical
- Capital-gain: - continuous
- Capital-loss: - continuous
- Hours-per-week: - continuous
- Native-country: Nationality - categorical

Figure 11.2 shows a decision tree for four of these variables:

- Capital-gain
- Age
- Marital Status
- Educations Yrs.

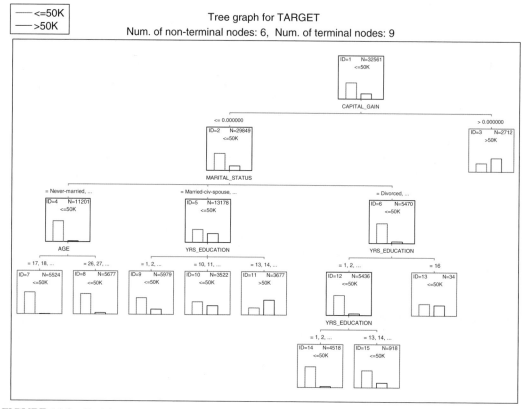

FIGURE 11.2   Decision tree structure.

From the decision tree shown in Figure 11.2, we can induce general rules to predict who is likely to have income:

1. Persons with IRS capital gains deductions are likely to have an income > $50,000/yr.
2. Married persons with no capital gains deductions and who have greater than 12 years of education (college work) are likely to have incomes > $50,000/yr.
3. All other persons are likely to have incomes <= $50,000/yr.

We can drill down a little deeper to evaluate how accurate the predictions are by constructing a classification matrix (Table 11.1). For the sake of this analysis, the >$50,000/yr category is considered the "positive" prediction.

According to the model evaluation criteria presented in Chapter 6, the Sensitivity of the model is 21,929/(21,929 + 2,791) × 100 = 88.71%, and the Specificity of the model is 3,232/(3,232 + 4,609) × 100 = 41.22%. Remember, the Sensitivity of the model measures how well it predicts incomes > $50,000/yr., and the Specificity of the model measures how well it predicts incomes <= $50,000/yr. Ideally, the Specificity value of 41.22% should be closer to the Sensitivity value of 88.71%. This means that the model predicts high

TABLE 11.1 The Classification Matrix for the Prediction of Incomes > $50,000/yr. Using Age, Marital Status, Capital Gains, and Years of Education

Classification matrix 1 (adult_train_data.sta) Dependent variable: TARGET Options: Categorical response, Analysis sample				
	Observed	**Predicted <=50K**	**Predicted >50K**	**Row Total**
Number	<=50K	21929	2791	24720
Column Percentage		87.15%	37.72%	
Row Percentage		**88.71%**	11.29%	
Total Percentage		67.35%	8.57%	75.92%
Number	>50K	3232	4609	7841
Column Percentage		**12.85%**	62.28%	
Row Percentage		**41.22%**	58.78%	
Total Percentage		9.93%	14.15%	24.08%
Count	All Groups	25161	7400	32561
Total Percent		77.27%	22.73%	

incomes better than it does low incomes. Nearly half (41.22%) of the incomes predicted as low were actually high. This model might work well if we wanted only to identify persons with a particularly high probability of having a high income. This model has only four predictor variables (Age, Capital Gains, Marital Status, and Years of Education). It appears that our short-list of variables is too short to do a good job of predicting low incomes. Maybe we should add more variables.

Table 11.2 shows the classification matrix for all the potential predictor variables in the data set (except for the Final_Weight variable).

TABLE 11.2 The Classification Matrix for the Prediction of Incomes > $50,000/yr. Using All Potential Predictor Variables

Classification matrix 1 (adult_train_data.sta) Dependent variable: TARGET Options: Categorical response, Analysis sample				
	Observed	**Predicted <=50K**	**Predicted >50K**	**Row Total**
Number	<=50K	22502	2218	24720
Column Percentage		85.98%	34.72%	
Row Percentage		91.03%	8.97%	
Total Percentage		69.11%	6.81%	75.92%
Number	>50K	3670	4171	7841
Column Percentage		14.02%	65.28%	
Row Percentage		46.81%	53.19%	
Total Percentage		11.27%	12.81%	24.08%
Count	All Groups	26172	6389	32561
Total Percent		80.38%	19.62%	

The Sensitivity value for the model with all predictor variable of 85.98% is slightly lower than the 88.71 figure for the model with four variables, but the Specificity value of 46.81% is slightly higher. *Which model is better?* The best rule of thumb to follow in this situation is Occam's Razor: the simplest model is the best. Therefore, we might select the model with four predictor variables.

The lift chart for the simplest model is shown in Figure 11.3.

The values in the cumulative lift chart shown in the figure have been normalized around the expected value of 1.00 (50:50 probability of high income). This form of the lift chart suggests that the model predicts better than random expectation (50:50) in the first eight deciles (up to 80% of the values).

CHAID

The acronym CHAID stands for Chi-square Automatic Interaction Detector. It was proposed by Kass (1980). Unlike C&RT, CHAID uses multiway splits instead of binary splits, where more than two splits can occur from a single parent node. When a categorical response variable has many categories (like car, truck, classic, motorcycle, etc.), the

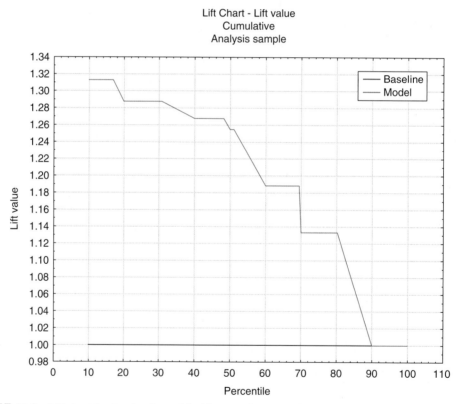

FIGURE 11.3   Lift chart for the simple model of four predictor variables.

algorithm will build many multiway frequency tables. This property has made CHAID very popular in research involving market segmentation studies.

Both CHAID and C&RT will construct trees, where each (nonterminal) node identifies a split condition to predict either a continuous or categorical response variable. Therefore, both algorithms can be applied to both classification and estimation (regression) problems.

CHAID relies on the Chi-square test to determine the best next split at each step in a classification problem. In F-regression-type problems, the algorithm uses the F-test in place of the Chi-square test.

Basic Steps in CHAID Processing

Preparing predictors. The first step is to create categorical predictors out of any continuous predictors by dividing the respective continuous distributions into a number of classes with an approximately equal number of observations. Another name for this process is *binning*.

Merging classes. The next step is to cycle through the predictors to determine for each predictor the two classes (or bins) that have the least relationship with the dependent variable. For classification problems, the Chi-square test is used to determine the significance of the relationship. For regression problems, the F-test is used. If the respective test for a given pair of predictor classes is not statistically significant as defined by a predetermined setting, the classes will be merged into a category and continue to analyze the next two classes. If the test shows significant difference, a new category is erected, and the next two classes are evaluated. Probability values (*p*-values) are calculated for each merged category.

Selecting the split variable. The next step is to choose the category with the smallest *p*-value; if it is greater than the predetermined threshold, no further split will be made along that branch and that category becomes a terminal node. This process is continued until no further splits can be made along any of the branches.

Exhaustive CHAID

A modification to the basic CHAID algorithm, called *Exhaustive CHAID*, does not check the *p*-value against a predetermined threshold value but performs a more thorough merging and testing of predictor variables. Consequently, this technique requires more computing time. The merging of classes continues (without reference to any threshold value) until only two categories remain for each predictor. The algorithm then selects from among the predictors the one that yields the most significant split. For large data sets and with many continuous predictor variables, Exhaustive CHAID may require significant computing time.

Issues with CHAID

- CHAID trees in regression problems can become very large. With modern computers, this is not so much a computational problem, but rather it makes the trees less comprehensible to the user.
- The basic algorithm can accommodate both categorical and continuous predictor variables (using a different significance test for each). But in practice, both variable types are combined often for analysis of covariance experimental designs. CHAID cannot handle experimental designs like that.

Random Forests and Boosted Trees

The random forests algorithm was proposed formally by Brieman (2001). It combines the concept of random subspaces and bagging (discussed further in Chapter 13). The random forests algorithm trains a number of trees on slightly different subsets of data (bootstrap sample), in which a case is added to each subset containing random selections from the range of each variable. This group of trees are similar to an *ensemble* (also discussed further in Chapter 13). Each decision tree in the ensemble votes for the classification of each input case.

Following are the steps in tree growth in random forests:

1. A random sample of the number of cases is taken. Subsequent samples for other trees are done with replacement (no case is left out, even the ones included in building the previous tree).
2. A subset of variables (the number of which is represented by the term *m*) is chosen, being much less than the number of variables, and the best split (based on the Gini score) is determined on this subset of variables. Increasing the value of *m* increases the correlation between trees (bad) and increases the predictive power of the tree (lessens the error rate), while decreasing the value of *m* does the opposite for both. There is a fairly wide "happy medium" in between values of *m* that are too low and too high. The *m*-number is the *only* setting in the algorithm to which the model is sensitive.
3. About one-third of the cases are used to create the out-of-sample testing data set. This testing data set is used to compute the prediction error rate. The average error rate is calculated from all trees built.
4. Variable importance is calculated by running the out-of-sample data set down the tree, and then the number of votes for the predicted class are counted. Then the values in each variable are randomly changed and run down the tree independently. The number of votes for the tree with the changed variable is subtracted from the number of votes for the unchanged tree to yield a measure of effect. Finally, the average effect is calculated across all trees to yield the variable importance value.

Advantages of Random Forests

- The random forests algorithm has relatively high accuracy among algorithms for classification.
- It can handle very large data sets with hundreds (even thousands) of variables.
- It provides an estimate of variables' importance, like neural nets.
- It has a robust method for handling missing data. The most frequent value for the variable among all cases in the node is substituted for the missing value.
- It has a built-in method for balancing unbalanced data sets (one class much rarer than other classes). We will revisit this concept in Chapter 16.
- It runs fast! Hundreds of trees with many thousands of cases with hundreds of variables can be built in a few minutes on a personal computer.

Boosted Trees

The boosted trees algorithm is an adaptation of the Stochastic Gradient Boosting technique of Friedman (1999).

The steps in the process are as follows:

1. In the first iteration, decision trees are built for each class of the categorical dependent variable.
2. Then a subsample of the training data is drawn at random (without replacement) and used to build a very simple tree (only one or two splits). This is called the *base learner*. The random draw is the stochastic elements of the algorithm.
3. Predicted values of the training set model and the base learner model are transformed with a Logistic function and used to calculate residuals.
4. A prediction update factor is calculated from the residuals and applied to generate a final prediction for the second tree.

This process is continued until the sum of the residuals declines below a threshold value. The effect of the application of the update factor in successive iterations of trees is to generate an error surface that declines in slope (forms a downward gradient). The overall effect of the algorithm is to create an adaptive weighted expansion of decision trees, which can produce an excellent fit of the predicted class values to the observed class values.

As is common in other machine learning techniques, this adaptive weighted expansion of trees can easily overfit the training data set. An out-of-sample data set can be used to calculate the testing error, which can be used to track the error through iterations of the algorithm. Figure 11.4 shows how one boosted trees algorithm in *STATISTICA* Data Miner tracks this testing error through the iterations of the model.

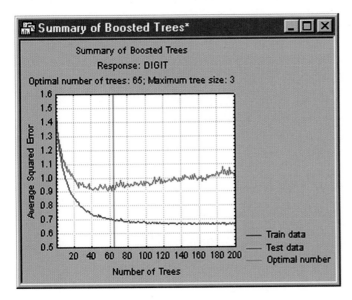

FIGURE 11.4 Error plot of training set and the testing set in a boosted trees model.

Notice that the error of the testing data set flattens out after about 35 trees are built and then begins to climb after about 65 trees are built. The optimum tree, therefore, is identified at tree 65.

Logistic Regression

Regressions will be discussed in greater detail in Chapter 12. But logistic regression is used in classification rather than numerical prediction. Therefore, we will include it here. The general form of the regression equation generated by the analysis is

$$Y = a + b_1X_1 + b_2X_2 + b_3X_3 + \ldots + b_nX_n \qquad \text{(Eq. 2)}$$

where a is the Y value, X is equal to 0 (the intercept), b_1 is the coefficient for X factor #1 (X_1), b_2 is the coefficient for X factor #2 (X_2), and so forth through the last X variable (X_n).

Instead of setting Y = the target variable in the data set, logistic regression uses the Logistic function to express Y as follows:

$$f(y) = \frac{1}{(1 + e^{-y})} \qquad \text{(Eq. 3)}$$

Figure 11.5 shows a plot of Equation 3.

Figure 11.5 describes the classical growth curve and is a suitable expression of many exponential relationships in nature. But many business data distributions are the mirror image of Figure 11.5. The Logistic function can be configured to express that data pattern also by removing the negative sign in Eq. 3. This transformed logistic data pattern is sometimes called the *inverse logistic* curve, as shown in Figure 11.6.

Logistic regression is used to model the nonlinear relationship between Y and the combined effects of the independent variables. This relationship is used to model the

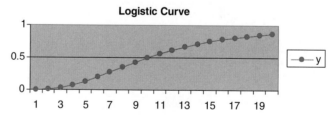

FIGURE 11.5   The Logistic curve.

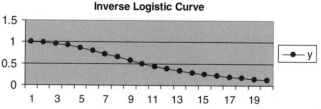

FIGURE 11.6   The Inverse Logistic function.

probability of an event's occurrence (a binary variable, like Yes/No or 1/0), using either categorical or numerical predictors. This algorithm has seen wide usage in business to predict customer attrition events, sales events of a specific product or group of products, or any event that has a binary outcome.

Neural Networks

Artificial neural nets are a crude attempt to mimic the function of human neurons. Neuron cells receive electrical impulses from neighboring cells and accumulate them until a threshold value is exceeded. Then they "fire" an impulse to an adjacent cell. The capacity of the cell to store electrical impulses and the threshold are controlled by biochemical processes, which change over time. This change is under the control of the autonomic nervous system and is the primary means by which we "learn" to think or activate our bodies.

Artificial neurons in artificial nets (Figure 11.7) incorporate these two processes and vary them numerically, rather than biochemically. The aggregation process accepts data inputs by summing them (usually). The activation process is represented by some mathematical function, usually linear or logistic. Linear activation functions work best for numerical estimation problems (i.e., regression), and the Logistic activation function works best for classification problems.

Artificial neurons are connected together into an *architecture* or processing operations. This architecture forms a network in which each input variable (called an input *node*) is connected to one or more output nodes. This network is called an artificial neural network (neural net for short). When the input nodes with summation aggregation function and a Logistic activation function are directly connected to an output node, the mathematical processing is analogous to a logistic regression with a binary output. This configuration of a neural net is a powerful classifier. It has the ability to handle nonlinear relationships between the output and the input variables, by virtue of the Logistic function shown in Figure 11.8.

The Logistic function fits many binary classification problems and can express much of the nonlinear effects of the predictors.

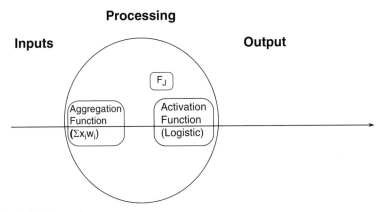

FIGURE 11.7   Architecture of a neuron.

FIGURE 11.8   A plot of the Logistic function.

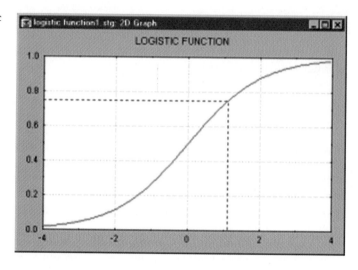

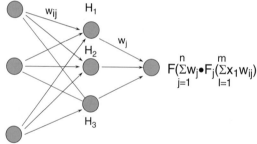

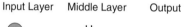

FIGURE 11.9   Architecture of a neural net.

The most interesting property of a neural net comes into view when you intercalate a middle layer of neurons (nodes) between the input and output node, as shown in Figure 11.9.

Weights (W_{ij}) are assigned to each connection between the input nodes and middle layer nodes, and between the middle layer nodes and the output node(s). These weights have the capacity to model nonlinear relationships between the input nodes and output node(s). Herein lies the great value of a neural net for solving data mining problems. The nodes in the middle layer provide the capacity to model nonlinear relationships between the input nodes and the output node (the decision). The greater the number of nodes in the middle layer, the greater is the capacity of the neural net to recognize nonlinear patterns in the data set. But as the number of nodes increases in the middle layer, the training time increases exponentially, and it increases the probability of overtraining the model. An over-trained model may fit the training set very well but not perform very well on another data set. Unfortunately, there are no hard and fast rules of thumb to define the middle layer nodes to put into your net. The general guideline to follow is to use more nodes when you have a lot of cases to use for training and use fewer nodes with fewer cases. If your classification problem is complex, use more nodes in the middle layer; if it is simple, use fewer nodes. See Chapter 12 for some additional rules of thumb.

The neural net architecture can be constructed to contain only one output node and be configured to function as a regression (for numerical outputs) or binary classification (yes/no or 1/0). Alternatively, the net architecture can be constructed to contain multiple output nodes and function as a clustering algorithm.

The learning process of the human neuron is reflected (crudely) by performing one of a number of weight adjustment processes, the most common of which is called backpropagation, shown in Figure 11.10.

Neural nets can be particularly effective in classification problems using predictor variables forming nonlinear relationships with the solution. The most common classification type is the binary classification. Logically, binary classification is defined as the XOR case, or an exclusive-OR operation. The XOR logical case permits one outcome or another outcome, but not both. The outcomes are mutually exclusive, which evaluate to "Yes" or "No," "1" or "0."

Neural Net Architecture

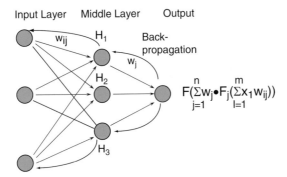

$$F(\sum_{j=1}^{n} w_j \cdot F_j(\sum_{l=1}^{m} x_1 w_{ij}))$$

FIGURE 11.10   A feed-forward neural net with backpropagation.

Naïve Bayesian Classifiers

In the general overview of Bayesian analysis in Chapter 1, the statement was made that Bayesian prediction follows patterns of human thinking more closely than does classical statistical analysis. The serious drawback of this fact is that two humans may (and often do) disagree in the decisions they make as a result of this thinking. Sir R. A. Fisher could not abide by this diversity in the decision-making process for medical purposes. That is why he developed the standard practices in statistical analysis in 1921. But there are many other situations in which the Bayesian approach to truth is much more appropriate and may even be better.

When we are faced with the need to classify some entity in the world around us, we include two general sources of evidence:

- Its similarity to each other based on some metrics;
- Past decisions on classifications of things like it.

Fisher excluded the second source of evidence from his analyses to calculate his probabilities from the former source only. Bayesians contend that in many cases that second source of evidence is critical to the proper classification of the entity. They integrate these sources of evidence by multiplying them to calculate the *joint* probability. To Fisherians, classification is a calculation involving simple probabilities; to Bayesians, classification is a judgment call based on joint probability. In many classification situations involving data attributes, we know relatively little about the entity we are classifying, and it may be acceptable to view the classification process as a judgment call. Following this logic, Naïve Bayesian classification has become accepted as a useful technique in data mining.

To demonstrate the concept of Naïve Bayesian classification, consider a group of objects classified according to their characteristics, as shown in Figure 11.11.

Given the past classification of the objects in the figure, our task is to classify new cases as they occur. Our approach is to decide to which class label they belong, based on the currently existing objects. Based on the fact that there are twice as many GREEN objects as RED, it is reasonable to believe that any new case is twice as likely to be a part of the GREEN group rather than the RED. In the Bayesian analysis, this belief is known as the *prior probability*. Prior probabilities are based on evidence from previous classifications, in this case the percentage of GREEN and RED objects, and can be used to predict the classification of new objects.

These prior probabilities can be expressed as

Prior probability for GREEN objects α (# GREEN objects/# TOTAL objects)
Prior probability for RED objects α (# RED objects/# TOTAL objects)

(Note: The α, or alpha symbol, means "is proportionate to.")

Since there is a total of 60 objects, 40 of which are GREEN and 20 RED, our prior probabilities for class membership are

Prior probability for GREEN α (40/60)
Prior probability for RED α (20/60)

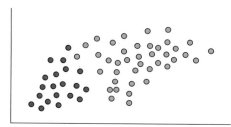

FIGURE 11.11   Objects classified in two groups, RED or GREEN, plotted in an analysis space defined by two axes of similarity (two metrics).

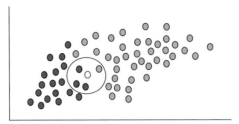

FIGURE 11.12   Position in the analysis space of a new object (white ball).

Now, we can consider the classification of a new object, indicated in Figure 11.12 as a white ball.

Having formulated our prior probability, we are now ready to classify a new object X (small WHITE circle). Since the objects are well clustered, it is reasonable to assume that the more GREEN (or RED) objects in the vicinity of X, the more *likely* that the new cases belong to that particular color. To measure this likelihood, we can consider a region around X (depicted by the larger circle), which encompasses a number (to be chosen *a priori*) of points irrespective of their class labels. Then we calculate the number of points in the circle belonging to each class label. From this, we calculate the likelihood as follows:

Likelihood of X given GREEN α (# GREEN in region/Total # GREEN cases in region)
Likelihood of X given RED α (# RED in region/Total # RED cases in region)

With the preceding information, it is clear that Likelihood of X given GREEN is smaller than Likelihood of X given RED, since the circle encompasses 1 GREEN object and 3 RED ones. Thus,

Probability of X given GREEN α (1/40)
Probability of X given RED α (3/20)

Although evidence of the prior probabilities suggests that X may belong to GREEN (given that there are twice as many GREEN as RED), the evidence from analysis of the region around it (likelihood) suggests that the class membership of X is RED. In the Bayesian analysis, the final classification is produced by *combining* both sources of evidence (prior probability and the likelihood) to form a *joint posterior probability* following Bayes' rule (see Chapter 1).

Posterior probability of X being GREEN α (Prior probability X Likelihood)
= (4/6) × (1/40) = 1/60

Likewise,

Posterior probability of X being RED = (4/6) ×(3/20) = 1/10

As a result, we classify X as RED since its class membership achieves the largest posterior probability. This joint posterior probability is also known as the *conditional probability*.

Despite its simplicity, Naïve Bayesian can often outperform more sophisticated classification methods.

Naïve Bayesian classifiers assume that the predictor variables are independent in their effects on the classification. This assumption is rather naïve in the face of reality, hence the origin of the name. But in spite of this rather too-strong assumption, it performs rather well with many data sets. This classifier can accept any number of either continuous or categorical variables. In fact, the Naïve Bayesian classifier technique is particularly suited when the number of variables (the *dimensionality* of the inputs) is high. Although the assumption that the predictor variables are independent is not always accurate, it does simplify the classification task dramatically, since it allows the class conditional densities to be calculated separately for each variable; i.e., it reduces a multidimensional task to a number of one-dimensional ones. Furthermore, the assumption does not seem to greatly affect the posterior probabilities, especially in regions near decision boundaries, thus leaving the classification task unaffected. In effect, Naïve Bayesian classifiers reduce a high-dimensional density estimation task to a one-dimensional kernel density estimation. This kernel function can be modeled in several different ways including normal, lognormal, gamma, and Poisson density functions.

WHAT IS THE BEST ALGORITHM FOR CLASSIFICATION?

If you have the time to use all the algorithms discussed in the preceding sections to classify your data sets, you will find that the best algorithm to use to classify one of your data sets may not work well for other data sets. In other words, different algorithms work best for different data sets. Using a diversity of algorithms is best. A good example is provided in the results of a study of performance of 10 data mining algorithms by Kalousis et al. (2004). They compared algorithm performance on 80 data sets available from the UCI Machine Learning Repository (http://archive.ics.uci.edu/ml/). Each data set was characterized by a number of criteria:

- Error correlation between two classifiers using the data set (EC)

  $EC = P(i)[Covariance(X_1, X_2)]$, summed for $target(i) = 1$ and $target(i) = 0$,
  $$EC = \sum_{i=1}^{i=0} P(i)[\mathrm{cov}(error_1, error_2)]$$

  where $P(i) =$ the prior probability of each target class (proportion $= 1$ and 0), summed for cases where $target = 1$ and $target = 0$.

  All data sets were clustered on their matrices of error correlation and grouped into two classes: relatively low EC and relatively high EC.
- Log of the total number of cases/number of attributes
- Sum of the logs of the total number of cases/number of cases where $target = 0$ and total number of cases/number of cases where $target = 1$
- Log of the total number of cases/number of cases where $target = 1$

The data sets with high EC and low EC were further characterized by other variables, as shown in Table 11.3.

TABLE 11.3   Comparison of Data Set Groups with Relatively Low and High EC and Other Data Sets Criteria

Data Set Criteria	Low EC	High EC
# Target classes	High	Low
Target class distribution	Relatively balanced	Relatively unbalanced
Total number of cases	High	Low
Total number of attributes	High	Low
Average number per class	High	Low

Ten algorithms were used in the analysis of each group of data sets: C5.0 rules, C5.0 decision tree, C5.0 boosted tree, Clementine automated neural net, Clementine Radial Basis Function net, Naïve Bayes classifier, Nearest Neighbor, multivariate decision tree, linear discriminant analysis, and a specialized rule induction engine. Results showed that different algorithms performed differently on the two different data set groups. The relative performance of the algorithms was related to differences among the data sets, in terms of:

- Data availability (number of cases, number of attributes, etc.);
- Class distribution (imbalance between the occurrence of target = 1 and target = 0);
- Information content (uncertainty coefficient of attributes and classes).

The bottom line to keep in mind is that different algorithms perform differently on different data sets. At the beginning of the analysis, you don't know which algorithms among those available to you will do the best job on your data set. This is why it is a good idea to use multiple algorithms to model a single problem and use the predictions of each algorithm as votes, with majority ruling the final classification for a given case. This approach is called *bagging*, which we will explore in more detail in Chapter 13.

Now that we have introduced several classification techniques and discussed some of the challenges related to them, you could continue our journey through the data mining process using the Recipe Interface of *STATISTICA* Data Miner. This new interface conducts you through the complex task of building a data mining model, similar to the way Turbo-Tax Interview software option guides a user through building a tax return. This interview process builds tax return "model" that minimizes income tax to be paid by the user. In an analogous way, the *STATISTICA* Data Miner Recipe Interface builds a data mining model that minimizes prediction error among an ensemble of prediction algorithms.

POSTSCRIPT

In this chapter, we have introduced several classification techniques and discussed some of the challenges related to them. Now, we can consider another set of data mining applications, in which the target variable is not a set of categories, but rather is a continuous number. These numerical methods require very different algorithms to process data.

References

Brieman, L., Friedman, J. H., Olshen, R. A., & Stone, C. J. (1984). *Classification and Regression Trees*. Boca Raton, FL: Chapman & Hall.

Brieman, L. (2001). Random forests. *Machine Learning, 45*, 5–32.

Friedman, J. H. (1999). Stochastic gradient boosting. http://www-stat.stanford.edu/~jhf/ftp/stobst.ps

Kalousis, A., Gama, J., & Hilario, M. (2004). On data and algorithms: Understanding inductive performance. *Machine Learning, 54*, 275–312.

Kass, G. (1980). An exploratory technique for exploring large quantities of categorical data. *Applied Statistics, 29*, 119–127.

Numerical Prediction

PREAMBLE

Modern humans have always been fascinated with numbers. The Industrial Revolution was founded on Aristotelian logic and numerical relationships between movements and actions and the things that cause them. World War II was at the same time the most traumatic conflict in the history of the world and the impetus that drove us into the age of high technology. The single most important influence in modern technology is the exponential rate at which we have been able to "crunch" numbers with computers.

The personal computer, or PC, is arguably the single most influential technological force in the world today. The $\sim$3 GHz PCs of today are over 600 times faster than the original

4.77 MHz IBM PC in 1981. These PCs can execute about 100–200 million instructions per second, and all of them are numeric. The human brain has about a trillion cells, and each one has many connections with each other. Each one of the brain cells operates with numbers also (in terms of nerve impulse strengths). The "thinking processes" of both humans and computers involve numerical analysis. Thus, it is fair to say that numerical analysis is the most basic function in both the carbon-based human world and the silicon-based computer world.

It is not the intention of this book to teach you how to do numerical analysis or even to understand how modern analytical algorithms formalize it. So far, you have seen very few equations in this book; that motif will be relaxed slightly in this chapter. The equations presented provide convenient display objects to refer to in subsequent discussions. But the presentation of these equations does not constitute a formal definition of the algorithms discussed. Rather than present algorithms in a formal mathematical format, we seek to explain how to *use* these algorithms to solve problems, the most basic of which involve numerical prediction. Even the classification of things is based on numerical operations.

In this chapter, we will explore the concepts of *linear* and *nonlinear* relationships between a given response variable (that which is predicted) and those things that control it (predictor variables). Each type of relationship requires different analytical techniques to express it. For linear relationships, we will review the assumptions of the Parametric Model of reality (introduced in Chapter 1) and relate them to some classical methods of numerical prediction. It is not our purpose to present an exhaustive treatment of numerical prediction algorithms, but rather to discuss common examples found in most data mining tool packages. Also, we will revisit briefly several of the algorithms described for classification in Chapter 11 (C&RT, boosted trees, and neural nets) because they can be configured for use also in numerical prediction.

LINEAR RESPONSE ANALYSIS AND THE ASSUMPTIONS OF THE PARAMETRIC MODEL

The goal in linear analysis is to find a set of predictor variables (X_1 to X_n), in which changes in each predictor variable cause a change in the response variable (Y) as a multiple of the change in the predictor variable. This type of change is called a *geometric* progression. A geometric progression follows a straight line of increase when plotted on a graph (Figure 12.1).

The relationship shown in Figure 12.1 follows a simple geometric progression of increase; Y increases 3 units for each unit of increase in X. The defining elements of this relationship are the *slope* (3.0) and the *intercept* (where the trend line crosses the y-axis; zero in this case). Linear relationships were assumed by Sir R. A. Fisher in his parametric methods of analysis.

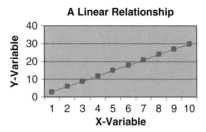

FIGURE 12.1   A typical linear relationship: $Y = 3X$.

PARAMETRIC STATISTICAL ANALYSIS

Parametric statistical analysis was introduced in Chapter 1, in terms of a number of assumptions that underlie it. It was Sir R. A. Fisher's purpose in proposing his radical methods in 1921 (see Chapter 1) to bring some consistency in the analysis of data in medical studies. He was concerned that different Bayesian medical researchers could come to different conclusions from the same experimental data because they each brought a different set of past experiences and knowledge to the study (and included it). He decided to restrict his analysis to only those relationships present in the experimental data in a given study. He reasoned that the two most important aspects of a variable in a data set are its central tendency and the distribution of data values around this point. He chose the average (mean) value as the measure of central tendency and the average difference (deviation) of each data value from the mean for that variable. Of course, he found that even if the data values were all positive, about half of the deviations were positive (for data values greater than the mean), and half of the values were negative (for data values less than the mean). Their arithmetic average was zero! But what he really wanted to express was the deviation itself, not its sign. So he just squared them (to remove the negative sign), added them up, and divided by the number of values. Fisher recognized that representing the data values by the mean eliminated details in the data, so he decided to follow a convention of subtracting 1 from the number of values for each mean calculated in his statistical methods that ensued. This subtraction increased the average deviation slightly to account for the increased uncertainty of the mean as a measure of the data values in the data set. The result was termed the *variance*. The square root of the variance was calculated to convert the variance value back to the original scale, which in turn yielded the *standard deviation*. The final formula for the standard deviation is

$$\sqrt{\frac{\sum (X - \overline{X})^2}{N - 1}} \tag{Eq. 1}$$

where X is a data point, N is the total number of data points, $\overline{X}$ is the mean, and Σ is the symbol for summing all of the squared differences from 1 to N.

The rest of Fisher's statistics (standard error, correlation coefficient, etc.) are just elaborations of these two *parameters* (mean and standard deviation). Hence, we see the origin of the term *parametric statistical analysis*. The next step was to derive a standard table of probability (the F-distribution) based on his definition of *likelihood*. The F-statistic was used to determine the significant difference between two data sets.

The final step for Fisher was to develop a scheme for analyzing the variance in his experimental data set to determine if there was a significant difference between one medical treatment in his experiment and another treatment (based on his tables of probability). This analysis of variance (called ANOVA) became the basis for his statistical conclusions about which treatments were *significantly* effective for each of several cases of a medical problem and which treatments were not. This is the basic pattern of analysis for all of Fisher's parametric statistical procedures. Granted, this is a very simplistic explanation of the elements of Fisher's landmark paper in 1921. But it serves to set the stage for evaluating how to apply his methods in other experimental areas.

ASSUMPTIONS OF THE PARAMETRIC MODEL

To achieve this analytical standardization, Fisher had to make a number of assumptions about the data he used. These assumptions were introduced in Chapter 1, and their relevance to numerical is discussed below.

The Assumption of Independency

Fisher was fortunate to be able to have a medical laboratory at his disposal, in which he could conduct very *controlled* experiments. It was necessary for all methodological and environmental effects be held constant, varying only the actual treatment of the patient. Other studies could examine the effects of varying methods or environmental conditions for a given medical treatment. For example, he could study the effects of room temperature on the activity of a drug used in a given treatment. To do this, he had to hold all other variables constant (including the treatment) and vary only the room temperature. Then he could do likewise for humidity and so forth. This experimental approach assured that the recorded effects of each variable were *independent* of each other. This variable independency is assumed in the mathematics Fisher followed to define standard deviation in Eq. 1.

The Assumption of Normality

An even more basic assumption than that of variable independence is the assumption of *normality*. Fisher's combination of deviations from the right of the mean with those from the left of the mean assumed that both sides of the distribution were similar. His probability tables also assumed such a distribution. The distribution this situation describes is called the *normal distribution,* shown in Figure 12.2.

The normal curve shown in Figure 12.2 is displayed with units on the x-axis graduated in terms of standard deviation units. The curve represents the frequency of values for any point along the x-axis. All of the area between the curve and the x-axis represents 100% of the values in this distribution. The area under the curve between $X = -1$ (one standard

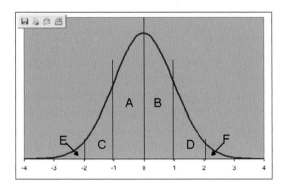

FIGURE 12.2 The standard normal (bell-shaped) curve. (*Source:* http://www.tushar-mehta.com/excel/charts/normal_distribution/)

deviation unit below the mean) and $X = +1$ (one standard deviation unit above the mean) represents about 68% of the area under the curve (areas A plus B). The area included by ±2 standard deviations is about 95% (areas A+B+C+D), and the area included by 3 units is about 99.5% (areas A+B+C+D+E+F). Fisher's mathematics assumes that the distribution of values in each variable of the data set follows a normal distribution around the mean value. If an analyst uses any classical parametric statistical procedure, he or she is making the assumptions of normality and independency, however unconsciously. Significant departures from a normal distribution can lead to spurious conclusions inferred from the application of normal parametric. Significant departures from independence (effects of some predictor variables are strongly related to each other) are even more problematic.

Fixes for Non-Normality

A common treatment of non-normal distributions is to transform the data with various utilities in the data mining or statistical tool package. Common transforms include

- Beta distribution
- Gamma distribution
- Binomial distribution

If the distribution of your data set fits a common distribution transform available in your tool package, you can create (in effect) a normal distribution from a non-normal distribution.

Normality and the Central Limit Theorem

The Central Limit Theorem states

A group of means of samples drawn from a nonrandom distribution is normally distributed.

Naïve practitioners often misinterpret this theorem. Careless reading of the theorem leads students to believe that all you have to do to get a normally distributed data set is to sample a non-normal population of data, and voila, the sample is normally distributed. That is not the case! The theorem states that the *group* of means taken from a non-normal distribution is normally distributed. That means if you take 100 samples of a non-normal distribution and calculate the mean for each of them, the distribution of the 100 mean values is normally distributed. This attribute of sampling can be applied when you take multiple samples of a population and submit the data to linear regression analysis. We will discuss this further in the following paragraphs.

Don't look for quick ways out of this assumption. If your data set distribution is significantly different from normal, it can ruin any analysis based on parametric statistical methods. And the most painful part of this problem is you may not know when you are wrong! Testing for a normal distribution is a recommended step and is available in most statistical and data mining programs. If you don't verify the normality of your data set, all the statistical tests may show a strong relationship in your model, but that relationship may fail miserably when you try to apply it.

We will discuss some fixes for nonindependency in the section titled "Linear Regression."

The Assumption of Linearity

The third major assumption inherent in classical parametric procedures is that the variables have a *linear* effect on the response variable (the target). This means that a plot of the relationship of any variable to the response variables is a straight line. Examples of common statistical procedures that make these three assumptions are ANOVA and linear regression. Many fixes exist for handling nonlinear variables in these linear analyses. We will look at them next.

LINEAR REGRESSION

Linear regression was first proposed by Sir Francis Galton (1822–1911). Galton coined the term *regression* to describe the observation that the majority of very tall fathers had sons who were shorter, and most very short fathers had sons taller than them. The trend of this progression in height was toward the average (or mean) height. This phenomenon was termed *regression to the mean*. His analysis of this effect became known simply as regression.

The major objectives of linear regression are to

- Determine if a relationship exists between one variable and another (or a set of others);
- Describe the nature of this relationship, if it exists;
- Quantify the accuracy of this relationship;
- Evaluate the relative contributions of each variable, if multiple variables are used.

Linear regression makes all three assumptions described previously. But you can appeal to the Central Limit Theorem to correct for non-normality by taking multiple samples and working on the means rather than the original data. That is, you can generate a group of data for a given variable by drawing a group of samples for the population, finding the mean, drawing another group of samples, finding the mean, and so forth until you have a group of means representing that variable. That group of means will be normally distributed. Engineering control charts rely on that property by making many samples of a process flow and creating this group of means. The control charts are based on the group of means, not the underlying data from which they were calculated.

The basic process of linear regression of a data set is to estimate parameters (coefficients) for each candidate predictor variable (X) to represent the effect that variable has on the response variable (Y). These effects are assumed to be linear and additive (that is, you add them up to get the total effect). The general form of the regression equation generated by the analysis is

$$Y = a + b_1X_1 + b_2X_2 + b_3X_3 + \ldots + b_nX_n \tag{Eq. 2}$$

In Eq. 2, the variable a is the Y value where $X = 0$ (the intercept), b_1 is the coefficient for X factor #1 (X_1), b_2 is the coefficient for X factor #2 (X_2), and so forth through the last X-variable (X_n). The coefficients reflect effects of two sources of relationship of the X-variable to the dependent variable (Y): the relative effect of X on Y, and the effect of differences in scale between X and Y. If variable Y and all of the X-variables have the same scale, then the coefficients reflect the relative predictability of each X-variable. But if the scales are different, the coefficients may not reflect much of the relative predictabilities. Usually, the scales of variables are quite different. One technique for overcoming the effects of scale is *standardization*. This is a process of transforming all variables to a common scale. One common standardization method (computing the z-score) is to subtract the mean from a value and divide by the standard deviation, which creates a scale in terms of standard deviation units. Another method used to create a common scale from 0 to 1.0 is to normalize the variable over its range as new_value = (old_value − minimum_value)/ (maximum_value − minimum_value).

Methods for Handling Variable Interactions in Linear Regression

The earliest approach to correct for variable interactions was to use factorial designs to separate "main" effects from "interaction effects." The combined interaction effect, termed as C, was added to the ANOVA analysis equations to "correct" for the interactions. Application of this approach to linear regression is not strictly appropriate because the calculation of the variable coefficients includes the effect of interactions with other variables. If an interaction between two variables is obvious, an additional variable can be derived as the product of the interacting variables. When these multiplicative terms are added to the regression equation, it may significantly increase the collinearity of the two variables.

Collinearity among Variables in a Linear Regression

When two variables are highly correlated to each other, the plots of these variables lie on nearly the same line. The total of all the collinearity between variable pairs is called *multicollinearity*. You can assess this effect by comparing the square of the sum of the simple correlation coefficients for all variables with the coefficient of determination (R^2). The R^2–value measures the combined effect of all the variables in explaining the variance in the dependent (response) variable. The Pearson simple correlation coefficients measure the degree of correlation between a single variable and the dependent variables. The degree to which the squared sum of the simple correlation coefficients exceeds the R^2–value is a measure of the amount of collinearity between the variables. Relatively high multicollinearity in a regression analysis will make it difficult if not impossible for the algorithm to find a single optimum solution. Because the interacting effects vary with the values in the variables, there may be several "optimum" solutions, depending on the relative frequencies of values among collinear variables. A good rule of thumb to follow in parametric statistical analysis is to eliminate one member of any pair of variables that is more than 80% correlated with the other. The other suggestion we can make is to limit the number of interaction variables to only those that are obvious.

The Concept of the Response Surface

Consider the problem of predicting one variable with two other variables. The plot of predicted points in a linear regression is a straight line (Figure 12.3).

This straight line is the best the algorithm can do to express the variation in the predicted values. The straight line shows the *response surface* of this linear regression. Another way to

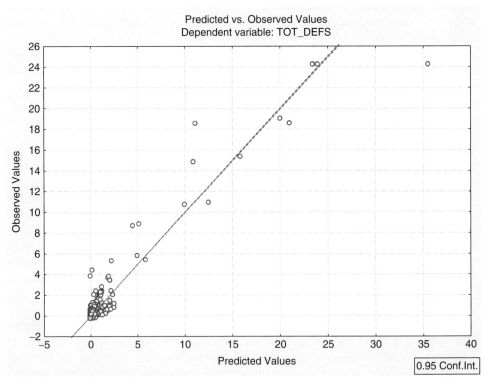

FIGURE 12.3   Plot of a two-factor regression.

look at the relationships among the variables is to look at a response surface in 3D space. Figure 12.4 shows the relationship as a plane, when a linear function is used to represent the data relationships.

Figure 12.5 shows the fit using a quadratic function rather than a linear function.

By fitting even more complicated functions to the three data values, you can conform the response surface even more closely to the data. The plot of the three-factor response surface using a negative exponential (an even more nonlinear function) is shown in Figure 12.6.

The departure of a data point from its predicted value is called the *residual*. One of the common reports of a multiple regression algorithm is a plot of the residuals versus the predicted values, as shown in Figure 12.7.

The ideal plot of residuals versus predicted values would be a long cluster parallel with the *x*-axis at $Y = 0$ (residual $= 0$). In Figure 12.6, you can see that most of the values are near the ideal. But there are several predictions that differ significantly from the raw data values (have a large residual). This plot is a visual form of model evaluation.

Another visual reflection of the strength of the model is the normal probability plot, shown in Figure 12.8.

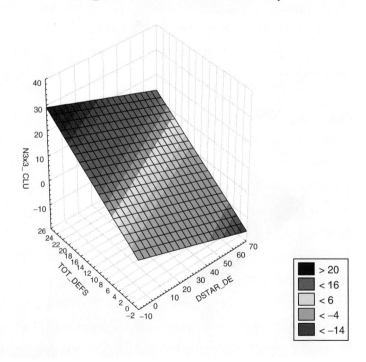

FIGURE 12.4   Fit of a linear three-factor response surface.

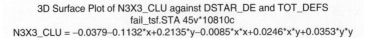

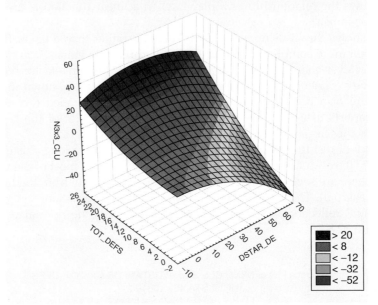

FIGURE 12.5 Plot of a quadratic fit of three variables.

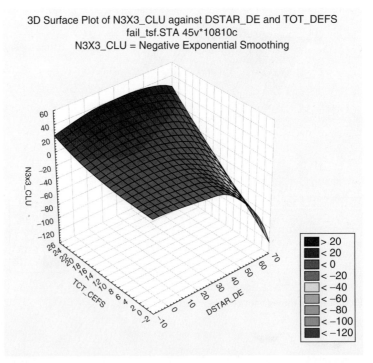

FIGURE 12.6 Plot of a negative exponential smoothing function applied to fit three variables.

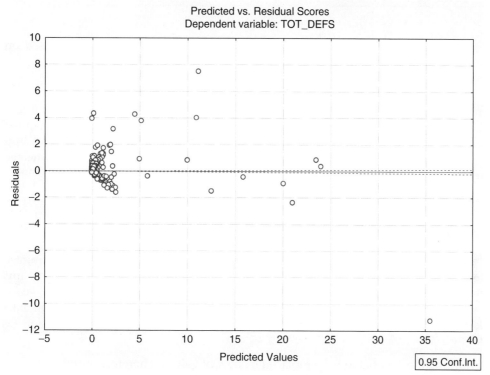

FIGURE 12.7   Plot of the residuals versus predicted values.

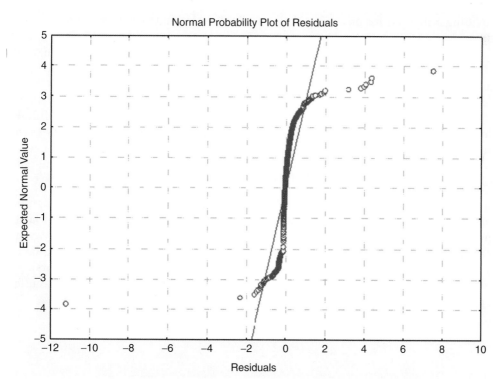

FIGURE 12.8   Normal probability plot of the three-factor regression.

The normal probability of a residual is the expected value of the residual based on the normal distribution, calculated as

$$P_i = i/(N + 1) \tag{Eq. 3}$$

where i is a residual value in a list, and N is the number in the list.

The ideal normal probability plot of residuals is a straight line. The red line on Figure 12.7 is the linear fit through the residual data points. This normal probability plot appears relatively close to normal, except for a few data points at the upper end. This means that the underlying regression that generated the residuals is likely to be valid.

GENERALIZED LINEAR MODELS (GLMs)

The multiple linear regression models suffered from a number of constraints and assumptions that could not be varied, lest the inferences from the model become invalid. Much work in the mathematical and statistical community over the past 20 years has provided a much more flexible framework for analyzing data in a regression context. This work has successively expanded and generalized the Multiple Linear Regression (MLR) model by

1. Permitting the Y-variable to be replaced by a *set* of values. Therefore, multiple dependent variables could be analyzed with the same solution techniques as single dependent variables;
2. Permitting each X-variable to be replaced by a *set* of values;
3. Providing a method for analyzing the sets of values by matrix algebra rather than standard arithmetic;
4. Providing for linear transforms of the Y matrix and the X matrix to perform high-order polynomial regressions;
5. Providing a number of methods for coding categorical predictors to accomplish the same goal as dummy variables, without increasing the number of variables;
6. Providing a method for overcoming the linear independency assumption of MLR, but allowing a solution of normal equations with a *generalized inverse* operation. A normal matrix inversion is very restrictive (you can do it only one way). But a generalized inverse operation can be done many ways, leading to many possible solutions;
7. Providing methods for handling redundant predictors.

The combination of these provisions of the GLM approach removes the most significant limitations of MLR. The matrix architecture of the analysis operations permits full-factorial designs ($N \times N$ variable analyses) to incorporate *all interactions* between predictor variables to be evaluated for effects on the dependent variable. These *interaction effects* can be combined with the *main effects* (attributable to the independent effects of the predictors).

A wide variety of experimental designs can be accommodated by GLMs. Examples of these designs include the following:

- Classical ANOVA designs can be used to assess the $N \times N$ effects of factorial experiments.

- ANCOVA (analysis of covariance) designs with both continuous and categorical predictors can be evaluated. ANCOVA designs permit the evaluation of the *significance* of the interactions in factorial designs.
- Mixed models of ANOVA and ANCOVA can be accommodated.

Many more experimental designs can be evaluated by GLM analysis, making it the most flexible parametric procedure to use, with the fewest assumptions. If you have a normally distributed data set, a GLM analysis may be the best way for you to go. Data mining and machine learning methods (see following sections) may be able to handle more complex response surfaces, but the techniques for evaluating models are far fewer than those available for parametric techniques. See Chapter 13 on model evaluation and enhancement for a discussion of some of these techniques.

METHODS FOR ANALYZING NONLINEAR RELATIONSHIPS

In a nonlinear relationship, the trend line of Y plotted against an X-variable is *not* a straight line, but rather it is a curved line, as shown in Figure 12.9.

Figure 12.9 shows the relationship with Y is not a multiple of X (as it was in the geometric progression), but according to the natural logarithm (Ln) of X. Notice that the slope of the plotted line is not constant; it can be evaluated only for a given point on the curved line. Most relationships in nature and in the business world are intrinsically nonlinear rather than linear in nature.

NONLINEAR REGRESSION AND ESTIMATION

It is possible to fit a nonlinear function simply by replacing some of the variables with polynomial terms, which include that variable. This approach is called polynomial (or curvilinear) regression. For example, if we replace predictor variable X with X^2, or some higher-order polynomial, the regression equation can account for nonlinear effects in X. This procedure is usually a trial-and-error process because it is difficult to know ahead of time which polynomial to use. A GLM model can be configured to do this easily. Contrary

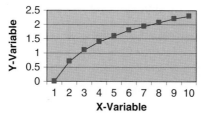

FIGURE 12.9   A plot of a nonlinear relationship ($Y = \mathrm{Ln}X$).

to what it appears to be, it still uses a linear function to fit the data, and we should not call this nonlinear regression. Techniques like this are referred to often as *intrinsically linear regression models.*

An intrinsically nonlinear regression model uses an arbitrary nonlinear function to replace one or more of the variables. This nonlinear function has no exact solution, but rather its parameters must be estimated. Hence, the better name for it is nonlinear estimation. These estimation procedures make a number of passes through the data set and minimize an error function along the way. Any one of a number of techniques can be used to find the "minimum" error (e.g., Least Squares, Maximum Likelihood, Quasi-Newton Method). There are a number of common nonlinear estimation techniques, which can be very useful in data mining applications. These techniques include

- Logistic (or Logit) Regression;
- Probit Regression;
- Poisson Regression;
- Piecewise Linear Regression.

Logit and Probit Regression

Logistic regression was introduced in Chapter 11 because it models binary outcomes that have only one of two possible values, which is a form of classification. Probit regression is similar to logit regression in that it too has only two possible outcomes, but there is a "fuzziness" associated with these outcomes. For example, many surveys use a multipoint scale to measure responses. A 5-point scale might be defined as follows: 5 = strongly agree; 4 = generally agree; 3 = neither agree nor disagree; 2 = generally disagree; and 1 = strongly disagree. Actually, this scale reflects a "feeling" about one of two possible outcomes: agree or disagree. The distribution of these responses can be transformed to reflect the appropriate area under a normal probability curve (assuming a normal distribution, of course), and they can be analyzed using the probit model in Equation 4:

$$NP(feeling) = NP(b_0 + b_1 * x_1 + b_2 * x_2 + \ldots) \tag{Eq. 4}$$

where NP is the normal probability, or space under the normal curve.

Poisson Regression

Poisson regression uses the Poisson distribution (rather than the normal distribution) to express data relationships. The Poisson distribution fits count data well, such as attendance counts on different days or for different events.

Exponential Distributions

The normal and Poisson distributions are types of exponential distributions because they include an exponential factor (representing a value with an exponent). These distributions can be classified according to two parameters: a *dispersion parameter* and an *index parameter.*

For a detailed discussion of these parameters and the distributions they express, see Jørgensen (1987). For our purpose here, we can classify a distribution by its index parameter p:

- $p = 0$ ... Normal distribution
- $p = 1$ ... Poisson distribution
- $p = 2$ ... Gamma distribution
- $p = 3$ ... Inverse Gaussian distribution

One of the common problems in data mining is modeling the occurrence of significant events. The significance of this occurrence is composed of two components: frequency of the event and severity of the event. Statistical analysis of these significant events requires the calculation of probabilities. The calculation of the probabilities for frequency follows the Poisson distribution, and that for severity follows a log of the normal distribution. You can calculate the probabilities easily enough according to the properties of each exponential distribution. The problem enters when you try to combine them. The calculated probabilities are *not* additive! You must find a way to combine inferences from the different probabilities. The most common application where this must be done is in the modeling of insurance credit risk.

There are three ways to model problems like insurance risk:

1. Model frequency and severity separately, using different algorithms, and then report on them separately.
2. Use a distribution, such as the Tweedie distribution (see Jørgensen, 1987), which has a p value of between 0 and 1. This is a compromise between the normal and Poisson distributions.
3. Use transform regression, a technique available in one data mining tool (IBM Intelligent Miner) to analyze a probability defined using elements of the mathematical expressions of both the normal and Poisson distribution (see Pednault, 2006).

Piecewise Linear Regression

Piecewise linear regression fits a linear regression on a number of portions of a nonlinear response curve. Piecewise linear regression carves up a nonlinear relationship into a number of linear ones. Consider the inverse logistic curve introduced in Chapter 11, with three linear functions fit to it (Figure 12.10).

Conceptually, we can see in Figure 12.10 how piecewise linear regression preprocesses the data. First, it determines appropriate breakpoints along the line at (a) and (b), and defines the three straight lines shown in the figure. Next, the algorithm fits an ordinary linear regression to each of the three lines and presents the results for the combined relationship. Essentially, this is the approach that C&RT follows to build a regression tree.

There are many other nonlinear estimation techniques available, although they are not included in this handbook. For more information on other common techniques, refer to Denison et al. (2003).

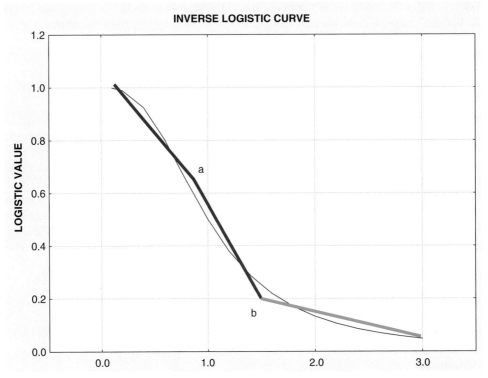

FIGURE 12.10   Piecewise linear segments expressing the nonlinear inverse logistic curve.

DATA MINING AND MACHINE LEARNING ALGORITHMS USED IN NUMERICAL PREDICTION

The most common machine learning algorithms used for predicting continuous response variables are

- C&RT;
- Neural nets;
- Decision trees;
- SVMs and other kernel techniques.

The material in the following sections may repeat some of the discussion in Chapters 6 and 7. We chose to accept this redundancy for the sake of coherency in this presentation. We hope that this restatement will aid your understanding, not bore you.

Numerical Prediction with C&RT

C&RT can be used for regression problems, as well as classification problems. In prediction, the continuous dependent (or response) variable Y is treated similarly to regression.

Predicted values are continuous numbers rather than categories. The continuous predictor variables are "binned"; that is, their ranges are divided into subranges using calculated split points. Each bin can participate in the formation of a number of if-then logical conditions. As was shown in Chapter 11, these if-then statements can be combined together to form a tree structure. The tree is grown along a particular branch using the Gini score as a split criterion (or some other metric) until the splitting process can't continue any further along that path because one of the stopping criteria was met.

The Tree Structure

A tree was built in *STATISTICA* Data Miner on an industrial failures data set, the first few nodes of which are shown in Figure 12.11.

The first variable (with the highest ranking) is split to separate those cases with values less than or equal to 9.86 and those with values greater than 9.86. Node 3 does not split any further because one of the stopping criteria has been met (the 15 cases in this node were less than the 1060 case minimum set in the tool). The SQL for the node is as follows:

```
/* Selecting cases related to Node 3 */
SELECT * FROM <TABLE>
WHERE ( ("RESP_DEF" > 9.86)
);
/* Assigning values related to Node 3 */
UPDATE <TABLE>
SET NODEID = 3, PREDVAL = 2.36, VARIVAL = 3.42)
WHERE ( ("RESP_DEF" > 9.86)
);
```

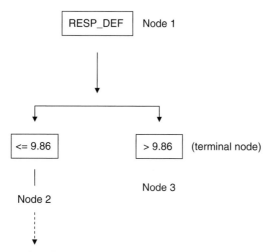

FIGURE 12.11   First three nodes of a decision tree, showing one terminal node.

From this SQL statement, a business rule can be induced: "If the value of the variable RESP_DEF is > 9.86, then assign the predicted value as 3.42" (compared to the observed value of 2.36). A similar (but more complex) business rule could be induced from this statement for terminal node 46.

The SQL assignment statement at Node 46 is as follows:

```
/* Assigning values related to Node 46 */
UPDATE <TABLE>
SET NODEID = 46, PREDVAL = −1.58286355732380e-001, VARIVAL = 1.86424905201287e-
   003
WHERE ( ("RESP_DEF" <= 9.85776807826836e+000)
And (DR3 <= 9.29624349632859e+000)
And ("PRE_L_DS1" > −7.37056366674857e-001)
And ("RESP_DEF" <= 6.67860053006884e-002)
And ("PRE_L_DS1" > 3.83034634488327e-002)
And ("RESP_DEF" <= −4.28291478552872e-002)
And ("PRE_L_DS1" > 3.63470956904821e-001)
And ("RESP_DEF" > −6.63181092458534e-002)
```

(*Note:* Values are not rounded in this example.)

Model Results Available in C&RT

Most of the report tables and charts are available from data mining tool packages providing C&RT. These features are described next.

Variable Importance Tables

The variable importance table will give you overall expression of the importance of a variable among all the splits in the tree (Table 12.1). The variable with the highest

TABLE 12.1   Variable Importance Table Generated by C&RT

Predictor importance 1 (fail_tsf.STA) Dependent variable: TOT_DEFS Options: Continuous response, Tree number 1

	Variable – rank	Importance
DR3	100	1.000000
PF_DS	93	0.930011
PF_AOL	93	0.928640
RESP_DEF	90	0.899691
PRE_L_DS1	89	0.886784
RESP_AVE	84	0.842368
PF_SR	78	0.783173
DR2	49	0.493977
PF_IC	44	0.438273
PF_PRE	41	0.409875

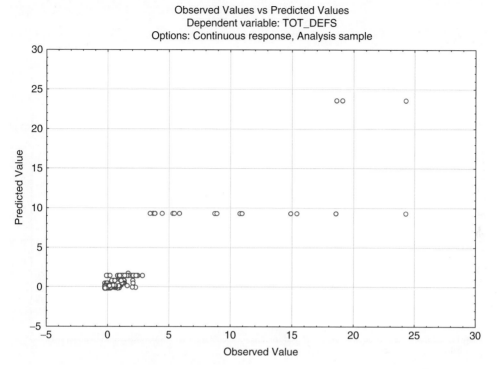

FIGURE 12.12   Observed versus predicted values.

importance value (DR3) is not the variable used for the first split. It draws its importance from its participation in many splits in the tree.

Observed Versus Predicted Plots

Figure 12.12 shows the observed versus the predicted values.

Normal Probability Plots of the Residuals

In Figure 12.13, the normal probability plot of the residuals shows that the large majority of the residuals are well behaved; that is, they fall near a straight line. Some cases shown on the lower left and upper right of the plot are anomalous, but in general the plot suggests that the model is valid.

ADVANTAGES OF CLASSIFICATION AND REGRESSION TREES (C&RT) METHODS

Following this approach, C&RT can produce accurate predictions based on a number of if-then conditions, and the results of the model have many advantages over many alternative techniques.

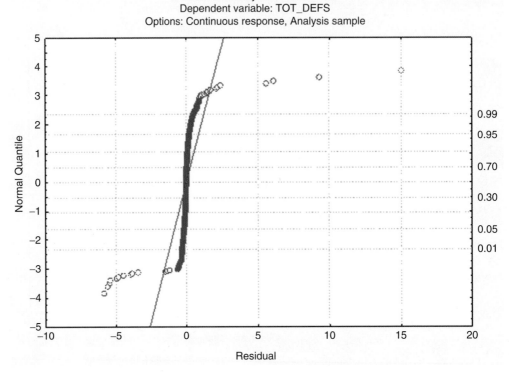

FIGURE 12.13 Normal probability plot.

Comprehensibility of the results. The simple architecture of the process permits rapid development of predicted values. This approach is much faster than calculating matrices and performing mathematical operations on all possible combinations of input variables. The decision tree process used in C&RT (and other decision tree algorithms) follows a winnowing process to separate the important predictors from the unimportant ones. Not only is the computation simpler, but the models are often simpler as well. But the most powerful feature of decision trees is the ease of understanding of the models by business people, particularly management. Models will not be accepted until managers understand them *in terms of their own business concepts.* Many data mining models have languished on the shelf because management did not understand them enough to trust them.

Handling of missing values. Most C&RT methods will handle missing values by suggesting a surrogate split to use in the case of missing values. This feature is nice in many applications with missing data.

Decision tree methods do not make parametric statistical assumptions. Predictions can be presented in a few logical if-then conditions at the terminal nodes. No implicit assumptions are made of a normal data distribution, or linear relationships among the variables and the response variable. Decision tree methods are well suited for data mining tasks, where the analyst does not know ahead of time which variables are important predictors. Thus, decision tree methods may uncover relationships and express them in a few decision rules, which might be masked by other more computationally intensive methods.

General Issues Related to C&RT

Multilevel splits. When there is one obvious split point in the range of a variable, C&RT does well. But if there are potentially multiple split points, the binary splitting approach may oversimplify relationships between variables. See Briemann et al. (1984) for more details and challenges of determining the best binary split point. Also, an excellent discussion of both decision trees and neural nets in general is provided in Ripley (1996).

The danger of overfitting. Theoretically, a decision could keep on splitting until it creates terminal nodes for every case. In that case, the tree will keep splitting until not only the signal pattern is modeled, but also the noise in the data is modeled perfectly. The prediction accuracy would be perfect, but the model would probably fail miserably on other data sets (the generality is low). The challenge in building a useful tree is determining when to stop splitting, thus creating a less predictive model that is more general. This issue is an expression of the general machine learning tendency to overtrain an algorithm.

The easiest way to address this issue is to impose one or more stopping rules on the training process. Common stopping methods are

- Less than a minimum number of cases is included in the split;
- Maximum number of terminal nodes (leaves) has been reached.

After the tree building has been stopped, many algorithms begin "pruning back" the tree, by iteratively evaluating the "sensitivity" of the solution of the elimination of variables one at a time. The goal in pruning decision trees is to find the simplest model within a specified range of the highest accuracy, one that is equally as accurate (or nearly so) in predicting new cases.

Model testing. Many decision tree algorithms provide an option to split the input data stream into a training set and a testing set. If this option is enabled (and we strongly suggest that you do so), the algorithm iterative builds a number of candidate trees with the training set and tests each tree against the testing data set to measure the generality of the prediction. Then the algorithm can choose which tree has reasonable accuracy and good generality. If this facility is not available in your decision tree algorithm, you should split the trees outside the tool and test each candidate tree manually.

Resampling. If the form of testing described in the preceding paragraph appears to be a good idea, then you can understand the value in doing it many times on different random samples of the data set. The variation in the predictions among trees built on different resampled data sets is an expression of the model error, or the error that is due not to the noise in the data but rather is caused by the effect of sampling a particular set of cases. One random sample of cases may have a significantly different "view" of the response signal than another sample set. Various resampling methods will be discussed in Chapter 13.

Large trees are problematical. Decision trees built on complex patterns in large data sets can become quite large (unless controlled by a maximum-size stopping function). With modern computers, this size is not so much a computation problem as it is a comprehensibility problem. Complex trees are difficult to present to the "consumers" of the project results.

APPLICATION TO MIXED MODELS

C&RT is useful for analyzing both categorical and continuous predictor variables. But it is also quite flexible in analyzing multiple response variables in full-factorial experiments. Some data mining tool packages provide for coded ANCOVA designs to separate the main effects from the interaction effects, similar to a GLM algorithm.

NEURAL NETS FOR PREDICTION

The operation of neural nets for classification was introduced in Chapter 11. Operation of this algorithm for prediction is very similar, except the prediction is not converted to a category at the end. Older algorithms require you to set a number of parameters. Usually, the parameters have default values, but you can modify them. In Chapter 11, we introduced the parameters of learning rate and momentum. In addition, you can change the network architecture (the number of middle, or "hidden," layers and the number of neurons to be used in each hidden layer). Finally, you may be able to modify the rate at which the learning rate degrades between iterations of the model, permitting a more thorough search over the response surface to find the solution with the lowest (global) minimum error. Therefore, modeling with neural nets is much more of an art than a science. Of course, academics and researchers will twiddle with these settings to optimize a given behavior of the neural net. But the business user will be quite happy to use the default settings most of the time. Why? The reason is that neural nets can produce a model for classification or prediction with default settings that are among the best models possible, and they can do it rather quickly.

Manual or Automated Operation?

Neural net implementations in several common data mining packages provide an automatic operation to select the optimum network architecture (e.g., SPSS Clementine, SAS-EM, and *STATISTICA* Data Miner). This optimization of network architecture is a huge benefit to the data mining practitioner. Algorithm implementations of this sort permit the user to spend less time on configuring the algorithms and spend more time on model enhancements (see Chapter 13). The *STATISTICA* Data Miner Recipe interface will train models using multiple algorithms automatically, permitting the data miner to "view" patterns in the data from several mathematical perspectives. This "synoptic" view of data patterns is a powerful means to capture all aspects of the response signal in the model results.

Structuring the Network for Manual Operation

In this context, manual operation means that you set the parameters for the algorithm's function yourself. Given a learning rate and momentum, the operation of a neural net is controlled by the number of layers and the number of nodes (processing elements) in each layer of the network. There is no best architecture for any particular application.

There are only general rules of thumb developed by practitioners over time that may be used to guide users.

Rule One: As the complexity increases in the relationship between the predictor variables and the response variables (*Y*), the number of the processing elements in the middle layer should also increase.

Rule Two: If the response pattern being modeled is highly nonlinear, then one or two middle layers may be necessary to capture the nonlinear relationships. If the response pattern is not particularly nonlinear, the additional layers lead to overtraining. For example, phone call duration data conform well to an inverse logistic curve and can be modeled successfully with a logistic regression or a two-layer neural net with a logistic activation function (the equivalent of a logistic regression). If you add a middle layer, you have to be careful not to overtrain the model. Addition of a second middle layer will usually degrade the performance of the model on the validation data set.

Rule Three: The number of nodes to put in the middle layer(s) should be no more than 1/5 to 1/10 of the number of cases available in the training data set. A factor closer to 1/5 should be chosen as a maximum for data sets with more complex patterns, and a factor closer to 1/10 should be selected for data sets with simpler patterns. Too many nodes in the middle layer will increase the likelihood of overtraining and generate models with relatively low generality. Too few nodes in the middle layer may not be able to capture the nonlinear patterns in the data.

Modern Neural Nets Are "Gray Boxes"

It used to be said that neural nets were "black boxes"; that is, they did not provide much information on *how* the solution was created. Early neural net algorithms yielded predictions or classifications but no measures of variable importance or error. Modern neural net algorithms provide a measure of variable importance (see Chapter 11) that opens up the black box to permit the user to see some reflections of the operational details. Also, data mining packages can add model evaluation tools to neural net outputs, such as

- Coincidence matrices (for classification only);
- Lift charts;
- Observed versus predicted charts;
- Residual plots;
- Metrics of prediction accuracy (more on this subject in Chapter 13).

Example of Automated Neural Net Results

STATISTICA Data Miner provides a very powerful automated neural net (SANN) algorithm. Try this algorithm (available on the CD-DVD) on any prediction or classification problem. The results from an SANN run on the Failures data set generated the report shown in Table 12.2.

The SANN algorithm performs an 80:20 randomized split of the data set, trains on the 80% portion, and tests the models on the 20% portion. Notice that SANN trained five neural nets, each with 10 predictors, and one output (prediction), and numbers of nodes

TABLE 12.2   Summary Report of the Five Best Networks Generated by the SANN Algorithm

Summary of Active Networks (fail_tsf.STA)

Index	Net. Name	Train Perf.	Test Perf.	Training Error	Test Error	Training Algorithm	Error Function	Hidden Activation	Output Activation
1	MLP 10–10–1	0.978341	0.968836	0.000039	0.000036	BFGS 23	SOS	Exponential	Exponential
2	MLP 10–5–1	0.971171	0.967780	0.000056	0.000035	BFGS 20	SOS	Tanh	Tanh
3	MLP 10–7–1	0.979169	0.967803	0.000037	0.000036	BFGS 32	SOS	Tanh	Identity
4	MLP 10–11–1	0.973139	0.967721	0.000052	0.000035	BFGS 13	SOS	Identity	Tanh
5	MLP 10–4–1	0.976285	0.967110	0.000042	0.000036	BFGS 31	SOS	Exponential	Identity

in the middle layer varying from 4 to 11 (see the Net. Name column). The term *MLP* stands for *multilayer perceptron*, another name for a neural net. The five nets also varied in the activation functions used to pass data through a node. Data accumulate (according to the accumulation function) and pass to the next node via a "firing" (activation) function. Different activation functions will generate slightly different solutions to the model for a given case. The model judged best by the algorithm report is model #1. An alternative approach would be to define the best model as the one with the highest evaluation of the ratio between the performance on the testing set (Test perf.) and the lowest testing error (Test error). If we use the second approach, model #2 has the highest score of 27,651, compared to 26,912 for model #1.

SUPPORT VECTOR MACHINES (SVMs) AND OTHER KERNEL LEARNING ALGORITHMS

The automated operation of some neural nets is eclipsed by the almost total automation of an SVM. Early SVM algorithms (e.g., SVM-light) required that inputs be scaled from −1 to +1. Modern SVM algorithms include some data preprocessing routines that standardize the inputs properly for the algorithm. The SVM algorithm in *STATISTICA* Data Miner is a good example of this automated operation (see Table 12.3).

The correlation value of 0.96275 is just the Pearson Product-Moment (simple) correlation coefficient between the observed and predicted values in the testing data set (25% hold-out sample). The evaluation of the other numbers in Table 12.3 would have meaning only when comparing two SVM models using different settings. The kernel function for this model runs as a Radial Basis Function (RBF). Other kernels available are linear, polynomial, and sigmoid. Some output plots are provided by this SVM. Figure 12.14 shows the plot of observed versus predicted values for this model.

TABLE 12.3   Performance and Error Report of the *STATISTICA* Data Miner SVM

Regression summary (Support Vector Machine), Test sample (fail_tsf.STA) SVM: Regression type 1 (C=10.000, epsilon=0.100), Kernel: Radial Basis Function (gamma=0.100) Number of support vectors= 25 (7 bounded)

	TOT_DEFS
Observed mean	−0.00999
Predictions mean	1.01321
Observed S.D.	0.98360
Predictions S.D.	0.77431
Sum of squared error	1.14743
Error mean	−1.02320
Error S.D.	0.31708
Abs. error mean	1.04233
S.D. ratio	0.32236
Correlation	0.96275

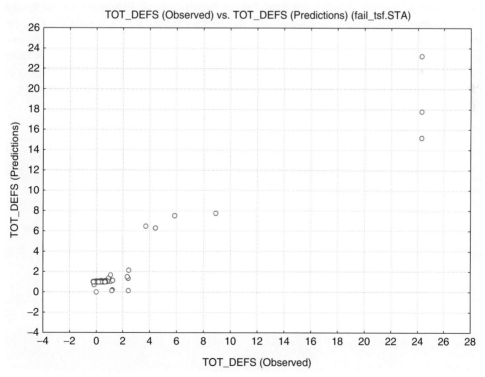

FIGURE 12.14   Observed versus predicted values for an SVM.

Compare the plot in Figure 12.14 with that of Figure 12.12 (generated by C&RT). The distribution of predicted values on Figure 12.14 falls closer to a straight line (the ideal) than do the values in Figure 12.12. The correlation coefficient of 0.96+ for the SVM is slightly better than that of 0.92 for C&RT. Does this prove that the SVM model is better than the C&RT model? No, the answer is not as simple as that. We must do more work to determine that.

To evaluate models, we must look at a number of other factors. Neither the C&RT model nor the SVM model included any assessment of the model error. We can assess this model error by performing a *V*-fold cross-validation operation in the models. We will discuss this method of resampling further in Chapter 13. If we had included the cross-validation operation in the modeling process, we might have seen different models selected as the best model for each algorithm. For now, suffice it to say that we can make no judgment of which model is best among those generated by SANN, C&RT, or the SVM.

POSTSCRIPT

The introduction to classification problems in Chapter 11 and to numerical prediction problems in this chapter provides a foundation for designing the appropriate analytical approach to most prediction problems you might face. Chapter 13 builds on this framework to show you how to evaluate and refine models after they have been trained. We place this chapter here in the sequence because it pertains to all models you will create in any of the general application areas discussed afterward. In addition, this information will help you understand and proceed through the tutorials in Part III of this book.

References

Brieman, L., Friedman, J. H., Olshen, R. A., & Stone, C. J. (1984). *Classification and Regression Trees*. Boca Raton, FL: Chapman & Hall.

Denison, D., Hansen, M., Holmes, C., Mallick, B., & Yu, B. (Eds.). (2003). *Nonlinear Estimation and Classification*. New York: Springer.

Jørgensen, B. (1987). Exponential dispersion models (with discussion). *J. R. Stat. Soc. Ser. B Stat. Methodology, 49,* 127–162.

Pednault, E. (2006). Transform regression and the Kolmogorov superposition theorem. *Proc. 6th SIAM Int. Conf. on Data Mining*. Society for Industrial and Applied Mathematics, 35–46.

Ripley, B. (1996). *Pattern Recognition and Neural Nets*. Cambridge, UK: Cambridge University Press.

Model Evaluation and Enhancement

PREAMBLE

One of the most common questions asked by beginning data miners is "How do I know when my model is any good?" This chapter will introduce you to a number of model metrics that you can use to measure the "goodness" of your model. But this process of modeling and evaluation is not a linear process; it is *iterative*. It is very rare that the best model will be trained initially. Often, the evaluation process will point out some issues that can be resolved by making some changes in the data preparation or modeling process. These changes may help to enhance the predictability of the model. Thus, the complete sequence of operations is

Modeling $\longrightarrow$ Evaluation $\longrightarrow$ Enhancement

This process may consist of many iterations; thus, we must view them as a single integrated operation. That is why the activities of model evaluation and enhancement are included in the same chapter.

INTRODUCTION

In this chapter, we will explain ways to tell how well your model is doing and then give a checklist of actions you can employ to improve its performance. Using a reliable technique for model assessment is essential. With that, you can search through the vast space of possible data transformations and model options to find the best model and have confidence in its out-of-sample performance. We'll explain some useful principles and shortcuts but emphasize that no method is as valuable as careful resampling (also known as cross-validation, leave-one-out, bootstrap, and jackknife; see Lachenbruch and Mickey, 1968; Lachenbruch and Goldstein, 1979; Efron, 1982; Rothpearl, 2008) that assesses the modeling algorithm over multiple splits of the data.

Before the checklist, we'll recap clustering and the five most popular algorithms for prediction and classification: linear regression, linear discriminant analysis, decision trees, neural networks, and nearest neighbors. These and other algorithms can be split into two broad groups, defined by what they do with the training data:

- **Consensus:** The data are used to create a model summarizing its information and then are thrown away (e.g., regression).
- **Contributory:** The data (or some of the data) are retained and potentially employed to evaluate a new case (e.g., nearest neighbors).

A key difference is that contributory methods require severe reduction of the number of dimensions to keep the input space as densely populated as possible. (*In high-dimensional space, no other point is really nearby.*) Consensus methods are much more tolerant of large numbers of features. However, the contributory methods (overlapping with what are normally called *nonparametric* methods) can be more flexible and adaptable to unusual situations, allowing their complexity to be contained in the retained data, rather than an equation.

Experience with many projects and software tools has revealed that most major techniques have strengths that can contribute to a solution, and that it's rare for one algorithm to dominate even one other in all properties. It is best to fit multiple models—especially ones of different types. Not only can insights be "fused" from multiple diverse models, but their estimates can be combined to (very often) yield an incremental improvement in performance, as shown near the end of the chapter.

Lastly, we end with some advice for the survival and success of the engagement of which the model is a key part—that is, with some practical words about consulting with clients (internal or external) and for developing your career skills as a data mining analyst. You can think of this as *modeler* evaluation and enhancement.

MODEL EVALUATION

How accurate is your model? The first clue is the training performance—how it does on the cases it is shown during training using the variables available then. If this accuracy is too low, the model is "underfit," and you must work hard to find useful relationships.

But the greatest danger in data mining is the opposite problem—overfit—where you fit the noise, as well as the signal, of the data. Overfit models look good while training but fall apart on evaluation when used on new data. Often analysts have brought a model all the way to implementation, feeling quite proud of their accomplishment, only to have reality hit at such a late stage that great damage is done. This is why overfit is so dangerous; the stakes are often high when it is discovered.

Splitting Data

The essential first step in any modeling task is to split off an evaluation set.[1] This should come before *any* other step (with the exception of examining the data just enough to stratify the training sample, if the minority cases are rare enough). Often, researchers don't do this early enough, so the lessons they learn—about which variables are useful, what the valid ranges of the data are, what outliers look like, etc.—are all tainted by looking at *all* the data, and the evaluation results aren't truly out-of-sample. It is essential that the data a model is presented on evaluation are completely new to the model (and the process that generated the model, including your own examinations). Only then is it a true evaluation.

Training error is obviously not our metric to optimize; if it were most important, a lookup table would be the ideal algorithm ("Let's see, case 17, . . . the answer is . . ."). For a model to be used, going forward, it needs to generalize to new data. Recall learning material from a textbook: some problems have answers (output labels) at the back (the training data) and similar but new problems on the test (evaluation). To do well on the test, a good learner induces the lessons (relationship between the questions and the answers) from the specific cases where the label (answer) is known and then can handle questions of the same type in the future.

But how do you split the data? As for proportion, it is usual to set aside 20–30% for evaluation. There is a trade-off; you want as much data as possible to drive learning, but also enough to comprehensively test the model. The practical consensus is to put more data on the training than evaluation side. As for case selection, the default method is to randomly split the data. It is too dangerous to split it by the order in the file, as there may be some information embedded in how the data was assembled. You may end up training on the credit applications that first arrived, for instance, and testing on those that last arrived, to find that they have very different behavior. The exception is if the model itself is time-based, e.g., as in stock-market investing. Then the best (toughest) metric is to train on the oldest data and evaluate on the newest. This reflects how the real world will challenge the model. Otherwise, you might not notice how vulnerable the model is to sudden

---

[1] Some like to call it a "test" set, but, as you often have to name files or variables to keep track of data, it's very useful to label the out-of-sample (evaluation) data with a different first letter than the in-sample (training) data, e.g., Tdata and Edata, etc.

changes in tax law, inherent market volatility, political regime, etc.[2] and side-step a problem in testing that you wouldn't be able to do in practice.

In classification, there is often one rare and important class, such as fraud, or cancer, or a prospect who actually becomes a customer. When you are splitting the data, it is important to have these influential cases fairly represented in each set—a process called "stratifying the sample." If the overall response rate to a mailing is 1%, for example, then separately split the 1 and 0 response cases (randomly) so that 1% of training and 1% of evaluation is a responder. Sometimes it can be important to stratify the input variables as well. For instance, political pollsters can do very well at extrapolating the preferences of about 1000 interviewees to gauge the sentiment of 120 million voters if they pay careful attention to categorizing whom they're interviewing (male/female, old/young, Black/White/Hispanic/Asian, land-line/cell-phone, past/new voter, etc.) and scale up their responses according to the estimated proportion in the full population. So, when stratifying on multiple variables, do the following:

1. Divide the data into sets according to the stratification variables. (If one group is too small, perhaps drop a variable from the stratification set.)
2. Randomly remove cases from each set according to your evaluation proportion.
3. Join the training cases of each group together to form your training sample. All other cases are your evaluation sample.

Avoiding Overfit Through Complexity Regularization

As you add terms to a model, the training error will go down (or at least stay flat). But, with more terms, the evaluation error will eventually go up, as shown in Figure 13.1.

There, the evaluation error tracked training well until a dramatic explosion upward on the ninth point (an 8th-order polynomial) where the model failed miserably. The data points and the estimation curves of the models of increasing order and flexibility are shown in Figure 13.2.

The data points (shown as diamonds in the figure) roughly rise from left to right, and this is reflected in the simplest model—a line. As you add more and more powers of X to the candidate input pool, the models get more flexible, until they eventually can exactly fit all the data points, when the number of coefficients in the model equals the number of constraints (data points) in the data. The final model is the most excessive in its

---

[2] This chapter is being written (Fall 2008) as the stock market goes through gut-wrenching drops (second in magnitude only to the Great Depression) with accompanying multiples of previously normal volatility, and when a change in political administration portends dramatic shifts in tax and spending policy. If a model can't weather such a change, it should at least signal "don't know"—i.e., that it is being asked a question that it's not qualified to respond to (hasn't seen that value or combination of values before in its inputs). See Mistake 8 in Chapter 20.

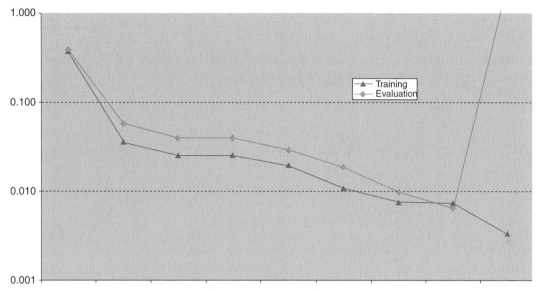

FIGURE 13.1 Training and evaluation error versus order (highest power) of a polynomial linear regression model.

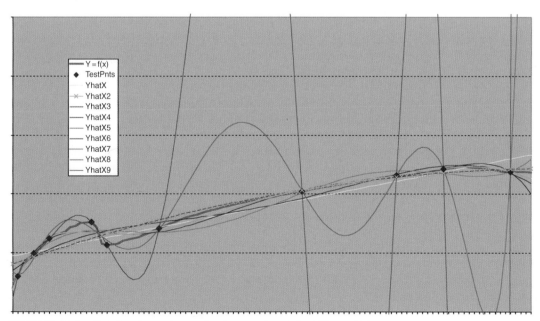

FIGURE 13.2 Training data points and estimates of models of Figure 13.1.

estimates (in fact, its error is off the scale in Figure 13.1) and the second-to-last model is also unsuitable for prediction—especially extrapolating outside the bounds of the data, but even interpolating between known points.

Statisticians have long known of this relationship between complexity and accuracy, and one way to avoid overfit is to regulate the complexity of the model. This is done in four main ways:

1. Reserving data and stopping when the evaluation performance starts to worsen;
2. Penalizing complexity according to number of parameters of the model;
3. Penalizing the "roughness" (e.g., integrated second derivative) of the estimation surface;
4. Penalizing the total magnitude of the estimated parameters, e.g., the sum of the absolute values of the regression weights.

In methods 2–4 shown here, the analyst minimizes a weighted sum of error and the penalized quantity to crown the "best" model.

Method 1, reserving data, will get our greatest attention in this chapter and book. Method 3, penalizing roughness, is rarely used but is more precise in its goals than Method 2, penalizing number of terms (which is very common), as exemplified by criteria such as Cp (Mallows, 1973), Akaike's information criterion (AIC; Akaike, 1973), and Minimum Description Length (MDL; Rissanen, 1978; Barron, 1985; and explained in Elder and Finn, 1991). These methods essentially boil down to minimizing the formula

$$error + \theta^* K$$

given the number of model terms, K, and complexity trade-off parameter, θ. The different criteria provide theoretical values for the latter, but most engineers (like one of the authors here) tend to experiment with that simple knob.

This approach (complexity penalty) does allow you to use all the data for training a model and can be used very effectively to protect against overfit. But there are several problems with defining complexity as the number of parameters, as is explained in Chapter 18, which details how to measure true complexity.

Method 4, parameter shrinkage, is exemplified by *Ridge Regression* (Hoerl and Kennard, 1970), where some "mass" is added to the diagonal values (the ridge) of the matrix to be inverted for linear regression, causing the overall parameters to be less extreme. This trick has been discovered also in other contexts, such as "optimal brain surgery" in neural networks (where weight magnitudes are penalized to keep them lower and under better control) and in Bayesian methods such as shrinking, where a prior parameter estimate of zero is combined with the measured parameter values to draw the final values lower. We are big fans of such Bayesian ideas but find the proper use of reserved data (see cross-validation in "Cross-Validation to Estimate Error Rate and Its Confidence") to be the most useful method of all.

Error Metric: Estimation

For estimation, the metric almost always used is minimum squared error (MSE). You set parameters to minimize the sum of the squared difference between the true and estimated values[3] over all cases:

$$\Sigma (truth - estimate)^2$$

While this is the best metric theoretically when the errors are normally distributed, the real reason for its use is its extreme mathematical tractability. With linear regression, for instance, you can near-instantaneously find the ideal joint set of weights for a large number of inputs in a single fast matrix inversion step. If you used another metric, it would take vastly longer. For instance, least absolute error

$$\Sigma \, |truth - estimate|$$

can be minimized, it can be shown, by a linear program. This is guaranteed to converge to an optimal value but takes several iterative steps. A more custom metric (such as dollars saved) requires a global search algorithm, which can take orders of magnitude longer and is not guaranteed to converge to the best possible value under the constraints of the parameters and data (unlike with squared error and absolute error).

However, even with its speed, MSE is rarely the preferred metric to employ. Its implicit normal distribution means that extreme errors are expected to be very unlikely. Thus, when an outlier occurs, it has a huge influence on the model selected. Also, the errors are nonlinear. A profit loss of $2 is not often four times worse, to the end user, than a loss of $1, but that's what MSE "believes." Even more important, one direction of error (e.g., positive: true value > estimated) could be a good thing—e.g., reflecting profits made—yet be counted as just as bad as its opposite. A custom error metric can better reflect the true trade-offs in a model's estimates. We have found that it is often worth the effort, on very challenging problems, to design a custom error metric and employ a global search algorithm (such as random search or GROPE; Elder, 1993) to minimize that metric over your model. This effort can often yield the key incremental improvement in performance that makes a model worthwhile where none was before.[4]

Error Metric: Classification

Percent correct (PC) is the default performance metric reported for classification but is rarely the best metric to use. The reason is that errors of one type (e.g., false dismissal) are often much more costly than errors of the opposite type (false alarm); yet PC counts them as equal. You must assess the relative costs of different kinds of errors to get the best balance. If the software tool allows, you can efficiently record this information in a cost

---

[3] Always define error as *true – estimate*. Satellites, for instance, have spun out of control because two groups used opposite definitions. It doesn't matter to calculate MSE, but does when you take corrective action due to the implemented model.

[4] See Mistake 3 in Chapter 20 for an example.

TABLE 13.1   Sample *Cost Matrix* for Credit Scoring Problem

Cost	Predicted Default	Predicted Good
True Default	0	7
True Good	1	0

matrix, as shown in Table 13.1. [A 2×2 matrix is shown, but the same concept works for K classes, where a $K×K$ matrix would require $K*(K-1)$ off-diagonal relative misclassification costs to be estimated and plugged in.]

In the credit scoring application of Table 13.1, analysts have estimated that a default is seven times worse than a miss. That is, giving credit to an applicant who will eventually default swallows the profit made from seven good customers. (In reality, risk is along a continuum, but to make it a two-class problem, default has to be defined to be something concrete like "more than 90 days late at least once in a 2-year period.") The diagonals have no cost, as that is where the predictions are correct. Note that the costs can be actual amounts or can be normalized so the smallest cost is 1.0; it is only their relative magnitude that matters. (Note also, some software tools require negative numbers; some positive.) Now, the algorithm can find the minimum cost rather than the maximum PC solution. This will be much more useful.[5]

The results can also be reported in a matrix, known as a confusion matrix, as shown in Table 13.2.

There, 10% of the data cases are defaults, but 19% are predicted as defaults. This is the result of defaults being more costly to miss, so there is a preference for false alarms over false dismissals. Overall, 15% are errors (3% false dismissals with a relative cost of 7 each, and 12% false alarms with a cost of 1 each), for a total cost of 0.33%. [Note that the overall cost is just the dot product of the cost and confusion matrices, and that it is a relative cost here, so only the relative (to total) population needs to be tracked in the confusion matrix.]

If the cost matrix (for a two-class problem) is hard to estimate and only a range of likely relative costs is known, you can refer to a Receiver Operating Characteristic (ROC) curve of

TABLE 13.2   Sample *Confusion Matrix* for Credit Scoring Problem (Cells Record % of Total Cases)

Confusion (% Total)	Predicted Default	Predicted Good	True Total
True Default	7	3	10
True Good	12	78	90
Predicted Total	19	81	100

[5] The model form, especially for a structurally unstable method like decision trees, is often very sensitive to small changes in the cost matrix. This is due to so many alternative models being explored by the data mining search method, and is normal, albeit unsettling.

the classifier. First employed at the dawn of radar, these curves show the trade-offs over the continuum of cutoff values (relative costs) of the two competing goals; that is, they reveal how changing cutoffs affect the classifier's accuracy versus precision, or false alarm versus false dismissal rate, or Type I versus Type II error, or false positive versus false negative rate, etc. One receiver (classifier) dominates another in a cost region if it is better by both criteria. The cost region of interest is that closest to the estimated relative cost. Thus, it is possible for an ROC curve to reveal that one classifier is better than another without anyone being able (or willing) to exactly specify the relative costs.

Note that each point along the ROC curve corresponds to a separate confusion matrix. And choice of a particular confusion matrix as best implies selection of a relative cost trade-off (e.g., cost matrix). Sometimes an analyst can present alternative solutions to domain experts in this "backwards" way to discover what underlying cost trade-offs match their intuition. Intuition is exceedingly powerful—a psychologist friend Dan Elash calls it "unarticulated lessons from experience"—so uncovering it by eliciting preferences between alternative solutions can be a very useful technique to know.

If a tool does not allow for cost matrices and only maximizes PC, you can give important classes the proper weight by manipulating the data. *Undersampling* is when the majority class is sampled and its other excess cases are ignored. This is tough for many analysts to be willing to do, given the great value of data, unless the data are very plentiful. More common is *oversampling*, where the rare cases are simply duplicated. In the previous example, then, six more copies of the default cases would be made so they can have seven times their original influence on the overall model. Be very careful when sampling either way, however, as problems can ensue (as explained in Mistake 9 of Chapter 20). Especially remember to oversample only the cases in the training data, and make sure that none of their copies appear in the evaluation data.

For estimation problems, you can also duplicate the cases of important classes to increase their influence on the resulting model.

Error Metric: Ranking

The error metrics described in the preceding sections calculate a global score, over all the data, but some problems really require a local score. That is, they may need to pay attention only to cases near a boundary. For example, a model to estimate political leanings might need to work well only at the boundary between two parties, where one could go either way, if it is to be used to identify persuadable constituents where time or money spent reaching out to them might affect their affiliation. (Hence, the influence of "battleground" states when a winner-takes-all-delegates score function is used, and the ignoring of "safe/lost" states by both parties.) On the other hand, if you are trying to identify the peaks of the curve (e.g., who might donate or might volunteer for campaiging), then an extreme is the most interesting region. Still, in both cases, the exact accuracy of the rest of the population is not important.

To optimize an estimation problem of this form, you would have to define a custom cost (or score) function and use a global search algorithm (and a lot of time) to find the best model parameters (for a given model structure). But if your data mining tool allows multiclass cost matrices, you could approximate a complex and flexible cost function through quantizing the output into multiple classes (along a continuum) and specifying the misclassification costs in the matrix. Then, to show that a particular kind of mistake doesn't matter, use a zero in the cost matrix. Look for natural or sensible breaks in the output when defining the classes. A slight loss will be the inherent ordering of real numbers; translated into classes, it will not be "remembered" that "medium" comes between "low" and "high." A decision tree, for instance, will not hesitate to put opposite extremes together in a node if the sample populations so dictate.

A simpler (albeit less optimal) way to get a feel for how your model is doing in the region of interest to you is to use a lift or gains chart, as shown in Figure 13.3.

In the lift chart, the percent of the total responses is the vertical axis, and the percent of the total population contacted is the horizontal axis. The diagonal line represents the baseline or random case, where mailing $X\%$ of the prospects will get you $X\%$ of the dollars. The line "lifted" above that is the result of using the model to order the cases (estimate the likelihood of response, $Y = 1$). The triangular knots on the model are the boundaries of the different regions. They are few in number because this lift chart was built using a decision tree, and each knot represents a leaf node. The leaf is heavily populated if the span between knots is large, and lightly populated if small. All cases within a given leaf node have the same probability, so ordering within a node is random, and this is represented by a straight line between knots. Note that the model does fairly well, as the line has lift (is above the baseline). In particular, the highest-ranked (first) 10% of the population provides 20% of the return (a lift of 2.0), and this is made up for by the bottom 40% of the population being needed to provide the same 20% return. (The curve always comes back

FIGURE 13.3 Sample lift chart for customer response application: % Total Responses versus % Population Contacted.

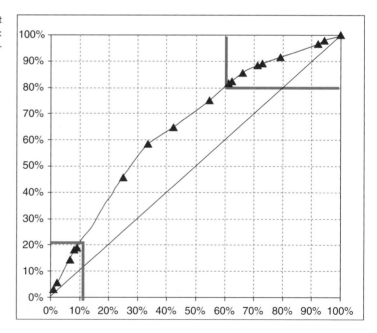

to the corner point in this form of the lift chart.) If you are interested in how the top quarter of the cases perform, the chart above says they provide over 45% of the return.

To build a lift chart using a decision tree, follow these steps:

1. Build the tree from the training data.
2. Order its leaf nodes by the response rate *in training order*[6](best first).
3. Run the model on the evaluation data, scoring each leaf node.
4. Graph evaluation response % versus solicited % using the training order of the leaf nodes from step 2.

If all nodes are well ranked, the slope of the line between successive knots (moving left to right) will decrease slightly. Note that the model represented by Figure 13.3 has some nodes whose ranking does not hold up on evaluation, as there are "dents" in the lift curve. A similar chart from a more continuous type of model, such as linear regression, will likely have many more knots, and it is possible (if the inputs are varied enough) that every predicted value will be distinct. This will make the lift curve smooth, though it can still be dented. Smoothness allows for much more variable cutoffs than can easily be had with the few knots of a tree's lift curve.

Cross-Validation to Estimate Error Rate and Its Confidence

What if, after splitting the data into training and evaluation, we are concerned that we just got lucky (or unlucky) in how the model performed? The evaluation result is just one number. How confident in it can we be? The best way to find out is to split the data multiple times and validate the original number. If we organize our data bookkeeping in such a way as to split the data V different times so each case is in the evaluation data exactly once, we have V-fold cross-validation, as described here and shown in Figure 13.4:

1. Separate the data into V subsets of equal size (stratifying if necessary). A rule of thumb is $V = 5$ or 10.
2. Train V models, leaving out a different data subset for each.[7] The training data have proportionate size $(V - 1)/ V$.
3. Test each model on the data held out for it (the gray blocks of Figure 13.4).
4. Accumulate these test results to get a distribution of out-of-sample accuracies.

With cross-validation (CV), an analyst gets a distribution of error results, which is a better estimate of true accuracy in two ways:

1. The mean is more accurate than a single experiment; and
2. The spread of the distribution provides an idea of the confidence in that mean value.

---

[6] We found a serious error in a widely disseminated tool from a major vendor—which the vendor fixed immediately upon our telling it privately—where it had instead ordered the nodes of the evaluation tree by its own ranking, rather than using the ranking set from training. This look-ahead error led all users to think they had better results than they really did!

[7] In the limit where $V = N$, the number of cases, you get the famous "leave-one-out" estimation method.

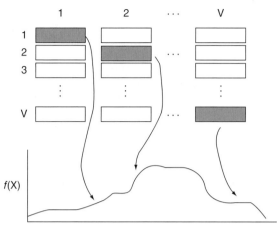

FIGURE 13.4   *V*-fold cross-validation.

To obtain the final model—since at this point you have *V* models that disagree on details—analysts often refit a model on the whole data set and impute the error accuracies of the CV models to it. Another option is to combine the CV models, which is in the spirit of the powerful ensemble modeling approach described in "Ensembles of Models: The Single Greatest Enhancement Technique."

Note that what you are testing when you perform CV is what you hold constant between the "plies." That is, some things change in response to the different data sets, such as the parameter values (e.g., *a*, *b*, *c*) estimated. But others won't, such as (in one example) the form of the model being fit (e.g., $Y = a^*x + b^*x^2 + c^*x^3$). So you are testing the overall accuracy of a model with that form on that data through the CV procedure. At the end, you know how well such a model performs, but you don't yet have the model itself; instead, you have *V* submodels. So one of these two approaches (refit or combine) can provide you the final model to use.

Bootstrap

Another powerful way to estimate accuracy with resampling is the bootstrap (largely legitimized by Efron, 1979; Efron and Gong, 1983). With CV, you have to trade off the size of the evaluation set and the number of evaluation error points. Use a plentiful set of errors to make a good distribution, and the sets may be too small to provide an accurate error estimate on their own. Bootstrapping is a much easier way to get a similarly good estimate:

1. Repeatedly copy[8] cases until you have a *replicant* data set of the size of the original.[9] (Approximately $1/e$ of the cases in the original data set will be left out.)

---

[8] Statisticians call this, less concisely, "repeatedly sample with replacement."
[9] Theory shows it really needs to be only half the size of the original.

2. Do this as many times as you wish (e.g., 50 or 100).
3. Fit a model to each replicant data set and *evaluate those models on the original data*.
4. Measure the mean and sigma of the resulting distribution of errors, as with CV.

Each replicant data set is like the original, but also weirdly unlike it. Some cases appear many times; others not at all. If the original were an animal, a replicant would have, say, one eye and five legs—each would resemble the original animal but over- and underemphasize different aspects of it. Together, the results provide a very useful and robust estimate of accuracy.

When should you use cross-validation over bootstrapping? Bootstrapping compared to CV is easier to program, does not trade off evaluation set size and number of sets, and produces twice the information (according to theory), for the same effort. For these reasons, many analysts who are comfortable with the randomness of resampling prefer it over the "fussy" bookkeeping of CV. But sometimes you need strict control over the cases making up the sets. The three most important situations in which you should use CV instead are if

1. You want to compare algorithms. (Data quality is usually more influential than algorithmic power on results. Controlling the data is necessary, though not sufficient, to make comparisons fair.
2. You want to combine models (See "Ensembles of Models: The Single Greatest Enhancement Technique"), so they will all be trained on the same subsets.
3. Cases are not independent, or they come in sets. For instance, the rows of a medical database may represent doctor visits. If you are to examine treatment efficacy, patients should be assigned to one group or another (training or evaluation), and all their cases should stay together.

Target Shuffling to Estimate Baseline Performance

What if you are performing discovery rather than fitting a model? How can you protect yourself against the "vast search effect" where you're bound to find something because you look at so many possibilities?

One of us has developed a useful approach, called target shuffling, which is simply to

1. Randomly shuffle the output (target variable) on the training data to break the relationship between it and the input variables;
2. Search for combinations of variables having a high concentration of interesting outputs (e.g., diseased patients);
3. Save the "most interesting" result and repeat the process many times;
4. Look at a distribution of these bogus "results" to see how much apparent results can be extracted from random data;
5. Evaluate where on this distribution your actual results stands;
6. Use this as your "significance" measure.

For example, it may appear, for students in a class, that the scores on the first homework are positively correlated with age. You could come up with a good reason for this. But look at enough candidate inputs, and the finite membership of the class will lead to something looking correlated. (It is likely that, presented with a result that the darker the hair, the better the grade, we would be eager to explain exactly why! We humans love stories and will look for ways to believe results that appear credible.) So randomly assign grades to the wrong students and repeat your search. Then you know there is no true relationship, and you can note the apparent relationship's strength. One hopes that the best result on the actual data is much stronger than the average best result on the shuffled data.

This idea of target shuffling was suggested to a vendor of data mining tools, Salford Systems, a company that improves, maintains, and sells a very useful suite of software programs (CART, MARS, Random Forests, and TreeNet) whose core code was written by Jerry Friedman and Leo Breiman—extremely well-respected leaders in our field. Salford's CART has a mode where it can run a battery of tests, so they called this idea "Battery Monte Carlo Test (MCT)." Figure 13.5A shows the procedure's result for a sample modeling problem (explained in "Examples: Decision Tree Surface with Noise" in Chapter 18), where $Y = f(x_1, x_2) + noise$, and there are 8 additional (useless, but distracting) random candidate

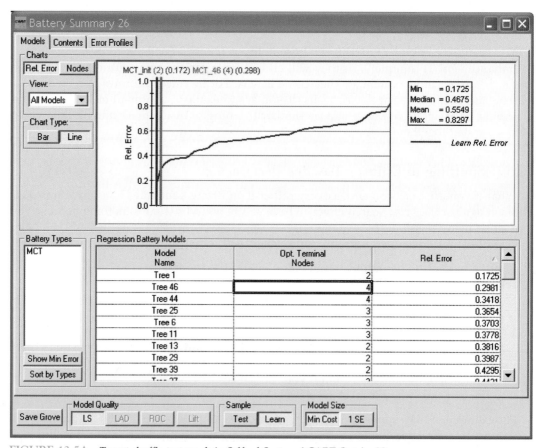

FIGURE 13.5A   Target shuffle test result in Salford Systems' CART (beta)—20 cases.

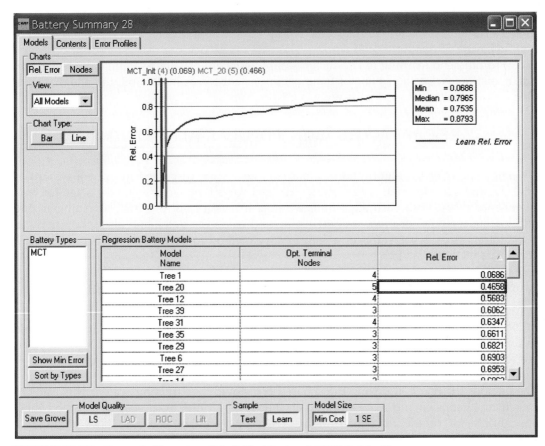

FIGURE 13.5B   Target shuffle test result in Salford Systems' CART (beta)—50 cases.

inputs and only 20 cases. The resulting tree with the real output labels (Tree 1) has a relative error of 0.172 (and two leaf nodes), whereas the best of 50 trees built on the target-shuffled data has an error of 0.298 (using four leaf nodes). (A perfect fit would be reflected by a relative error of 0, and the fit of a constant by a relative error of 1.) Thus, the best (of 50) random trees was able to achieve almost 6/7ths of the performance of the real tree, bringing strong doubt into the reliability of the stability of the real tree's training accuracy estimate.

It is hard to build much of a model with only 20 cases, especially given 10 candidate inputs. Indeed, if we increase the number of training cases to 50, as shown in Figure 13.5B, the result is more reliable. There, the relative improvement of the real tree is 0.069 and that of the best target-shuffled tree is 0.466, so the real tree shows almost twice the improvement (i.e., movement from 1 to 0).

The metric version in Figures 13.5A and 13.5B is in beta form (using a readily available chart form not designed for it), so it undoubtedly needs to be made more clear. Still, we have found the most useful way to use target shuffling is to calculate the distribution of results of the shuffled-Y models (mean and standard deviation) and note, for the real model's results, how many standard deviations above the shuffled mean it is. This is a very good measure of quality.

We recommend learning to use such resampling techniques (cross-validation, leave-one-out, bootstrap, jackknife) as one of the most important topics in data mining and statistics. One useful resource to explore is www.statistics.com.

RE-CAP OF THE MOST POPULAR ALGORITHMS

Some aspects of the algorithms summarized next were discussed in previous chapters (7, 8, 11, and 12). We briefly highlight the distinct ways they work to set the stage for how to enhance them. With each algorithm, we note if it is a *consensus* method (summarizing and tossing the data) or *contributory* (keeping and using the data), and also whether it can select variables. It is very rare for a contributory method to be variable-selecting, but most consensus methods can be.

Linear Methods (Consensus Method, Stepwise Is Variable-Selecting)

Methods traditionally used in statistical analysis often contribute significantly to a data mining effort, at the very least providing a baseline against which to compare more modern techniques. Linear regression (LR) predicts a response (dependent) variable by a weighted sum of predictor (independent) variables. The estimation surface is a plane in the space of candidate variables, though those variables can be nonlinear functions of the original ones. The plane is the one minimizing the squared error over the "training" cases. This leads to elegant mathematics and fast computation, but is rarely the exact scoring function most appropriate for estimation penalization. Linear discriminant analysis (LDA) predicts a categorical response variable by creating a discriminating plane separating the groups of the response variable. A quadratic extension allows for nonlinear boundaries but requires estimating covariance matrices for each class.

Decision Trees (Consensus Method, Variable-Selecting)

Decision trees (DTs) are the most popular inductive method in current use. DTs are often built in two stages. When *growing*, the algorithm finds at each node (subset of data) the best feature discriminating between the classes and then splits the data into two new nodes based on that feature. This is applied recursively to the resulting data subsets until a class assignment can be made for each leaf. The second stage of *pruning* works by clipping off (reabsorbing) the least useful branches of the tree to best balance the accuracy on the training data with the complexity of the model. (A simpler model is generally more *robust*, i.e., more accurate on new data.) The final tree partitions the feature space in a number of labeled regions—typically in the form of axis-parallel hyper-rectangles.

Independently developed in multiple fields (including statistics, computer science, artificial intelligence, and psychology), DTs are heavily used in the newer disciplines of machine learning and data mining. Early psychology researchers believed that such trees were a good model of how humans build classifications. The three most important and widely used decision tree implementations are CART (Breiman et al., 1984), CHAID, and C4.5 (Quinlan, 1991). There are literally dozens of tree algorithms, including one developed by one of the authors which looks for two-step splits, but expert use of a subset of such tools is sufficient to explore their utility.*

While trees tend to be the least accurate of the major methods (when used alone), they have many other strengths, such as effective handling of missing data, and the ability to work with both numerical and symbolic data. Also, with methods of combining DT models, their accuracy can be improved to a competitive level.

Neural Networks (Consensus Method)

Artificial neural networks (ANNs) have received much attention for the past two decades, due initially to their hypothetical analogy to neurons in brains. ANNs consist of layers of nodes, each implementing a linearly weighted sum of its inputs with a bounded sigmoidal (S-shaped) output transformation ("squashing" function). The outputs of each node on a layer feed into every node on subsequent layers. With the backpropagation weight adjustment procedure, cases are fed through the network, and their errors are fed back to adjust the weights of the node to a degree proportional to their contribution to the error. Recursively, weights for nodes which feed into that node are similarly adjusted back to the first layers. Initial weights are usually set randomly.

Though ANNs have been overpromoted and only weakly use their parameters, they are surprisingly robust in practice. The sigmoidal transfer functions in each node serve to dampen the problem of overfit from extrapolation (like a logistic transformation). Also, the networks operate primarily in a linear region unless the training data, represented for many cycles, push nodes into more flexible nonlinear regions, so the potential for overfit is muted.

ANNs do not naturally select variables, but use, to some degree, all the candidates provided. So we have found that variable reduction and transformation can strongly impact ANN performance. Accordingly, we give great attention to variable selection and successively eliminate raw and derived data features having low influence (as measured by their linkage weight paths). This is similar to the "backwards elimination" method of variable selection often done in regression, but only rarely with ANNs.

Nearest Neighbors (Contributory Method)

Nearest neighbor algorithms classify a test example by finding its closest neighbors in a multidimensional feature space populated by known examples from a reference (training) data set. The class prediction is estimated to be that of the nearest neighbor, or by a weighted average of the classes of the k nearest neighbors. The distance between two data records depends critically on the variables used in the feature space, some on the metric used to scale numerical variables, and, when applicable, on how distances between

Texas Two Step: www.datamininglab.com/PRODUCTS/tabid/56/Default.aspx

symbolic variables are determined. One of the authors has developed a custom method of searching for combinations of dimensions (data features) for nearest neighbors, leading to strong predictive performance.*

Clustering (Consensus or Contributory Method)

Cluster analysis divides a heterogeneous group of records into several more homogeneous classes, or *clusters*. These clusters contain records that are similar in their values on particular variables. Unlike the classification and estimation techniques discussed in the preceding sections, clustering can be performed on *unlabeled* cases—that is, where the output value or class is unknown in the training set. Sometimes, clustering is useful even when labels are available—by illuminating groupings of cases or suggesting data features to be calculated to help other methods.

Like nearest neighbors, clustering depends on a distance metric, such as Euclidean distance (root sum squared of distances along each feature) or Manhattan distance (sum of absolute feature differences). For both algorithms, it is a serious challenge to incorporate symbolic features alongside numeric ones. As with nearest neighbors and neural networks, clustering also uses all dimensions provided, making feature reduction essential. Some clustering algorithms are consensus methods—such as *k*-means, which summarizes the clusters as *k* normal-shaped blobs—and some are contributory, such as single-link (SLINK), which adds the closest unclustered point to each cluster.

Knowing how each algorithm basically works—consensus or contributory, and variable-selecting or not—will help guide which enhancement suggestions, listed next, to try.

ENHANCEMENT ACTION CHECKLIST

After training and evaluating your model, you may want or need to improve its results. Here is a checklist of 10 practical steps that usually help so consider trying:

1. *Transform real-valued inputs to be approximately normal in distribution.* Regression, for instance, behaves better if the inputs are Gaussian; extremes have too much influence on squared-error. For variables that are typically log-normally distributed, like income, this involves transforming the variable via a logarithm or the more general Box-Cox function.
2. *Remove outliers.* Note the extremes of each variable and investigate any that are too many standard deviations from the mean. (This is iterative, as outliers have an undue effect on the calculation of the deviation itself.) Don't necessarily assume it is an error (see Mistake 6 in Chapter 20), but set them aside.
3. *Reduce variables.*
   a. *Correlation:* Variables are often very similar to others, so those with high (say, 99%) correlation with others may be redundant. Beware, however, of outliers (again) which can mask or create correlation. For instance, we once had a data set where

---

Nearest Neighbor: www.datamininglab.com/PRODUCTS/tabid/56/Default.aspx

we discovered (painfully) that missing values were sometimes miscoded with a large positive number (e.g., 99999). This led to chaos in the correlation checks, where the value reported depended only on whether or not missing cell values in two variables were on the same case.

   b. *Principal components (PC):* You may retain the vast majority of the variance in the input data by replacing the original variable, X, with a smaller subset of PCs. These are linear transformations of X designed to be mutually orthogonal and span as much of the input space as possible. For instance, you might be able to represent 90% of the space covered by cases in 150 variables using only 20 PC dimensions. Note, however, that PCs don't consider the output variable when being discovered, so they may not be the best vocabulary for classification; note also that they still require measuring all of X to be calculated.

   c. *Follow the choices of variable-selecting algorithms:* Many algorithms, such as neural networks or nearest neighbors, don't themselves select variables. But you can first run different algorithms that do—such as stepwise regression, decision trees, or polynomial networks (Elder and Brown, 2000)—and follow their lead. Try using only the superset of variables that they pick up. There is no guarantee this will be the best set for your algorithm, but the approach often proves useful in practice.

4. *Divide and conquer.* Many simple models may be more useful than a single complex one. You can remove from training any simple subset of the problem you can clearly define and focus modeling energies on the hard part. For instance, if all patients with a certain symptom are to be recommended for immediate treatment, take that out of the data as a known situation and train on the rest. Slice the data enough this way and many aspects of the problem will change, leading to the possibility of a novel discovery due to the novel perspective of what the problem really is.

5. *Combine variables to create higher-order features.* Don't try to "build a critter from pond-scum"; use higher-order components. For instance, on a trajectory estimation problem, calculate where the craft will land without any complex effects, such as earth's rotation and air resistance, and estimate the shortfall instead of the full distance.

6. *Impute missing data.* The easiest way to handle missing data on training is to not allow it. This means deleting either a case or variable for which a cell is empty, so this can vastly reduce your data. To get the benefit of "holey" cases, try filling in the data with different alternatives:

   a. Mean of known cases for variable
   b. Last value (if cases are in order)
   c. Estimated value from other known input values (best, but most complex)
   d. If categorical, the label "missing"

7. *Explode categorical variables to allow use of estimation routines.* A categorical variable can't be used directly in an estimation algorithm like regression or neural networks. But you can "explode" a C-category variable into C binary variables where each holds a 1 if the category is its value for that case.

8. *Merge categories if there are too many.* Each value of a categorical variable is usually allowed to have its own parameter by an algorithm, so overfit is very likely if there are

too many values. For instance, State can have over 50 values (given Washington, DC and Puerto Rico, etc.) but the important (well-populated) states in a database may be many fewer. Keep the large ones and merge the remaining into "Other" or divide by regions. Always look for differences among categories and merge them if they aren't significant enough.

9. *Merge variables with similar behavior.* If you have real-valued variables that are strongly correlated, you can average them to create one candidate variable. For binary variables, you could examine their expected value conditioned on the output in the training data set and create a super-variable that is the sum of those which tend toward one of the classes of the output. For instance, people might have flags (bits) reflecting their interest (or lack thereof) in NASCAR racing, vegetarian cooking, *Field and Stream* magazine, progressive rock, etc. You could find which are correlated with voting patterns for each party over the training data and create one variable for each party that holds the sum of these bits. The union of those bits will not be as empty for the super-variable.

10. *Spherify data.* Many algorithms prefer the variables to be on the same scale and be independent (for instance, nearest-neighbor distances). But this is rare in practice. A person's weight might be in pounds and height in feet. Those are likely correlated. And a unit step in one variable (height) is much more significant than such a step in the other (weight). A first step would be to normalize the data by transforming each variable by $Z = (X - \mu)/\sigma$ where μ and σ are the mean and standard deviation of X. The next step up would be to also take care of the correlation. We recommend using the *Mahalanobis Distance* function for this when possible.

ENSEMBLES OF MODELS: THE SINGLE GREATEST ENHANCEMENT TECHNIQUE

The most useful and powerful result of statistical research over the past decade has come to be known as *ensemble modeling* (also known as "bundling," "model fusion," or "committee of experts").[10] The idea is to employ multiple models to do better than a single one—often even the retrospective best of the individual models. The key requirements of the models to be joined are that they must each meet some minimum standards of accuracy and diversity (difference from one another in behavior). It is still not well known how these qualities interact exactly.

Most researchers and practitioners build multiple versions of models of the same type (e.g., decision tree), but we have found it most useful to join models of completely different types (Elder and Lee, 1997). Still, we emphasize that the concept has helped either way. In particular, we've never seen a problem in which a single decision tree beats an ensemble of decision trees on new data.

---

[10] Our opinion, but also shared by prominent data mining statisticians Jerry Friedman, Trevor Hastie, and Rob Tibshirani (1998), who wrote "Boosting [a type of ensembling] is one of the most important recent developments in classification methodology."

Bagging

Breiman (1996) created a simple but powerful algorithm he called *bagging*, for *b*ootstrap *agg*regating:

1. Create M bootstrap replicates of the data set (copy cases randomly from the data to build M other same-size sets).
2. Fit a model to each of the replicates.
3. Average (or vote, for a classification problem) the predictions of the M models.

The final model is a compromise of its component models, and for decision trees, the resulting surface transitions less suddenly than it does for a single tree (as shown later, in "Examples: Decision Tree Surface with Noise" in Chapter 18).

Figures 13.6 and 13.7 (from Elder and Ridgeway, 1999) show how the out-of-sample error for a suite of regression and classification problems, respectively, was reduced on *all* of the problems attempted when going from a single tree to a bagged ensemble model.

Boosting

Freund and Schapire (1996) came up with the idea of *boosting*, in which variety is created from weighting cases according to which ones were easier or harder to model correctly.

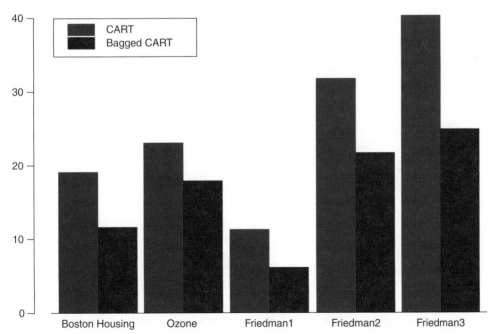

FIGURE 13.6 Bagged tree better than single tree for all five *estimation* test problems (Elder and Ridgeway, 1999).

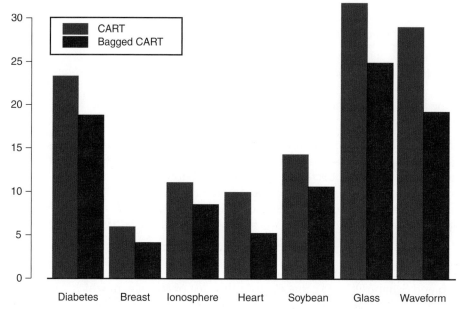

FIGURE 13.7   Bagged tree better than single tree for all seven *classification* test problems (Elder and Ridgeway, 1999).

The harder cases get more weight (influence) in subsequent modeling passes, and the easier cases, less:

1. Equally weight the observations (cases).
2. For j in 1 to M:
    a. Using the case weights, fit a classifier $f_j(x)$.
    b. Up-weight the poorly predicted cases; e.g., multiply their weight by $1 + \alpha$.
    c. Down-weight the well-predicted cases; e.g., divide by $1 + \alpha$.
3. Merge $f_1(x), \ldots f_M(x)$ to form the boosted classifier, giving most weight to the earliest models and least to the (somewhat obsessed) last models.

This algorithm is very popular because it works well over a wide swath of applications for a number of different modeling approaches.

Ensembles in General

To build ensembles, you need to construct varied (and accurate enough) models and combine their estimates. There are five ways to generate variability, by modifying

1. Case weights, as with boosting (real-valued weights) and bagging (integer weights);
2. Case values—adding noise to the input or output variable values;
3. Variable subsets (as with *random forests*, where each tree being built only considers for splitting a random subset—say, 10%—of the potential candidate inputs at any one node);

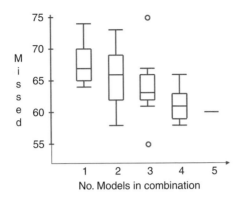

FIGURE 13.8   Median error of ensemble reduces as more models are combined.

4. Guiding parameters (such as running an algorithm with different settings);
5. Modeling technique (e.g., tree, regression, MARS, neural network).

There are three ways to combine estimates, by using

1. Estimator weights (perhaps based on estimated or training accuracy);
2. Voting (when the problem is predicting the best class);
3. Partitions of the design space (as with *model gating*, where different models take control depending on which input space region we are in).

In our estimate, the most common combination used is cross-validation and averaging, especially with decision trees. That is, building a model of one type on V different overlapping folds of the data and then averaging the resulting estimates. In our experience this (and other variations) are much more likely to help, rather than hurt, performance. Further, as shown in Figure 13.8, for a credit-scoring example, the more (good enough) models are combined, the lower the median out-of-sample error of the ensemble.

Surprisingly (as shown in Chapter 18), ensembling has also been shown to be simpler (in behavior, not form) than the use of just a single model.

HOW TO THRIVE AS A DATA MINER

Now that we've studied a lot of useful ways to assess and improve a model, let's step back and look at how to ensure success for the larger project of which the modeling is a part. The key is to make whomever hires you more money than you cost! Keep the business problem—not the (usually more interesting) technical issues—at the forefront of your thoughts.

Big Picture of the Project

Let's take as a focusing example the problem of fraud detection—one of the data mining problems akin to finding "needles in a haystack." (We'll focus on fraud detection in detail in Chapter 19, but for now it'll serve as a motivating challenge.) You may be required to scour vast amounts of data yet have only very few known instances of fraud, so they won't be well represented as a class in the training phase. There are also typically fraudulent cases that escaped detection, which adds to the challenge of training a model to distinguish between fraud and nonfraud, as those are mislabeled as valid cases. Further, while statistical outliers in the data—which are signs of unusual activity—are good places to search for fraud, a typical "alert" system built only on these will have far too many false alarms to be useful in practice.

Before you begin modeling, successful fraud detection requires looking at the entire business process and identifying where fraud can originate. Begin the project with a careful evaluation of the client's existing business process. Then collect cases of fraud that have been found by auditors or others through the existing manual processes. From knowledge of the business process and these known cases, design metrics for measuring fraud and work with the client to automate their calculation. Finally, develop the detection models. This process delivers value to clients at each stage.

The return on investment (ROI) in fraud detection data mining can be extremely impressive. Take the time to develop metrics of success and the baseline ("before") performance so that you can score your impact (the "after" picture, which should look better). For example, on a large project led by one of our colleagues, the client had an alert system for its enormous data processing task whose warnings turned out to be fraud only 1% of the time (very inefficient, though far better than random). With the data mining solution, however, the hit rate improved to an astonishing 25%. In another fraud detection project, our colleagues were able to achieve a savings of over $20 million on an engagement that took less than a staff-year to complete and deliver. Wouldn't it have been great if, in the business negotiations, the confident and expert team had been able to negotiate a share of the success? Even $5\% \times$ (after–before) is serious compensation.

Project Methodology and Deliverables

The authors have found that an investment in teamwork and creative problem solving early in a project pays off down the road. It is essential to cooperate and communicate closely (both with team members and the client) throughout a project to successfully implement the solution on time. You will need client experts to convey business understanding and requirements throughout the project, so get to know them and treat them with respect. Join client domain expertise with your expertise in business analysis, systems engineering, and technical skills in predictive modeling.

We recommend operating in a rapid-prototyping framework, where a baseline solution is developed quickly, to discover the technical and interface issues the final system will face. Then strengthen the system by iteratively improving components—often possible in parallel. This allows the path-critical components to be identified and improved as time and budget allow, and provides decision makers with better estimates of the trade-offs involved. Include regular administrative and technical updates and on-site meetings at critical junctures. A client's great fear is that it will launch analysts on a project which will become open-loop, with researchers chasing down rabbit-holes and not reporting back when necessary. Frequent (e.g., weekly) detailed meetings and intermediate reports are painful (from the point of view of a technical person eager to try the next research idea) but essential for the survival of the project.

At the end (either problem or budget runs out), submit findings in a complete report (a short paper, headed by an executive summary, with model and methodology details in appendices) and a presentation, deliver the results in the form of computer code or logical rules that can be converted to the best format for the client, and work closely with those responsible for implementation to ensure that the models integrate seamlessly into the existing systems environment. Always remember to suggest valuable next steps suggested by your findings; you'll want to make it easy for the client to ask you to continue your valuable work!

Professional Development

Continual learning is vital. The latest software tools are powerful, able to tremendously augment the productivity of an analyst—but only one who knows in what direction to push. That experience is hard to accumulate in faster than real time, but a single trait—humility—helps greatly. Usually, the data mining expert is the person in the room who knows the least about the specific problem being faced by a client. No matter how expert you become in the mining craft, you must be open to learning the business constraints, metrics, and vocabulary of the client as quickly as possible. Clients have great tolerance for (in their mind) stupid questions in the first couple of weeks of an engagement, but they aren't encouraged if they have to explain things more than once. Fortunately, many problems that seem completely unrelated become clean data mining problems with the right level of abstraction. We have used data mining to successfully

- Select profitable stocks;
- Detect fraudulent claims;
- Score an applicant's credit;
- Discover cross-selling opportunities for products;
- Quantify drug efficacy;
- Forecast new product sales;
- Discover new customers;
- Recognize objects in images;
- Verify identify through biometrics; and
- Anticipate shifts in market sectors for hedge funds.

The data miner, with experience, gains a breadth of perspective that enables him or her to find creative and effective solutions to a great variety of challenges. It's this variety, and the "detective" nature of the work, that can be most satisfying.

Though the data miner has the least domain-specific knowledge, he or she brings an analytic expertise that is rare and necessary to the team. Statistical thinking is hard—it's been shown to be the "least caught" discipline taught at universities—and uncommon.

Just handling the data right, with validation and true out-of-sample data set aside, can be a critical contribution. But it's also the trained analytic mind, seeing the client's problem for the first time, that provides a valuable outside perspective. Often, a group has been addressing a problem in just one way for so long that the group members act as if it's the only way, and an outsider committed to understanding the details of their challenge can make great contributions just by asking intelligent, probing questions.

To increase your effectiveness, seek training. There are excellent short courses on data mining, which are quicker for many to digest than books. Also look for opportunities to give presentations. If you are not one of those for whom public speaking is terrifying,[11] count yourself lucky; it can be a distinguishing skill.[12] Conferences are very eager for industrial contributions; they are dominated by academics, due to the score function for academics that requires the writing of papers and self-promotion. It is hard to attract industrial practitioners, who often have to write papers on their own time, and whose management is rightly fearful that industrial trade secrets will be revealed. (And it is astonishing to newcomers how slight a deviation from standard practice is jealously guarded as a valuable intellectual property.) But, when a good industrial paper is delivered at a conference, it is eagerly listened to and appreciated. The real world has distinct advantages over toy problems in imparting useful lessons. So learn to present and speak; this skill is rare and thus valuable. Conveying the value of data mining to management will ensure the needed backing for your analytics. Again, it is only by helping others succeed at their goals that we get the chance to succeed at ours.

Three Goals

To conclude, we want to share three goals that will contribute to your data mining successes:

1. *Maintain a diversity of projects.* Whether a project is military, medical, or monetary, it hinges on whether you and the client can successfully learn useful patterns from historical data. A mix of applications will expose you to the collected wisdom of expert practitioners in each domain, teaching you much. For instance, a data feature similar to one useful in drug efficacy studies has, for one team, contributed to fraud risk estimation, and techniques originally applied to high-performance aircraft have improved investment and credit projects.

---

[11] It is said that a majority of Americans rank their fear of public speaking above even their fear of death! Perhaps this is due to it being much easier to imagine yourself speaking than dying. Nevertheless, the willingness and ability to speak well is worth cultivating; it is valued due to its scarcity.

[12] We have found *Toastmasters* to be a useful way for those without practice or comfort speaking in public to become aware of the components of different kinds of useful talks—from jokes and introductions, to technical talks and "elevator speeches" (30-second summaries of what you're working on in case the big boss asks you in an elevator)—and gain practice and advice on how to improve their presentation skills.

2. *Employ a diversity of algorithms.* Unlike most practitioners, you should not be wedded to, or limited by, a single data mining technology. Though some of the authors have spent years of our early careers developing particular competitive algorithms, we have learned that each problem presents its own spectrum of challenges and requirements and that a toolbox of powerful techniques is needed.

3. *Combine models.* Learn and use the breakthrough technique of combining multiple models to achieve estimates more robust and accurate than possible through any single model—sometimes even the *retrospective* best of the individual models. This simple idea contains subtleties in its execution, yet practice can make it a powerful tool in your arsenal. Chapter 18 explains new research showing how an ensemble can even be *less complex* than any of its components, helping to explain its superior performance.

POSTSCRIPT

Now, we are ready to apply data mining technology to four major areas: medical informatics (Chapter 14), bioinformatics (Chapter 15), Customer Relationship Management (Chapter 16), and fraud detection (Chapter 17).

References

Akaike, H. (1973). Information theory and an extension of the maximum likelihood principle. In P. N. Petrov & F. Csaki (Eds.), *Proceedings of the Second International Symposium on Information Theory* (pp. 267–281). Budapest: Kiado Academy.

Barron, A. R. (1985). *Logically Smooth Density Estimation.* Ph.D. Thesis, Dept. Electrical Engineering, Stanford University.

Breiman, L., Friedman, J., & Stone, C. *Classification and Regression Trees.* Monterey, CA: Wadsworth & Brooks.

Breiman, L. (1996). Bagging predictors. *Machine Learning, 26*(2), 123–140.

Efron, B. (1979). Bootstrap methods: Another look at the jackknife. *Annals of Statistics, 7,* 1–26.

Efron, B. (1982). The jackknife, the bootstrap, and other resampling plans. SIAM monograph No. 38, Society for Industrial and Applied Mathematics.

Efron, B., & Gong, G. (1983). A leisurely look at the bootstrap, the jackknife, and cross-validation. *American Statistician, 37,* 36–48.

Elder, J., & Brown, D. (2000). Induction and polynomial networks. In M. Fraser (Ed.), *Network Models for Control and Processing,* Chapter 6. Bristol, UK: Intellect.

Elder, J. F. (1993). *Efficient Optimization through Response Surface Modeling: A GROPE Algorithm.* Dissertation. Charlottesville: School of Engineering and Applied Science, University of Virginia.

Elder, J. F., & Finn, M. T. (1991). Creating 'optimally complex' models for forecasting. *Financial Analysts Journal,* Jan/Feb, 73–79.

Elder, J. F., & Lee, S. S. (1997). *Bundling Heterogeneous Classifiers with Advisor Perceptrons.* ERI and University of Idaho Technical Report, available at www.datamininglab.com under presentations.

Elder, J. F., & Ridgeway, G. (1999). *Combining Estimators to Improve Performance: Bundling, Bagging, Boosting, and Bayesian Model Averaging.* Tutorial at 5th International Conference on Knowledge Discovery and Data Mining, San Diego, CA (August 15, 1999).

Freund, Y., & Schapire, R. E. (1996). *Experiments with a new boosting algorithm. Machine Learning: Proceedings of the 13th International Conference.* July, 148–156. San Francisco: Morgan Kaufmann.

Friedman, J., Hastie, T., & Tibshirani, R. (1998). *Additive Logistic Regression: A Statistical View of Boosting*. Technical Report, Stanford University.

Hoerl, A. E., & Kennard, R. W. (1970). Ridge regression: Biased estimation for nonorthogonal problems. *Technometrics, 12*, 55–67.

Lachenbruch, P. A., & Goldstein, M. (1979). Discriminant analysis. *Biometrics, 35*, 69–85.

Lachenbruch, P. A., & Mickey, M. R. (1968). Estimation of error rates in discriminant analysis. *Technometrics, 10*(1), 1–11.

Mallows, C. L. (1973). Some comments on Cp. *Technometrics, 15*, 661–675.

Quinlan, R. (1991). *C4.5: Programs for Machine Learning*. San Mateo, CA: Morgan Kaufmann.

Rissanen, J. (1978). Modeling by shortest data description. *Automatica, 14*, 465–471.

Rothpearl, A. (2008). The jacknife technique in statistical analysis. *Chest, 1989, 95*, 940, http://chestjournal.org

14

Medical Informatics

PREAMBLE

The first general application area we will explore is *medical informatics*. The reason for treating it first is that in no other discipline has statistical and data mining applications been applied for a longer time, and it is an area in which analytical results can have such profound effects in both positive and negative directions. We will consider what medical informatics is and how data mining and text mining are related to medical informatics and 3D medical information. We will also discuss some journals and associations in medical informatics and look at an example of a medical informatics study.

WHAT IS MEDICAL INFORMATICS?

In a very simple sense, we might think of the new area of medical informatics as operations in which the computer meets medicine. Medical informatics has at least three areas specific to medical and health care management, research, and delivery. These three general areas may be thought of as (1) things related to a patient/doctor relationship like prescriptions and records, involving the accumulation of considerable data; (2) the area of three-dimensional imaging like PET and MRI scanning and all the image processing data; and (3) medical literature review and compilation. All three areas involve the accumulation of large amounts of data, and thus the possibility/need of analyzing these data to learn how to be more efficient and more accurate in management and diagnosis, etc.

We might categorize the patient/doctor type of medical informatics data coming from the following activities:

- Electronic prescribing
- Personal health records
- Computerized practitioner order entry
- Identity management of patients
- Electronic health records
- Good project management and software selection (for the preceding to be really successful)

Because of the large number of research papers and amount of data in text format (over 5,000 papers a month between medical informatics and bioinformatics), medical informatics requires *text processing* in addition to standard data mining methods. Thus, text mining algorithms must be added to data mining algorithms in the arsenal of tools needed to make sense out of all these data.

Additionally, because of the use of many 3D imaging methods in medicine, medical informatics requires analytical methods for image and structural informatics. Visual (and even auditory) data mining has not yet reached a plateau in its potential and possibilities; however, it is increasing in importance for diagnosis and decisions on actions to take in treatment (see Chapter 21).

The model in Figure 14.1 shows where medical informatics as a discipline fits into the fields of biological sciences, clinical and health services, and information technology and analysis.

The field of medical informatics is large, as is bioinformatics (the topic of the next chapter in this book), and it is not the primary purpose of this book to cover this topic comprehensively. So, only a brief summary of the field will be presented in this chapter; however, some of the most recent and important volumes written in this field will be provided in the References section, for any readers wanting to pursue this area in more detail.

HOW DATA MINING AND TEXT MINING RELATE TO MEDICAL INFORMATICS

Knowledge management, data mining, and text mining have come into the mainstream of business during the past 10 years, and their full implementation into health care delivery is crucial to bring the efficiencies in cost and accuracy that are so badly needed. Medical informatics data are usually structured, factual, numeric, and historical. These data contain textual data referred to as "unstructured," in that they do not consist of numbers or codes that can be contained in a database. But these data are factual and are every bit as important

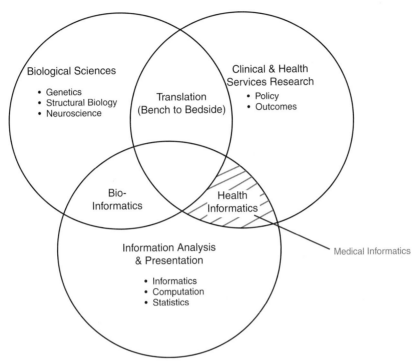

FIGURE 14.1   An informatics-centric view of the intersections and overlap among the biological sciences, health services research, and information analysis and presentation. It shows that bioinformatics includes a core of biomedical informatics techniques that can be applied to both biological science and health services research, together with its own applied computation and statistics techniques. Similarly, "clinical" or "health" informatics reuses that core, while adding clinical epidemiology, etc. The common core leads naturally to the common information foundation needed for rapid translation between bench and bedside. A "networked" or "web-like" organization structure is needed to provide a critical mass of people to nurture growth in each of the core disciplines of biomedical informatics, statistics, and computation, while at the same time providing separate concentration on application to the biological sciences, health services research, and translation between the two. From http://www.mc.vanderbilt.edu/dbmi/informatics.html (Vanderbilt University Medical Center Department of Biomedical Informatics).

as data structured in databases. With the advent of text mining algorithms, this textual "knowledge" material can be converted into numbers to function just like variables such as the values of blood pressure, enzyme levels, LDL and HDL lipid levels, etc.

Knowledge gathered from both data mining and text mining methods is needed to, at the very least, formulate new hypotheses, and at its highest level, provide the foundation for making decisions and taking action (which may be needed "immediately" in life-and-death health situations). While standard data mining algorithms can be used to discover patterns or knowledge that were previously unknown to medical science, text mining is needed to extract needed information for document classification, document clustering, entity extraction, information extraction, and summarization.

Data volumes in medical informatics and bioinformatics are being generated faster than researchers can handle using traditional methods of the past century. Thus, new methods, e.g., in data mining and text mining, and in other database searching methods are needed. In the medical informatics field, three methods of data or information retrieval stand out:

1. The **MedBlast system**, making use of BLAST (which will be defined and discussed in Chapter 15), allows researchers and clinicians to search for articles of interest among the plethora of research discovered each month.
2. The **HelpfulMed system** is used in medical informatics to retrieve documents from different databases, which are then clustered using a self-organizing map algorithm.
3. The **NLM's Visible Human** produces three-dimensional representations of the normal male and female human bodies by obtaining transverse CT, MR, and cryosection images of representative male and female cadavers. The data provide a good test-bed for medical imaging and multimedia processing algorithms. The Visible Human project has been applied to various diagnostic, educational, and research uses.

Data mining has a valuable *predictive* power to enable clinicians to determine, with some measure of accuracy, the proper dosage or treatment protocol. Classification is the most widely used technique in medical data mining, using things like *decision trees* (see one of the Medical Informatics Tutorials included on the DVD accompanying this book for examples, such as Tutorial Z). It is difficult to select in advance which algorithms will be best for a particular study; thus, the authors of this book always recommend doing a "competitive evaluation" of data mining algorithms to see which performs best. For example, Dreiseitl et al. (2001) compare five classification algorithms for the diagnosis of pigmented skin lesions. Their results show that *logistic regression, artificial neural networks*, and a *Support Vector Machine* performed comparably, while *k-nearest neighbors* and *decision trees* performed worse. But with other situations, other diseases, decision trees can perform quite adequately (again, see one of the tutorials accompanying this book, on the DVD). Another example is Acir and Guzelis (2004), who applied a Support Vector Machine algorithm in automatic spike signal detection in electroencephalograms (EEGs), which can be used in diagnosing neurological disorders related to epilepsy.

And, finally, a third example is Kandaswamy et al. (2004), who used an artificial neural network to classify lung sound signals into six different categories to assist diagnosis.

The amount of information in medicine/bioinformatics is so vast that clinicians and researchers cannot begin to read all the literature but must have methods of seeking what they need. Google's Page Rank algorithm can do well in finding popular web sites, but it just cannot meet the need of these clinicians/researchers to find very specific needed information. The following sections describe a couple of examples of document retrieval systems that medical informatics have been using in recent years.

XplorMed

XplorMed is a system developed by Perez-Iratxeta and colleagues (Perez-Iratxeta et al., 2001, 2002, 2003) for browsing the MEDLINE literature database. Given a set of MEDLINE abstracts, the system computes the words with the strongest relations with a set of categories, and then extracts keywords and computes their relationship to each other.

ABView: HivResist

ABView: HivResist is another method that focuses on a small set of the MEDLINE literature. It was originally developed to study HIV drug resistances and their associated mutations. Using such a focused set of literature reduces the ambiguity of terms.

For the future, nothing can substitute for a full-fledged text mining and data mining analysis of the medical literature. Such methods are, of course, the main topic of this book.

3D MEDICAL INFORMATICS

What Is 3D Informatics?

Three-dimensional medical informatics is the application of data analysis to images, volume data, and other dimensional data in addition to text-based metadata associated with imaging. Medical storage repositories are filling rapidly with this type of information, e.g., nonprint data in the form of audio recordings, films of X-rays and 3D imaging, and video recordings. The tools to index all of these nontext medical information are somewhat rudimentary today, so more sophisticated/focused systems are needed; no doubt data and text mining will play an important part here. These types of data may include more than three dimensions; e.g., 3D informatics may also include things like position, time, scale, and also multichannel data rather than just one dependent dimension. Object recognition may be very important to 3D medical informatics; object recognition is discussed in Chapter 21 as one of the "developing new areas of data mining."

Listing of types of scans done in medicine:

- X-rays (two-dimensional)
- CT scans—ray-computed tomography (three-dimensional)
- PET scans (3D)
- Magnetic Resonance Imaging (MRI)

All of these new three-dimensional scanning methods, with an increasing deluge of data, motivated the development of the following tools:

- An industry of Picture Archiving and Communications Systems (PACS)
- The creation of standards for DIgital COmmunications in Medicine (DICOM)

Three-dimensional methods in medicine are not just in body imaging, as there is a relatively new area of "surgical templates," e.g., 3D templates, created for each patient, an example being the screws that need to be put into a patient's vertebrae in certain kinds of back/vertebrae surgery, where there is a need for precise placement to avoid disturbing nerves or blood vessels. Computer-aided/analysis methods are being used to make decisions on such types of things.

Future and Challenges of 3D Medical Informatics

- Three-dimensional medical informatics can provide new frontiers in medical education in anatomy, physiology, molecular biology, and other areas.
- Already, research in 3D techniques has made an impact on computer-assisted surgical planning and image-guided interventions, and we'd expect that the number of these will increase in the future.
- Computer-aided diagnosis/early detection may become one of the most used tools in medicine, provided from data mining analysis of patterns, and thus the discovery of predictable patterns.

This discussion is short for three reasons:

1. Medical informatics is a relatively new field.
2. There are recent extensive volumes/books on the subject that can serve well. Those who want to explore medical informatics more thoroughly can order a four-volume set that was published in September 2008 (at a price tag of about $3,000): Joseph Tan (ed.). (2008). *Medical Informatics: Concepts, Methodologies, Tools, and Applications*; published by Medical Information Science Reference; ISBN-10: 1605660507
3. The focus of this book is *practical* data mining, and the purpose of this chapter on medical informatics is just to briefly make you aware of this domain. There are several medical informatics tutorials included within the printed pages of this book, on the accompanying DVD, and also on the companion Web site, where additional tutorials will be added from time to time.

Journals and Associations in the Field of Medical Informatics

- American Medical Informatics Association (http://www.amia.org/)
- Journal of the American Medical Informatics Association (http://www.jamia.org/)

POSTSCRIPT

With this brief introduction to the large field of medical informatics, you are ready to do some tutorials associated with this chapter. See Diabetes Clinical Management (on the included CD-DVD) for an example of how data mining can be applied to diagnosis and managing patient care. You will find several other tutorials on medical informatics on the CD-DVD. The next chapter follows the initial biological theme to focus on bioinformatics, which is concerned primarily with DNA microarray processing.

References

Acir, N., & Guzelis, C. (2004). Automatic spike detection in EEG by a two-stage procedure based on support vector machines. *Computers in Biology and Medicine, 34*, 561–575.

Dreseitl, S., Ohno-Machado, L., Kittler, H. et al. (2001). A comparison of machine learning methods for the diagnosis of pigmented skin lesions. *Journal of Biomedical Informatics, 34*, 28–36.

Kandaswamy, A., Kumar, C. S., Ramanthan, R. P. et al. (2004). Neural classification of lung sounds using wavelet coefficients. *Computers in Biology and Medicine, 34*, 523–537.

Perez-Iratxeta, C., Bork, P., & Andrade, M. A. (2002). Association of genes to genetically inherited diseases using data mining. *National Genetics, 31*, 316–319.

Perez-Iratxeta, C., Bork, P., Pérez, A. J., & Andrade, M. A. (2003). Update on XplorMed: A web server for exploring scientific literature. *Nucleic Acids Research, 31*, 3866–3868.

Perez-Iratxeta, C., Bork, P., & Andrade-Navarro, M. A. (2008). K2D2: Estimation of protein secondary structure and CD spectra. *BMC Structural Biology, 8*, 25.

Bibliography

Everitt, B. S. (2003). *Medical Statistics: From A to Z: A Guide for Clinicians and Medical Students*. UK: Cambridge University Press.

Greenes, R. A., & Shortlife, E. H. (1990). Medical informatics: An emerging academic discipline and institutional priority. *JAMA, 263*(8), 1114–1120. http://www.mc.vanderbilt.edu/dbmi/informatics.html; http://www.sigkdd.org/explorations/issue.php?issue=current

ISO Press Medical Informatics Publications. http://www.iospress.nl/html/mit.html

Kuhn, K. A., Warren, J. R., & Leong, T.-Y. (2007). *Proceedings of the 12th World Congress on Health {Medical} Informatics* (Vol. 129). Studies in Health Technology and Informatics. http://www.iospress.nl/html/shti.php

Mantas, J. (Ed.). (2000). *Health and Medical Informatics Education in Europe* (Vol. 57). Studies in Health Technology and Informatics. Amsterdam: IOS Press.

Ong, K. R. (Ed.). (2007). *Medical Informatics: An Executive*. Healthcare Information and Management Systems Society.

Shortliffe, E. H., & Perreault, L. E. (2001). *Medical Informatics: Computer Applications in Healthcare and Biomedicine* (2nd ed.). New York: Springer Verlag.

Shortliffe, E. H., & Cimino, J. J. (Eds.). (2006). *Biomedical Informatics: Computer Applications in Health Care and Biomedicine*. New York: Springer.

Shortliffe, E. H., Perreault, L. E., Wiederhold, G., & Fagan, L. M. (1990). *Medical Informatics: Computer Applications in Health Care*. Boston: Addison-Wesley Longman.

Spiegelhalter, D. J., Abrams, K. R., & Myles, J. P. (2004). *Bayesian Approaches to Clinical Trials and Health-Care Evaluation*. West Sussex, England: John Wiley & Sons Ltd.

Tan, J. (Ed.). (2008). *Medical Informatics: Concepts, Methodologies, Tools, and Applications*. Hershey, PA: Medical Information Science Reference.

Bioinformatics

PREAMBLE

This chapter is designed to give you a very brief exposure to the burgeoning field of bioinformatics. We will not present a comprehensive exposure to how data mining is applied in bioinformatics; instead, we will present a simple outline of major operations in bioinformatics and a short discussion of the direction in which the field is moving in the future.

Bioinformatics is the application of information technology to the field of molecular biology. Bioinformatics entails the creation and advancement of databases, algorithms, computational and statistical techniques, and theory to solve formal and practical problems arising from the management and analysis of biological data. Over the past few decades, rapid developments in genomic and other molecular research technologies combined with developments in information technologies to produce a tremendous amount of information related to molecular biology. Bioinformatics, then, is the name given to these mathematical and computing approaches used to glean understanding of biological processes.

The primary goal of bioinformatics is to increase our understanding of biological processes. What sets it apart from other approaches is its focus on developing and applying computationally intensive techniques (e.g., data mining and machine learning algorithms) to achieve this goal. Major research efforts in the field include sequence alignment, gene finding, genome assembly, protein structure alignment, protein structure prediction, prediction of gene expression and protein-protein interactions, and the modeling of evolution.

The following lists provide an outline of points that a bioinformatics scholar or researcher will have an understanding of or at least an "acquaintance" with in regards to the part they play in the overall field. Then we will discuss various aspects of bioinformatics. Bioinformatics is such a large field that to do this topic justice, we'd have to write several volumes, each with thousands of pages. But it is not the purpose of this practical book to give the new user a comprehensive exposure to bioinformatics data mining and text mining. Our purpose here is to simply give you a simple outline of what is involved in bioinformatics and a short discussion of where the field is headed in reference to making better use of data mining and text mining tools in the future.

Areas of study in bioinformatics:

1. Sequence analysis of DNA/RNA structures
2. Evolutionary biology
3. Gene annotation
4. Biodiversity
5. Gene expression
6. Regulation of genes
7. Protein expression and its regulation
8. Predictions of protein structures
9. Biological systems modeling

Fields related to bioinformatics:

1. Biophysics
2. Biocybernetics
3. Biomedical informatics
4. Computational biology
5. Medical informatics
6. Genomics
7. Computational and mathematical biology and biomodeling
8. Proteomics
9. Pharmacogenomics
10. Pharmacogenetics
11. Chemoinformatics
12. Molecular and metabolic networks modeling
13. Artificial intelligence
14. Neuroinformatics
15. Statistics
16. Data mining and information analysis

Figure 15.1 shows where bioinformatics fits into the total picture of biology, medicine and health care services, and information technology.

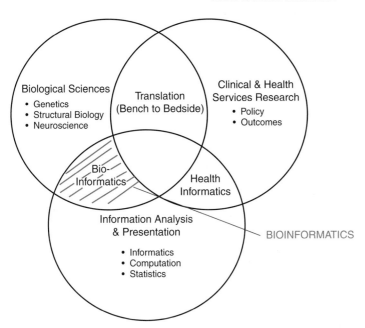

FIGURE 15.1   From http://www.mc.vanderbilt.edu/dbmi/informatics.html (Vanderbilt University Medical School Department of Medical Informatics).

The bioinformatics field was greatly enlarged by the full sequencing of the human genome in the late 1990s; Figure 15.2 shows a map of the human chromosome, resulting from the Genome Project.

WHAT IS BIOINFORMATICS?

Bioinformatics is the activity of biologists using computers to find information related to the way the genetic blueprint in the genes unfolds, makes us who we are, and regulates our lives. We might say that bioinformatics is the study of the gene and all of its consequences.

Today, bioinformatics involves biologists, molecular geneticists, pharmacologists, protein chemists, other scientists and mathematicians, statisticians, and, yes, data miners. The impact of the advances in bioinformatics is felt across an even wider field of professionals such as lawyers, judges, and congress people who often have to consider DNA evidence in legal cases and the writing of bills in Congress. Thus, bioinformatics invades our everyday lives, whether or not we realize it!

Bioinformatics involves (1) analyzing DNA sequences, (2) analyzing RNA sequences, and (3) analyzing protein sequences. But it involves more than this. It also involves finding what are called redundant sequences, aligning sequences, aligning multiple sequences, finding structure, and then finding 3D structure of these molecules. In one sense, we might say that bioinformatics involves working with the entire genome. Those involved in bioinformatics seek to understand what it *is* by breaking it apart to understand the pieces (and then putting it back together again, as they come to understand these pieces consisting of RNA, DNA, and protein sequences). Then they seek to understand what it *does* by studying

FIGURE 15.2 Map of the human X chromosome (from the National Center for Biotechnology Information (NCBI) website; www.ncbi.nlm.nih.gov). Assembly of the human genome is one of the greatest achievements of bioinformatics.

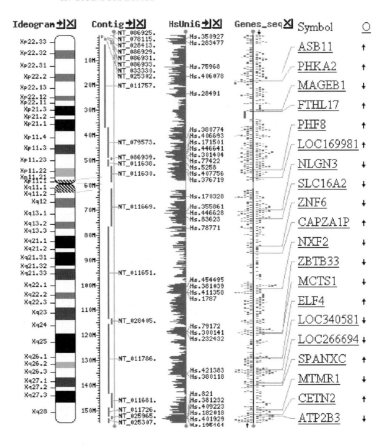

the production of proteins and how these proteins work to make all body structures and all body processes, including vision, feel, movement, cognition, behavior, and other things, all of which involve amino acids to proteins to organelles to cells to tissues to organisms. Finally, they seek to understand how the genome is related to the ability in humans to create things and processes far beyond what they were born with, even to build computing systems and robots that can begin to mimic the abilities of humans.

Some of the terms and concepts you initially need to be aware of in bioinformatics are as follows:

- A *DNA sequence* is always defined as a sequence of nucleotides, these nucleotides being the four biochemical structural components that Watson and Crick discovered in the 1950s: adenine, guanine, cytosine, and thymine.
- *Palindrome sequences* are nucleotide sequences that read the same way in both directions, as these play important biological roles.
- *RNA molecules* are the stranded structure coming from the DNA of the nucleus that goes out into the cell and does the work, e.g., making proteins that are either structural or regulate other chemical and structural pathways.

- *Genomics* is the study of organism-wide genes, "working with all the genes at the same time," e.g., considering the entire "blueprint."
- *PubMed* is a listing of the literature published on biology and genetics (from 1965 onward) that allows researchers in the field to find what they need quickly so that they can see what has already been researched and thus determine what new directions of research will be of value. See www.ncbi.nlm.nih.gov/entrez/query.fcgi?DB=pubmed.
- *BLAST* is a database mining program, used extensively in bioinformatics, to explore databases of information constructed from what has already been discovered and cataloged about DNA, RNA, and protein structures. BLAST will take a sequence and compare it to a database of already-analyzed sequences to see whether there are matches and thus help in identifying a protein or sequence. See www.ncbi.nlm.nih.gov/blast.
- ClustalW2 is a data analysis program that looks for multiple sequence alignments to see how a sequence that a researcher has just found fits into what is already known. See http://www.ebi.ac.uk/Tools/clustalw2/index.html.
- *FASTA* is another sequence alignment and database scanning program created in 1988, although ClustalW2, listed previously, is most widely used.
- *GenBank* is a repository for storing sequences of nucleotides and protein sequences so that researches can check their newly discovered sequences against ones already known. See www.ncbi.nlm.nih.gov/entrez/query.fcgi?db=Nucleotide.
- Entrez/Gene is one of the newer gene banks that is more "gene centric"; e.g., it allows questions/queries about a specific gene; some of these can be found at http://www.ncbi.nlm.nih.gov/sites/entrez/gene.
- The *Ensemble Project* has concentrated on the human genome in addition to other vertebrates. See www.ensembl.org/.

Let's look at a specific example of how the knowledge gained from bioinformatics is used in a practical way today. We will do this by looking at a specific technique called *PCR*, which stands for Polymerase chain reaction. This technique involves mixing the following in a test tube:

- The DNA template
- Primers, a mixture of nucleotides and other biochemicals
- A heat-resistant enzyme called DNA polymerase

What is the purpose of conducting this DNA polymerase chain reaction? The answer is to amplify sequences of DNA so that there is enough to work with in other biochemical tests. How can this procedure be used practically? For example, in forensic science a small DNA sample—from fingerprints (for example) or a licked stamp—will provide enough of the DNA to identify the person from whom this DNA came. See the following web site for more information: http://nature.umesci.maine.edu/forensics/p_intro.htm.

After sequence identification, alignment, 3D structure visualization, and other procedures to generate bioinformatics data, the vast opportunity for structured data mining analyses comes into play. But before we discuss the possible future of data mining for bioinformatics, let's look at a list of the tools and computer software available and in use today in this field.

TABLE 15.1　Free Resources for Doing Protein Amino Acid Sequence Analysis (e.g., Gene Sequence Analysis)

Name	Site URL	What It Works With
ExPASy	www.expasy.org/tools	Proteins
PSSMs	www.ncbi.nlm.nih.gov/blast/blastcgihelp.shtml#ps;pssm	Domains*
PIR	http://pir.georgetown.edu	Proteins
CBS	www.cbs.dtu.dk/services	Proteins
Hits	http://hits.isb-sib.ch/	Proteins
InterPro	http://www.ebi.ac.uk/Tools/InterProScan/	Domains*

*Domains = a portion of a protein that can "keep its shape" if you remove it from the rest of the protein; it consists of at least 50 amino acids; in one sense domains are like building blocks of a protein; each of the domains can function as a single type function and exist separately, but added together into a complete protein, there is a functioning unit (e.g., like an office; each person in each office cubicle can do a function, and in one sense exist separately; but when all of the people in all of the cubicles of an office are added together, they produce a complete product of that office or organization).

The Internet contains a variety of free resources for doing sequence analysis. Table 15.1 lists some of the more usable sites.

DATA ANALYSIS METHODS IN BIOINFORMATICS

BLAST, which stands for Basic Local Alignment and Search Tool, is the first and most popular data mining tool for DNA/protein sequences. Prior to the development of this tool, biologists had to search a database of published sequences, print them out, hang them on the wall, and look at them for hours to try to distinguish any patterns and make sense out of them. Also, they had to determine if any new sequence discovered was really new or existed already in the literature. BLAST does this automatically.

Other computer programs search databases like BLAST, but BLAST has been the most popular one. Some of the alternatives to BLAST are

- **Smith-Waterman (SSEARCH):** This is considered more accurate than BLAST but much slower in operation.
- **FASTA:** This is more accurate for DNA comparisons but slower than BLAST.
- **BLAT:** This can locate cDNA rapidly and also find close proteins (e.g., mouse versus man, as both are mammals).

These additional database search tools can be found at the sites shown in Table 15.2.

ClustalW2: Sequence Alignment

After sequences are found, these sequences have to be aligned, in an attempt to see the whole picture, to see what is important, and to determine what may just be extraneous or redundant sequences (there is a lot of, in fact considerable, redundancy in DNA). Several

TABLE 15.2 Some of the Most Used Genomic Database Search Tools

Region of World	Program Name	URL
USA	FASTA	http://Fasta.bioch.Virginia.edu/fasta
Europe	SSEARCH	www.ch.embnet.org/software/GMFDF_form.html
USA	BLAT	www.genome.ucsc.edu

tools can do this, with names like ClustalW2, Tcoffee, and MUSCLE, but ClustalW2 is the one most often used.

ClustalW2 is a progressive algorithm that uses a little trick to build up an alignment of sequences: it compares sequences two by two and eventually clusters them into what looks like a phylogenetic tree or, what statisticians call in traditional cluster analysis, a *dendogram*. ClustalW2 is among the most cited references in the entire history of biology. The *W* in ClustalW2 stands for weights; every sequence receives a weight proportional to the amount of new information it contributes to the overall alignment and genome. ClustalW2 can be found at the following site: www.ebi.ac.uk/Tools/clustalw2/index.html.

Once the sequences have been identified and a protein identified, the next thing that many bioinformatics researchers want to do is to look at the sequence in 3D format. This can be done by going to another NIH governmental web site: www.ncbi.nlm.nih. gov/Structure/. We will not go into detail here in this book but provide this site for the overall sketch of what bioinformatics is all about.

After looking at the 3D structure, a researcher may want to find proteins with similar shapes; this can be done at another NIH web site, known as VAST service: http://www. ncbi.nlm.nih.gov/Structure/VAST/vast.shtml.

Searching Databases for RNA Molecules

We will not go into detail on RNA molecules in this book, but we will give you the terminology, the references at the end of this chapter, and the web sites to get you started. There are several subtypes of RNA; they are listed with corresponding web sites in Table 15.3.

Databases abound in the bioinformatics field and become most useful resources, avoiding reinventing of a research protocol and thus making for efficiencies in this field of research. Some of the most practical databases are listed in Table 15.4.

WEB SERVICES IN BIOINFORMATICS

SOAP and REST-based interfaces have been developed for a wide variety of bioinformatics applications, allowing an application running on one computer in one part of the world to use algorithms, data, and computing resources on servers in other parts of the world. The main advantages lie in the end user's not having to deal with software and database

TABLE 15.3　RNA Web Sites, for RNA Types and Subtypes

Type/Name of RNA	URL	Description of the RNA Type
Micro RNAs, called miRNAs	www.ambion.com/techlib/resources/miRNA/index.html	Definition, visual overview, and database of miRNAs
Micro RNAs, called miRNAs	http://microrna.sanger.ac.uk/sequences/	miRNAs at the Sanger Center in the UK, a very extensive listing
Ribosomal RNA, called rRNA, the "larger unit"	www.psb.ugent.be/rRNA/1su/	A database on the "larger" of the two ribosomal units; also has online software
Ribosomal RNA, called rRNA, the "smaller unit"	www.psb.ugent.be/webtools/rRNA/ssu/	A database on the "smaller" of the two ribosomal units
Noncoding or nontranslating regions of genes	www.hsls.pitt.edu/guides/genetics/obrc/rna/sequences/URL1101850039/info http://www.ncbi.nlm.ni.gov/pubmed/9399823?dopt=Abstract	Small noncoding RNAs; these are the untranslated regions of genes
tmRNAs = function as both transfer and messenger RNAs	www.indiana.edu/~tmrna/	Fairly recently discovered type of RNA, the tmRNAs
Generic RNAs = all types of RNA	www.imb-jena.de/RNA.html	"RNA World," a very complete RNA web site

TABLE 15.4　Important Bioinformatics Databases to Help You Start Your Bioinformatics Studies or Research

Name of Database	URL	Type of Data Found in This Database
GenBank / DDB / EMBL	www.ncbi.nlm.nih.gov/entrez/query.fcgi?db=Nucleotide	Nucleotide sequences of RNA and DNA
PubMed	www.ncbi.nlm.nih.gov/entrez/query.fcgi?DB=pubmed	Literature references
NR	www.ncbi.nlm.nih.gov/entrez/query.fcgi?db=Protein	Nonredundant protein sequences (e.g., proteins that are unique and perform known functions)
OMIM	www.ncbi.nlm.nih.gov	Genetic diseases

maintenance overhead. Basic bioinformatics services are classified by the European Bioinformatics Institute (EBI—http://www.ebi.ac.uk/services/) into three categories: Sequence Search Services (SSS), Multiple Sequence Alignment (MSA), and Biological Sequence Analysis (BSA). The availability of these service-oriented bioinformatics resources demonstrate the applicability of web-based bioinformatics solutions and range from a collection of

TABLE 15.5   Important Software Programs Used in Bioinformatics*

Category	Name	URL	Function
Prediction	GenScan	http://genes.mit.edu	Prediction of genes, e.g., DNA
Prediction	PsiPred	http://bioinf.cs.ucl.ac.uk/psipred/	Prediction of protein structure
Prediction	Mfold	www.bioinfo.rpi.edu/applications/mfold/	Prediction of RNA structure
Visualization	Logos	http://weblogo.berkeley.edu	MSA (=Multiple Sequence Alignment) visualization
Visualization	Rasmol	www.umass.edu/microbio/rasmol/	Visualization of structure

*BLAST and ClustalW2 are also very important but were discussed earlier in this chapter.

standalone tools with a common data format under a single, standalone, or web-based interface, to integrative, distributed, and extensible bioinformatics workflow management systems.

The field of bioinformatics, with its development of many kinds of freeware, has been rather chaotic. One explanation for this problem is that bioinformatics is a very large field, including many different knowledge domains, each with its own practitioners working in isolation from each other. It is only recently that these diverse areas have been brought together under the same umbrella of bioinformatics. Later, we will suggest that we are currently in a stage of "sifting out" what is really needed and that future years will bring a more or less standard data analysis methodology, involving data mining and text mining. But for now, Table 15.5 provides a list of some of the software programs that are being used in the field of bioinformatics. BLAST, which is the major database search tool, and ClustalW2, which is the major multiple alignment of sequences tool, were discussed previously so are not included in the table.

HOW DO WE APPLY DATA MINING METHODS TO BIOINFORMATICS?

There is a multitude of open source and freeware computer software sources on the Web, which provide programs written by various bioinformatics researchers. Additionally, some have written books explaining how users can write their own programs. One of these is with the use of Perl, available as an open source code base. Tisdall (2001) wrote *Beginning Perl for Bioinformatics*, in which sequences examples are emphasized. Perl is a relatively easy computer language to learn, and we mentioned it in Chapter 9 on text mining. Tisdall (2001) shows how to build an interface for the NIH GenBank database, which is mentioned earlier in this chapter. He also provides an introduction on how to use Perl to interface with the BLAST sequence alignment tool, mentioned earlier in this chapter. It allows the user to develop skills in using Perl to parse annotations in GenBank and BLAST output.

Some of the other algorithms and concepts that can be found in more technical literature in the field of bioinformatics are described in the paragraphs to follow.

Gollery (2008) and many other sources describe Hidden Markov Models (HMM) and related methods, like the HMMER package, Sequence Analysis Method (SAM), and the PSI-BLAST algorithm. This area has its own set of HMM databases with names like Pfam, PANTHER, COG, and metaSHARK. We won't discuss these programs here; we mention them only to whet your appetite, in case bioinformatics is an area you would like to study further. Googling any of these terms will provide a plethora of information.

Shui Qing Ye (2008) describes some additional DNA and genome analysis tools, some for phylogenetic analysis, SNP analysis, haplotype analysis, and regulation of gene expression. SAGE is one of the tools described. SNP, pronounced "snip," stands for Simple Nucleotide Polymorphism. A *polymorphism* is a substitution in one base of the code of DNA and is what makes some people have blue eyes and other people have brown eyes, for example, among many examples of polymorphisms we could list. One of the co-authors spent his doctoral candidate years looking for polymorphisms in blood enzymes of mice; this was years ago, shortly after Watson and Crick discovered the DNA molecule. So, you can see that basic genetics and polymorphisms, DNA, and RNA sequences are still in active service in the twenty-first century as we continue to work out the details of bioinformatics. Haplotype analysis involved looking at the haploid genotype. Complete genetic complements of most organisms contain two of each type of chromosome. A set of one of each chromosome is called a haploid, with one haploid contributed by the male and other by the female. The complements of genes in a haploid are called a haplotype. Haplotypes are inherited as units; they consist of a combination of alleles at different markers along the same chromosome.

Serial Analysis of Gene Expression, or SAGE, was developed in the mid-1990s. It provides an overview of a cell's complete gene activity. The technique captures RNAs and then allows a quantitative analysis of the transcripts of information that are made from these RNAs. It has the potential of developing a catalog of not only the mRNAs present in a cell, but also their prevalence.

For those who would like to gain a practical understanding of what bioinformatics is all about, we recommend Shui Qing Ye's book *Bioinformatics: A Practical Approach*, as it contains not only some unique biocomputing tools including use of Perl and R languages, but also useful web sites and database listings. Particularly, we refer you to "Tutorials" in Ye (2008), like the one on page 352, which takes you step by step through analyzing a protein sequence, and the one on page 456, which again takes you step by step through the process of creating a Perl script.

In her book on bioinformatics, Parida (2007) clearly presents the opinion that most bioinformatic success stories require algorithmic and statistical ingenuity. As such, she discusses and develops a considerable number of algorithms for better pattern discovery in the bioinformatic field. These include such unique names as

- Prim's algorithm
- Fitch's algorithm
- Discovery algorithm

- Pattern statistics algorithms, such as
  - Trees—Counting Binary Trees
  - Trees—Enumerating Unrooted Trees (Prufer Sequence)
  - Bayes' theorem with multiple events
  - Probability spaces
  - Discrete probability distributions (Binomial, Poisson)
  - Continuous probability space
- Parikh mapping-based algorithm
- Naïve algorithm
- Uno-Yagiura RC algorithm

We won't discuss these algorithms in detail because this chapter provides only an overview of bioinformatics. As mentioned before, the field is so large that it is a topic of several volumes in itself. The purpose here is only to acquaint you with this rapidly developing field and give an idea of how data mining and text mining will play an increasingly important part in it.

Our final reference is Mitra et al. (2008). In this source, the authors discuss the development and application to bioinformatics of such machine learning methods as electron density map interpretations, bi-clustering, and their application to such practical things as cancer tumor treatment. Some of the specific machine learning and other statistical methods they discuss are listed here to illustrate another group of terms that is part of this very large field. They illustrate that both traditional–frequentist methods and Bayesian and machine learning methods will be needed to solve all the issues in bioinformatics. These methods include

- Frequentist statistical inference
- Bayesian inference
- Unsupervised Learning methods, such as
  - Principal components analysis
  - Multidimensional scaling
  - Cluster analysis
- Fuzzy Sets (FS)
- Evolutionary Computing (EC)
- Rough Sets (RS)
- Network inference
- Bi-clustering, including
  - Multi-object bi-clustering
  - Fuzzy possibilistic bi-clustering
- 3D protein image analysis, using
  - ARP/WARP
  - RESOLVE
  - TEXTAL
  - ACMI
- Practical tumor classification using Bayesian machine learning methods, using various types of Support Vector Machines, called
  - Reproducing Kernel Hilbert Space (RKHS) based classification

Using this last approach to cancer tumor classification, Mitra et al. (2008) provide practical examples of and benchmark comparisons to earlier less-accurate methods of classification (see pages 316+), to such cancers as leukemia and colon tumors.

POSTSCRIPT

Statistical and data mining analysis in bioinformatics is still in somewhat of a chaotic state of trial and error. Researchers in this field are still debating the value of this or that method; the field does *not* yet have a standard methodology of analyzing its very large and complex databases. As Parida (2007) pointed out in the introductory sentences of her book:

> Major scientific discoveries have been made quite by accident: however, a closer look reveals that the scientists were intrigued by a specific PATTERN in the observations; an excellent example is Edward Jenner and the discovery of how milkmaids in England developed immunity to smallpox, followed by the development of the smallpox vaccination, and the eventual eradication, a few years ago, of smallpox on planet Earth. (page 1; paraphrase)

Pattern discovery is the operation that data mining and text mining excel in doing. These tools are necessary in the study of any complex bioinformatics phenomena. Ideally, in the future, a data mining/pattern recognition system will have standardized components, bringing bioinformatics and DNA microarray analysis into a more mature field, and providing practical ways of making accurate decisions in many areas, including disease diagnosis with accompanying action plans with a known probability of success.

Tutorial Associated with This Chapter on Bioinformatics

On the accompanying DVD packaged with this book, go to the Tutorial section and select the "Tutorial—CancerGene," where a *K*-nearest neighbors Bayesian model is used to show a 99% accuracy prediction of breast cancer.

A "Text Mining in Bioinformatics & Medical Informatics" white paper is included on the DVD; go to the TUTORIAL – PDF_PPT_ETC folder to find this paper; it contains a good discussion of the issues surrounding text mining in the bioinformatics field.

Books, Associations, and Journals on Bioinformatics, and Other Resources, Including Online

Associations and Organizations

- Bioinformatics Organization (Bioinformatics.Org): The Open-Access Institute (http://www.bioinformatics.org/)
- EMBnet (http://www.embnet.org/)
- European Bioinformatics Institute (http://www.ebi.ac.uk/)
- European Molecular Biology Laboratory (http://www.ebi.ac.uk/)
- The International Society for Computational Biology (http://www.iscb.org/)

- National Center for Biotechnology Information (http://www.ncbi.nlm.nih.gov/)
- National Institutes of Health home page (http://www.nih.gov/)
- Open Bioinformatics Foundation: umbrella nonprofit organization supporting certain open source projects in bioinformatics (http://www.open-bio.org/wiki/Main_Page)
- Swiss Institute of Bioinformatics (http://www.isb-sib.ch/)
- Welcome Trust Sanger Institute (http://www.sanger.ac.uk/)

Major Journals

- *Algorithms in Molecular Biology* (http://www.almob.org/)
- *Bioinformatics* (http://bioinformatics.oxfordjournals.org/)
- *BMC Bioinformatics* (http://www.biomedcentral.com/bmcbioinformatics)
- *Briefings in Bioinformatics* (http://bib.oxfordjournals.org/)
- *Evolutionary Bioinformatics* (http://www.la-press.com/journal.php?journal_id=17)
- *Genome Research* (http://genome.cshlp.org/)
- *The International Journal of Biostatistics* (http://www.bepress.com/ijb/)
- *Journal of Computational Biology* (http://www.liebertpub.com/products/product.aspx?pid=31)
- *Cancer Informatics* (http://www.la-press.com/cancer-informatics-journal-j10)
- *Journal of the Royal Society Interface* (http://publishing.royalsociety.org/index.cfm?page=1058)
- *Molecular Systems Biology* (http://www.nature.com/msb/index.html)
- *PLoS Computational Biology* (http://www.ploscompbiol.org/home.action)
- *Statistical Applications in Genetic and Molecular Biology* (http://www.bepress.com/sagmb/)
- *Transactions on Computational Biology and Bioinformatics - IEEE/ACM* (http://portal.acm.org/toc.cfm?id=J954)
- *International Journal of Bioinformatics Research and Applications* (http://www.inderscience.com/browse/index.php?journalcode=ijbra)
- List of bioinformatics journals (http://www.bioinformatics.fr/journals.php)

Other Sites

- Human Genome Project and Bioinformatics (http://www.ornl.gov/sci/techresources/Human_Genome/research/informatics.shtml) *Excellent resource site—Co-Authors.*
- List of Bioinformatics Research Groups (http://www.bioinformatics.fr/laboratories.php)
- List of Bioinformatics Research Groups at the Open Directory Project (http://www.dmoz.org/Science/Biology/Bioinformatics/Research_Groups/)

References

Gollery, M. (2008). *Handbook of Hidden Markov Models in Bioinformatics*. Boca Raton, FL: Chapman & Hall/CRC Press.

Mitra, S., Datta, S., Perkins, T., & Michailidis, G. (2008). *Introduction to Machine Learning and Bioinformatics*. Boca Raton, FL: Chapman & Hall/CRC Press.

Parida, L. (2007). *Pattern Discovery in Bioinformatics: Theory & Algorithms*. Boca Raton, FL: Chapman & Hall/CRC Press.

Tisdall, J. (2001). *Beginning Perl for Bioinformatics*. Sebastopol, CA; Cambridge, MA: O'Reilly Media, Inc.

Bibliography

Baldi, P., & Brunak, S. (2001). *Bioinformatics: The Machine Learning Approach* (2nd ed.). Cambridge, MA: MIT Press, A Bradford Book.

Claverie, J.-M., & Notredame, C. (2007). *Bioinformatics For Dummies*. New Jersey and Indiana: Wiley Publishing, Inc.

Cristianini, N., & Hahn, M. W. (2007). *Introduction to Computational Genomics: A Case Studies Approach*. Cambridge, MA: Cambridge University Press.

Draghici, S. (2003). *Data Analysis Tools for DNA Microarrays*. Boca Raton, FL: Chapman & Hall/CRC.

Gentleman, R. (2009). *R Programing for Bioinformatics*. Boca Raton, FL: Chapman & Hall/CRC Press.

Knudsen, S. (2002). *A Biologist's Guide to Analysis of DNA Microarray Data*. Hoboken, NJ: Wiley-Liss.

Lesk, A. M. (2002). *Introduction to Bioinformatics*. Oxford, UK and New York: Oxford University Press.

Majoros, W. H. (2007). *Methods for Computational Gene Prediction*. Cambridge, MA: Cambridge University Press.

Parmigiani, G., Garrett, E. S., Irizarry, R. A., & Zeger, S. L. (2003). *The Analysis of Gene Expression Data: Methods and Software*. Berlin, Germany, and Hoboken, NJ: Springer.

Schena, M. (2003). *Microarray Analysis*. Hoboken, NJ: Wiley-Liss.

Simon, R. M., Korn, E. L., McShane, L. M., Radmacher, M. D., Wright, G. W., & Zhao, Y. (2004). *Design and Analysis of DNA Microarray Investigations*. Berlin, Germany, and Hoboken, NJ: Springer.

Tramontano, A. (2006). *Introduction to Bioinformatics*. Boca Raton, FL: Chapman & Hall/CRC Press.

Xiong, J. (2006). *Essential Bioinformatics*. Cambridge, MA: Cambridge University Press.

Ye, Shui Quig. (2008). *Bioinformatics: A Practical Approach*. Boca Raton, FL: Chapman & Hall/CRC.

Customer Response Modeling

PREAMBLE

Most organizations, whether for profit or nonprofit purposes, exist to develop and promote some things or ideas related to their organization. One of the major activities of these organizations is to appeal to people outside their organizations to join them, support them, or purchase their goods or services. Traditional means of doing this included offering goods and services in storefronts, by advertisements in appropriate venues, and by contacting a broad spectrum of people by phone or mail. These methods are rather passive. The philosophy was to build it, show it, advertise or promote it, and customers would come.

Since the early 1990s, many businesses have taken a more active approach by using various technological approaches to identify specific prospective customers and going after their business, rather than waiting for them to respond to the passive appeals. The key issue in this process is identifying *which* prospects are most likely to respond to the appeals. The activity of identifying prospects and quantifying their likelihood to respond is one of earliest applications of data mining technology to business.

EARLY CRM ISSUES IN BUSINESS

One of the early business issues to be addressed with data mining technology in the mid-1990s was Customer Relationship Management (CRM). CRM systems were built to manage how a business relates to its customers. Customer-facing systems were built to manage call centers and to inform marketing and sales efforts. In support of the marketing and sales channels, analytical modeling systems were built by pioneers in data mining technology. NCR built some of the earliest analytical CRM product suites in 1998 in the form of Churn*Sentry* (for customer retention modeling) and Growth*Advisor* (for cross-sell and up-sell modeling). Both of these products included a data discovery tool, a model manager, numerous canned reports (via Cognos), and a campaign management system. Soon, other CRM systems were built, notably by Siebel and Vantive, to serve sales force automation, but later extended to cover call centers and some front-end office operations.

On the analytical side of CRM, the major foci were

- Customer response modeling with predictive analytics for
  - Customer acquisition
  - Customer retention
  - Customer up-sell (selling an enhanced product or service)
  - Customer cross-sell (selling a different product or service)
- Customer Lifetime Value (LTV) modeling

The trend in marketing with analytics was to move from a broadcast marketing operation to a one-to-one marketing operation. Naturally, the key in this activity was predicting which products or services a particular customer was likely to respond to. The most common approach used to do this was to model customer actions in the past and use the model to predict actions in the future. This is a form of human behavioral modeling.

KNOWING HOW CUSTOMERS BEHAVED BEFORE THEY ACTED

To be competitive in today's markets, we must capture and leverage information from historical detail records describing what our customers did in the past. This information can be very useful in defining patterns in the behavior of customers leading up to the decision to leave the company. For a given customer, the decision to leave the company did not happen in a vacuum. Many factors contributed to this decision, such as dissatisfaction with service, perception of the greater value of competitive goods and services, and changes in business needs. Some of these factors, such as customer satisfaction, can be tracked through customer care programs. However, most factors that contribute directly to attrition cannot be captured and stored in corporate databases. The only way to reflect these attrition variables is to relate them to customer behavior patterns that can be tracked from data in the data warehouse. The pattern of historical information of customers who have left the company can be used to predict which present customers have a high probability of leaving in the near future. How is this possible?

Transforming Corporations into Business Ecosystems: The Path to Customer Fulfillment

Ever since the Industrial Revolution, Western society has tended to view the world as a *machine*, composed of components that functioned like cogs, wheels, and springs.

Newton formalized this approach in science. However, it worked only within the range of Newton's instruments. Later discoveries by Einstein (relativity) and quantum physicists caused the Newtonian concept of the world to fall to pieces!

Business also picked up on this metaphor in the Industrial Revolution. The automobile assembly line of Henry Ford was viewed as the paragon of efficiency.

As long as the product was relatively simple in organization, this metaphor appeared to work. An efficient business became defined in terms of

- A "well-oiled machine"
- "Having momentum"
- "Gaining steam"
- "Firing on all eight cylinders"

The primary business unit became the *corporation*. The prevailing attitude was "Us against Them." Only the strong competitors survived. For these corporations, the primary business activity was *production.* It was expected that revenue would be maximized as production was optimized. Generations of operations research practitioners sought to optimize processes that would maximize business revenue.

With the advent of fast computers, flexible communications, and the Internet, a new business paradigm has emerged: the *business ecosystem* (Inmon et al., 1998). Moore (1996) maintains that real competition in these business ecosystems is not dead (actually, it is intensifying); it has just changed its expression. The old expression of competition pitted offers and markets against each other. The products improved as companies listened to customers and made the products fit their desires. The problem with this approach is that it ignores the *environment* and the *system* in which those offers and markets are embedded. It also ignores the great benefit that can come with co-evolution with other "competitors" to satisfy customers more than if they operated separately. Moore stresses the importance of the environment and the system in which our businesses are enmeshed. This emphasis points also to the need to consider systems' effects in our analyses of customer behavior.

As businesses became more complex, the machine metaphor began to break down. In both science and business, it became increasingly obvious by the 1980s that we had to begin to look at the world in a different way. In these increasingly complex systems, there seemed to be important properties that did not emerge until the system was complete and operating as a whole. These *emergent* properties often controlled the major responses of the system. These influences are causing a profound shft in science and business toward viewing the world as *organism!*

Petzinger (1999) remarks that the key characteristic of modern civilization is that of *economizing,* and that our genes are programmed for business.

This view of business as "organism" flowed out of the the central concept proposed by Rothschild (1990):

There is a parallel between the response of natural systems to rapid environmental change and the response of business systems to rapid technological change. (p. xiii)

From this principle, it is argued that our view of the world-as-machine greatly hinders us from economizing very well in this age of rapid technological change. Why? Because the rules keep changing faster than our machine-like business systems can accommodate. Perhaps it is time for a "new" science to help us understand life in the midst of rapid change (Rothschild, 1990).

CRM IN BUSINESS ECOSYSTEMS

In the freewheeling business of today, companies try to build Customer Relationship Management programs that aim to create the same kinds of relationships with their customers. To build these relationships, companies must learn to understand their customers. To understand their customers sufficiently to build effective customer relationships, marketers must

- Learn how to identify the right set of customers to do business with (segmentation);
- Learn how to identify valued customers;
- Learn how to recognize danger signals in their data relating to customer behavior that, if unchecked, might lead to decisions to leave the company;
- Use segments defined by attrition probability algorithms to strengthen and maintain relationships with valued members of the existing customer base.

The key principle in this approach is that the most powerful predictors of customer behavior in the future are customer behavior patterns in the past. Other customer characteristics are important also in defining patterns of customer behavior (i.e., demographic and firmographic information). However, unless we include in our models of customer behavior the patterns of past customer behavior related to their future actions, they will not be very powerful predictors of what customers actually do. When these patterns are combined with the more static customer information gathered by businesses in their day-to-day operations (e.g., the date a business started business), companies can take a quantum leap forward in understanding the customer and improving customer loyalty.

Differences Between Static Measures and Evolutionary Measures

The key difference between historical behavior patterns and relatively static characteristics of customers is that historical patterns enable us to track the *development* of the decision to leave rather than just the decision itself. These evolving behavior patterns are very organic in nature and are driven by a number of significant nonlinear events (NLEs). Farrel (1998) maintains that bursts of customer demand (or "anti-demand" like attrition) are driven by these NLEs. The evolutionary nature of these NLEs renders them much richer in predictive value than static characteristics alone because they can capture the mood of the customer, preferences, attitudes, and many clues that help you to understand why the customer did what he did. Some static characteristics are certainly related to the attrition decision, but they tell only a part of the story. To see the other half of the story, we must add variables that express this development of the decision to leave the company. This is a very organic view of customer behavior similar to the way biologists view the complementary effects of intrinsic (organism-based) and extrinsic (environmental) influences on organism response. This viewpoint represents a dramatic shift in mindset from the traditional way that many companies view their data.

How Can Human Nature as Viewed Through Plato Help Us in Modeling Customer Response?

If human nature is a common basis for human action, then to predict the action of customer response, we must model human nature. We must focus on variables available to us in our databases that reflect some aspect of human nature that leads to the response. These variables might include

- Historical customer care data
- Historical use of company services
- Historical billing revenue data
- Historical contract data
- Selected demographic data

How Can We Reorganize Our Data to Reflect Motives and Attitudes?

The key to successful customer response modeling is to associate with each customer a historical time-series of fields (selected from those listed in the preceding section) that in some way *reflect* motives and attitudes that *caused* the customer decision. These motives and attitudes flowing out of our human nature are the reality behind the "shadows" of the action. To see the deeper reality of what causes these shadows, we must turn around, so to speak (like those in Plato's cave), and look at the data in a different way. We must abstract information from the time-series of customer response in a form that is related to the customer action to be modeled. These abstractions are called *temporal abstractions*.

The use of temporal abstractions has attracted widespread interest in medical and pharmaceutical informatics for predicting patient responses (Kahn et al., 1991; Haimowitz and Kohane, 1996; Kattan et al., 1997). Temporal abstractions are one type of data abstraction used to map data elements to some context environment. Data abstractions can be classified into four groups (Lavrac et al., 2000):

- **Qualitative abstraction:** A numeric expression is mapped to a qualitative expression. For example, in an analysis of teenage customer demand, compared to that of others, customers with ages between 13 and 19 could be abstracted as a value of 1 to a variable "teenager," while others are abstracted to a value of 0.

- **Generalization abstraction:** An instance of an occurrence is mapped to its class. For example, in an analysis of Asian preferences, compared to non-Asian, listings of "Chinese," "Japanese," and "Korean" in the Race variable could be abstracted to 1 in the Asian variable, while others are abstracted to a value of 0.
- **Definitional abstraction:** One data element from one conceptual category is mapped to its counterpart in another conceptual category. For example, when combining data sets from different sources for an analysis of customer demand among African-Americans, you might want to map "Caucasian" in a demographic data set and "White Anglo-Saxon Protestant" in a sociological data set to a separate variable of "Non-Black."
- **Temporal abstraction:** A variable in a time domain with one reference is mapped to a time domain with a different reference.

The first three types of data abstractions are usually referred to by data miners as forms of *recoding*. The fourth type, temporal abstraction, is not commonly used. However, the methodologies of several data mining tool vendors have (or had) forms of temporal abstractions integrated into their design:

- **SAS Enterprise Miner:** Ability to define "lag" variables
- **Orchestrate–PreludePLUS™** by Torrent Systems (now owned by IBM)
- **Churn*Sentry*™** and **Growth*Advisor*™** by NCR
- **KXEN™:** Knowledge Extraction Engine

What Is a Temporal Abstraction?

Modeling customer behavior with temporal abstractions involves rearranging all the modeling variables to more clearly reflect patterns of change in the customer response variable. Then the modeling tool can easily recognize the pattern that exists between the response variable and various states of predictor variables in the past with respect to the response variable. These time-series representations of each predictor variable are a form of temporal abstraction. See Figure 16.1.

Figure 16.1 displays fields in six customer records from a telecommunications company lined up like beads on an abacus. The data on the left abacus represent information stored for monthly call duration extracted from multiple records in the database. This arrangement is similar to the format of the data extracted from databases into flat files to be submitted to the modeling tool for analysis. In the default configuration (left abacus), the yellow beads (a given state in the time-series) are scattered all over the abacus. The diagram on the right of Figure 16.1 shows the rearranged data. Now, the yellow beads are lined up. The pattern emerges to the physical senses of our eyes and likewise to the mathematical senses of the modeling tool.

The same approach can be used to model customer fraud, or propensity-to-buy to serve cross-selling and up-selling campaigns. In Chapter 1, we showed that we must include both Aristotelian and Platonic approaches to truth to model a complex system. Customer behaviors in the context of the business ecosystems within which they operate can be modeled successfully using this combined approach. This approach to customer behavior modeling will permit us to see the "shadows" of customer behavior (following Aristotle) and

How Temporal Abstractions Work

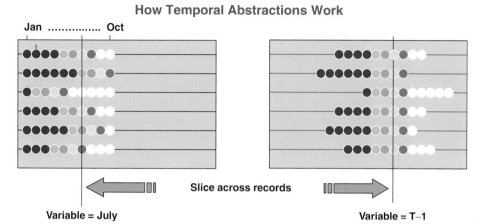

FIGURE 16.1 Pattern emergence facilitated by a temporal abstraction.

reflections of the causes (the deeper reality) of this behavior (following Plato). Such a combined perspective on the nature of customer behavior can provide much more powerful models than those based on one perspective alone.

Example

We can see the relative contributions of temporal abstraction and static variables in modeling voluntary attrition (disenrollment) among customers of a large insurance company (see Nisbet, 2004). These events were modeled separately with each of two variable sets: one using temporal abstraction variables and one set without them. The temporal abstractions were created by taking quarterly snapshots of policy records for a given household. The snapshots represent temporal objects in a temporal database (Jensen et al., 1996). The temporal abstractions represent keys of this derived temporal database in which the temporal tuples are the response quarter and a given quarter prior to the response. These snapshot temporal abstractions follow Snapshot Dependency Theory as extended by Wijsen et al. (1993) and formalized by Wijsen (2001), and they represent keys for a sequence of snapshot relations in the household insurance policy history indexed in reference to the response quarter.

Sir R. A. Fisher designed his statistical tools for use in the medical world to permit different researchers to analyze the same data and get the same results. Previous (Bayesian) statistical methods with their subjective "priors" did not lend themselves well to that end. To make these methods work, scientists had to perform controlled experiments, holding all variables constant and varying the treatment of one variable at a time. Results were compared to a "control" group with no treatments. Laboratory conditions of temperature, light, moisture, etc., often had to be held constant because the physics of variable response might be affected by the environment. These highly controlled conditions are almost never found outside a laboratory, but business analysts used these methods anyway.

Machine learning technology (particularly, neural nets) developed in the AI community was not based on calculation of "parameters" like standard deviation. Modern neural nets do not depend on data drawn from a distribution of any particular kind (e.g., normal distribution). Patterns in data sets can be modeled directly in the form of weights assigned to each input variable.

The tool chosen for the analysis of the insurance disenrollment event was an automated backpropagation neural net in a prior version of SPSS Clementine (Version 5.1). Data preparation of the temporal abstraction variables was done with a C-program outside the data mining tool (at that time, no data mining tool could do that). A Clementine stream was designed to input data, train the neural net, and score the holdout data set with the trained model (Figure 16.2).

A second Clementine stream was used to aggregate and decile the scored list, and to create the lift curves (Figure 16.3).

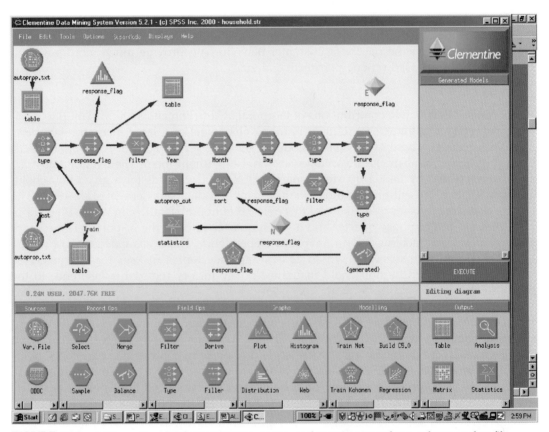

FIGURE 16.2   A Clementine visual programming stream used to train a neural net and score a data file.

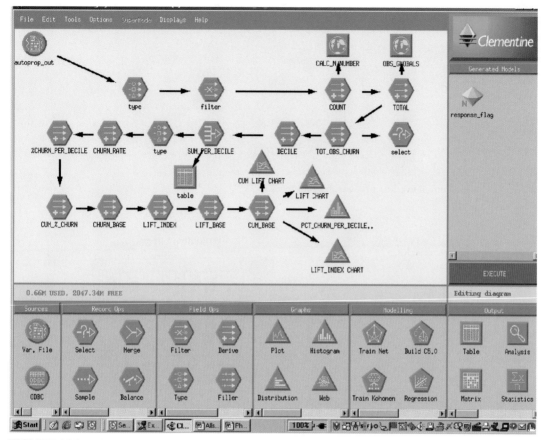

FIGURE 16.3   Clementine stream used to create the lift curves.

Results

The cumulative lift diagram (Figure 16.4) is created by plotting the cumulative response(s) along with the cumulative response that you would expect from random selection. The random number expected for customer response in each decile is 10% of the total. The figure shows that the random expectation for customer response increases by 10% each decile (shown by the diagonal red line). The difference between the response line (blue) and the random expectation line (red) reflects the "lift" that the model gives to the predicted response rate for a given decile. The total area between the lift curve and the random line represents the total effect of the model for increasing the total customer response across all deciles of the scored list.

Comparison of Static and Temporal Effects

The lift curve (Figure 16.4) was calculated with holdout data scored by the model, which used both static and temporal abstraction variables. Another model was trained using only

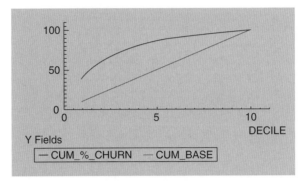

FIGURE 16.4   Cumulative lift curve for disenrollment.

FIGURE 16.5   Lift curve for static plus
temporal abstraction variables.

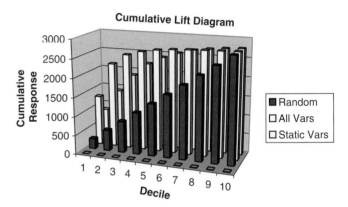

the static variables, and the results plotted together with those for the model using all of the variables. Figure 16.5 shows that only about 60% of the lift (extension of the bar above random for a given decile) was due to the static variables. The rest of the lift is provided by the temporal abstraction variables.

CONCLUSIONS

The biological metaphor appears superior to the machine metaphor in helping us understand the causes of customer behavior in business. Customers are biological entities that respond in a biological manner. This manner is the result of the complex interaction of many factors that operate as a system to influence the nature of the response. It seems reasonable to expect that the only way to create highly predictive models of customer behavior is to express some degree of this complex interaction in the design of the modeling methodology.

From this perspective, it is easy to see why static variables alone will detect only part of the signal of customer response. Most attempts at modeling customer response are confined to analyzing historical variables and their transforms as if they are time-independent influences on customer response. Occasionally, time-series analysis is applied to capture time-dependencies. However, time-series analyses (like all parametric methods, as represented in Chapter 1) suffer from many assumptions that are not satisfied in business data (e.g., linear additivity, variable independence, membership in a specific distribution such as a normal distribution). Therefore, many analyses of historical data treat the data elements as if they were static variables with no temporal attributes. For example, call durations for March, April, and May are part of a time-series of historical data for a given customer, but they are usually submitted to the modeling tool without regard to their sequence or relationship to the point in time of the customer response. This relationship constitutes a time-dependency between the response date and the sequence in the state of each variable prior to that date.

Temporal abstractions permit expression of this time-dependency between an event and the influences affecting its occurrence. The new variables derived from these temporal expressions provide a rich source of predictability for customer response. They provide insights into why the customer acted and represent the "other half" of the story of customer response that we can see in our databases. Temporal abstractions (and trend variables calculated from them) permit the data mining tool to capture much of the signal of customer response resident in the time-series of the historical data.

POSTSCRIPT

In the preceding discussion of Customer Relationship Management models, we restricted our focus to just those issues related to building the business base, from which increased profitability could ensue. But there are other issues related to profitability that are not related to building the customer base, but to shrinking it. One such issue is the incidence of fraud. Chapter 17 will explore some of the issues and challenges of modeling the exceedingly rare (but potentially devastating) effects of fraud.

References

Farrel, W. (1998). *How Hits Happen*. New York: Harper-Business.

Haimowitz, I. J., & Kohane, I. S. (1996). Managing temporal worlds for medical trend diagnosis. *Artificial Intelligence in Medicine, 8*(3), 299–321.

Inman, W. H., Imhoff, C., & Sousa, R. (1998). *Corporate Information Factory* (pp. 1–11). Hoboken, NJ: Wiley Computer Publishing.

Jensen, C. S., Snodgrass, R. T., & Soo, M. D. (1996). Extending existing dependency theory to temporal databases. *IEEE Trans Knowledge and Data Engr, 8*(4), 563–581.

Kahn, G. M., Fagan, L. M., & Sheiner, L. B. (1991). Combining physiologic models with symbolic methods to interpret time-varying patient data. *Meth Inform Med, 30*, 167–178.

Kattan, M. W., Oshida, H., Scardino, P. T., & Beck, J. R. (1997). Applying a neural network to prostate cancer survival data. In N. Lavrac, E. Keravnou, & B. Zupan (Eds.), *Intelligent Data Analysis in Medicine and Pharmacology* (pp. 295–306). Boston: Kluwer Academic Publishers.

Lavrac, N., Keravnou, E., & Zupan, B. (2000). *Intelligent Data Analysis in Medicine*. White Paper. Slovenia: Faculty of Computer and Information Sciences, University of Ljubljana.

Moore, J. F. (1996). *The Death of Competition: Leadership and Strategy in the Age of Business Ecosystems* (pp. 1–23). New York: Harper-Business.

Nisbet, B. (2004). Temporal abstractions model customer behavior in business ecosystems: Insightful data mining. *PC-AI Journal, 16*(6), 34–41.

Petzinger, T., Jr. (1999). *The New Pioneers: The Men and Women Who Are Transforming the Workplace and Marketplace* (p. 23). New York: Simon & Schuster.

Rothschild, M. (1990). *Bionomics—Economy as Ecosystem*. New York: Henry Holt.

Wijsen, J. (2001). Trends in databases: Reasoning and mining. *IEEE Trans Knowledge and Data Engr, 13*(3), 426–438.

Wijsen, J., Vanderbulcke, J., & Olivie, H. (1993). Functional dependencies generalized for temporal databases that include object-identity. *Proc. Int'l Conf. Entity-Relationship Approach* (pp. 100–114). Arlington, TX.

Fraud Detection

PREAMBLE

Fraud can be defined as a criminal activity, involving false representations to gain an unjust advantage (*Concise Oxford Dictionary*). Fraud occurs in a wide variety of forms and is ever changing as new technologies and new economic and social systems provide new opportunities for fraudulent activity. The total extent of business losses due to fraudulent activities is difficult to define. One estimate claims that financial losses range from $100–150 billion per year. The Association of Certified Fraud Examiners estimates that U.S. organizations lose about 7% of their revenues to fraud. If this were to hold true for all organizations contributing to the Gross Domestic Product of about $14 trillion for 2007, fraud losses could be as high as $1 trillion.

This discussion of fraud detection is not intended to be inclusive of all types of fraud, nor is it definitive of even the types discussed in the following sections. The purpose of this chapter is to introduce you to fraud detection, give you a simple example of how to build a fraud model, and direct you to additional references to broaden and deepen your knowledge of the vast scope of fraud detection.

ISSUES WITH FRAUD DETECTION

Fraud Is Rare

Fraud is usually a rare event and often exceedingly so. Identifying fraud is very difficult because of its rarity and because its very nature is stealthy. This stealthy action is directed against an external individual or organization (public or private) for the purposes of some sort of gain. The vast majority of the records (i.e., 99.9%) may be legitimate. Only 0.1% of the records may be fraudulent. It may be relatively easy to build a fraud model on these records that is 99% accurate (overall). For other modeling problems in business, this accuracy would be exceeding high. But this model would miss 9 out of 10 fraudsters! Much more time must be spent to identify many more of the 9 fraudsters that would be missed. Often, the extra accuracy is associated with higher cost, but the cost of *not* doing so may be much higher.

Fundamentally, fraud is a form of human response that can be modeled in ways very similar to customer response in business. But because of its rare and stealthy nature, the fraud signal is very diffuse and must be detected with much more rigorous methods than the more conventional responses of attrition and cross-sell/up-sell discussed in Chapter 16 on customer response modeling.

Fraud Is Evolving!

Fraudsters may adapt quickly to many fraud detection methods, by devising novel and increasingly subtle ways to get away with it. Also, fraud detection schemes must evolve also to try to keep up with (and get ahead of) fraudsters. This process is very much like the way flu viruses evolve. Flu vaccine designers try to craft new vaccines not just to confer immunity to strains of flu viruses they know, but to get ahead of the next epidemic. Fraud detection is a lot like that.

Large Data Sets Are Needed

Large credit card issuers like Capital One may process billions of transactions per year. Even a very small percentage of fraud among these billions of transactions can result in proportionately large losses. AT&T processed almost 300 million telephone calls *each day* in 1998 (Cortes and Pregebon, 1998). Phone fraud was one of the major incentives that prompted AT&T Bell Labs to develop Hancock, a large database computer system capable of analyzing huge volumes of call detail records. In addition to the fast computer systems, you must use fast and efficient algorithms to process all these data in time to make actionable any information related to fraud.

The Fact of Fraud Is Not Always Known During Modeling

Sometimes you can identify fraudsters, and sometimes you can't. When you can "tag" a certain group of records as fraudulent, the analyses to model them are called *supervised*. The training of the model is supervised by the known identity of the fraudster. If you can't identify the fraudulent records up front, the analyses are called *unsupervised*. In either event, Bolton and Hand (2002) suggest that we should view the fraud predictions as *suspicion scores*.

When the Fraud Happened Is Very Important to Its Detection

The temporal dimension of fraud provides a rich source of information related to fraud. The occurrence of a fraud event at a given time may be highly related to the pattern of events that happened in the past. These historical data are the most important source of attributes needed to sufficiently define the fraud signature in the data set. Many derived variables can be constructed with various time dimensions (e.g., time since the last transaction). These variables are forms of temporal abstractions we met in Chapter 16. The same principles that apply to customer behavior in response models also apply to behavior of fraudsters. We might even expect that many of the most powerful predictor variables in fraud models are temporal in nature, as is the case in customer response models.

Fraud Is Very Complex

Fraud events involve much complexity. In addition to the data complexity listed in the preceding sections, the series of events associated with the fraud event may be quite complex. This complexity is partly due to the fraudster's need for stealth and secrecy, and partly due to the intentional obfuscation of the trail of evidence indicating fraud.

Fraud Detection May Require the Formulation of Rules Based on General Principles, "Red Flags," Alerts, and Profiles

Fraud modeling requires the construction of reference objects based on relationships that have been drawn in the past between various conditions and the incidence of fraud. Examples of such rules that suggest fraud include

- **General principle:** The incidence of fraud is more likely when the opportunity is high and the potential gains are large.
- **A "red flag":** A large number of accidents or claims is made by one individual.
- **A "red flag":** The same professional service person is involved with the claim (e.g., a doctor).
- **An alert:** A new product is introduced before fraud management systems are put in place.

Fraud profiles will be discussed separately later in the chapter.

Fraud Detection Requires Both Internal and External Business Data

Most companies have some sort of internal data describing their business events (selling things or providing services). But the forms of data gathered for internal purposes most often are related to billing and account service purposes. Many potentially predictive variables are not gathered by internal systems (e.g., years in business), but must be gathered from external sources. Information can be gathered from various data providers to enhance the corporate database, including

- Demographic data (available from Axciom, Experion, Equifax, Lexis-Nexis, etc.);
- Firmographic data (e.g., Dun & Bradstreet data and other business data sources);
- Psychographic data (inferences and classifications of people according to various measures of attitudinal and philosophical views).

Very Few Data Sets and Modeling Details Are Available

There is good reason for the lack of data sets and modeling details. You would not want potential fraudsters to learn how to defeat your detection strategies. Fraud data sets and modeling methodologies are tightly kept secrets. A company like Fair Isaac (generator of the FICO credit scores) has a huge library of predictor variables it won't share with anyone. In academia, fraud researchers share their methods in very formal and general terms that only experts can understand, read "between the lines," and relate to detailed instructions. Fraud modelers may be technical experts in a given business and would love to have access to detailed methodological presentations.

Very few fraud data sets are available in the public domain. Following are the only two that the authors are aware of:

1. A relatively small data set of Spanish automobile insurance claims (a Research Paper in Economics, or RePEc, data set. See http://repec.org/ and Artis et al. (1999).
2. The KDD Cup 1999 Network Intrusion Detection data set (http://kdd.ics.uci.edu/databases/kddcup99/kddcup99.html). (*Note:* This data set will be used in the example described later.)

HOW DO YOU DETECT FRAUD?

The basic approach to fraud detection with an analytical model is to identify possible predictors of fraud associated with known fraudsters and their actions in the past. The most powerful fraud models (like the most powerful customer response models) are built on historical data.

If the fraud response can be identified, it can be used to characterize the behavior of the fraudster in the specific fraud act and in historical data. The application of the term *supervised* is drawn from the broader discipline of classification (see Chapter 11 for an introduction to the terms *supervised* and *unsupervised*). Supervised classifications are based on some measure of true class membership of a given entity. According to Bolton and Hand (2002), supervised modeling has the drawback that it requires "absolute certainty" that each event can be accurately classified as fraud or nonfraud. In addition, the authors note that any models of fraud can be used to detect only types of fraud that have been identified previously.

Unsupervised methods of fraud modeling rely on detecting events that are abnormal. These abnormal events must be characterized by relating the events to symptoms associated with fraudulent events in the past. Statistical classification as fraud by unsupervised methods does not prove that certain events are fraudulent, but only suggests that these events should be considered as probably fraud suitable for further investigation.

Link analysis is the most common unsupervised method of fraud detection. The process of performing link analysis is known as link discovery (LD). This discipline has its origin in discreet mathematics, graph theory, social science, and pattern analysis. The object of LD is to find hidden links among patterns that appear to be unrelated. The approach is to relate groups and activities to some behavior, such as fraud. LD is related in a broader context to the recent emergence of social network analysis.

In traditional data mining, entities modeled are variables, which may be correlated (linked) to other variables in their effect on a target variable. In LD, entities are not variables, but rather are relationships between entities. LD evaluates the likelihood that a given pattern in a data set (expressible in a specific graphic data structure) matches some target pattern. In this regard, LD is very "Platonic" in its search for truth, compared to the more Aristotelian approach of supervised methods of fraud detection.

Another common unsupervised method is the application of Benford's Law to detection of fraudulent financial reports. Benford's Law states that in numerical lists involving real-life processes and events, the leading digit is not distributed in a uniform manner (Benford, 1938). The digit 1 appears about a third of the time, and the digit with the lowest frequency is 9. This principle is attributed to Benford, but it was published earlier by Newcomb (1881). As Director of the Nautical Almanac Office, Newcomb observed that pages of logarithm books were unevenly worn. Logarithms were used extensively in the calculation of nautical chart values. The earlier pages of the logarithm books were more worn than the later pages. This observation led him to form the general principle that any list of numbers taken from any set of data will contain numbers beginning with the digit 1 more frequently than any other number. Benford's Law can be applied to check the "normalcy" of street numbers, bill amounts, stock prices, or expense reports. This principle was derived from observations in the real world, but it remained unproven mathematically until Hill (1996) offered a formal proof. Checks against the relative frequencies of initial digits presented by Benford (1938) can be used to flag suspicious numerical lists. If the frequency of initial digits in a list is significantly different from the frequencies listed by Benford, then the list can be flagged as probable fraud.

Despite the wide range of unsupervised methods of fraud detection in use today, in the interest of parsimony, we will consider only supervised methods of fraud detection in this chapter.

SUPERVISED CLASSIFICATION OF FRAUD

Several elements are crucial to the successful production and deployment of any supervised fraud model:

- The fraud event and the relationship of that event to specific transactions or responses of the fraudster must be accurately identified.
- Historical data of past transactions or responses must be available to derive powerfully predictive variables.

- Profiles of the past behavior and actions of both the fraudsters and the nonfraudsters must be built and employed in the modeling methodology.

Fraud can occur in many aspects of business:

- **Credit card fraud:** Stealing or counterfeiting credit card numbers, or nonpayment of accounts.
- **Charge-back fraud:** Transaction reversals after an item is shipped.
- **Check fraud:** Taking advantage of the "float" in time between writing the check and payment by the bank. In one form, the fraudster writes a check he knows is bad to delay payment until the check clears ("kiting") or withdraws money from an account fed by a bad check and then abandons the account.
- **Application fraud:** Untrue statements on a credit application, leading to assignment of an artificially low credit risk.
- **Merchant fraud:** Involves the collusion of a merchant with another fraudster. One scheme is "white plastic fraud," in which a merchant sends fraudulent sales drafts to a bank and pockets the sales draft payment by the bank.
- **Claim fraud:** Submitting inflated or false claims.
- **Life insurance:** False or "engineered" death claims.
- **Health care fraud:** False billings by health care providers.
- **Automobile:** Includes "soft" fraud of filing multiple claims and "hard" fraud of engineering accidents.
- **Property:** Includes arson and destruction of unsold property.

HOW DO YOU MODEL FRAUD?

There are three general approaches to modeling fraudulent events depicted in Figure 17.1.

Early fraud models employed expert systems to detect fraudulent events. An expert system is a collection of expert opinions on a number of decision criteria. Instead of sifting out mathematical patterns in a data set, these systems induced rules from the responses of a group of experts in the field. These rules can be coordinated into a flow chart leading to a decision. The problem with expert systems is that they are based on subjective inputs that may be contradictory. Subsequent fraud detection systems used automated rule induction engines, based decision tree technology, and fuzzy logic. Some of these fraud detection systems are still marketed today (iPrevent by Brighterion).

The most comprehensive fraud detection systems were developed by HNC Systems in the late 1990s (now owned by Fair Isaac & Co.). The Fair Isaac fraud detection systems Falcon Fraud Manager, eFalcon, and LiquidCredit Fraud Solution are built around a sophisticated system of predictive variables derived from extensive historical customer data. These predictors have been selected by many years of modeling fraud in many companies. The variables are submitted to a powerful backpropagation neural net developed by HNC Systems.

Types of Fraud Detection Systems

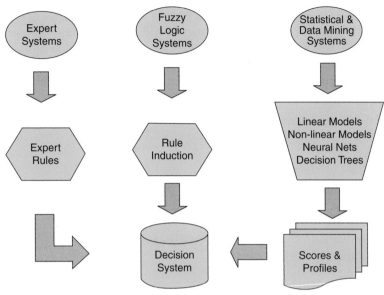

FIGURE 17.1   Types of fraud models.

HOW ARE FRAUD DETECTION SYSTEMS BUILT?

Credit card fraud gets the most press coverage, and investment fraud may cause the biggest financial "hits." But application fraud is viewed by some as the most common type of fraud. The initial problem with application fraud is that there are probably a large number of fraudulent applications that are never caught. Application fraud can occur in many situations in which a customer fills out an application. Credit applications (including credit cards) are particularly vulnerable to fraudulent information, which can cause the credit risk associated with the application to be significantly underestimated.

Successful fraud detection requires looking at the entire business process and identifying where fraud can originate. A successful fraud detection team begins each project with a careful evaluation of the client's existing business process. Then the team collects cases of fraud that have been found by auditors or others within the existing manual processes. From knowledge of the business process and these known cases, team members design metrics for measuring fraud and work with the client to automate their calculation. Finally, they develop the detection models. This process delivers value to clients at each stage.

The return on investment (ROI) in fraud detection data mining can be extremely impressive. On one of the authors' projects, the client had an alert system for its enormous data processing task whose warnings turned out to be fraud only 1% of the time (very inefficient, though better than random). With the data mining solution, however, the hit rate improved to 25%. In another fraud detection project, the analysts were able to achieve a savings of over $20 million on an engagement that took less than 12 staff months of effort to complete and deliver.

The most successful application fraud detection systems are based on extensive customer historical data. Patterns of both fraudsters ("bads") and nonfraudsters ("goods") are created, based on many variables. These variables include not only the information from the application form, but also information from a number of other sources. Some of the most predictive variables are those derived from combinations of variables based on domain knowledge.

Some of the sources of information include

- Near real-time access to credit bureau data;
  Names and addresses;
  Employer data;
  Banking and credit data.
- Characteristics of the applicant extracted from other external data sources (e.g., ZIP code lists by city, county, and state). Checks will be made to see whether names and addresses match among different sources of information for an applicant. Other checks may include
  The phone number is in the list for a given ZIP code;
  The phone number is valid or invalid;
  The SSN is valid or invalid;
  SSN was never issued;
  Date of birth is valid or suspicious;
  Aliases were used in the past.
- Checks will be made for duplicates among specific services and for missing services that are related to existing services for an applicant.
- Many temporal abstraction variables are based on
  Time since a specific action occurred, like a late payment;
  Time since last loan charge-off;
  Number and balances of charge-offs during the last time period.

The application fraud modeling system may be embedded in a system that incorporates the checks listed here and may operate on all data gathered during all phases of data checking. There are many fraud management systems based in general on this approach. Included in these systems are

- The Fair Isaac: Falcon Fraud Manager
- Agilis International: NetMind
- SAS: Fraud Management
- Neural Technologies: Minotaur
- 41st Parameter: Fraud Management Solutions
- SAP: Biometric Fraud Mitigation Solution

Some of these products include a number of optional modules that contain various kinds of checks, powerful modeling algorithms, and a complex scoring system. Some of these systems can be put in place to analyze credit card applications with a near real-time response rate.

INTRUSION DETECTION MODELING

Business network intrusion is a major problem in our digital age. Sometimes people hack into business and governmental systems just for the fun of it. Other times, the intrusion is malicious, seeking information that can be used for fraudulent purposes in a commercial or military context. This type of fraud has led to the development of sophisticated countermeasures to assure network security. To this end, the KDD Cup 1999 Network Intrusion Detector data set was created during the 1998 DARPA Intrusion Detection Evaluation Program, hosted by MIT Lincoln Labs. A data set was generated by collecting 9 weeks' worth of raw TCP dump data from a local area network (LAN) simulating a real LAN in a U.S. Air Force environment. The simulated LAN was hit by many simulated intrusion attempts. The TCP data consisted of about 5 million connection records in the main data set intended for model training and about 2 million connection records in the test data set. Each connection consisted of a number of TCP packets associated with a start time and end time flowing from a start IP address to a destination IP address. A packet is a short burst of data sent over a network; it is quality checked at the destination system with various forms of cyclic redundancy checks (CRCs). If the CRC at the destination is different from that of the source, the packet is retransmitted. Each data record (line) in the packet was labeled as a binary attack versus nonattack variable and as a categorical variable with one of 24 attack types.

The data set contains three sets of predictor variables:

1. Basic features;
2. Content features suggested by domain knowledge;
3. Network traffic features using a 2-second time window (one type of time-based feature).

Stolfo et al. (2000) defined additional time-based traffic features of the connections. These high-level variables included "same host" features, which were calculated for connections with the same destination host in the past 2 seconds. Similar "same service" features were calculated. A similar set of time-based features was built using a "connection" window of 100 connections. These high-level derived variables are likely to be quite predictive of different patterns of intrusions, similar to the temporal abstraction variables used to train the churn model described in Chapter 16.

COMPARISON OF MODELS WITH AND WITHOUT TIME-BASED FEATURES

The time-based features presented in the KDD Cup data set and those generated by Stolfo are forms of temporal abstractions in which the base is the time of the connection and the abstraction is drawn from data within 2 seconds previous to the connection time.

The importance of temporal abstractions for predicting churn in the insurance industry was demonstrated in Chapter 16. Similar time-based derived variables are also very important predictors of fraud. Analyses of all predictor variables in the KDD Cup 1999 data set were submitted to the Variable Selection feature in *STATISTICA* Data Miner. The most important predictors of network intrusion (regardless of its type) are shown in Figure 17.2.

Notice that except for the variable *Logged_in*, all of the predictor variables are time-based. This result indicates how important time-based variables will be in supporting any model built on this data set. The data set was constrained to the variables listed in Figure 17.2 and submitted to the Data Mining Recipes module in *STATISTICA* Data Miner.

The next step is to load the data set and select the target(s) and the set of predictor variables to use in building the model (Figure 17.3).

Step 2 in the Recipe calculates the variable statistics shown in Figure 17.4.

In Figure 17.4 notice that the Selected Testing Sample box was clicked, and the default 20% sample was chosen, which caused the notation "Selected" to appear on the screen.

Step 3 in the Recipe looks for redundant variables (Figure 17.5). Either the Pearson's Product-Moment Correlation Coefficient (simple parametric correlation) or the Spearman's Rank correlation coefficient (nonparametric) can be selected as the criterion for judging whether any two variables are correlated at a sufficient level to be redundant.

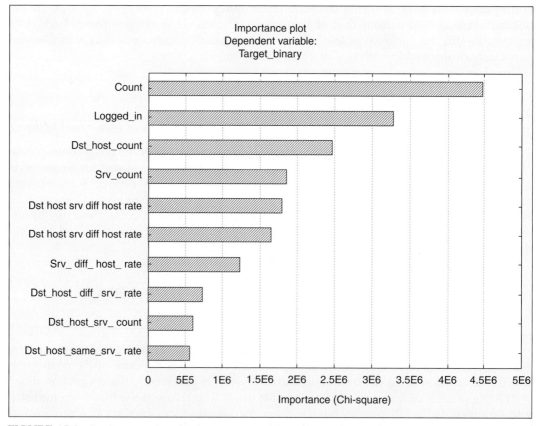

FIGURE 17.2   Importance values for the most powerful predictors of network intrusion.

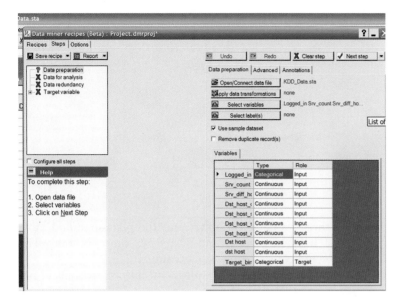

FIGURE 17.3   *STATISTICA* Recipes step 1: variable selection.

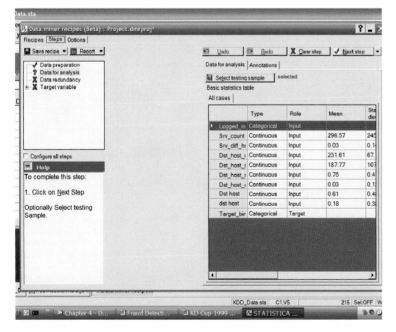

FIGURE 17.4   Descriptive statistical data.

FIGURE 17.5 Selection of the criterion to use for redundancy checking.

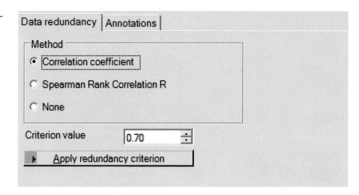

Redundancy was found between three pairs of variables, and the recommended variables to delete were Srv_count, Dst_host_srv_count, and Dst_host_same_srv_rate. The variable list was amended, and the Recipe construction was continued.

Step 4 in the Recipe builds models for C&RT, boosted trees, and automated neural net (Figure 17.6).

All models in Figure 17.7 marked as TRUE will be evaluated in Step 5 of the Recipe.

For this data set, the boosted trees model (rightmost lift curve through the first five deciles) performed the best among the models. The total area between the curve and the baseline reflects the total predictive power, and is largest for the boosted trees model.

FIGURE 17.6 Selection of the models to train.

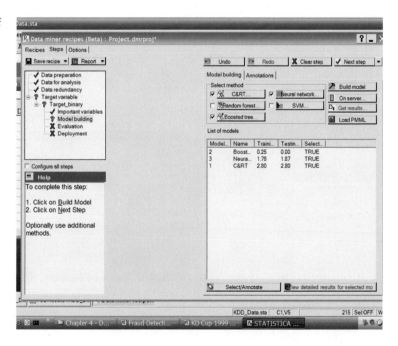

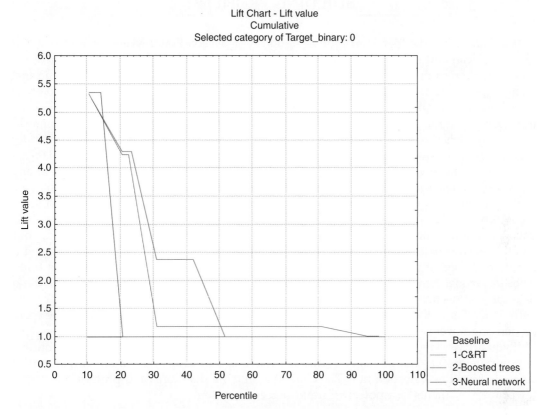

FIGURE 17.7   Lift index of the testing data set centered around 1.00 (random performance).

These lift index curves reflect how far down the scored list sorted on prediction probability a fraud analyst can go before reaching the point of randomness in the prediction of attack. Even though the model produces a lift over random selection for only the top half of the data set, the classification of any record as attack or normal is still much more accurate than random selection.

The preceding models illustrate

1. Many variables collected to assess fraud detection are not related to the fraud action at all (only one basic variable had enough predictive power to be included in the model).
2. Most of the final predictor variables were time-based.
3. Time spent in deriving time-based variables can pay off with big returns in model performance.
4. Models for other kinds of fraud detection can be built similarly.

BUILDING PROFILES

Fair Isaac offers the Merchant Profiles option to its Falcon Fraud Manager. It claims that this option can add up to 50% more detections of merchant fraud. When implemented, the Merchant Profiles option provides a score for each merchant to combine with the normal modeling score from Falcon Fraud Manager. This is a good example of how you can combine model predictions and profiles to create a more powerful fraud detection system, as depicted in Figure 17.1. Similar profiles can be built for customers in a commercial or credit context.

Many models can be built following this example. Each model could predict fraud under slightly different conditions. For example, the target variable in the KDD Cup 1999 data set included 24 categories of fraud. For the sake of illustration, the occurrence of fraud in any form was modeled in the earlier example. We could have restricted the model to just one of the 24 categories of fraud. Each of the models could be used to generate a fraud score for each type of fraud. These scores, plus the rules of thumb, demographic and firmographic data, and information from other external sources can be composed into a profile. From your score data, you could build multiple profiles that pertain to different types of fraud and different conditions of fraud (male, female, age, etc.). Potential predictor variables for a fraud detection model may come from data elements listed in the earlier section on how fraud detector systems are built. In addition to those variables, many time-based variables can be derived, similar to the ones used for the KDD Cup Network Intrusion model. The time spent on deriving novel variables is the most effective way to increase fraud detection rates.

If you are working on a fraud detection project in which the fraudsters can be identified, appropriate profiles can be built for various customer segments and combined with model scores to boost the detection rate. The combination of model scores and profiles constitutes the primary elements of a powerful fraud detection system.

DEPLOYMENT OF FRAUD PROFILES

These profiles can be loaded into real-time systems, and credit card applicants (for example) can be matched relatively quickly to known fraud profiles. Model scores and elements of profiles can be composed into business rules and programmed into SQL or some other production system interface. For example, some business rules resulting from this composition in a credit card environment might include

1. If the ZIP code on the application is not in the known list of the phone number area code $\rightarrow$ Fraud (a "red-flag" fraud indicator)
2. If the fraud model score is > 0.60 and the customer demographic profile matches that of a group of known fraudsters at the 85% level $\rightarrow$ Fraud

These business rules can be generated directly from rule induction engines, indirectly from decision tree algorithms, or inferred from combinations of neural net predictor variables with relatively high importance values.

POSTSCRIPT AND PROLEGOMENON

You might not have come this far in the book before trying one of the tutorials. But if you resisted that temptation, you are much better prepared to work on the tutorials in Part III. Tutorials included in the printed pages of Part III include those that the authors judged to be most pertinent to the interests of the wide audience of our readers. Other tutorials are included on the enclosed DVD.

References

Artis, M., Ayuso, M., & Guillen, M. (1999). Modelling different types of automobile insurance fraud behavior in the Spanish market. *Mathematics and Economics, 24,* 67–81.

Benford, F. (1938). The law of anomalous numbers. *Proc Amer Phil Soc, 78*(4), a551–572.

Bolton, R. J., & Hand, D. J. (2002). Statistical fraud detection: A review. *Statistical Science, 17.3,* 235–249.

Cortes, C., & Pregibon, D. (1998). Giga-mining. In R. Agrawal & P. Stolorz (Eds.), *Proceedings of the Fourth International Conference on Knowledge Discovery and Data Mining* (pp. 174–178). Menlo Park, CA: AAAI Press.

Hill, T. (1996). A statistical derivation of the significant-digit law. *Statistical Science, 10,* 354–363.

Newcomb, S. (1881). Note on the frequency of use of the different digits in natural numbers. *American Journal of Mathematics, 4*(1), 39–40.

Stolfo, S., Lee, W., Prodromidis, A., & Chan, P. (2000). Cost-based modeling for fraud and intrusion detection: Results from the JAM project. In C. Panel (Ed.), *Proceedings of the 2000 DARPA Information Survivability Conference and Exposition* (pp. 130–144). Hoboken, NJ: Wiley-IEEE Press.

TUTORIALS—STEP-BY-STEP CASE STUDIES AS A STARTING POINT TO LEARN HOW TO DO DATA MINING ANALYSES

If a picture is worth a thousand words, a good tutorial can be worth this whole book. We have packaged a number of tutorials in these printed pages, which cover a wide range of applications, using one the three data mining tools introduced in Chapter 10. Some readers may charge ahead directly guided by these tutorials; other readers will go through some or all of the preceding chapters before tackling a tutorial. Whichever approach you took, you are in now in the "meat" of this book. Parts I and II were designed to lead up to the tutorials in Part III. As you go through these tutorials, you may remember some factoid about a subject presented in one of the previous chapters. If so, use the tutorials as a springboard to jump into your own application area, but also let the tutorials point you back to important topics presented previously in this handbook.

Guest Authors of the Tutorials

David P. Armentrout, PhD, Practicing Clinical Psychologist, Faculty IHI Family Practice Residency Program, and Family Medical Care Clinics, Tulsa, OK

Ira L. Cohen, PhD, Chairman, Department of Psychology, NYS Institute for Basic Research in Developmental Disabilities, Staten Island, NY

Dursun Delen, PhD, Associate Professor, Management Science and Information Systems, William S. Spears School of Business, Oklahoma State University, Tulsa and Stillwater, OK

Dalton Ellis, MD, IHI Medical Resident, Tulsa, OK

Ashley Estep, DO, IHI Medical Resident, Tulsa, OK

Cheryl Flynt, RD, MPH, LD/N, Manager, Center for Nutritional Excellence, Florida Hospital, Orlando, FL

Joseph M. Hilbe, JD, PhD, NASA Solar System Ambassador Jet Propulsion Laboratory, California Institute of Technology, Pasadena, CA; Software Reviews Editor, *The American Statistician*; Adjunct Professor, Statistics, Arizona State University; Emeritus Professor, University of Hawaii; served as Law Professor; served on FDA Statistics Group; served on editorial boards of six statistics journals, including *Statistics* and *Sports Medicine*; and latest endeavor astrostatistics with International Statistics Institute

Thomas Hill, PhD, Vice President for Development, StatSoft, Inc., Tulsa, OK

Karen James, RN, BSN, CHPN, Private Practice, Tulsa, OK

Wayne S. Kendal, MD, PhD, FRCSC, FRCPC, Division of Radiation Oncology, The Ottawa Hospital Regional Cancer Centre; University of Ottawa, Ottawa, Canada

Stacey Knight, MStat, Department of Biomedical Informatics, University of Utah, Salt Lake City, UT

Ivan Korsakov, StatSoft, Inc., Tulsa, OK

Mourad Krifa, PhD, Department of Fiber Science, School of Human Ecology, University of Texas at Austin, Austin, TX

Sachin Lahoti, Data Mining Consultant, StatSoft, Inc., Tulsa, OK

Walter L. Larimore, MD, Clinical Instructor, IHI Medical Family Residency; Assistant Professor Department of Family Medicine, University of Colorado Health Services Center, Denver, CO

Juan S. Lebron, Graduate Student, Department of Physics, University of Puerto Rico, Rio Piedras, Puerto Rico

Kiron Mathew, Senior Analyst (DB Marketing), Key Bank—Key Corporation, Cleveland, OH

Allen Joseph Mednick, PhD Candidate, Department of Education, University of Pennsylvania, Philadelphia, PA

Ronald Mellado Miller, PhD, Department of Psychology, Brigham Young University-Hawaii

Mary A. Millikin, PhD, Director of Institutional Research, Tulsa Community College, Tulsa, OK

Linda A. Miner, PhD, Professor and Director of Academic Programs, Southern Nazarene University, Tulsa, OK

Jose F. Nieves, PhD, Department of Physics, University of Puerto Rico, Rio Piedras, Puerto Rico

Hanu R. Pappu, PhD, Department of Plant Pathology, and Center for Integrated Biotechnology, Washington State University, Pullman, WA

Timothy Potter, MD, IHI Medical Residency, Tulsa, OK

Mollie R. Poynton, PhD, APRN, BC Assistant Professor, University of Utah, Salt Lake City, UT

David Redfearn, PhD, Woodland Hills, CA

Stephanie Rick, Sales and Marketing Manager, Florida Hospital Publishing, Florida Hospital, Orlando, FL

Greg S. Robinson, PhD, Challenge Quest, LLC, Founder & President, Tulsa, OK

James Ross, MD, Private Practice, Tulsa, OK

Enis Sakirgil, MD, IHI Medical Residency, Tulsa, OK

Jessica Sieck, Statistical Consultant, StatSoft, Inc., Tulsa, OK

Alan Stolzer, PhD, Embry-Riddle Aeronautical University, Daytona Beach, FL

Shirley C. Strum, PhD, Department of Anthropology; University of California, San Diego; and Director, Uaso Ngiro Baboon Project; Nairobi, Kenya

David Vergara, SPSS Inc., Chicago, IL

Nephi Walton, Department of Biomedical Informatics, School of Public Health; University of Utah, Salt Lake City, UT

Chamont Wang, PhD, Professor of Mathematics and Statistics, The College of New Jersey, Ewing, NJ

Charles G. Widmer, DDS, MS, Coordinator of Graduate Orthodontic Research; Head, Division of Facial Pain, Department of Orthodontics, University of Florida, Gainesville, FL

How to Use Data Miner Recipe *STATISTICA* Data Miner Only

Gary Miner, Ph.D.

The new Data Miner Recipe (DMRecipe) will be presented in this tutorial because it is the easiest way for you to actually "do" a data mining project, automatically get results, and thus get an overall feel about what data mining can do with your data if you're new to data mining. The process can involve as few as four to six mouse clicks and is so easy that you could even write these short steps on a sheet of paper, leave it on your assistant's desk, asking him or her to run this analysis the next day while you are away on a business trip or a meeting across town. The fastest way to understand this process is to present some illustrations of what you will see on your PC screen, followed by a full explanation of what DMRecipe can do for you.

The DMRecipe is really the first method that you, whether a new user or experienced data miner, should use to look at a new data set because it is the most rapid, simplest way to get a feel for a new data set, and the resulting analysis may be all that is needed, thus negating further interactive or DMWorkspace analysis, resulting in a real savings of time and energy.

Before you start working with DMRecipe, if you don't have a data set already open on the PC screen, click on the File menu at the top of the *STATISTICA* screen, and select either Open or Open Examples. Find a data set of interest, opening it on the PC screen. Then follow these steps:

1. Click on the Data Mining menu. When the menu drops down, select the first option, Data Miner Recipes, as illustrated in Figure A.1.

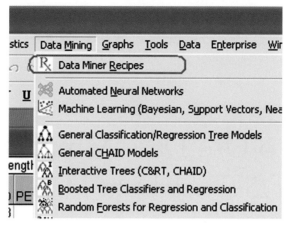

FIGURE A.1   Data Miner Recipe is found on the Data Mining pull-down menu.

    After a few seconds, this Data Miner Recipe dialog will pop up, as shown in Figure A.2.

2. Click on the DMRecipe dialog, to bring it to the topmost screen, and then click the New button.
3. Click the Open/Connect Data File button (see the upper-center part of the dialog in Figure A.3) and select the variables of interest. (In this example, the famous Irisdat.sta data set is being used; it contains data on three species of iris, the three species listed as one variable, the dependent variable in this example, and several independent/predictor variables that are continuous measures of the flower petal length, width, etc.)
4. Click the Select Variables button on the DMRecipe dialog, where indicated in Figure A.4.
5. Select the variables shown in Figure A.5 and click OK to accept them and place in the DMRecipe analysis.
6. Click on the down arrow icon in the upper-right corner of the DMRecipe dialog and release on the Run to Completion selection, as shown in Figure A.6.

    Alternatively, you could click on the Next Step button, which is located just to the left of the down arrow, but then you would have to sit at your PC and click on each step of the process—although in some cases this may be to your advantage.

Go away and have lunch, take a walk, or go home if you know you have a large data set that may take hours for the complete data mining computations to take place or before useful results workbooks are shown. When you come back, the results workbook shown in Figure A.7 will be open on the screen, with the competitive analysis of algorithms as the topmost screen.

As you can see from the results in Figure A.7, boosted trees had the lowest error rate, which can be converted into the accuracy rates. The accuracy rate is the primary way of

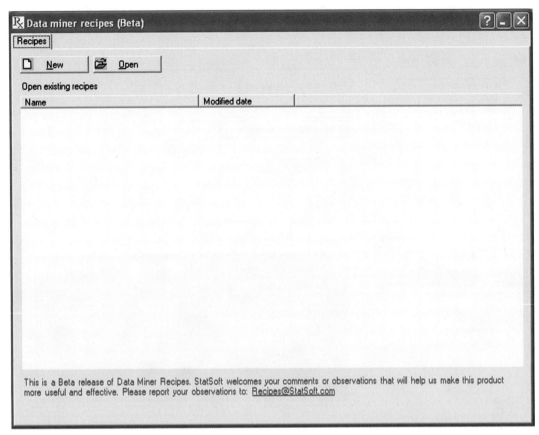

FIGURE A.2   DMRecipe beginning dialog.

expressing the "correctness" of data mining algorithms; it is equal to (100 − error rate). Thus, the accuracy rate for boosted trees is 99.37%, almost a 100% accuracy in predicting which plant species a particular specimen is, based on the dimensions of the plant's flower petals in this example. Since boosted trees had the highest accuracy rate, it was selected for the deployment model, which can be used on future data sets.

The model developed for Figure A.7 has very high accuracy rates; thus, any of the three algorithms—e.g., neural networks, C&RT (i.e., decision trees), or boosted trees—could have been reasonably used as the deployment model.

However, you will run into data sets where there are some differences in accuracy scores among the models, with maybe none of the algorithms providing the hoped-for "95% or higher" accuracy. In these cases, *hybrid models* can be made, usually called *consensus models*; these are discussed in Chapter 9 and in some of the other tutorials.

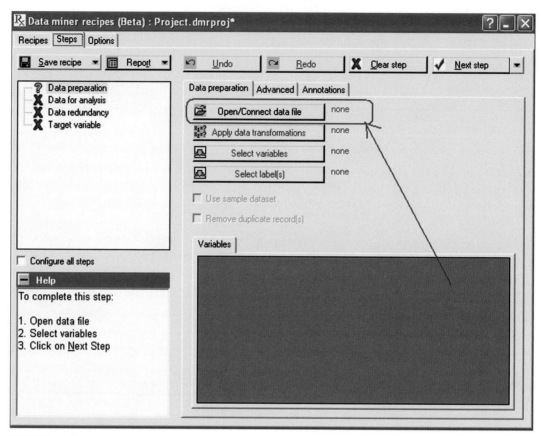

FIGURE A.3   The Open/Connect data file button used to connect a data file to the DMRecipe.

If the DMRecipe workspace you used was an old one or a partially constructed one, where you had to click on the Open button (instead of the New button) at the beginning of the process, you may have needed only two or three clicks to run the project.

The following text is taken from the online help of *STATISTICA*; this discussion provides an overview of the basic concepts in the DNRecipe approach; StatSoft, Inc. (2008). STATISTICA (data analysis software system), version 8.0. www.statsoft.com.

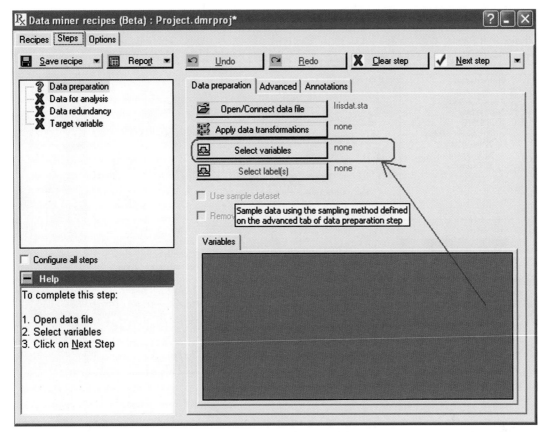

FIGURE A.4   Select variables button; click to open "Select variables" dialog.

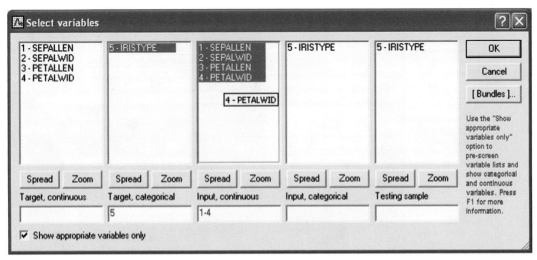

FIGURE A.5   Select variables dialog.

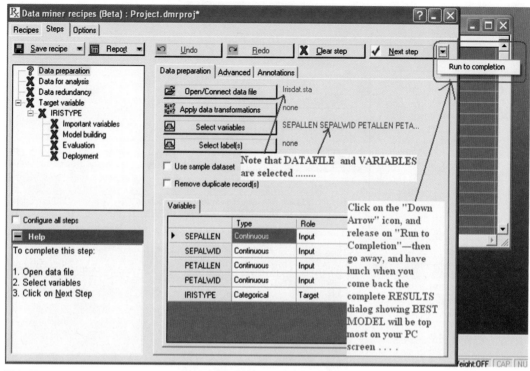

FIGURE A.6   DMRecipe dialog with variables selected and ready to run.

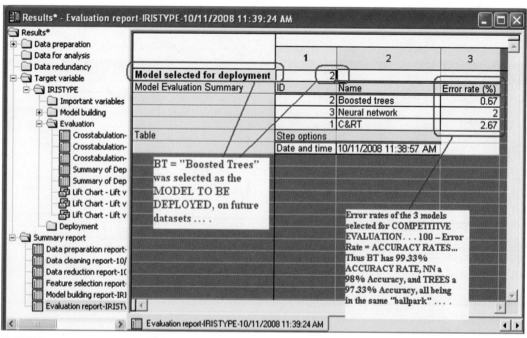

FIGURE A.7   Results workbook presented at the completion of DMRecipe project computations.

WHAT IS *STATISTICA* DATA MINER RECIPE (DMR)?

STATISTICA Data Miner Recipe (DMR) provides a systematic method for building advanced analytic models to relate one or more target (dependent) quantities to a number of input (independent) predictor variables. The target variables can be continuous or categorical. Continuous target variables are usually associated with regression tasks, and categorical variables are used in classification problems. *STATISTICA* DMR is capable of handling both types of variables and, thus, capable of building predictive models for tackling regression and classification problems.

STATISTICA DMR is a complete solution that makes the process of predictive model building and data mining a systematic and step-by-step process. Model building in DMR starts with preliminary data analysis, preprocessing the analysis variables, dimensionality reduction, and elimination of any redundancy that might exist in the data set. Once data definitions and preparation are complete, DMR can create various predictive models (such as neural networks, Support Vector Machines, trees, etc.) for modeling the values of the target data from the input variables. This step is followed by model evaluation and, finally and most important of all, model deployment in which the predictive models can be used for making predictions on unseen (new) data (e.g., for "scoring" of databases).

In addition to a recipe-like user interface for building predictive models, *STATISTICA* DMR also supports the off-loading of computationally demanding tasks. With DMR, you also can save projections and reload them in the future for further deployment.

CORE ANALYTIC INGREDIENTS

At the heart of *STATISTICA* Data Miner Recipe is a step-by-step recipe consisting of many analytic ingredients, which together create a self-contained tool for creating and deploying predictive models. The individual analytic methods and steps that need to be applied in a particular order may be available in various other scientific or academic domains, but the unique combination of analysis methods and steps in DMR has created an analytic workflow that can satisfy the needs of beginners and advanced users of data mining tools. *STATISTICA* DMR can be run as a nearly single-step data mining process (just specify variables or fields and then run-to-completion) or can be used by experienced data mining practitioners to "host" sophisticated and "fine-tuned" data mining models (with data preprocessing, transformations, etc.) for deployment.

The various ingredients of *STATISTICA* DMR are implemented in unique and specific steps, and the results are supported by various spreadsheets and graphs that are designed to aid us in drawing conclusions and interpreting the results.

1. **Data preparation**. Essentially, in this first step, you prepare the data for modeling. Specific data cleaning and transformations procedures can be implemented to eliminate specific unusual and duplicate cases. In addition, a "blind-holdout-sample" can be selected to be used later for validating models. Also, the target (dependent) and input

variables to the process are specified; targets are the variables or outcomes of interest that are to be predicted using the inputs (independent variables).

For example, suppose the task is to identify an accurate model for predicting two important outcomes related to credit risk: default probability and direct profit/loss over the lifetime of the loan. In this case, there would be two target variables (in the training data, i.e., a sample of individuals who had previously taken out a loan): (1) whether a respective individual defaulted on the loan and (2) how much profit/loss overall accumulated over the lifetime of the loan. Typical predictors might be the credit rating of each person (at the time the loan was originated), average income, etc.

2. **Data analysis**. In this stage, you can conduct statistical analyses of your variables including the targets and the inputs. You can review various statistics of the data such as mean, standard deviation, skewness, kurtosis, and observed minimum and maximum. You can also review the variable roles (inputs or targets) and their types (continuous or categorical). For regression analysis, the target variables are invariably continuous. For classification tasks, only categorical variables are chosen as target variables.

3. **Data redundancy**. Often, a number of variables can carry, to some degree, the same information. For example, the height and weight of people might in many circumstances carry similar information because the two variables are correlated. In other words, from the height, you can predict the weight with some accuracy. Thus, it may not be necessary to use both height and weight in the same analysis. Intuitively, the exclusion of variables that may carry useful information may sound like a bad idea, since the height and weight can never be related with perfect correlation, but in fact it is a consequence of the curse of dimensionality. It demonstrates that the benefits gained by reducing the curse of dimensionality can actually outweigh the loss of some information that might be incurred as a result.

In this step, a simple correlation test is applied to identify redundant inputs and remove them from further consideration for modeling. Note that the data redundancy scheme applies only to continuous inputs.

4. **Dimension reduction**. One of the important functionalities available in *STATISTICA* DMR is the ability to identify a small number of important inputs (for predicting the target) from a much larger number of available inputs (and is effective in cases when there are more inputs than cases or observations). Even after the preceding step has been applied (data redundancy), usually a large number of inputs remain for model building. While many methods exist—and are typically in use—to "screen" inputs to identify those that appear to be related in some way to the target variable of interest, a major analytic challenge is to find the interactions between inputs that predict the target of interest.

For example, predicting (modeling) the performance of a boiler requires a particular "configuration" of settings of different multiple inputs; also, the most simple monotone relationships (the more of X, the more of Y) are usually known already. The accurate prediction of credit or insurance risk (from inputs beyond commonly known and obvious risk factors) also usually requires the identification of a specific "configuration" of demographic and other variables that are related to risk. Interactions between inputs often cannot be explicitly screened entirely, because even with as few as 100 inputs, there can be more than 160,000 interactions between, for example, 3 specific inputs that can be arranged out of 100 inputs. *STATISTICA* DMR uses tree-based algorithms for finding important input predictor variables and interactions among them.

5. **Model building**. In this step, the actual models for predicting the targets from the inputs are built. Traditionally, building predictive models often falls under the domain of data mining or statistics. In *STATISTICA* DMR, the goal is to largely "automate" the process of generating good (accurate) predictive models; thus, by default the program will automatically search a specified number of different predictive models such as various tree models, Support Vector Machines, and neural networks. For the latter models such as neural networks, DMR automatically chooses good "candidate models" for further consideration. These computations can be time consuming and, hence, can be off-loaded to the server (from the desktop computer), where results can be picked up later or even the next day. Also, a large number of graphical displays are available to the Model Builder to review how well the different models predict the targets of interest. However, model building and selection are mostly automated to empower subject experts in the domain of interest (e.g., engineers who serve in the Model Builder role) rather than statisticians or data mining professionals to quickly and effectively build accurate predictive (e.g., neural networks) models.

6. **Model evaluation**. By the time you reach this step, you already have built your predictive models. Like any tool, your predictive models need to be tested on data that were not presented to the models during their training. This is also very similar to quality control, which needs to be applied to items coming out of production lines to ensure they meet certain specifications and standards. To do this, you test your models with data sets that were unseen before. In this case, the validation data set can help. The aim here is to see how well your models will perform on future data during the later and most important stage of deployment. The ability to predict new data is known as *generalization*. If your models did not generalize well on the validation data, it is recommended that you investigate the conditions and settings under which they were built and try creating more models that meet your needs.

7. **Deployment**. After building your data mining models using *STATISTICA* DMR, you can put your predictive models to "use" for predicting future or "new" data as needed. The process of using predictive models for predicting data that were not used in training the model is known as *deployment*. Deployment is by far the most important state of predictive modeling, and it is indeed the ultimate goal of the Model Builder. It is also the stage where your predictive models face the test of the real world. A good predictive model is one that predicts unseen data with the desired accuracy.

STATISTICA Data Miner Recipe provides a direct interface to *WebSTATISTICA* Enterprise to "attach" fully trained data mining models (data miner recipes) to data configurations for automated scoring of new data (e.g., new credit applications) in a Web-based solution and predict expected outcomes (e.g., for a continuous manufacturing process, and to track prediction residuals in standard QC charts).

Obviously, this DMRecipe method should be the first choice of most new users in looking at a new data set.

Data Mining for Aviation Safety
Using Data Mining Recipe "Automatized Data Mining" from *STATISTICA*

Alan Stolzer, Ph.D.

The following tutorial will use the Federal Aviation Administration's (FAA's) Service Difficulty Report (SDR) database to explore factors that lead to undesirable events called *unscheduled landings* and determine which data mining (DM) model or models appear to be the most predictive of that event.

It should be noted from the outset that this is a greatly simplified exemplar study. As will be described later, the database used is very complex, and much of it is in text form. Since this study is limited to a demonstration of DM tools in a particular application, highlighting the level of automation available, we will hand-select a small subset of the variables available to us. Selecting other variables may require text mining and methods to cope with the large number of categorical responses in some of the variables. While the software package we chose is capable of both data and text mining, we are using only DM methods for this study.

To achieve our objectives with this tutorial, we will accomplish the following:

- Briefly discuss airline safety and data mining's importance in improving safety;
- Introduce and describe the SDR database;
- Prepare the data for our study;
- Describe our DM approach using *STATISTICA* by StatSoft;
- Determine which DM algorithm appears to produce the most accurate results in predicting unscheduled landings based on error rate.

AIRLINE SAFETY

Commercial airline travel is statistically one of the safest modes of transportation in the world, but it wasn't always that way. In the early decades of air travel, accidents were not uncommon. The best-fit regression line in Figure B.1 reveals a substantial decline in the fatal accident rate from 1950 through about 1980, even though the actual number of departures was increasing just as dramatically.

But that's only part of the story. Sometime around 1980 there appears to be a leveling off of the accident rate along with a corresponding reduction in the variability of the data.

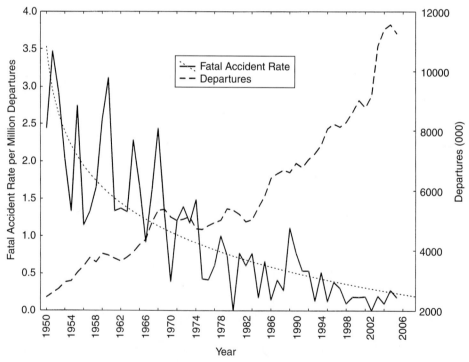

FIGURE B.1   Fatal accident rates per million departures—U.S. air carriers operating under 14 CFR 121—1950 through 2006 (ATA, 2007).

That dramatic improvement in safety in earlier decades coincided with the change from the old model of "wait for a crash, fix the problem, then fly again" to more of a proactive, system safety approach. Many experts believe that the leveling off of the accident rate (our current state) suggests that we may have gained about as much as we can using these methods and that we need to employ new, more scientific methods if we want to drive the accident rate down even further. There are many methods that could be discussed in this context, but one of particular interest is data mining.

There is a wealth of data being collected in the aviation industry; however, not much of it is being used very effectively. For instance, extensive data are collected on a type of report called *Service Difficulty Reports (SDRs)*. An SDR is required to be filed by an operator following an event with an aircraft, such as a malfunction, defect, failure, or other occurrence. FAA Form 8070-1 is used to collect the pertinent information about the event. SDR information is stored in a database and can be used to observe trends and reveal patterns of failures, design deficiencies, and other anomalies.

Of course, it will not be a casual observation of the data that will reveal these trends and patterns; rather, it will be a rigorous, methodical, well-structured search using data mining algorithms that will produce actionable results for the aviation safety professional.

One example of countless studies that might be undertaken with this large, complex database is to explore the occurrences of "unscheduled landings" at airlines. In the context of this database, unscheduled landings occur when a commercial flight fails to land at its intended destination due to a problem with the aircraft; rather, the landing occurs at an airport that is not included in the flight plan. These events pose several problems for air carriers, including safety and economic considerations.

Air carriers incur significant costs as a result of unscheduled landings. The obvious cost is the additional fuel necessary for another takeoff and climb, which is the phase of flight when an airliner consumes the most fuel. If the unscheduled landing becomes necessary early in the flight, the crew may even have to jettison or dump fuel. Other costs can include personnel, landing fees, maintenance costs associated with transporting company maintenance personnel and/or contracting services, transporting and housing passengers, and the costs related to an out-of-service airplane. The sum of these costs makes unscheduled landings a very unwelcomed event for air carrier management.

But safety is also an issue. Sometimes these unscheduled landings occur at airports served by the air carrier; sometimes not. While flight crews are trained to handle unusual situations, the safety margin can certainly decline when a landing is made at an unfamiliar airport and accomplished in an airplane with a problem significant enough to require an unscheduled landing to begin with.

SDR DATABASE

The SDR database can be found at http://av-info.faa.gov/dd_sublevel.asp?Folder= \SDRS. The data are stored in downloadable files in either fixed width delimited or tab delimited, and in either a zipped (.zip) or text (.txt) format. The data fields are shown in Table B.1 as detailed in the file named filelayout.txt on the SDR web site. The files appear to be updated weekly.

TABLE B.1   Service Difficulty Report: File Layout

Columnname	Columntype	Length	Description
c5	Char	13	Date and sequence number
c10	VarChar	8	Date on which the report was received
c12	Char	5	Unique sequential number assigned on date
c14	Char	4	Region code
c15	Char	4	Year
c16	Char	2	Month
c17	Char	2	Day
c18	Char	13	Operator control number
c20	Char	1	Segment code 1 = aircraft, 2 = engine, 3 =
c25	VarChar	8	Date of occurrence
c35	Char	1	A = Air Carrier; G = General Aviation
c40	Char	4	Air Transport Association (ATA) code
c90	Char	16	Manufacturer's part number
c100	Char	16	Descriptive name of part
c110	Char	12	Component manufacturer's name
c120	Char	12	Component manufacturer's model designation
c130	Char	6	Aircraft manufacturer's name
c140	Char	6	FAA assigned code to identify aircraft group
c150	Char	12	Aircraft manufacturer's model number
c152	Char	7	Aircraft make and model sequence number
c160	Char	4	Region responsible for aircraft
c170	Char	6	Engine manufacturer's name
c180	Char	6	FAA assigned code to identify engine group
c190	Char	12	Engine manufacturer's model number
c192	Char	5	Engine make and model sequence number
c200	Char	4	Region responsible for engine certification
c210	Char	6	Propeller manufacturer's name
c220	Char	6	FAA assigned code to identify propeller group
c230	Char	12	Propeller manufacturer's model number
c240	Char	4	Region responsible for propeller certification
c250	Char	16	Location on aircraft of the defective or m
c260	Char	16	Text reflecting condition of failed part
c270	Char	1	Submitter code
c280	Char	1	Report submitted
c290	Char	1	Alert code
c300	Char	4	Air carrier operator code
c310a	Char	1	1st occurrence – Precautionary Procedures
c310b	Char	1	2nd occurrence – Precautionary Procedures
c310c	Char	1	3rd occurrence – Precautionary Procedures
c310d	Char	1	4th occurrence – Precautionary Procedures
c314a	Char	24	1st occurrence – Precautionary Procedures
c314b	Char	24	2nd occurrence – Precautionary Procedures
c314c	Char	24	3rd occurrence – Precautionary Procedures
c314d	Char	24	4th occurrence – Precautionary Procedures
c320a	Char	1	1st occurrence – Nature condition code (Fo
c320b	Char	1	2nd occurrence – Nature condition code (Fo

Continued

TABLE B.1   Service Difficulty Report: File Layout—Cont'd

Columnname	Columntype	Length	Description
c320c	Char	1	3rd occurrence – Nature condition code (Fo
c324a	Char	26	1st occurrence – Nature condition text (Fo
c324b	Char	26	2nd occurrence – Nature condition text (For
c324c	Char	26	3rd occurrence – Nature condition text (Fo
c330	Char	2	Stage of operation code
c332	Char	15	Stage of operation text
c340	Char	1	Report status
c350	Char	2	Microfilm roll number of original report
c360	Char	4	Microfilm frame number of original report
c370	Char	2	Region receiving report
c380	Char	2	District office receiving report
c390	Char	5	Aircraft registration or tail number
c400	VarChar	5	Total time accumulated on part regardless
c410	VarChar	5	Total time accumulated on part since overh
c420	Char	1	Severity factor, the higher the number the
c430	Char	12	Manufacturer's serial number of component
c440	Char	12	Manufacturer's serial number of aircraft
c450	Char	12	Manufacturer's serial number of engine
c460	Char	12	Manufacturer's serial number of propeller
c490	Char	4	1st Code - Up to 2 three-digit ATA codes
c510a	VarChar	120	1st line of remarks (For use in report pri
c510b	VarChar	120	2nd line of remarks (For use in report pri
c510c	VarChar	120	3rd line of remarks (For use in report pri
c510d	VarChar	120	4th line of remarks (For use in report pri
c510e	VarChar	120	5th line of remarks (For use in report pri
c510f	VarChar	120	6th line of remarks (For use in report pri
c602	Char	1	Aircraft weight class code
c604	Char	1	Aircraft wing type code
c606	Char	1	Aircraft power class code
c608	VarChar	2	Number of engines
c610	Char	2	Design characteristic code for typical engine
c612	Char	1	Engine power class code
c614	Char	1	Engine type code
c616	Char	2	Landing gear code
c620	Char	8	FAA type certificate control number for ai
c640	Char	8	FAA type certificate control number for en
c652	Char	1	Propeller power code
c654	Char	1	Propeller type code
c660	Char	8	FAA type certificate control number for pr
Eof	VarChar	1	NULL

In addition to the data files, there are three files that provide information about the database. As mentioned previously, one is filelayout.txt. A second file is sdrcodes.doc; this file provides a description of most of the codes included in the database. A third file is not relevant for this tutorial.

PREPARING THE DATA FOR OUR TUTORIAL

For this study, we will concern ourselves only with air carrier-related events; general aviation events will be omitted. In addition, we will restrict our examination to events that occurred in the calendar year 2007.

Microsoft Excel will be used to compile the data. On the SDR web site (address provided previously), download and save the files named sdr2007a.txt and sdr2007g.txt (under "Tab Delimited Data").

Select File and then Open, point to the *xxxx*.txt file you want to open, and select it. That action automatically opens the Text Import Wizard. Select Delimited (Step 1), click Next, and then under Delimiters (Step 2), select Tab, click Next, and then Finish. Once the file has loaded into Excel, sort the data by C25 (column J) in ascending order. Ensure that only 2007-dated events are included in the file. Accomplish this procedure for both files (i.e., sdr2007a.txt and sdr2007g.txt) and combine the contents of the two files into a single Excel file named 2007.xls.

It is worth spending some time examining this file in some detail to familiarize yourself with the contents and to check for obvious missing, miscoded, or otherwise erroneous data. Some have estimated that data cleaning and exploration constitutes as much as 80% of the DM effort (Dasu and Johnson, 2003, p. ix). This tutorial will not go into detail on this step, but excellent references are widely available on the subject.

The file named 2007.xls contains 51,711 records, but recall that we are concerned only about air carrier-related events. Thus, sort by C35, A = Air Carrier, G = General Aviation, and delete those pertaining to general aviation; this leaves a total of 48,849 records.

When we examine Table B.1, it is clear that the SDR database contains some rather large text fields. Since this exercise is limited to data mining, eliminate text fields C510a through C510f, each of which contains text fields of up to 120 characters in length.

Finally, since our objective is to predict unscheduled landings, we need a discriminator. Therefore, insert a column after c314d, "4th occurrence – Precautionary Procedures Text (For use in report printout)," and label the field Unsched Lndg. Sort the data by C314a in descending order; this places UNSCHED LANDING occurrences at the top. There are 3,204 incidences of UNSCHED LANDING in C314a. C314b, C314c, and C314d reflect second, third, and fourth occurrences of UNSCHED LANDINGs, so these should be sorted successively to ensure that all occurrences of interest are at the top of our spreadsheet. All totaled, there are 3,405 incidences of UNSCHED LANDINGs in the 2007.xls data set. For those records that have UNSCHED LANDING in C314a, C314b, C314c, or C314d, place a YES in the column that was inserted after C314d. For those that do not have UNSCHED LANDING in C314a through C314d, place a NO in the column. While this procedure sounds tedious, it can be accomplished in just a few minutes using copy and paste functions.

The Excel spreadsheet is now ready to import into our data mining package: *STATISTICA* by StatSoft. *STATISTICA* was selected due to its excellent, industry-leading automated data mining capability.

To begin, launch *STATISTICA*, select File and then Open, select All Files from the drop-down list under Files of Type, click on 2007.xls, and then select Open. This will load the Excel datasheet into *STATISTICA* for the project.

Although *STATISTICA* can determine on its own what type of data is contained in the spreadsheet, it is often helpful for the user to specify the type of data. By reviewing the listing in Table B.1, we can see that we would want the software to view all of the data fields as categorical, except for C400 and C410, "total time accumulated on part regardless of overhaul" and "total time accumulated on part since overhaul," respectively. To ensure that the program treats these variables in the way intended, we would normally (see next paragraph) double-click in the variables header and then select All Specs . . .; this opens the Variable Specifications Editor dialog. Change the Type and Measurement Type as appropriate; then click OK.

Given the complexity of this data set, however, we decided to concern ourselves with only four variables for this study. These variables are C14, "Region code"; C130, "Aircraft manufacturer's name"; C160, "Region responsible for aircraft"; and C330, "Stage of operation code." There are many more variables in this data set that may be interesting to study, including some rich text fields, but this would require a much more complex project as well as text mining that is well beyond the scope of this tutorial. Data mining is often highly dependent on the subject matter expertise of the analyst, and choices must routinely be made about the cost versus benefit of including certain data. Use of these four variables will suffice in demonstrating the kind of study that can be accomplished with readily available data sets and automated data mining.

DATA MINING APPROACH

While there are many algorithms to choose from, we decided to select a new *STATISTICA* feature, still in Beta, called Data Miner Recipes (DMRecipe) to guide us through the exploration of this complex database. According to the *STATISTICA* Quick Reference (StatSoft, 2008) manual, the *STATISTICA* Data Miner Recipe (SDMR) approach is intended to provide an intuitive graphical interface to enable those with limited data mining experience to follow a recipe-like process to achieve results. SDMR guides the user through all phases of the data mining project, from querying databases through deployment of a model. SDMR was selected to demonstrate that extensive data mining experience, while desirable, is not required to explore complex databases and formulate predictive models. Depending on the performance of the models tested using SDMR, we could also easily look at other methods built into *STATISTICA*.

First, select Data Mining, click on Data Miner Recipes and, under Recipes, select New. Under the Data Preparation tab, we must first connect our data to the project by selecting Open/Connect Data File and clicking on our spreadsheet from the drop-down menu (see Figure B.2). Click on the Select Variables tab and identify 45 – UNSCHED LNDG as the target categorical variable, and 4, 17, 21, and 52 (C14, C130, C160, C330, respectively) as the input, categorical variables, as shown in Figure B.3. Then click OK. Select the Advanced tab, select

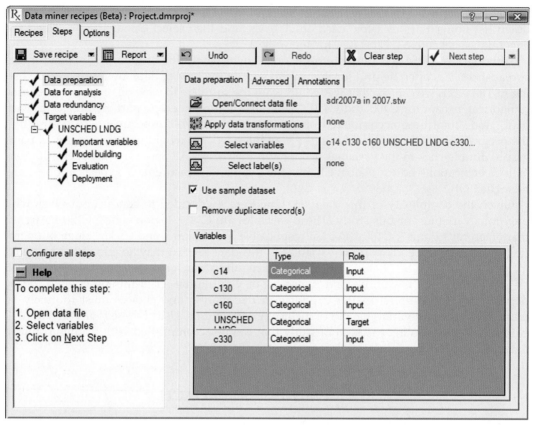

FIGURE B.2   Data preparation.

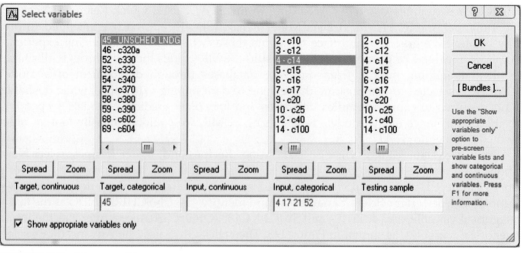

FIGURE B.3   Select variables.

Use Sample Data, and click on the Stratified Random Sampling radio button. Click on More Options, Strata Variables. Then select 45 – UNSCHED LNDG and click OK. Click Next Step. This action will cause the program to selectively sample from the database using the UNSCHED LNDG discriminator we included under the data preparation step.

The Data for Analysis dialog appears next. Choose Select Testing Sample, then % of Cases, opting for the default of 20%, Click OK. Finally, click Next Step.

The Data Redundancy dialog appears next. Since we're not using any continuous predictor variables, leave the Method as None. Click Next Step.

The Important Variables dialog appears next. This function is where the program attempts to identify the variables, or features, most important to the project. Since we've selected only four variables for this study, select None and click Next Step.

The Model Building dialog appears next. Since we desire to know which model among *STATISTICA*'s many options best predicts unscheduled landings, we selected all methods, i.e., classification and regression trees, random forest, boosted trees, neural networks, and support vector machine models. Click on Build Model (see Figure B.4).

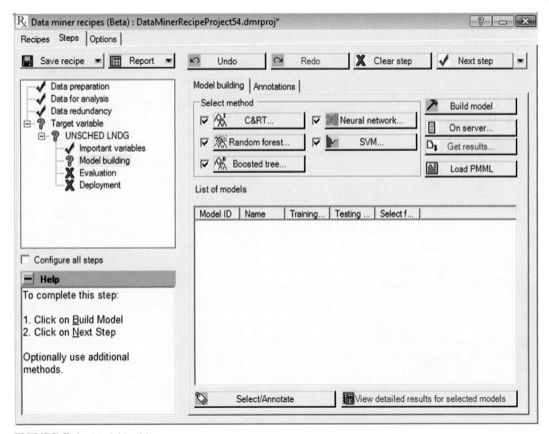

FIGURE B.4 Model building.

DATA MINING ALGORITHM ERROR RATE

The Evaluation dialog appears next. In our case, the boosted trees model performed the best with a 7.08 error rating, followed closely by neural networks with a 7.10 error rating (see Figure B.5).

After clicking Next Step, we now arrive at the Deployment tab. Select Data File for Deployment, and select another file from the FAA's SDR database for deployment. We chose sdr2008a.txt. Following the same procedure as before, we can save this file to Excel for importing into *STATISTICA*. This time, however, knowing that we are interested in only four variables, we can eliminate all others from the spreadsheet. Click Next Step. This will deploy our boosted trees model on a new data set.

STATISTICA Data Miner Recipe produces an array of reports automatically. By clicking on Summary Report, we can examine reports on data preparation, data cleaning, data reduction, feature selection, model building, evaluation, and deployment. For example,

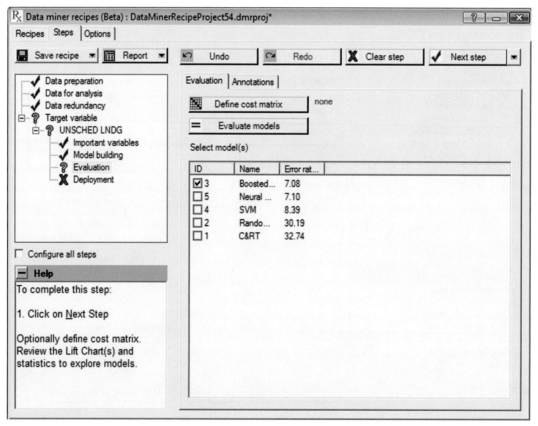

FIGURE B.5   Evaluation.

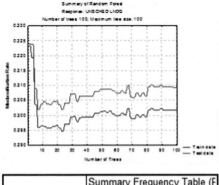

	Summary Frequency Table (F Table: UNSCHED LNDG(2) x	
	UNSCHED LNDG	Model-2-P YE
Count	YES	
Column Percent		
Row Percent		
Total Percent		
Count	NO	
Column Percent		
Row Percent		
Total Percent		
Count	All Grps	

	Summary Frequency Table (F Table: UNSCHED LNDG(2) x	
	UNSCHED LNDG	Model-2-P YE
Count	YES	
Column Percent		
Row Percent		
Total Percent		
Count	NO	
Column Percent		
Row Percent		
Total Percent		
Count	All Grps	

FIGURE B.6   Random Forest summary.

Figure B.6 shows a summary for the Random Forest model, with a chart indicating misclassification rate of the training and testing data, and partial views of frequency tables.

For our data, we know that the boosted trees model performed the best. We are not surprised by the results, since research on DM algorithms has indicated that for some difficult estimation and prediction tasks, boosted trees can yield better models than other methods, including neural networks (StatSoft, 2003). Although neural networks are a close second and are often a very effective model, they suffer from some drawbacks, including lack of explicitness. As several researchers have observed, the process neural networks goes through is largely hidden and left unexplained (Wang, 2003, p. 233). Therefore, we prefer the boosted trees model for this project.

CONCLUSION

We can conclude a number of things from this simple study. First, the SDR database contains a wealth of information that can be explored through data mining techniques to search for patterns and trends. Thanks to Data Miner Recipe, in just a few clicks of a mouse, we trained and tested a subset of the variables using several popular DM algorithms.

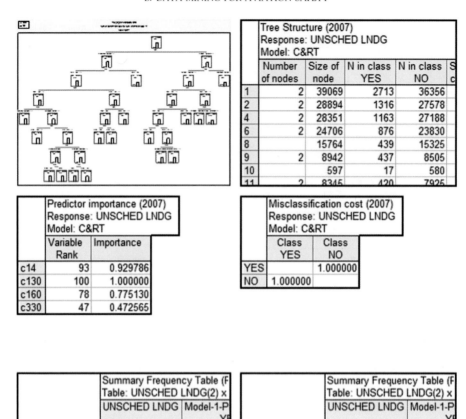

Tree Structure (2007)				
Response: UNSCHED LNDG				
Model: C&RT				
Number of nodes	Size of node	N in class YES	N in class NO	Sc
1	2	39069	2713	36356
2	2	28894	1316	27578
4	2	28351	1163	27188
6	2	24706	876	23830
8		15764	439	15325
9	2	8942	437	8505
10		597	17	580
11	2	8345	420	7925

Predictor importance (2007)		
Response: UNSCHED LNDG		
Model: C&RT		
Variable	Rank	Importance
c14	93	0.929786
c130	100	1.000000
c160	78	0.775130
c330	47	0.472565

Misclassification cost (2007)		
Response: UNSCHED LNDG		
Model: C&RT		
	Class YES	Class NO
YES		1.000000
NO	1.000000	

| Summary Frequency Table (F |
| Table: UNSCHED LNDG(2) x |
| UNSCHED LNDG | Model-1-P |
| YF |

| Summary Frequency Table (F |
| Table: UNSCHED LNDG(2) x |
| UNSCHED LNDG | Model-1-P |
| YF |

FIGURE B.7   C&RT results.

By examining the C&RT results shown in Figure B.7, we can determine that the aircraft manufacturer's name is the most important variable from the four selected, and stage of operation is the least important. For this demonstration study, this conclusion is not profound; for a full-scale study, that information would be very informative and might inspire other studies to determine the extent of the relationship between who made the airplane and the incidences of unscheduled landings.

Finally, we determined that both boosted trees and neural networks produced models with very low error rates (misclassification) for this exemplar tutorial. The models were built by the software and can be deployed in new data sets to predict when unscheduled landings might occur.

We hope it is evident through this tutorial that aviation safety has much to gain through the use of automated data mining tools and existing databases. The ultimate goal is to use the information DM can yield to drive the airline accident rate even lower than it is today.

References

Air Transport Association [ATA]. (2007). *Annual Traffic and Ops: U.S. Airlines*. Retrieved December 3, 2007, from http://www.airlines.org/economics/traffic/Annual+US+Traffic.htm.

Dasu, T., & Johnson, T. (2003). *Exploratory Data Mining and Data Cleaning*. Hoboken, NJ: Wiley & Sons.

StatSoft. (2003). STATISTICA *Data Miner*. Tulsa, OK: StatSoft, Inc.

StatSoft. (2008). STATISTICA *Quick Reference*. Tulsa, OK: StatSoft, Inc.

Wang, J. (2003). *Data Mining: Opportunities and Challenges*. Hershey, PA: Idea Group Publishing.

Predicting Movie Box-Office Receipts
Using SPSS Clementine Data Mining Software

Dursun Delen, Ph.D.

INTRODUCTION

Predicting box-office receipts (i.e., financial success) of a particular motion picture is a difficult and challenging problem. According to some domain experts, the movie industry is the land of hunches, and the wild guesses due largely to the difficulty associated with predicting the product demand.

In their highly publicized research, Sharda and Delen (2007) explored the use of a variety of data mining models in predicting the financial performance of a motion picture at the box office before its theatrical release. In their system, they converted the forecasting problem into a classification problem. That is, rather than forecasting the point estimate of box-office receipts, they classified a movie based on its box-office receipts in one of nine categories ranging from "flop" to "blockbuster," making the problem a multinomial classification problem.

DATA AND VARIABLE DEFINITIONS

The data were drawn (partially purchased) from ShowBiz Data Inc. The data set contains 2632 movies released between the years 1998 and 2006. The variable of interest in this study is the box-office gross revenue. The dependent variable, the box-office gross revenue, is discretized into nine classes (i.e., bins) using the following breakpoints:

Class No.	1	2	3	4	5	6	7	8	9
Range (in Millions)	<1 (Flop)	>1 <10	>10 <20	>20 <40	>40 <65	>65 <100	>100 <150	>150 <200	>200 (Blockbuster)

A summary of the decision variables along with their specifications is given in Table C.1. For descriptive details and justification for inclusion of the independent variables, refer to Sharda and Delen (2007).

Now that the data are briefly defined, in the next section, we will explain a step-by-step process to develop a number of different classification models. Specifically, this tutorial will introduce you to the Clementine toolkit for data mining and show you how to develop prediction models for the movie forecasting project. The first part provides a tour of the workspace of the Clementine toolkit. The second part is a step-by-step guide to data mining in Clementine using the movie forecasting data set.

TABLE C.1   Summary of Independent Variables

Independent Variable Name	Number of Values	Possible Values
MPAA Rating	5	G, PG, PG-13, R, NR
Competition	3	High, Medium, Low
Star value	3	High, Medium, Low
Genre	10	Sci-Fi, Historic Epic Drama, Modern Drama, Politically Related, Thriller, Horror, Comedy, Cartoon, Action, Documentary
Special effects	3	High, Medium, Low
Sequel	1	Yes, No
Number of screens	1	Positive integer

GETTING TO KNOW THE WORKSPACE OF THE CLEMENTINE DATA MINING TOOLKIT

Clementine uses a visual workspace approach to data mining that provides an intuitive way to develop data mining applications. This visual workspace approach is also adapted by other popular data mining toolkits in the market, including SAS-Enterprise Minder, *STATISTICA* Data Miner, and RapidMiner. Each process in Clementine is represented by an icon (or node) that can be connected to each other to form a stream representing the flow of data through a variety of processes. Although it may take some time to get used to this paradigm of developing a data mining application, you will soon find it simple, user friendly, and exceedingly powerful.

When you first start Clementine, the workspace opens in the default view with an empty stream (see Figure C.1). As shown in Figure C.1, the area in the middle is called the *stream canvas*. This is the main area you will use to build your data mining models in Clementine.

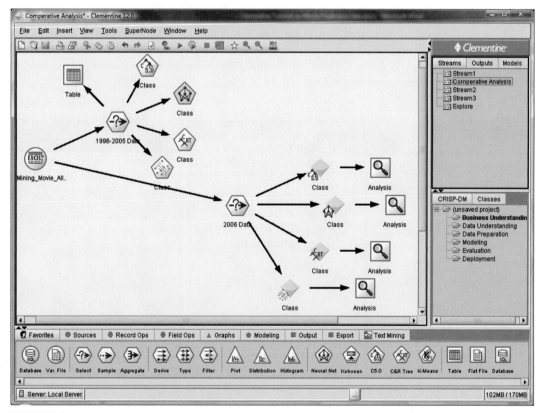

FIGURE C.1   The Clementine workspace.

Most of the data and modeling tools in Clementine reside in palettes, the area below the stream canvas. Each tab contains groups of nodes that are a graphical representation of data mining tasks, such as accessing and filtering data, creating graphs, and building models. To add nodes to the canvas, you can double-click on icons from the node palettes or drag and drop them onto the canvas. You then connect them to create a stream, representing the flow of data.

On the top right side of the window are the output and object managers. These tabs are used to view and manage a variety of Clementine objects. The Streams tab contains all streams open in the current session. You can save and close streams as well as add them to a project (Figure C.2A). The Outputs tab contains a variety of files produced by stream operations in Clementine (Figure C.2B). You can display, rename, and close the tables, graphs, and reports listed here. The Models tab is a powerful tool that contains all generated models (models that have been built in Clementine) for a session (Figure C.2C). Models can be examined closely, added to the stream, exported, or annotated.

On the bottom right side of the window is the projects tool, used to create and manage data mining projects. There are two ways to view projects you create in Clementine: Classes view and CRISP-DM view. The CRISP-DM tab provides a way to organize projects according to the Cross-Industry Standard Process for Data Mining, an industry-proven, nonproprietary methodology (Figure C.3A). For both experienced and first-time data miners, using the CRISP-DM tool would help to better organize and communicate the data mining effort. The Classes tab provides a way to organize the work in Clementine categorically by the types of objects created (Figure C.3B). This view is useful when taking inventory of data, streams, and models.

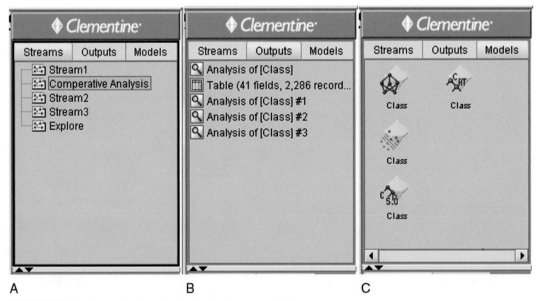

A   B   C

FIGURE C.2   Clementine's output and object manager tabs.

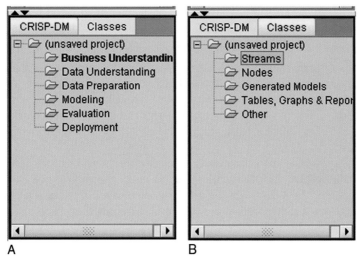

FIGURE C.3   Clementine's project tool tabs.

As an intuitive data mining toolkit, Clementine offers a strategic approach to finding useful relationships in large data sets. In Clementine, in contrast to more traditional hypothesis-driven statistical methods, you do not necessarily need to know exactly what you are looking for when you start the data mining application. You can explore your data, fit different models, and investigate different relationships until you find paths that lead to useful information.

Clementine provides templates for many of these data mining applications. Clementine Application Templates, also known as CATs, are available for the following types of activities:

- Web mining
- Fraud detection
- Analytical CRM
- Telecommunications analytical CRM
- Microarray analysis
- Crime detection and prevention

These templates provide an excellent starting point for a data mining project. With the flexible and powerful tools of Clementine, you can easily learn how to explore data in the various phases of a data mining project, including the following:

- **Visualization**, which helps you gain an overall picture of your data. You can create plots and charts to explore relationships among the fields in your data set and generate hypotheses to explore during modeling.
- **Manipulation**, which lets you clean and prepare the data for modeling. You can sort or aggregate data, filter out fields, discard or replace missing values, and derive new fields.

- **Modeling**, which gives you the broadest range of insight into the relationships among data fields. Models perform a variety of tasks such as predict outcomes, detect sequences, and group similarities.

The first step is to load the data file using a Variable File node. You can add a Variable File node from the palettes—either click the Sources tab to find the node or use the Favorites tab, which includes this node by default. Next, double-click the newly placed node to open its dialog box. Click the button just to the right of the File box marked with ellipses (...). This opens a dialog box for browsing to the directory in which the data set is stored. Open the data file.

Select Read Field Names from File and notice the fields and values that have just been loaded into the dialog box. Before clicking OK to close the dialog box, take a moment to look at the data using the other tabs on the Source node. Click the Data tab to override and change storage for a field. Note that storage is different than type or usage of the data field. The Filter tab can be used to remove any fields from the data that is brought into Clementine. Clicking on a field's arrow will mark it with a red X and filter it out. For this tutorial, though, we want to keep all fields. The Types tab helps you learn more about the type of fields in your data. You can also choose Read Values to view the actual values for each field based on the selections that you make from the Values column. This process is known as instantiation. (See Figures C.4A through C.4E.)

Now that you have loaded the data file, you may want to glance at the values for some of the records. One way to do this is by building a stream that includes a Table node. To place a Table node in the stream, either double-click the icon in the palette or drag and drop it on to the canvas. (*Hint*: Double-clicking a node from the palette will automatically connect it to the selected node in the stream canvas. However, you cannot connect to terminal nodes like tables and graphs.) Next, if the nodes are not already connected, you can use your middle mouse button to connect the Source node to the Table node. To simulate a middle mouse button, click the Alt key while using the mouse.

RESULTS

Now that you have built a stream, you must execute it in order to view its output. Click the green arrow button on the toolbar to execute the stream and view an output table showing all of the records in the data file. (See Figures C.5A through C.5C.)

During data mining, it is often useful to explore the data by creating visual summaries. Clementine offers several different types of graphs to choose from, depending on the kind of data that you want to summarize. For example, to find out what proportion of the success categories are represented in the data set, you use a Distribution node (see Figure C.6A). Place a Distribution node in the workspace and connect it to the Source node. Then double-click the Distribution node to open its dialog box and set the options for display. Select Class as the target field whose distribution you want to show. Then click Execute from the dialog box. The distribution graph helps you see the "shape" of the data distribution. It shows that the highest success categories (7, 8, 9) are less represented than the lower class categories. (See Figure C.6B.)

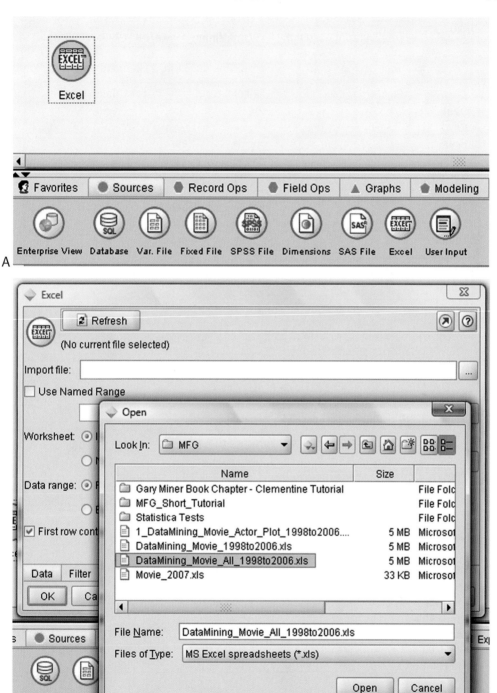

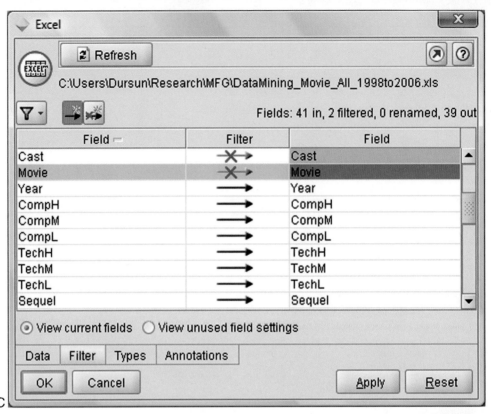

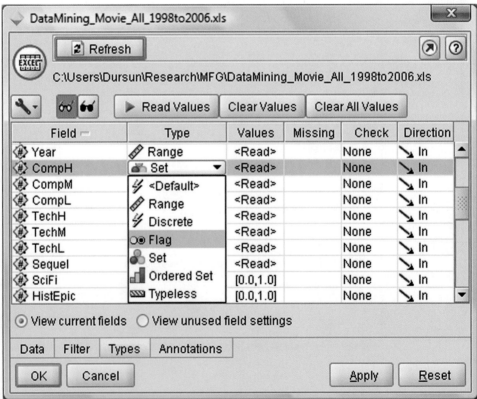

FIGURE C.4—Cont'd

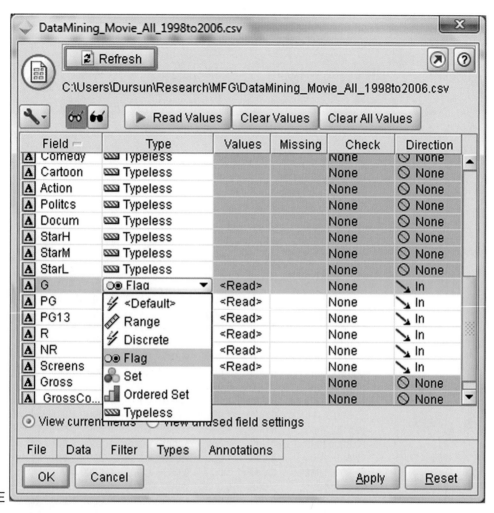

FIGURE C.4—Cont'd

In the remainder of this section, we describe and show the model building and testing steps. First, we split the data on two mutually exclusive sets using the Year variable. That is, we split and use the movies for the years 1998 to 2005 for mode building and we use the movies for the year 2006 for model testing. See Figures C.7 through C.10.

Executing a C5.0 modeling node produces a developed model represented by a diamond type icon. Click the Models tab of the Managers area (upper right) to view the already built models. To examine the patterns and/or rules generated by the model, right-click over the gem and choose Browse from the pop-up menu that appears.

In order to execute the model against some test data set (see Figure C.11), just drop it on the canvas and connect it to the data source, attach an Analysis node to it, and execute it.

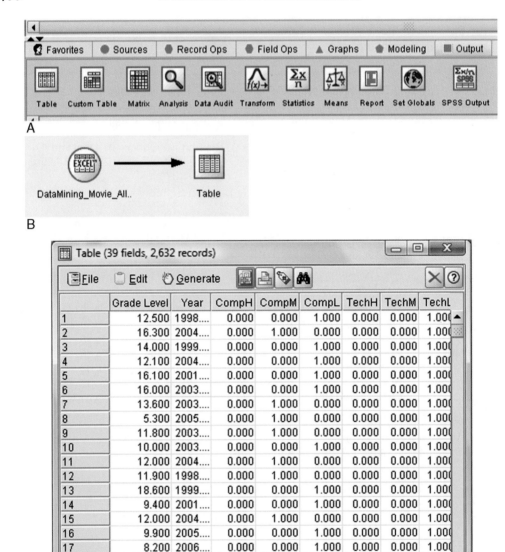

FIGURE C.5 Exploring the data using a Table node.

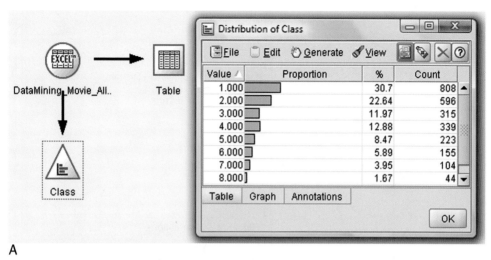

A

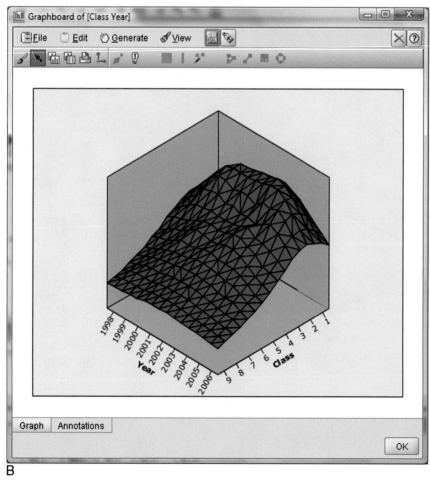

B

FIGURE C.6 A graph-board output for class distribution via the years.

III. TUTORIALS—STEP-BY-STEP CASE STUDIES

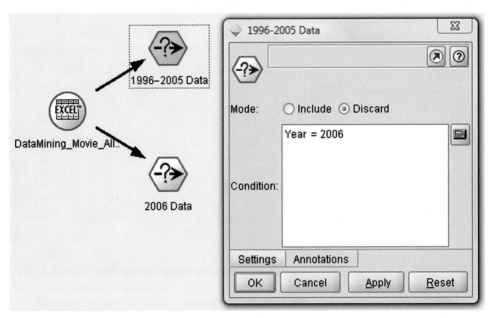

FIGURE C.7   Splitting the data into training and testing sets.

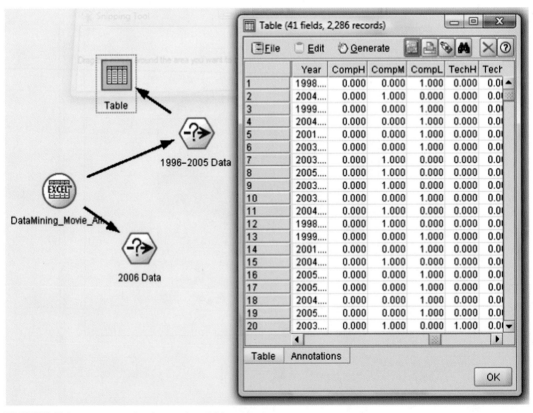

FIGURE C.8   Exploring the data with a Table node.

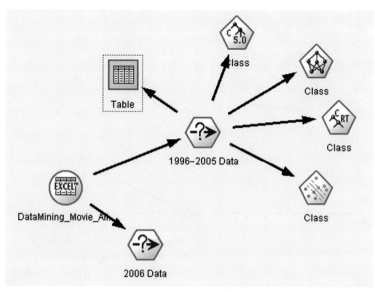

FIGURE C.9 Building prediction models for C5, ANN, C&RT, and SVM.

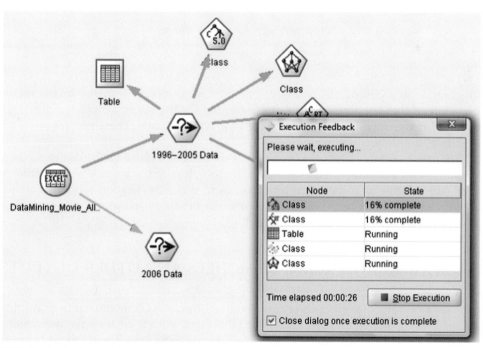

FIGURE C.10 Model building process.

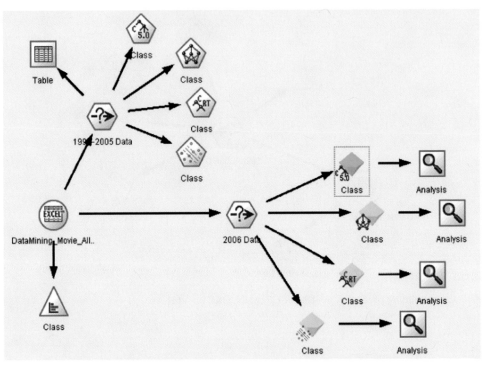

FIGURE C.11   Executing the developed models against the test data set.

It will generate analysis results in tabular as well as graphical formats. In the following section, some of these analysis results are presented for the current project.

In the test mode, only the movies from the year 2006 are used. These movies are not used for the model building.

The Analysis window shows the accuracy of the model (Figure C.12). With this data set, the model is 51.45% accurate; it predicted the correct class of the movies with a rather good accuracy (considering that there are nine classes in this classification model). You would use the Analysis node to help you determine if the model is acceptably accurate for your particular data set. (See Figures C.13 through C.21.)

PUBLISHING AND REUSE OF MODELS AND OTHER OUTPUTS

There are a number of ways to export individual objects in Clementine. For example, you can export output objects such as graphs or tables in a variety of formats, including JPG, PNG, and BMP. You can also publish the graph to the SPSS Web Deployment Framework by selecting Publish to Web from the output window's File menu.

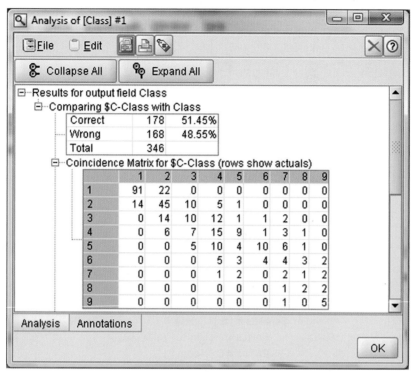

FIGURE C.12   The analysis results (confusion matrix) for C5.

You can also export model results as Predictive Modeling Markup Language (PMML) file or C code using the context menu options available from the Model tab of the Managers window.

Another deployment option is the Clementine Solution Publisher. Solution Publisher enables you to use streams independent of the standard Clementine environment. There are two steps to the Clementine Solution Publisher: publishing and executing.

• **Publishing** occurs in the stream canvas when you attach the Publish node to a stream, specify options, and execute.
• **Executing** occurs when someone runs the results that you published, re-creating the entire stream process built in Clementine without actually having Clementine on his or her machine.

For example, perhaps modeling in Clementine enabled you to learn the rules for predicting the financial success of a particular motion picture. By publishing the stream, you can deploy the stream to deliver decision recommendations for Hollywood managers (see Figure C.22). These decision makers can enter a hypothetical movie's parameters into an application or file and execute the published stream to make necessary data manipulations and run the model to receive a decision recommendation.

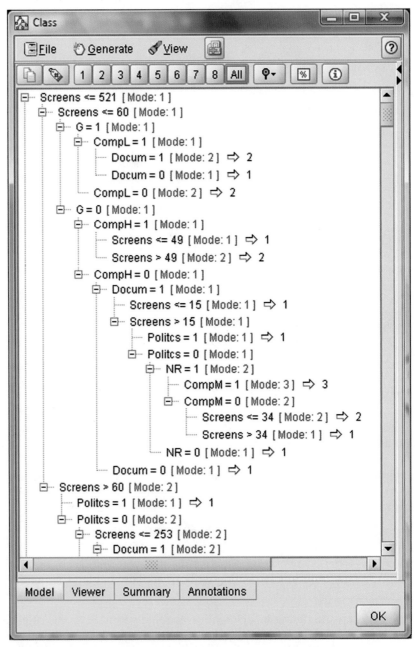

FIGURE C.13   A partial representation of the decision tree in indented list format for C5.

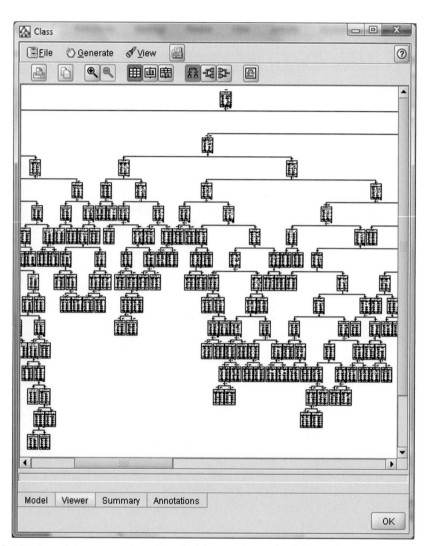

FIGURE C.14   A partial representation of the decision tree in graphical format for C5.

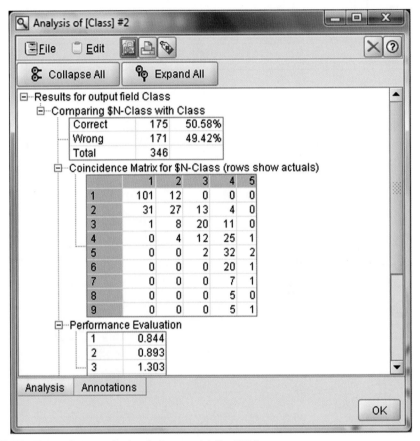

FIGURE C.15   The analysis results (confusion matrix) for ANN.

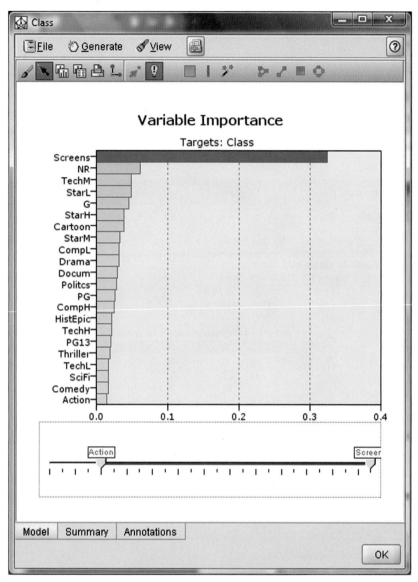

FIGURE C.16   The variable importance chart (based on sensitivity analysis) for ANN.

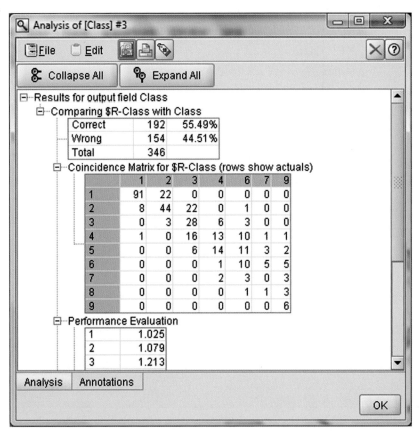

FIGURE C.17   The analysis results (confusion matrix) for C&RT.

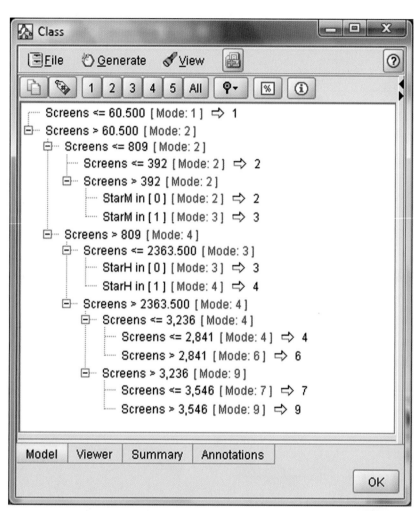

FIGURE C.18   A partial representation of the decision tree in indented list format for C&RT.

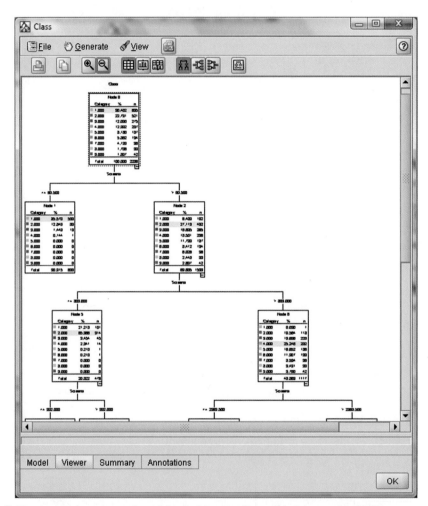

FIGURE C.19   A partial representation of the decision tree in graphical format for C&RT.

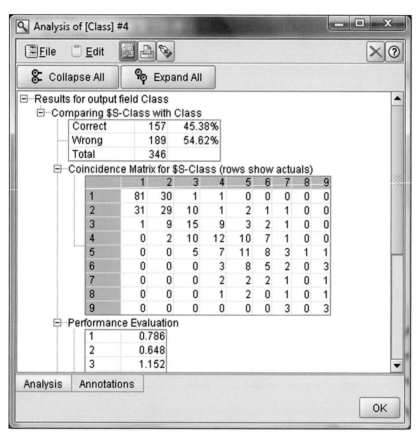

FIGURE C.20   The analysis results (confusion matrix) for SVM.

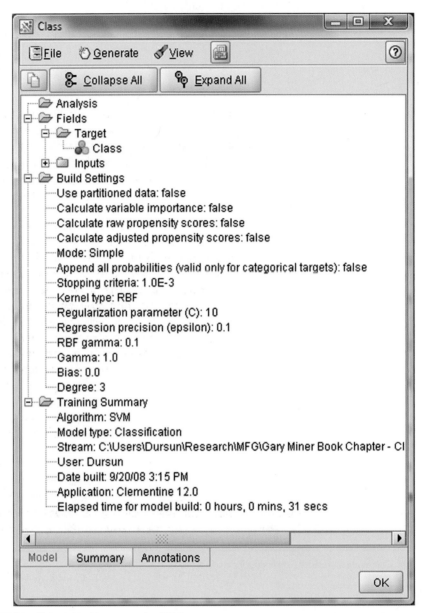

FIGURE C.21   The analysis summary for SVM.

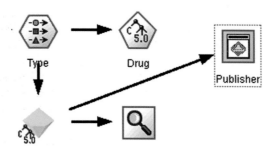

FIGURE C.22   Publishing the model from within a modeling canvas.

References

Sharda, R., & Delen, D. (2007). Predicting box-office success of motion pictures with neural networks. *Expert Systems with Applications, 30,* 243–254.

Detecting Unsatisfied Customers: A Case Study Using SAS Enterprise Miner Version 5.3 for the Analysis

Chamont Wang, Ph.D.

Note: The data set for this tutorial is not provided on the DVD that accompanies this book. Capital One, headquartered in Richmond, Virginia, requests that any readers of this book who want to use this data set contact Capital One directly; details on how to obtain this data set are provided on the DVD and/or the Elsevier companion Web page for this book.

INTRODUCTION

In this tutorial, we will use a specific data set to illustrate the application of SAS-EM 5.3 in two different scenarios of Customer Relationship Management (CRM). The tutorial is organized in the following manner:

- A Primer of SAS-EM Predictive Modeling
- Scoring Process and the Total Profit
- Oversampling and Rare Event Detection
- Decision Matrix and Profit Charts
- Micro-Target the Profitable Customers
- Import Excel File to the SASUSER library

The Data

The data have 19,991 cases with a binary target (satisfied versus unsatisfied customers) and 24 predictors. The data are commonly known as the *Capital-One Data*. A detailed description of the data can be found in the 2001 paper "Predicting Dissatisfied Credit Card Customers" by Zachary Arens and Dr. Wegman (http://www.galaxy.gmu.edu/stats/syllabi/inft979/ArensPaper.pdf).

The Objectives of the Study

Scenario 1

The first objective of the study is to provide a special program to detect and retain the unsatisfied customers for profit. The cost of the program is $15.00 per customer, and the benefit of retaining an unsatisfied customer is $278.52. The total profit is

$$P = 278.52^*M^*D - 15^*(D + S)$$

where

M = The rate of retention of dissatisfieds exposed to program = 25%;
D = Number dissatisfieds exposed to program (correct classifications);
S = Number satisfieds exposed to program (misclassifications).

The main results of our analysis are summarized in Table D.1 where the best model doubles the profit as reported in the Arens-Wegman paper.

TABLE D.1   Results

Model	Correct Classifications	Misclassifications	Predicted Positive	Total Profit	Comment
Tree	546 cases	1126 cases	1672 cases	$12,937.98	
Regression	899 cases	2257 cases	3156 cases	$15,257.37	
Neural Network	1175 cases	3102 cases	4277 cases	$17,660.25	The best

Scenario 2

A competitor has access to the data and plans to mount a campaign to lure away the unsatisfied customers. Assume that the profit is $278.52 for each defection and that the cost is $25.00 for luring away a customer. Further assume that the success rate of the campaign is 50%. A blanket effort to reach all of the 19,991 customers would result in the following profit:

$$\text{Total profit} = \$278.52^*2994^*0.5 - 25^*19991 = -\$82,830.60$$

$$\text{Average profit} = \text{Total profit}/19,991 = -\$4.14339$$

In other words, without any model, there will be a loss of $82,830.60. On the other hand, if we target the top 5% of the customers, then the total profit would be $36,313.65 as shown in the subsequent analysis. The difference of model or no-model would be $119,144.25.

The techniques in Scenario 2 are commonly used in micro-targeting, such as the famous 1998 KDD-Cup competition where a national veterans' organization seeks to better target its solicitations for donation (http://www.kdnuggets.com/meetings/kdd98/kdd-cup-98.html). The techniques should be equally effective in the campaigns for political donations and direct marketing.

SAS-EM 5.3 Interface

Figure D.1 shows the key components of the SAS-EM 5.3 interface.

- **Toolbars:** There are three rows of tools that can be activated by clicking or by dragging to the workspace. Move the cursor to a specific tool, and a small window will pop up that gives a brief description of the tool's functionality.
- **Project Panel:** To manage and view data sources, diagrams, and results.
- **Properties Panel:** To view and edit the settings of data sources, diagrams, and nodes.
- **Diagram Workshop:** To graphically build, edit, run, and save process flow diagrams.

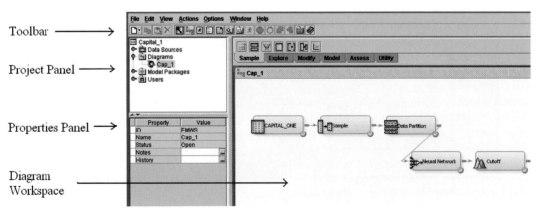

FIGURE D.1   SAS-EM version 5.3 interface.

A PRIMER OF SAS-EM PREDICTIVE MODELING

This section provides information on the construction of the following SAS-EM process flow, which contains only four nodes, as shown in Figure D.2.

The construction is sufficient for small- or medium-sized data sets. For large data sets, a sample node can be added with little effort and will be discussed at the end of this section.

1. Creating a Project:

Select File, New, Project from the main menu. Specify the project name, Capital_1, in the name field of the pop-up window, as shown in Figure D.3.

FIGURE D.2   SAS-EM process flow.

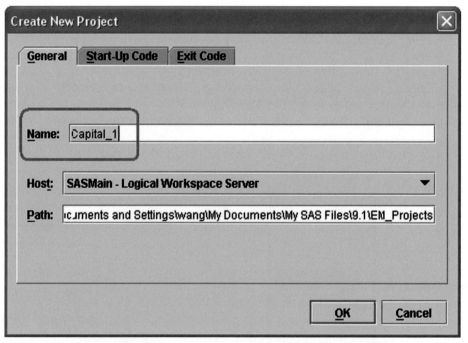

FIGURE D.3   SAS-EM Create New Project window.

2. Creating a Data Source:

Select File, New, Data Source to open the Data Source Wizard, shown in Figure D.4.

Click on the Next button to browse the folder of the Capital_One data, which resides in the SASUSER library (see "Appendix" on how to import Capital_One Excel data to the SASUSER library), as shown in Figures D.5 and D.6.

Click on OK, click on Next *four times,* and then on Finish to import the data.

3. Creating a Diagram:

Select File, New, Diagram and then type the name, Cap_1, in the name field of the pop-up window (see Figure D.7) and then click on the OK button:

4. Creating the Process Flow:

From the Project panel (upper-left corner), drag the Capital_One icon right under Data Source to the Diagram workspace to create the Data Sources node, as shown in Figure D.8.

5. Editing Variables:

In the workspace, right-click on the Data Source node, and then click on Edit Variables, as shown in Figure D.9.

In the pop-up window, identify the target variable, SATISF1. Change the model Role from Input to Target and change the Level from Nominal to Binary, as shown in Figure D.10. Then click on OK.

6. Data Partition node:

Drag the Data Partition icon (in the third toolbar) to the diagram workspace (see Figure D.11). Move the cursor to the right edge of the Data Source node until a pencil appears and then connect the Data Source and the Data Partition nodes.

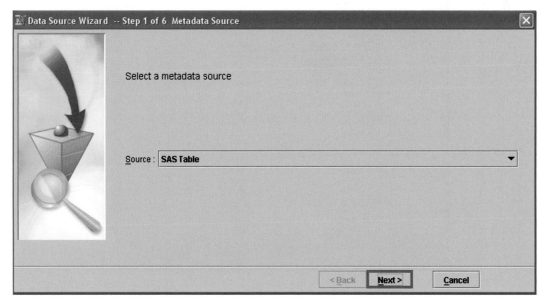

FIGURE D.4   SAS-EM Data Source Wizard window.

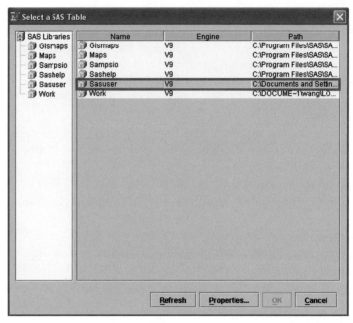

FIGURE D.5   SAS-EM Select a SAS Table path dialog.

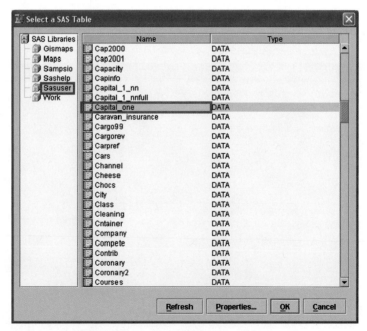

FIGURE D.6   SAS-EM Select a SAS Table data type dialog.

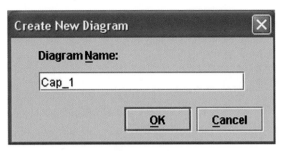

FIGURE D.7   SAS-EM Create New Diagram naming dialog.

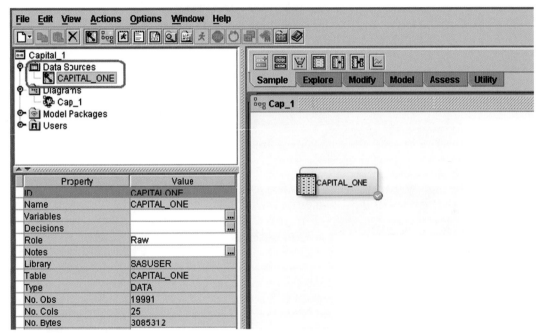

FIGURE D.8   SAS-EM version 5.3 interface with Data Mining Workspace window with the Capital_One data icon selected.

In SAS-EM 5.3, the default setting is 40% – 30% – 30% for the partition of the original data into Training, Validation, and Test data sets.

7. Neural Network node:

Click on the Model tab (in the third row of toolbars) to activate the Neural Network icon ![icon]. Drag the icon to the workspace and then connect the Data Partition node and the Neural Network node, as shown in Figure D.12.

Right-click on the Neural Network node and select Run. Click on Yes in the pop-up window. Wait for the next pop-up window and then click on Results. The next pop-up window contains a lot of information. This is useful in many other studies. In this case, we will skip these results and go straight to the Cutoff node for profit calculation.

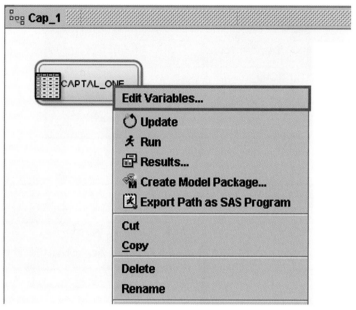

FIGURE D.9 Right clicking on Data Source provides flying menu of options.

Name	Role	Level	Report	Order	Drop	Lower Limit	Upper Limit	Type	Label	Format
ADB	Input	Interval	No		No			Numeric	ADB	
APPLY	Input	Interval	No		No			Numeric	APPLY	
APR	Input	Interval	No		No			Numeric	APR	
BALANCE	Input	Interval	No		No			Numeric	BALANCE	
CASH1	Input	Nominal	No		No			Character	CASH1	$3.0
CASH2	Input	Nominal	No		No			Character	CASH2	$3.0
CASH3	Input	Nominal	No		No			Character	CASH3	$3.0
CASH4	Input	Nominal	No		No			Character	CASH4	$3.0
CASH5	Input	Nominal	No		No			Character	CASH5	$3.0
CASH6	Input	Nominal	No		No			Character	CASH6	$3.0
CASHADV	Input	Interval	No		No			Numeric	CASHADV	
CBAL	Input	Interval	No		No			Numeric	CBAL	
CCRED	Input	Interval	No		No			Numeric	CCRED	
CREDIT	Input	Interval	No		No			Numeric	CREDIT	
EMPLOY	Input	Nominal	No		No			Character	EMPLOY	$3.0
HI	Input	Interval	No		No			Numeric	HI	
ICRED	Input	Interval	No		No			Numeric	ICRED	
JOINT	Input	Nominal	No		No			Character	JOINT	$3.0
LIMIT	Input	Nominal	No		No			Character	LIMIT	$3.0
PASTDUE	Input	Interval	No		No			Numeric	PASTDUE	
PURCH	Input	Interval	No		No			Numeric	PURCH	
PURTIME	Input	Interval	No		No			Numeric	PURTIME	
SATISF1	Target	Nominal ▼	No		No			Character	SATISF1	$5.0
TIME	Input	Unary	No		No			Numeric	TIME	
YEARS	Input	Binary	No		No			Numeric	YEARS	
		Nominal								
		Ordinal								
		Interval								

Variables - lds3

(none) □ not Equal to Apply Reset

Explore... OK Cancel

FIGURE D.10 SAS-EM Variables window for data set.

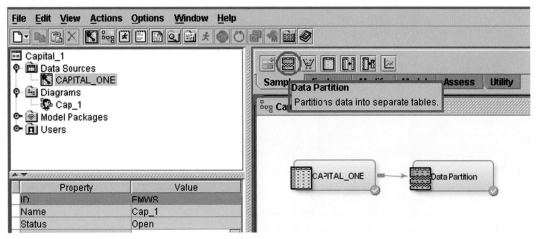

FIGURE D.11   Adding Data Partition to SAS-EM workspace.

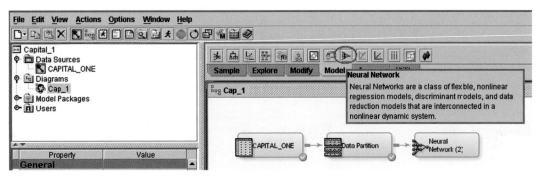

FIGURE D.12   Connect the Data Partition node and the Neural Network node.

8. Cutoff node:

Click on the Assess tab on the third row of the toolbars (right above the workspace) to activate a new row of tools (see Figure D.13).

Drag the Cutoff icon 🔲 to the workspace and connect this new node to the Neural Network node, as shown in Figure D.14.

Right-click on the Cutoff node and then click on Run. Wait for the window to pop up and then click on Results. In the next window, click on View, Analytical Results, Model Diagnostics, as shown in Figure D.15.

The next window, shown in Figure D.16, shows the variables that are used in the *Profit formula* at the beginning of this tutorial (Correct Classifications = Counts of True Positives, Misclassifications = Counts of False Positives, and Pred_Pos = Counts of Predicted Positives):

Note that different cutoff values (a.k.a., *threshold,* which ranges from 0 to 0.99, increment by 0.01) give different counts of True Positives, False Positives, and Predicted Positives. Our next task is to use Excel or SAS codes to find the optimal profit.

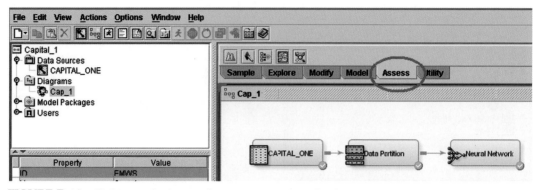

FIGURE D.13   Clicking on the Assess tab activates a new set of tools in SAS-EM 5.3 workspace.

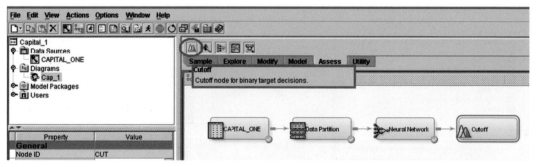

FIGURE D.14   Cutoff icon put into workspace.

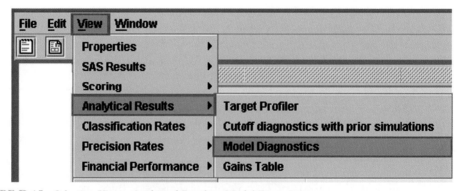

FIGURE D.15   Selecting View→Analytical Results→Model Diagnostics.

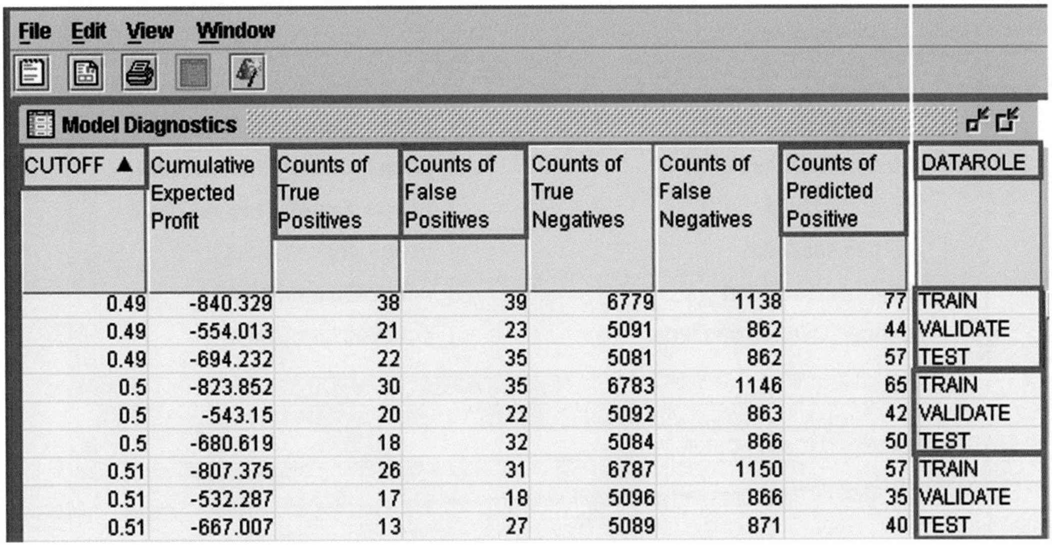

File Edit View Window

Model Diagnostics

CUTOFF ▲	Cumulative Expected Profit	Counts of True Positives	Counts of False Positives	Counts of True Negatives	Counts of False Negatives	Counts of Predicted Positive	DATAROLE
0.49	-840.329	38	39	6779	1138	77	TRAIN
0.49	-554.013	21	23	5091	862	44	VALIDATE
0.49	-694.232	22	35	5081	862	57	TEST
0.5	-823.852	30	35	6783	1146	65	TRAIN
0.5	-543.15	20	22	5092	863	42	VALIDATE
0.5	-680.619	18	32	5084	866	50	TEST
0.51	-807.375	26	31	6787	1150	57	TRAIN
0.51	-532.287	17	18	5096	866	35	VALIDATE
0.51	-667.007	13	27	5089	871	40	TEST

FIGURE D.16 Model Diagnostics showing variables used in the *profit formula*.

9. Excel or SAS codes for Profit Calculation:

In the Model Diagnostics spreadsheet window, click on File, Save As to save the spreadsheet in the folder of your choice (in our case, we saved the file in My SAS Files/9.1, which is the location of the SASUSER folder), and then assign the name, Cap1_Cutoff, to the file (see Figure D.17):

The following SAS codes can be used to calculate the Total Profit and Average Profit for different cutoff:

```
data Cap1_NN (Keep = Cutoff FP_CLASSIFS TP_CLASSIFS Pred_Pos Pred_Neg
     Total_Profit Average_Profit DataRole);
  Set sasuser.cap1_cutoff;
  Total_Profit = 278.52*.25*TP_CLASSIFS-15*Pred_Pos;
  If abs(pred_pos) < 0.000001 then Average_Profit = .;
  else Average_Profit = Total_Profit/Pred_Pos;
Run;
/* Remark:
  FP_CLASSIFS = False Positive
  TP_CLASSIFS = True Positive
*/
```

SAS-EM 5.3 has a Program Editor to run the preceding codes; see Figures D.18 and D.19.

After the codes are run successfully, click on the Explorer button to open the Work library and then to locate the Cap1_nn data set (see Figure 9.20).

Click on Cap1_nn to view the results. In the pop-up window, click on the DATAROLE tab to sort the TEST (HOLDOUT) data (see Figure D.21).

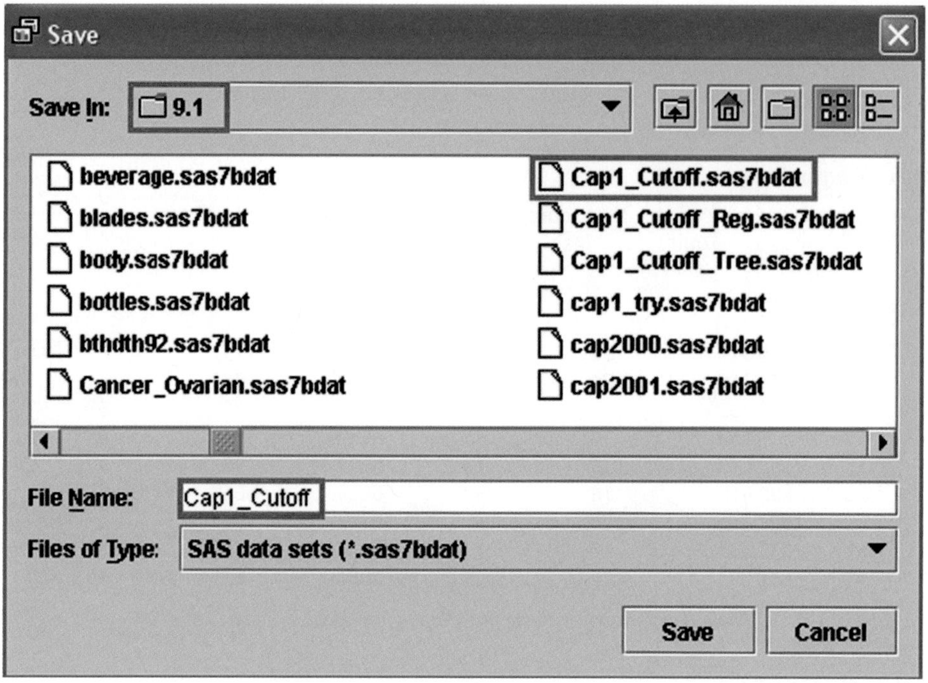

FIGURE D.17 Save window used to save the Model Diagnostics spreadsheet to a folder of one's choice.

FIGURE D.18 Program Editor used to run SAS codes.

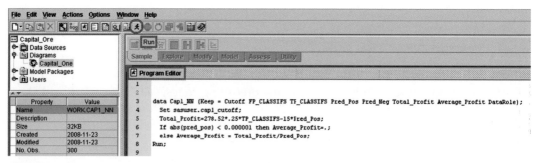

FIGURE D.19 Another view of SAS-EM Program Editor.

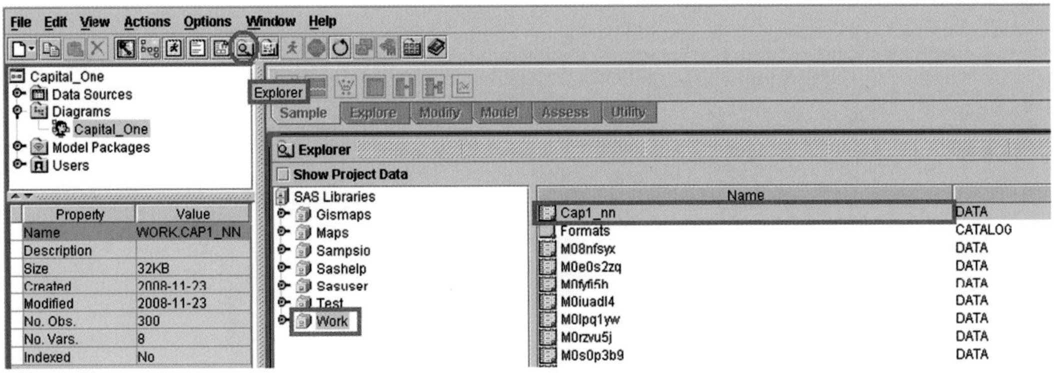

FIGURE D.20   Explorer button to open Work library and the Cap1_nn data set.

WORK.CAP1_NN

	CUTOFF	Co...	Cou...	Cou...	Counts...	DATAROLE	TOTAL_PROFIT	AVERAGE_PROFIT
41	0.4	48	113	161	5839	TEST	927.24	5.7592546584
42	0.41	47	99	146	5854	TEST	1082.61	7.4151369863
43	0.42	45	90	135	5865	TEST	1108.35	8.21
44	0.43	41	79	120	5880	TEST	1054.83	8.79025
45	0.44	37	69	106	5894	TEST	986.31	9.3048113208
46	0.45	30	59	89	5911	TEST	753.9	8.4707865169
47	0.46	27	54	81	5919	TEST	665.01	8.21
48	0.47	25	50	75	5925	TEST	615.75	8.21
49	0.48	25	42	67	5933	TEST	735.75	10.981343284
50	0.49	22	35	57	5943	TEST	676.86	11.874736842
51	0.5	18	32	50	5950	TEST	503.34	10.0668
52	0.51	13	27	40	5960	TEST	305.19	7.62975
53	0.52	13	23	36	5964	TEST	365.19	10.144166667
54	0.53	12	21	33	5967	TEST	340.56	10.32
55	0.54	9	20	29	5971	TEST	191.67	6.6093103448
56	0.55	8	18	26	5974	TEST	167.04	6.4246153846
57	0.56	7	13	20	5980	TEST	187.41	9.3705

FIGURE D.21   The Work.Cap1_nn data set.

This window shows that at Cutoff = 0.49, the *highest* average profit of the Test data set is $11.8747 per customer who is in the program. Sorting the data again on the Cutoff column reveals that the average profits from the Training and Validation data sets are substantially higher than that of the Test data (see Figure D.22). This is an indication of *model overfitting*.

WORK.CAP1_NN

	CUTOFF △	Co...	Cou...	Cou...	Counts...	DATAROLE △	TOTAL_PROFIT	AVERAGE_PROFIT
142	0.47	25	50	75	5925	TEST	615.75	8.21
143	0.47	49	48	97	7897	TRAIN	1956.87	20.173917526
144	0.47	26	31	57	5940	VALIDATE	955.38	16.761052632
145	0.48	25	42	67	5933	TEST	735.75	10.981343284
146	0.48	44	42	86	7908	TRAIN	1773.72	20.624651163
147	0.48	23	27	50	5947	VALIDATE	851.49	17.0298
148	0.49	22	35	57	5943	TEST	676.86	11.874736842
149	0.49	38	39	77	7917	TRAIN	1490.94	19.362857143
150	0.49	21	23	44	5953	VALIDATE	802.23	18.2325
151	0.5	18	32	50	5950	TEST	503.34	10.0668
152	0.5	30	35	65	7929	TRAIN	1113.9	17.136923077
153	0.5	20	22	42	5955	VALIDATE	762.6	18.157142857

FIGURE D.22   Another view of the data set.

Homework 1

Do the same by using Dmine Regression and Decision Tree as indicated by the process flow shown in Figure D.23.

The final results, shown in Table D.2, would include the Average Profit for the best cutoff for each model.

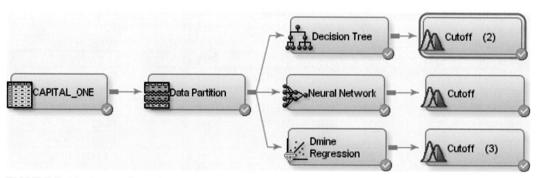

FIGURE D.23   Process flow of SAS-EM.

TABLE D.2   Results

Average Profit	Training	Validation	Test	Cutoff
Neural Network	$19.36	$18.23	$11.87	0.49
Dmine Regression	$26.78	$11.28	$11.67	0.58
Decision Tree (Gini)	$06.89	$03.85	$02.95	0.13 to 0.16

Discussions

a. From Table D.2, neural network appears to be the best model among the three, albeit a certain extent of overfitting.

b. If the purpose of the study is to *apply the model to a new data set*, then the criterion of the model selection should be placed on the performance of the model on the Test (holdout) data, as shown in the preceding.

c. The Arens-Wegman paper (2001, p. 10), on the other hand, used the entire data set as the basis for model comparison. The paper did not specify the cutoff value. The subsequent SAS-EM process flow shows that if you choose a different cutoff, it is possible to double the profit as reported in the Arens-Wegman paper.

d. In addition, the following process flow is a necessity in the scoring of new data sets.

e. Note that the default tree uses a chi-square criterion in the modeling-building process and will result in constant probability and hence no result. Click on the Tree node to activate the related Property Panel; click on the right of the Criterion tab and select Gini (see Figure D.24) or other Splitting Rule for the tree model.

Homework 2

Add the Sample node as shown in Figure D.25. Click on the Sample node to activate its Properties Panel. Move the cursor to the box next to Percentage and change 10.0 to 20.0:

The original data have 19,991 cases. The sampled data will have about 4,000 records, a number good enough to build a decent model.

Homework 3

Use Excel to reconfirm the previous results.

Note: To export the Captial_one_NN data to Excel, right-click on the file and then select View in Excel, as shown in Figure D.26.

You can then work on the file in the Excel window.

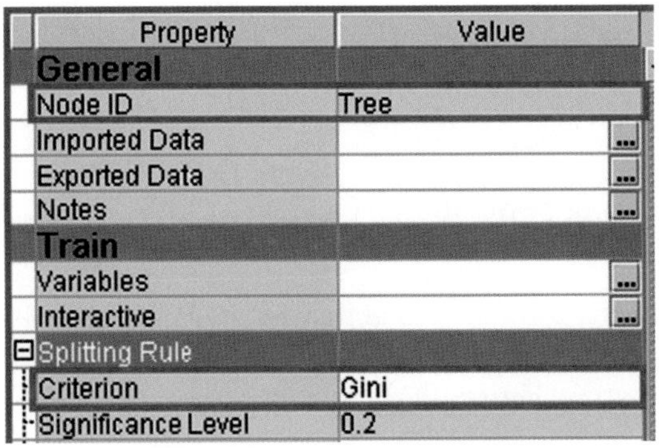

Property	Value	
General		
Node ID	Tree	
Imported Data		...
Exported Data		...
Notes		...
Train		
Variables		...
Interactive		...
⊟ Splitting Rule		
Criterion	Gini	
Significance Level	0.2	

FIGURE D.24 Property panel activated and Gini Splitting Rule selected.

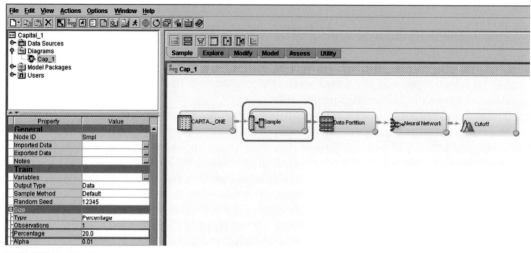

FIGURE D.25   Sample node added to workflow.

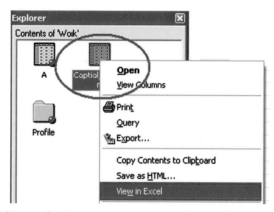

FIGURE D.26   Right-clicking on data icon opens menu where data can be viewed in Excel.

SCORING PROCESS AND THE TOTAL PROFIT

This section presents an SAS-EM process flow that can be used to score the new data set or to calculate decision consequences on an existing population (see Figure D.27).

1. Score node:

    Click on the Assess tab (on the third row of the toolbars) to activate a new set of tools (see Figure D.28).

    Drag the Score tab to the workspace as shown in Figure D.29.

FIGURE D.27   Process flow where Score icon can be used to score new data.

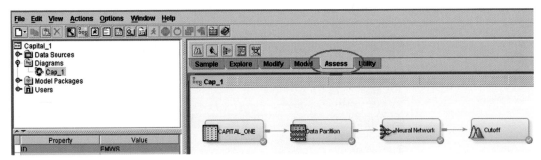

FIGURE D.28   Assess tab activates a new set of tools.

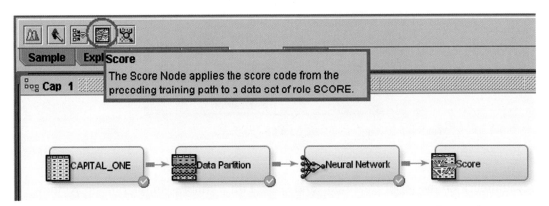

FIGURE D.29   The Score icon has been dragged from the Score button into the Process flow workspace.

2. Scoring data set:

Copy the Capital_One data node and paste it under the Neural Network node. Connect this input data node to the Score node, as shown in Figure D.30.

Change the role of the second Input Data from Raw to Score, as shown in Figure D.31. First click on the Input Data node to activate its Property Panel, then click on the right of the Role tab.

CAPITAL_ONE
Data Partition
Neural Network
Score
Input Data

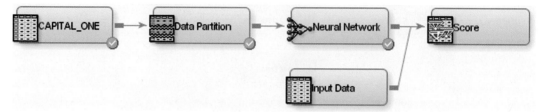

FIGURE D.30 Input Data node connected to Score node.

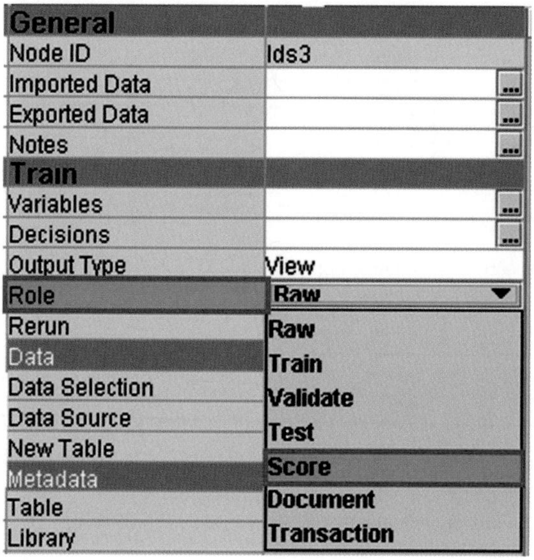

General
Node ID
Ids3
Imported Data
Exported Data
Notes
Train
Variables
Decisions
Output Type
View
Role
Raw
Rerun
Data
Data Selection
Data Source
New Table
Metadata
Table
Library
Raw
Train
Validate
Test
Score
Document
Transaction

FIGURE D.31 Window to change Input Data from Raw to Score.

Right-click on the Input Data node and select Edit Variables. In the pop-up window, double-click on Role to move the Target Variable (SATISF1) to the top, as shown in Figure D.32. Change the role of SATISF1 from Target to Rejected; then click on OK.

Run the Score node.

3. SAS Code node:

Click on the Utility tab (on the third row of the toolbars; see Figure D.33), drag the SAS Code node to the workspace, and then connect the SAS Code node with the Score node.

Click on the SAS Code node to activate its Property Panel. Then click on [...] at the right of Code Editor, as shown in Figure D.34.

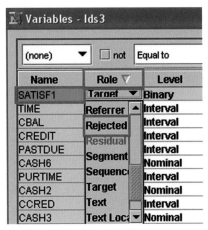

FIGURE D.32   Edit Variables window.

FIGURE D.33   Click on SAS coded node and drag to workspace.

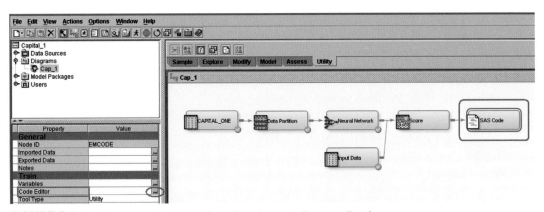

FIGURE D.34   Clicking on the SAS Code node activates its Property Panel.

In the pop-up window, paste and run the following SAS macro that is tailored for this specific problem. The macro can be modified with little effort for similar situations.

```
Data Cap1_NN_Pr (KEEP = P_SATISF1UNSAT Target);
        Set &EM_Import_Score;
        If SATISF1 = 'UNSAT' then Target = 1; else Target = 0;
Run;
%macro cutoff100;
%do
  n = 1% to 100;
    Data Cap1_NN_Pred_&n. ;
      set Cap1_NN_Pr;
      Cutoff = &n/100;
      If P_SATISF1UNSAT > Cutoff then D_target=1; else D_target=0;
      If D_target=1 and Target=1 then Tr_Positive=1;
        ELSE Tr_Positive=0;
      If D_target=1 and Target=0 then F_Positive=1; ELSE F_Positive=0;
    run;
    proc means data=Cap1_NN_Pred_&n. noprint;
      by cutoff;
      var Tr_Positive F_Positive;
      output out=Cap1_NN_Pred_&n._sum(drop=_type_
            _freq_) sum=Tr_Positive_sum F_Positive_sum;
    run;
    proc append base=all data=Cap1_NN_Pred_&n._sum; run;
%end;
%mend cutoff100;
%cutoff100;
data Cap1_NN_all;
  set all;
  Pred_Pos = Tr_Positive_sum + F_Positive_sum;
  Total_Profit=278.52*.25*Tr_Positive_sum-15*Pred_Pos;
  Average_Profit = Total_Profit/Pred_Pos;
Run;
Proc print;
```

After the codes are executed, click on OK and then click on Run, Results, as shown in Figure D.35, to view the results.

A new window appears with the results shown in Figure D.36.

So the best Total Profit is \$17,660.25 when Cutoff = 0.21, while the best Average Profit is \$17.2039 when Cutoff = 0.55, which is consistent with the earlier result in item 9 in "A Primer of SAS-EM Predictive Modeling."

FIGURE D.35 Click on Run and the Results to view results.

Obs	Cutoff	Tr_Positive_ sum	F_Positive_ sum	Pred_Pos	Total_ Profit	Average_ Profit
1	0.01	2943	17048	19991	-94943.91	-4.7493
2	0.02	2943	17000	19943	-94223.91	-4.7247
3	0.03	2933	16782	19715	-91500.21	-4.6411
4	0.04	2914	16365	19279	-86283.18	-4.4755
5	0.05	2874	15540	18414	-76093.38	-4.1324
6	0.06	2780	14177	16957	-60783.60	-3.5846
7	0.07	2664	12867	15531	-47470.68	-3.0565
8	0.08	2547	11643	14190	-35502.39	-2.5019
9	0.09	2430	10547	12977	-25454.10	-1.9615
10	0.10	2299	9549	11848	-17640.63	-1.4889
11	0.11	2184	8630	10814	-10138.08	-0.9375
12	0.12	2058	7838	9896	-5141.46	-0.5195
13	0.13	1964	7072	9036	1213.32	0.1343
14	0.14	1859	6380	8239	5857.17	0.7109
15	0.15	1741	5728	7469	9190.83	1.2305
16	0.16	1641	5218	6859	11377.83	1.6588
17	0.17	1542	4693	6235	13844.46	2.2204
18	0.18	1444	4243	5687	15240.72	2.6799
19	0.19	1340	3838	5178	15634.20	3.0194
20	0.20	1262	3464	4726	16983.06	3.5935
21	0.21	1175	3102	4277	17660.25	4.1291
22	0.22	1086	2799	3885	17343.18	4.4641
23	0.23	1015	2539	3554	17364.45	4.8859
24	0.24	946	2272	3218	17599.98	5.4692
25	0.25	869	2032	2901	16993.47	5.8578
26	0.26	808	1785	2593	17366.04	6.6973
27	0.27	745	1589	2334	16864.35	7.2255
28	0.28	676	1407	2083	15824.88	7.5972
29	0.29	623	1236	1859	15494.49	8.3349
53	0.53	83	101	184	3019.29	16.4092
54	0.54	78	95	173	2836.14	16.3939
55	0.55	74	86	160	2752.62	17.2039
56	0.56	63	77	140	2286.69	16.3335
57	0.57	55	65	120	2029.65	16.9138

FIGURE D.36 Results of the SAS-EM Run.

Homework 4

a. Build other models as shown in the process flow in Figure D.37. Click on the Control Point node at the end of the flow, go for a coffee break, and then come back to view the results.

b. To speed up the process, you can add a Sample node to use, say 20% (n ~ 4,000 cases), as discussed in "Homework 2."

Hint: To find the Control Point icon ▣, click on Utility tab, and the first icon in Figure D.38 will do.

The results are shown in Table D.3.

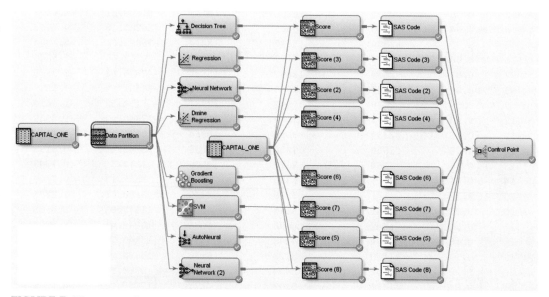

FIGURE D.37   Process flow with many additional models added.

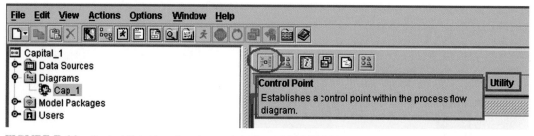

FIGURE D.38   Control Point icon location on interface of SAS-EM 5.3.

TABLE D.3   Results

Model	Cutoff	True-Positive Sum	False-Positive Sum	Predicted Positive	Total Profit	Average Profit
Tree	0.22	546	1126	1672	$12,937.98	$7.7380
Regression	0.23	899	2257	3156	$15,257.37	$4.8344
NN_MLP	0.21	1175	3102	4277	$17,660.25	$4.1291
DM_Reg	0.23	799	1852	2651	$15,869.37	$5.9862
Boosting	0.21	960	2391	3351	$16,579.80	$4.9477
SVM	0.23	1763	7824	9587	$−21,047.31	$−2.195
Auto_Neural	0.24	0	0	0	$0	$0
NN_RBF	0.22	830	2174	3004	$12,732.90	$4.2386

Discussions

a. The default neural network (multilayer perceptron) produced the best total profit. The tree model is the best in terms of average profit for each customer in the retention program (n = 1,672) but its total profit is about $5,000 less than that of the neural network model.
b. Auto-Neural and SVM, again, do not perform well.
c. The preceding exercise follows the Arens-Wegman approach to use the entire data set in the comparison of the models. The approach may be useful in certain applications, but a caution is that if the so-called best model overfits the data, then the resulting profit may be misleading.

OVERSAMPLING AND RARE EVENT DETECTION

In this section, we will first explore the distribution of the target variable to motivate the need of oversampling technique in the detection of rare event. We will then discuss the *profit matrix*, which is unique in SAS-EM among the data mining packages we have known. The process needs the addition of the Sample node to the previous flow, as shown in Figure D.39. In the first two nodes, we will have to make changes in the settings for rare event detection.

1. Explore the Target Variable:

   Right-click on the Data Source node and then select Edit Variables. In the pop-up window, click on the Role twice to sort the Target to the top of the column (see Figure D.40).

   Highlight the Target and then click on the Explore button (at the lower-right corner of the window) as shown in Figure D.40 to view the distribution of the variable. The result is shown in Figure D.41.

   The bar chart indicates that there are about 15% unsatisfied customers.

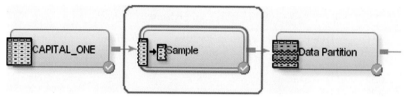

FIGURE D.39   Sample node added to process flow.

Name	Role ▽	Level	Report	Order	Drop	Lower Limit	Upper Limit	Type	Label	Format
SATISF1	Target	Binary	No		No			Character	SATISF1	$5.0
TIME	Input	Interval	No		No			Numeric	TIME	
CBAL	Input	Interval	No		No			Numeric	CBAL	
CREDIT	Input	Interval	No		No			Numeric	CREDIT	
PASTDUE	Input	Interval	No		No			Numeric	PASTDUE	
CASH6	Input	Nominal	No		No			Character	CASH6	$3.0
PURTIME	Input	Interval	No		No			Numeric	PURTIME	
CASH2	Input	Nominal	No		No			Character	CASH2	$3.0
CCRED	Input	Interval	No		No			Numeric	CCRED	
CASH3	Input	Nominal	No		No			Character	CASH3	$3.0
EMPLOY	Input	Nominal	No		No			Character	EMPLOY	$3.0
JOINT	Input	Nominal	No		No			Character	JOINT	$3.0
CASH5	Input	Nominal	No		No			Character	CASH5	$3.0
LIMIT	Input	Nominal	No		No			Character	LIMIT	$3.0
CASH1	Input	Nominal	No		No			Character	CASH1	$3.0
PURCH	Input	Interval	No		No			Numeric	PURCH	
CASHADV	Input	Interval	No		No			Numeric	CASHADV	
ICRED	Input	Interval	No		No			Numeric	ICRED	
APR	Input	Interval	No		No			Numeric	APR	
YEARS	Input	Interval	No		No			Numeric	YEARS	
ADB	Input	Interval	No		No			Numeric	ADB	
APPLY	Input	Interval	No		No			Numeric	APPLY	
BALANCE	Input	Interval	No		No			Numeric	BALANCE	
CASH4	Input	Nominal	No		No			Character	CASH4	$3.0
HI	Input	Interval	No		No			Numeric	HI	

FIGURE D.40   Target variable sorted to top of variable list.

2. The Sample node:

   Click on Sample tab (in the third toolbar), drag the Sample icon to the diagram workspace, and then connect the Sample node with the Data Source node, as shown in Figure D.42.

3. Oversampling and the Rare Event Detection:

   Click on the Sample node to activate its Property Panel; then edit the settings as shown in Figure D.43.
   - Click on the right of the Criterion tab to activate a drop-down menu and then select Level Based option.
   - Click on the right of the Level Selection tab and then select the Rarest Level.
   - Change the Percentage to 100.0 to use all rare cases in the sampled data.

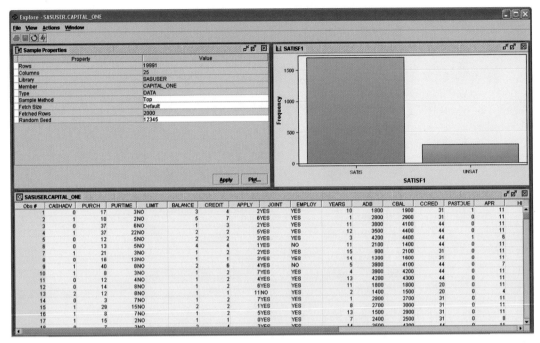

FIGURE D.41   Results showing distribution of the variable.

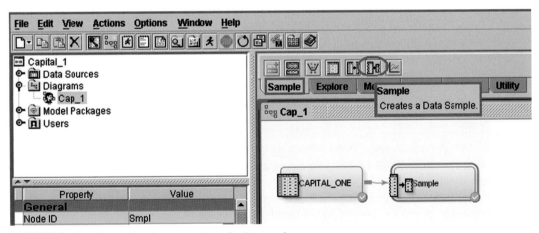

FIGURE D.42   Connecting Sample node to the Data node.

Connect the Sample node to the Data Partition node. Run the Data Partition node (which, by default, will also run the Sample node) and select Results to *verify the oversampling results* (2,943 satisfied and 2,943 unsatisfied customers, etc., as shown in Figure D.44).

Property	Value
General	
Node ID	Smpl
Imported Data	…
Exported Data	…
Notes	…
Train	
Variables	…
Output Type	Data
Sample Method	Default
Random Seed	12345
⊟Size	
┊Type	Percentage
┊Observations	1
┊Percentage	100.0
┊Alpha	0.01
┗PValue	0.01
Cluster Method	Random
⊟Stratified	
┊Criterion	Level Based
┊Ignore Small Strata	No
┗Minimum Strata Size	5
⊟Level Based Options	
┊Level Selection	Rarest Level
┊Level Proportion	100.0
┗Sample Proportion	50.0

FIGURE D.43   Editing the settings in the Property node.

Homework 5: The default of Sample Proportion is 50%, which would select equal cases of satisfied and unsatisfied customers. Move the cursor and change 50% to 75% and report the difference of the model results.

4. Rare Event Detection and Prior Probability:

Note that the model will be built by using the sampled data that has disproportions of the satisfied and unsatisfied customers (50%–50%), which does not represent the original proportion of 86.3%–13.7%. Consequently, you need to specify the prior probability to use the model in a proper manner when you score on the new data. To do this, click on the Data Source node to activate the Property Panel, as shown in Figure D.45.

In the Train subpanel, click on ▣ at the right of Decisions; then click on the Build button in the pop-up window to activate the Prior Probabilities tab, as shown in Figure D.46.

Data=DATA

Variable	Numeric Value	Formatted Value	Frequency Count	Percent
SATISF1	.	SATIS	2943	50
SATISF1	.	UNSAT	2943	50

Data=TEST

Variable	Numeric Value	Formatted Value	Frequency Count	Percent
SATISF1	.	SATIS	884	50
SATISF1	.	UNSAT	884	50

Data=TRAIN

Variable	Numeric Value	Formatted Value	Frequency Count	Percent
SATISF1	.	SATIS	1176	50
SATISF1	.	UNSAT	1176	50

Data=VALIDATE

Variable	Numeric Value	Formatted Value	Frequency Count	Percent
SATISF1	.	SATIS	883	50
SATISF1	.	UNSAT	883	50

FIGURE D.44   Results from oversampling to verify frequency counts of Satisfied versus Unsatisfied.

Click on the Prior Probabilities tab, check the Yes radio button, fill in the prior probabilities of {0.1472, 0.8628} in the Adjusted Prior field, and then click on OK (see Figure D.47).

In the process flow shown in Figure D.48, click on the Control Point node at the end of the flow, go for a walk, and then come back to view the results, as shown in Table D.4. Or you can add the Sample node to the Data Source node (with default Percentage = 10.0%) to speed up the process.

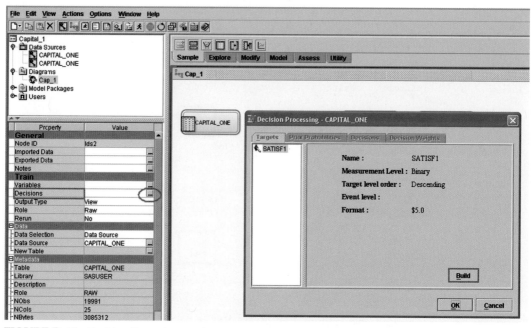

FIGURE D.45   Decision Processing window over SAS-EM interface.

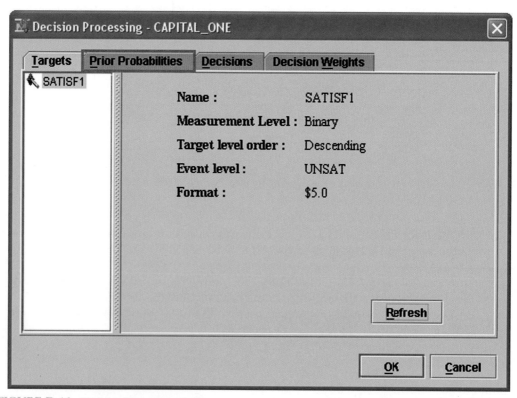

FIGURE D.46   Decision Processing window.

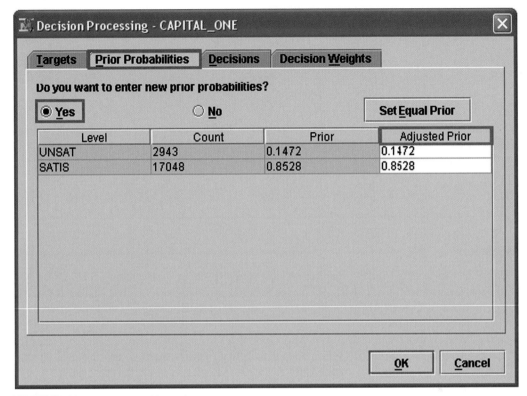

FIGURE D.47 Prior Probabilities tab.

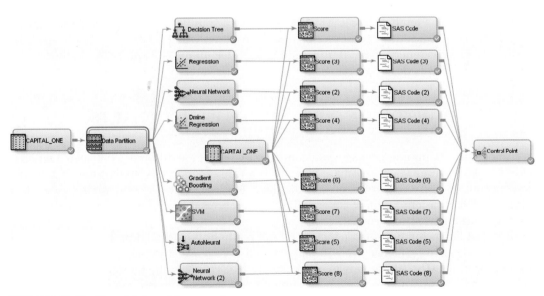

FIGURE D.48 Completed model process flow.

TABLE D.4   Results

| | | Oversampling | | | | |
Model	Cutoff	True-Positive Sum	False-Positive Sum	Predicted Positive	Total Profit	Average Profit
Tree	0.23	966	2755	3721	11447.58	3.07648
Regression	0.21	1132	3257	4389	12986.16	2.9588
NN_MLP	0.23	1006	2869	3875	11922.78	3.0768
DM-Regression	0.24	1008	2692	3700	14687.04	3.9695
Boosting	0.21	1023	2770	3793	14336.49	3.7797
SVM	0.25	1629	7038	8667	−16577.73	−1.91274
Auto_Neural	0.23	0	0	0	0	0
NN_RBF	0.22	966	2659	3625	12887.58	3.5552

Discussion

a. Table D.4 indicates that Dmine Regression gives the best total profit. Here, Dmine Regression computes a forward stepwise least-squares regression that includes two-way interactions of classification variables, binning of interval variables, and the grouping of classification variables.

b. Auto_Neural and SVM (Support Vector Machine) do not perform well in this example.

c. Oversampling is a necessity when the probability of rare event is about 4–6% or less. In this case study, the size of the data is 19,991 with 2,943 (14.72%) unsatisfied customers. The oversampling technique as mentioned in the Arens-Wegman paper turns out less than optimal in terms of profit.

d. In SAS-EM, oversampling is referred to *stratified random sampling*. The technique is very useful when the subpopulations of the good and bad customers are lopsided (say 95% versus 5%). In literature, some practitioners take a sample of the good customers and call the techniques downsampling or undersampling, while others may duplicate the data of bad customers 10 or 20 times and call the technique oversampling. In this study, what we did was downsampling.

DECISION MATRIX AND THE PROFIT CHARTS

Almost all data mining packages are confused by the subtle difference between a *misclassification matrix* and a *decision matrix*. SAS-EM is a rare exception. Note that the first matrix does not allow nonzero entries on the diagonal line, while the second matrix is able to accommodate different kinds of cost-profit considerations. The difference may seem small, but the consequence is enormous.

Note that the techniques in this section do not really fit the original problem statement of the Capital One study. Hence, we will make up a different scenario that is common in many business settings. The situation is mathematically equivalent to the famous 1998 KDD-Cup Competition where a national veterans' organization seeks to improve its solicitations for donation (http://www.kdnuggets.com/meetings/kdd98/kdd-cup-98.html).

Using the Capital One data, assume that a competitor has access to the data and would like to mount a campaign to lure away the unsatisfied customers. Assume that the profit is $278.52 for each defection and that the cost is $25.00 for reaching out to each of the 19,991 customers. Further assume that the success rate of the program would be 50%. A blanket effort to reach all of the 19,991 customers would render the following results:

$$\text{Total profit} = \$278.52^*2994^*0.5 - 25^*19991 = -\$82,830.60$$

$$\text{Average profit} = \text{Total profit}/19,991 = -\$4.14$$

In other words, without any model, there will be a *loss* of $82,830.60. However, if we target certain customers, then the total profit would be $24,031.17 or more. The difference between model and no-model is about $106,862.

The following steps show how to accomplish this with SAS-EM.

1. Data Source node:

Click on the Data Source node to activate its Property Panel. Then click on the right of Decisions, ▦. On the Decision Process pop-up window, click on Build to activate the decision menu (see Figure D.49).

Click on Decisions in the menu and select Yes (see Figure D.50).

Click on Decision Weights in the menu and enter weight values for the decision (see Figure D.51).

The numbers in the matrix are based on the assumptions that the profit is $278.52 for each defection, the program enjoys a 50% success rate, and that the cost for reaching out to each customer is $25.00. Hence the decision weight in the first cell of the matrix is $278.52*0.5 − $25 = $114.26, while the dollar amounts in all other cells are −$25.00.

2. Comparison of the Profit Charts:

Build the process flow shown in Figure D.52 to compare the profits of different models. To do so, first click on the Model tab to activate a number of predictive modeling tools.

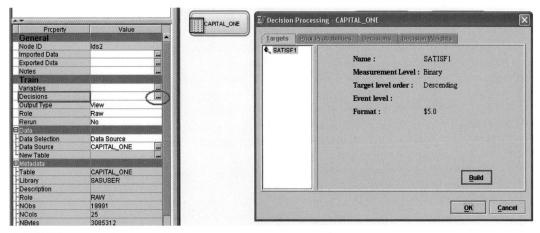

FIGURE D.49 Clicking on Build to activate Decision menu.

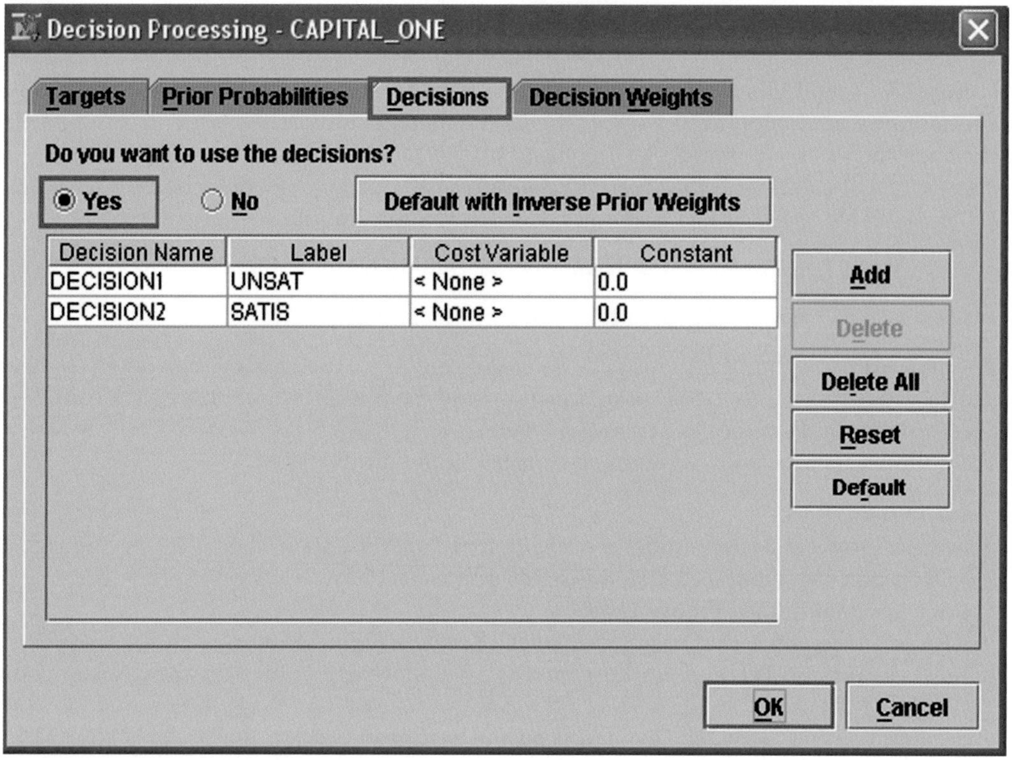

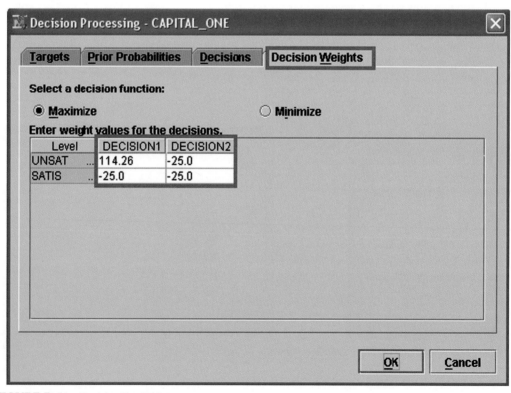

FIGURE D.50   Decision Processing.

FIGURE D.51   Decision Processing.

Then drag Dmine Regression and Decision Tree icons to the workspace (see Figures D.53 and D.54).

Next, click on the Assess tab to activate a number of new tools that contain a special icon called Model Comparison (see Figures D.55 and D.56).

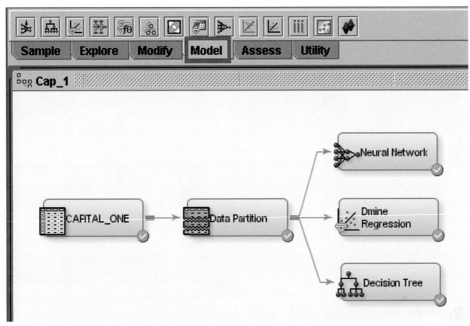

FIGURE D.52   Model flow.

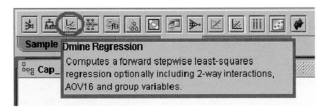

FIGURE D.53   Dmine Regression button.

FIGURE D.54   Decision Tree button.

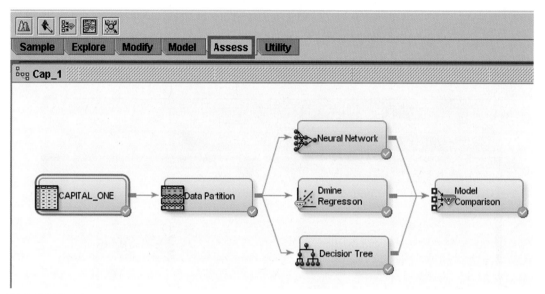

FIGURE D.55   Assess button.

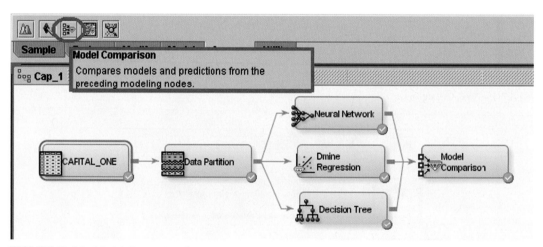

FIGURE D.56   Model Comparison button.

Run the Model Comparison. In the Results windows, go to the lower-left corner and click on the arrow of Cumulative Fit for a drop-down window (see Figure D.57). Select Expected Profit from the new window, as shown in Figure D.58.

Note that in SAS-EM 5.3, the Expected Profit gives the cumulative mean of the profit, while Total Profit gives the noncumulative profit. In this case, we will first use the cumulative mean.

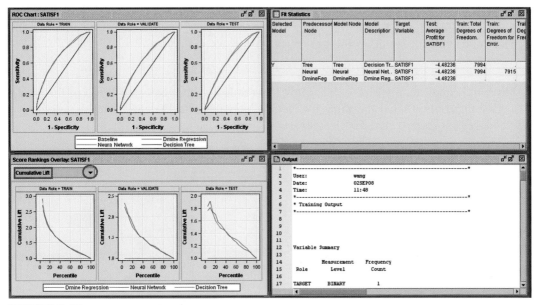

FIGURE D.57   SAS-EM Results screen.

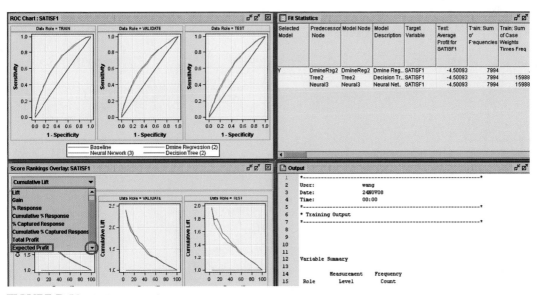

FIGURE D.58   SAS-EM Results screen.

Enlarge the Expected Profit window and click on the Dmine Regression option at bottom (see Figure D.59).

The Total Profit of selecting the top 5% of customers would be

$$0.05^*19,991^*\$36.33 = \$36,313.65$$

Move the cursor to any blue box to view Percentile and the Expected Profit of the model (see Figure D.60). Do the same for Neural Network and Decision Tree.

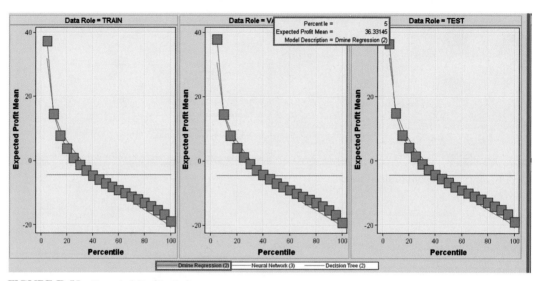

FIGURE D.59 Expected Profit window.

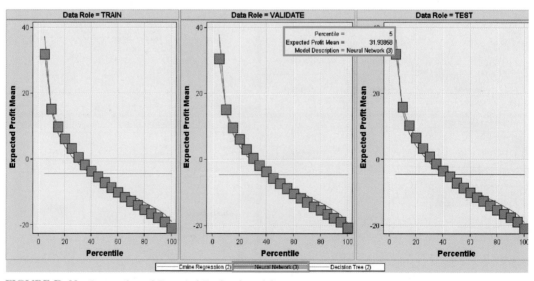

FIGURE D.60 Percentile and Expected Profit of model.

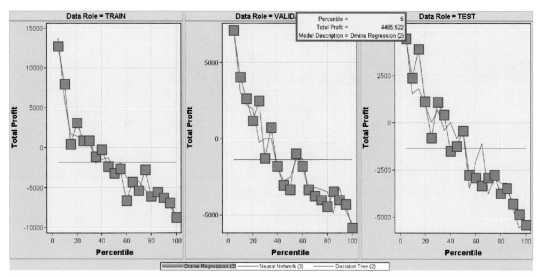

FIGURE D.61   Total Profit window.

The mean profit of Neural Network is substantially lower than that of the Dmine Regression. The default Trees of chi-square, Gini, and Entropy methods all rendered *a flat line* and hence will not be further discussed in this section. A homework problem is to compute the total profit at different cutoff values (10%, 15%, etc.). Our calculations indicate that the cutoff at 5% would be the best choice.

Discussions

a.   If we use Dmine Regression to select the top 5% of 19,991 customers, then the total profit would be $36,313.65.

b.   It is tempting to add up the mean profits of Training, Validation, and the Test data sets, and then calculate the total profit. A cautionary note is that this would distort the predictive power of the model in the event of *overfitting*.

c.   Note that the Test data consist of only 30% of the 19,991 customers. Hence, the total profit of the top 5% of the customers in the Test data should be

$$\$36,313.65*30\% = \$10,894.10$$

But if you click on the Total Profit tab, then you will get $4,465.52, as shown in Figure D.61.

We are still investigating the discrepancy.

MICRO-TARGET THE PROFITABLE CUSTOMERS

This section presents detailed steps on how to identify the customers that would be most profitable. Recall that neural networks is a competive model; hence, we will focus on this model. See Figure D.62.

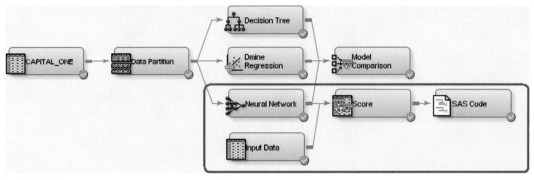

FIGURE D.62   Focus on NN model to identify customers that would be most profitable.

1. The Score node is shown in Figure D.63.
2. The Score Data are shown in Figure D.64.
3. The SAS Code node is shown in Figure D.65.

The code is as follows:

Data Customers;
    Set &EM_Import_Score;
    Customer_ID = _N_;
Run;
PROC Sort data = Customers;
    By descending P_SATISF1UNSAT;
Run;
Data good_customers;
    Set Customers;
    Obsnum = _N_;
    If Obsnum > 0.05*19991 THEN delete;
Run;
PROC Print data = good_customers noobs split = '*';
    VAR obsnum Customer_ID P_SATISF1UNSAT;
    LABEL P_SATISF1UNSAT = 'Predicted*Unsatisfied';
    TITLE "Credit Worthy Applicants";
Run;
Proc print;

The results are shown in Figure D.66.

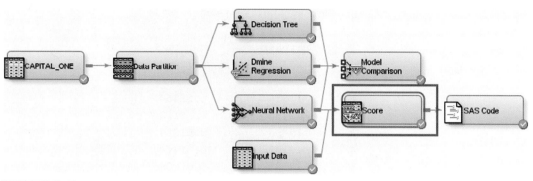

FIGURE D.63   Score node.

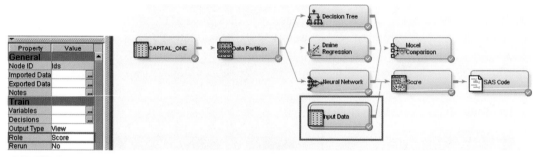

FIGURE D.64   Score data.

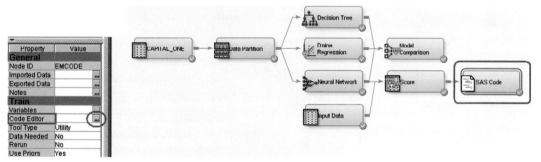

FIGURE D.65   SAS code.

APPENDIX

To import Capital_One Excel data to the SASUSER library, follow these steps:

1. Open base SAS. Select File, Import Data, as shown in Figure D.67.
2. Click on Next and then the Browse button to browse the location of the data (see Figure D.68).

"Credit Worthy Applicants"

Obsnum	Customer_ ID	'Predicted Unsatisfied'
1	13756	0.62254
2	9642	0.62138
3	19460	0.61927
4	19183	0.61744
5	17838	0.61606
6	3106	0.61306
7	8494	0.61106
8	14421	0.61075
9	12008	0.60931
10	16021	0.60894
11	9982	0.60738

FIGURE D.66   Results of Score node.

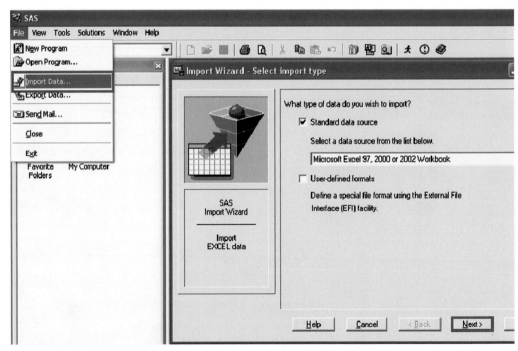

FIGURE D.67   Base SAS interface window.

FIGURE D.68   Window to locate data.

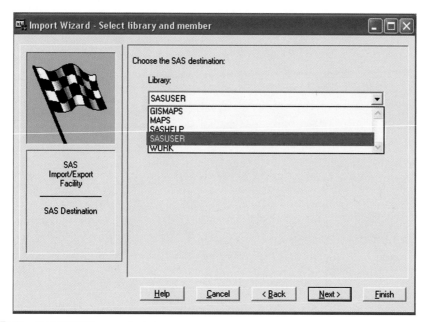

FIGURE D.69   SASUSER library selected.

3. Select the SASUSER library, as shown in Figure D.69.
4. Type the name of the file, click on Next, and in the next window, click on Finish to complete the data import (see Figure D.70).

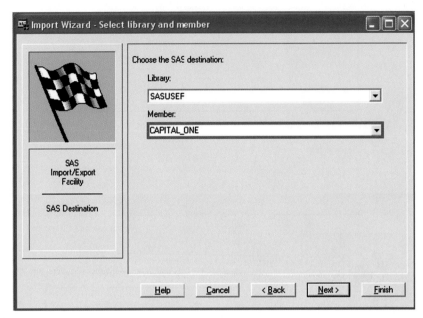

FIGURE D.70   Click on Finish to import data.

Reference

Arens, Z., & Wegman, D. (2001). *Predicting Dissatisfied Credit Card Customers.* http://www.galaxy.gmu.edu/stats/syllabi/inft979/ArensPaper.pdf

Credit Scoring Using *STATISTICA* Data Miner

Kiron Mathew
Edited by Gary Miner, Ph.D.

INTRODUCTION: WHAT IS CREDIT SCORING?

Over the last half of the twentieth century, lending to consumers exploded. Consumer credit has had one of the highest growth rates in any sector of the business. Consumers use credit to obtain goods and services now and then pay for them later. Many credit applicants are, in fact, good credit risks, but some are not. The risk for financial institutions comes from not knowing how to distinguish the good credit applicants from the bad credit applicants. One widely adopted technique for solving this problem is called *credit scoring*.

Credit scoring is the set of decision models and their underlying techniques that aid lenders in the granting of consumer credit. These techniques decide who will get credit, how much credit they should get, and what operational strategies will enhance the profitability of the borrowers to the lenders. Further, it helps to assess the risk in lending. Credit scoring is a dependable assessment of a person's creditworthiness, since it is based on actual data.

A lender must make two types of decisions: first, whether to grant credit to a new applicant and, second, how to deal with existing applicants, including whether to increase their credit limits. In both cases, whatever the techniques used, the vital point is that there is a very large sample of previous customers with their application details and subsequent credit history available. All the techniques use the sample to identify the connection between the characteristics of the consumers (annual income, age, number of years in employment with their current employer, etc.) and how "good" or "bad" their subsequent history is.

Typical application areas in the consumer market include credit cards, auto loans, home mortgages, home equity loans, mail catalog orders, and a wide variety of personal loan products.

CREDIT SCORING: BUSINESS OBJECTIVES

The application of scoring models has nowadays come to cover a wide range of objectives. The original idea of estimating the risk of defaulting has been augmented by credit scoring models at other aspects of the credit risk management: at the pre-application stage (identification of potential applicants), at the application stage (identification of acceptable applicants), and at the performance stage (identification of possible behavior of current customers).

Scoring models with different objectives has been developed. They can be generalized into the following four categories:

1. **Marketing aspect**:

   Purposes:
   Identify creditworthy customers most likely to respond to promotional activity to reduce the cost of customer acquisition and minimize customer dissatisfaction. Predict the likelihood of losing valuable customers and enable organizations to formulate effective customer retention strategy.

   Examples:
   Response scoring: The scoring models that estimate how likely a consumer would respond to a direct mailing of a new product.
   Retention/attrition scoring: The scoring models that predict how likely a consumer would keep using the product after the introductory offer period is over or change to another lender.

2. **Application aspect**:

   Purposes:
   Decide whether or not to extend credit, and how much credit to extend. Forecast the future behavior of a new credit applicant by predicting loan-default or poor-repayment behaviors at the time the credit is granted.

   Example:
   Applicant scoring: The scoring models that estimate how likely a new applicant of credit will become default.

3. **Performance aspect**:

   Purpose:

   Predict the future payment behavior of the existing debtors to isolate problem ones, to which more attention and assistance can be devoted, thereby reducing the likelihood that these debtors will become a problem.

   Example:

   Behavior scoring: Scoring models that evaluate the risk levels of existing debtors.

4. **Bad debt management**:

   Purpose:

   Select optimal collections policies to minimize the cost of administering collections or maximizing the amount recovered from the delinquents' account.

   Example:

   Scoring models for collection decisions: Scoring models that decide when actions should be taken on the accounts of delinquents and which of several alternative collection techniques might be more appropriate and successful.

The overall objective of credit scoring is not only to determine whether or not the applicant is creditworthy but also to attract quality credit applicants who can subsequently be retained and controlled while maintaining an overall profitable portfolio.

CASE STUDY: CONSUMER CREDIT SCORING

Description

In credit business, banks are interested in information regarding whether or not prospective consumers will pay back their credit. The aim of credit scoring is to model or predict the probability that consumers with certain characteristics are to be considered as potential risks.

The example in this case will illustrate how to build a credit scoring model using *STATISTICA* Data Miner to identify inputs or predictors that differentiate risky customers from others (based on patterns pertaining to previous customers) and then use these inputs to predict the new risky customers. This is a sample case typical for this domain.

The sample data set used in this case, CreditScoring.sta, has 1,000 cases and 20 variables or predictors pertaining to past and current customers who borrowed from a German bank (source: http://www.stat.uni-muenchen.de/service/datenarchiv/kredit/kredit_e.html) for various reasons. The data set contains various information related to the customers' financial standing, reason to loan, employment, demographic information, etc.

For each customer, the binary outcome (dependent) variable "creditability" is available. This variable contains information about whether each customer's credit is deemed good or bad. The data set has a distribution of 70% creditworthy (good) customers and 30% not creditworthy (bad) customers. Customers who have missed 90 days of payment can be thought of as bad risk, and the customers who have ideally missed no payment can be thought of as good risk. Other typical measures for determining good and bad customers

TABLE E.1   Variables

Category	Variables
1. Basic Personal Information	Age, Sex, Telephone, Foreign worker?
2. Family Information	Marital Status, Number of dependents
3. Residential Information	Number of years at current address, Type of apartment
4. Employment Status	Number of years in current occupation, Occupation
5. Financial Status	Most valuable available assets, Further running credits, Balance of current account, Number of previous credits at this bank
6. Security Information	Value of savings or stocks, Guarantors
7. Others	Purpose of credit, Amount of credit in DM

are the amount over the overdraft limit, current account turnover, number of months of missed payments, or a function of these and other variables.

Table E.1 provides the complete list of variables used in this data set.

In this example, we will look at how well the variables listed in Table E.1 (*Balance of current account*, *Value of savings or stocks*, etc.) allow us to discriminate between whether someone has good or bad credit standing. If we can discriminate between these two groups, we can then use the predictive model we built to classify or predict new cases where we have information on these other variables but do not know the person's credit standing. This information would be useful, for example, to decide whether or not to qualify a person for a loan.

Data Preparation

It is rather straightforward to apply DM modeling tools to data and judge the value of resulting models based on their predictive or descriptive value. This does not diminish the role of careful attention to data preparation efforts. Data are the central items in data mining; therefore, it should be massaged properly before feeding it to any data mining tool. Otherwise, it would be a case of Garbage In, Garbage Out (GIGO). Major strategic decisions are impacted by these results; therefore, any error might cost an organization in millions of dollars. Therefore, it is important to preprocess the data and try to improve the accuracy of the decisions made.

The following points were noted during this stage:

- Insight into data: Descriptive statistics were discovered.
- There are no outliers in the data.
- There are no missing values in the data.
- No integration is required.
- No transformations are required.
- Feature selection: Variables were reduced from 20 to 9.

Feature Selection

To reduce the complexity of the problem, we can transform the data set into a data set of lower dimensions. The Feature Selection tool available in *STATISTICA* Data Miner automatically found important predictors that clearly discriminate between good and bad customers.

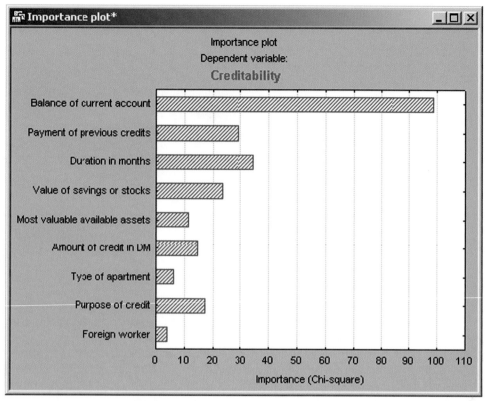

FIGURE E.1 Creditability bar chart.

The bar plot and spreadsheet of the predictor importance usually shed some light on the variables that are largely related to the prediction of the dependent variable of interest. For example, Figure E.1 is the bar plot of predictor importance for the dependent variable "Creditability".

In this case, the variables *Balance of current account*, *Payment of previous credits*, and *Duration in months* stand out as the most important predictors.

These predictors will be further examined using a wide array of data mining and machine learning algorithms available in *STATISTICA*'s Data Miner.

STATISTICA Data Miner: "Workhorses" or Predictive Modeling

The novelty and abundance of available techniques and algorithms involved in the modeling phase make this the most interesting part of the data mining process (Figure E.2). Classification methods are the most commonly used data mining techniques that are applied in the domain of credit scoring to predict the risk level of credit takers. Moreover, it is good practice to experiment with a number of different methods when modeling or mining data. Different techniques may shed new light on a problem or confirm previous conclusions.

FIGURE E.2 Techniques and algorithms involved in the modeling phase.

STATISTICA Data Miner – "Workhorses" for Predictive Modeling

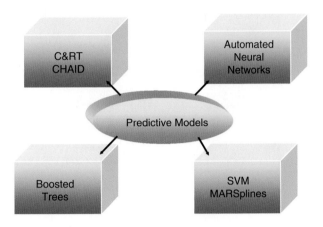

STATISTICA Data Miner is a comprehensive and user-friendly set of complete data mining tools designed to enable you to more easily and quickly analyze your data to uncover hidden trends, explain known patterns, and predict the future. From querying databases and drilling down to generating final reports and graphs, this tool offers ease of use without sacrificing power or comprehensiveness. Moreover, *STATISTICA* Data Miner features the largest selection of algorithms on the market for classification, prediction, clustering, and modeling, as well as an intuitive icon-based interface. It offers techniques from simple C&RT and CHAID to more advanced intelligent problem solver, boosted trees, Support Vector Machines and MARSplines.

Overview: *STATISTICA* Data Miner Workspace

The Data Miner Workspace depicts the flow of the analyses; all methods of *STATISTICA* Data Miner are available as icons via simple drag-and-drop.

Figure E.3 shows how the Data Miner Workspace should look.

Note the following regarding the *STATISTICA* Data Miner Workspace shown in the figure:

1. Split the original data set into two subsets; 34% of cases retained for testing and 66% of cases were used for model building;
2. Used the Stratified Random Sampling method to extract equal numbers of observations for both good and bad risk customers;
3. Used the Feature Selection tool to rank the best predictor variables for predicting the dependent variable creditability;
4. Took the best 10 predictors from a total of 20 variables based on feature selection for model-building purposes;

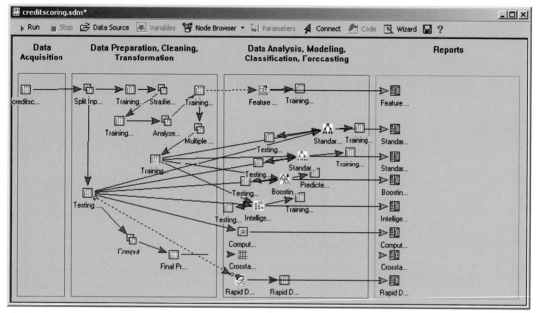

FIGURE E.3   The Data Miner Workspace.

5. Used different advanced predictive models (machine learning algorithms) to detect and understand relationships among words;
6. Used comparative tools such as lift charts, gains charts, cross-tabulation, etc., to find the best model for prediction purposes;
7. Applied the model to the test set kept aside (hold-out sample) to validate prediction accuracy.

ANALYSIS AND RESULTS

The following four models were selected from the arsenal of data mining techniques:

- Standard Classification Trees with Deployment (C&RT or CART)
- Standard Classification CHAID with Deployment
- Boosting Classification Trees with Deployment
- Intelligent Problem Solver with Deployment (Classification)

Decision Tree: CHAID

Decision trees are powerful and popular tools for classification and prediction (Figure E.4). The attractiveness of decision trees is due to the fact that, in contrast to neural networks, decision trees represent *rules* (they are easy to interpret).

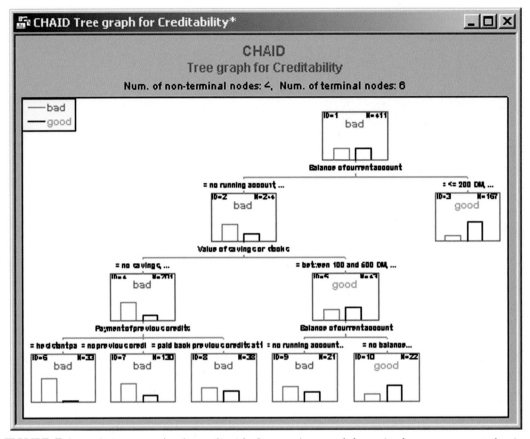

FIGURE E.4   A decision tree for the credit risk data set (excerpted from simple output generated using *STATISTICA* Data Miner).

The results you see on your screen might vary because of the different training samples generated each time the data source node is split for training and testing.

Each box in the tree in Figure E.4 represents a *node*. The top node is called the *root node*. A decision tree grows from the root node, so you can think of the tree as growing upside down, *splitting* the data at each level to form new nodes. The resulting tree comprises many nodes connected by *branches*. Nodes that are at the end of branches are called *leaf* nodes and play a special role when the tree is used for prediction.

In Figure E.4 each node contains information about the number of instances at that node, and about the distribution of dependent variable values (credit risk). The instances at the root node are all of the instances in the training set. This node contains 411 instances with equal proportion of customers from both the "good" and "bad" categories obtained using the Stratified Random Sampling feature from *STATISTICA* Data Miner. Below the root node (*parent*) is the first *split* that, in this case, splits the data into two new nodes (*children*) based on the predictor balance of current account.

The rightmost node resulting from this split contains 167 instances associated with good credit risk. Because most of the instances have the same value of the dependent variable (creditability), this node is termed *pure* and will not be split further. The leftmost node in the first split contains 244 instances. This node is then further split based on the predictor *value of savings or stock*, resulting in two more nodes and so on.

The order of the splits—*balance of current account, value of savings or stocks,* and then *payment of previous credits*—is determined by an induction algorithm

A tree that has only pure leaf nodes is called a *pure tree*, a condition that is not only unnecessary but is usually undesirable. Most trees are impure; that is, their leaf nodes contain cases with more than one outcome to avoid overfitting.

The rules for the leaf nodes in Figure E.4 are generated by following a path down the branches until a leaf node is encountered. For example,

IF Balance of current account = no running account, no balance
AND Value of Savings or Stocks = no savings, less than 100 DM
AND Payment of previous credits = hesitant payment of previous credits
THEN Creditability = bad

Classification Matrix: CHAID Model

The classification matrix can be computed for old cases as well as new cases. Only the classification of new cases (testing data set) allows us to assess the predictive validity of the model; the classification of old cases only provides a useful diagnostic tool to identify outliers or areas where the model seems to be less adequate.

The program computes the matrix of predicted and observed classification frequencies for the testing data set, which are displayed in a results spreadsheet, as well as a bivariate histogram, as shown in Figure E.5.

The classification matrix shows the number of cases that were correctly classified (on the diagonal of the matrix) and those that were misclassified as the other category.

In this case, the overall model could correctly predict whether the customer's credit standing was good or bad with 62.83% accuracy. Our main goal is to reduce the proportion of bad credits. The percent of correct predictions for the bad category is 66.30%. In other words, if there are 100 bad customers, our model will correctly classify approximately 66 as bad (which is far better than the law of chance).

COMPARATIVE ASSESSMENT OF THE MODELS (EVALUATION)

It is good practice to experiment with a number of different methods when modeling or mining data rather than relying on a single model for final deployment. Different techniques may shed new light on a problem or confirm previous conclusions.

The gains chart provides a visual summary of the usefulness of the information provided by one or more statistical models for predicting categorical dependent variables.

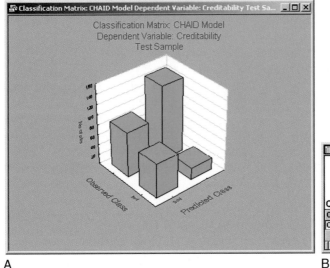

A   B

FIGURE E.5   Histogram and spreadsheet of results.

Specifically, the chart summarizes the utility that you can expect by using the respective predictive models, as compared to using baseline information only.

The overlaid gains chart (for multiple predictive models) based on models trained in *STATISTICA* Data Miner, shown in Figure E.6, is computed using the Compute Overlaid Lift Charts from All Models node.

This chart depicts that the Boosting Trees with Deployment model is the best among the available models for prediction purposes. For this model, if you consider the top two deciles, you would correctly classify approximately 40% of the all cases in the population belonging to the bad category. The baseline model serves as a comparison to gauge the utility of the respective models for classification.

Analogous values can be computed for each percentile of the population (loan applicants in the data file). You could compute separate gains values for selecting the top 30% of customers who are predicted to be among bad customers (hence no loan approval), the top 40%, etc. Hence, the gains values for different percentiles can be connected by a line that will typically ascend slowly and merge with the baseline if all customers (100%) were selected.

The lift chart (similar summary chart) shown in Figure E.7 depicts that the Boosting Trees with Deployment model is the best among the available models for prediction purposes.

If you consider the top two deciles, you end up with a sample that has almost 1.7 times the number of bad customers when compared to the baseline model. In other words, the relative gain or lift value due to using the Boosting Trees with Deployment predictive model is approximately 1.7.

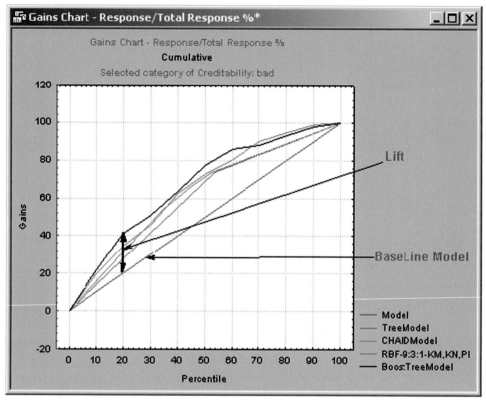

FIGURE E.6   Overlaid gains chart.

Classification Matrix: Boosting Trees with Deployment Model (Best Model)

The classification matrix for testing data set in Figure E.8 shows the number of cases that were correctly classified (on the diagonal of the matrix) and those that were misclassified as the other category.

In this case, the overall model could correctly predict whether the customer's credit standing was good or bad with 65.65% accuracy. Our main goal is to reduce the proportion of bad credits. The percent of correct predictions for the bad category is 73.91%.

DEPLOYING THE MODEL FOR PREDICTION

Finally, deploy the Boosting Classification Trees with Deployment model. In particular, save the PMML deployment code for this model and then use that code via the Rapid Deployment node in *STATISTICA* Data Miner to predict (classify) the credit risk of new loan applicants. If the bank could decide beforehand who would be more likely to default on a loan, this could save the bank money.

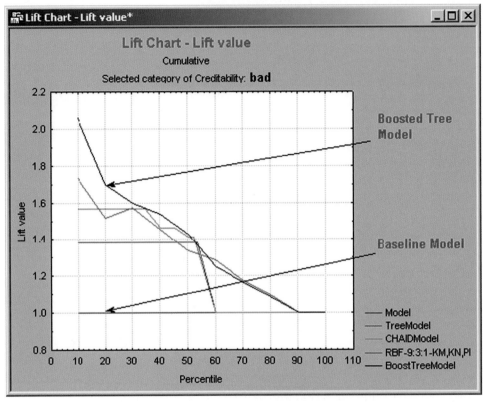

FIGURE E.7 Lift chart.

Data: 2-Way Summary Table: Observed Frequencies (Compute Best Pre...

Creditability	Classification Matrix: Test Sample (Boosted Tree		
	Predicted bad	Predicted good	% correctly predicted
Oberved:bad	68	24	73.91
Oberved:good	108	129	62.44
		Overall Hit Ratio	65.65

FIGURE E.8 Classification matrix.

CONCLUSION

The purpose of this example is to show how easily a large number of the most sophisticated methods for predictive data mining can be applied to data and how sophisticated ways for combining the power of these methods for predicting new observations become automatically available.

F

Churn Analysis
With SPSS-Clementine

Robert Nisbet, Ph.D.

In this exercise, we will use a generalized approach for creating customer response models. This tutorial will analyze churn patterns in a real telecommunications data set. Some fields of the data set have been recoded to preserve anonymity. The Temporal Abstracts have already been calculated with SQL in the database pivot process (see Chapter 4 for a discussion of temporal abstracts). The reference date is the date that records in the training set churned (disconnected service). This pivoting process operates directly on the database early in the Data Preparation phase of modeling. Your job is to select the variables that produce a model with a lift index of least a 3.0 in the first decile. There is an optimum set of variables that can be used, but some of the possible sets of variables may lead to net solutions with the minimum accuracy. See how much better than 3.0 lift index you can do!

Review the information on the Clementine interface in Chapter 10 to learn how to enter nodes onto the modeling canvas and link them together.

OBJECTIVES

1. Learn how to use the Derive node and Sample node.
2. Learn how to use the Distribution node and Generated Balance node.
3. Learn how to use a Type node to set the data type of variables.
4. Learn how to use a neural net to define a pattern of churned records in the database.
5. Learn how to create and use lift charts to evaluate the performance of a model.

STEPS

1. Paste a Variable File node (from the Sources palette) onto the upper-left region on the palette.
    a. Double-click the node to edit it.
    b. Click on the ellipsis at the end of the File box and choose the file named CH_10k.dat.
    c. In the Delimiters pane (lower left), click on the Tab box (to specify a tab-delimited ASCII file).
    d. Click on the Types tab at the bottom of the screen.
        i.   Also in the Types tab, click on the CR_CLASS (credit class) variable; then scroll down to the bottom of the field list, hold down the Shift key, and click the last field. This should highlight all the fields between the two clicks. Click in the Direction box of one of the highlighted fields and select In. This will set all those fields as inputs. (*Note:* You could set the types one at a time, but this way saves a lot of time for a long field list.)
        ii. Scroll down to the bottom of the field list and select CHURN_FL.
        iii. Click in the directions box for the CHURN_FL field and select Out. This will set that field to be the target variable.
        iv. Click in the Type box to the right of the field name CHURN_FL and select Discrete. This will set the data type of the target variable to be a discrete value. It must be typed as a discrete (categorical) value for the next operation.
    e. Click on the Filter tab and rename the field to CHURN by clicking on the right-hand file name and editing it.
    f. Click on the Filter tab.
        i. Notice the same list of field names with arrows between them. An arrow can be marked with an X by clicking on it to delete the variable for further processing. The following list of variables is described and listed with an X to delete them in this Filter tab. Note that many variables are listed in sets of three (e.g., DUR1, DUR2, DUR3). These are temporal abstractions. For example, DUR1 refers to the duration in minutes of use (MOU) one month before the churn month (the index month). If the customer did not churn, a random index month is chosen during the 4-month churn analysis window for this data set.
            1. CUST_ID—Customer ID (*X*)
            2. START_DT—Start date of the billing period (*X*)
            3. END_DT—End date of the billing period (*X*)
            4. LATE_ST—Late statement
            5. CREDIT_CLASS—Credit class
            6. INC_RANGE—Income range
            7. GENDER—Gender (M/F)
            8. AGE—Age in years
            9. INCOME—Income in dollars/mo
            10. CUST_TYPE—Customer type

11. INPUT_DT—Input date (X)
12. ACCT_TYPE—Account type (X)
13. AMT_CHARGED1, 2, 3—Amount charged in dollars one month, 2 months before, and 3 months before the index month
14. CALL_TP1, 2, 3—Calling type for the three temporal abstractions
15. NUM_SP1, 2, 3—Number of special services for the three temporal abstractions
16. DUR1, 2, 3—Cumulative duration of all calls for the three temporal abstractions
17. CALLS1, 2, 3—Number of calls for the three temporal abstractions
18. BAN_ST1, 2, 3—Unknown
19. CLOSED1, 2, 3—Closed dates (X)
20. DUE_DT1, 2, 3—Due date (X)
21. CHARGE1, 2, 3—Unknown (X) bad data
22. CH_BEG1, 2, 3—Charge begin date (X)
23. CH_END1, 2, 3—Charge end date (X)
24. BILL_ST1,2, 3—Billing statement (X) sparse date
25. LT_PMT1, 2, 3—Late payments (X) all zeros
26. ADJ1, 2, 3—Adjustments (X) sparse data
27. RECURR1, 2, 3—Recurrent charges
28. ONETIME1. 2. 3—One-time charges
29. P_METH1, 2, 3—Payment method
30. CR_DT1, 2, 3—Credit date (X)
31. PAY_VAL1, 2, 3—Payment
32. DUE_VAL1, 2, 3—Amount due
33. CHURN—Churn flag (1 = churn; 0 = no churn)
34. TEN_RAW—Tenure temporary transform or tenure in years (X)
35. TENURE—Normalized tenure transform

   To delete the variables marked with an (X), click on the arrow between the name lists. Also, you can change a field name simply by editing the right-hand name in the list.

2. Connect a Sample node (from the Record Ops Palette).
   a. Double-click to edit the node.
   b. Click on the Annotations tab.
   c. Name the node Training Set and click the Settings tab to return.
   d. Note that the first radio button in the Mode section selects Include sample. This means that all records selected in this node will be included in the output.
   e. Click on the 1-in-*n* radio button. The effect of the preceding two steps is to select every other record and include it in the training set. Later, we will set up another Sample node to select every other record and discard it from the second set. Those two sample nodes will function together to divide the data set into two equal parts.
   f. Click OK to exit the node.

3. Connect a Table node (from the Output palette) directly above or below the Filter node.
   a. Edit the Table node and execute the table by double-clicking on Execute in the Edit menu. Look at the CHURN field in the table to verify that it is now a set of numbers (0 and 1). Click on the *X* in the upper right.

4. Connect a Distribution node (in the Graphs palette) to the Training Set Sample node.
   a. Edit the Distribution node.
   b. In the Field box, click on the arrow to select the CHURN variable.
   c. Execute it.
   d. You should see a bar graph with unequal-length blue bars, showing the relative occurrence of 1 and 0. Neural nets require that there be a nearly equal number of 1s and 0s to train properly. Therefore, we will clone the 1s until they are approximately equal to the 0s. We will use the Balance node to do this. And we can use the Generate option in the Distribution bar chart to do this for us. (*Note:* We could balance the record set by deleting the number of 0s to equal the number of 1s. Balancing by boosting will retain all the information in the 0-records for use in training the model.)
   e. Click on the Generate option at the top of the Distribution chart.
   f. Choose the Balance Node Boost option.
   g. Click on the *X* in the upper right of the chart.
   h. The Generate option creates a Balance Node in the upper left of the palette screen. Remember, the Balance Node contains the factor by which the number of Churn records (CHURN = 1) will be increased ("boosted") to match the number of nonchurn records (CHURN = 0). This balanced pattern is necessary for proper pattern matching by the machine learning (neural net or rule induction) tool.

5. Drag the Generated Balance node to a position to the right and below the Training Set Sample node and connect it to the Generated Balance node.
   a. Rename the node to Boost by clicking on the node, selecting Annotations, and entering the new name.
   b. Right-click on the arrow connecting the Distribution node and delete the arrow link. Reconnect the Distribution node to the Boost node.
   c. Execute it.
   d. The blue bars should be about the same length now.
   e. Click OK to exit.
   Your screen should look like Figure F.1 now.

6. Connect three Derive nodes in a series to the right of the Type node. We will define three new variables with these three nodes.
   a. First Derive Node: Name it UNDR_PAY1 in the Field name.
      Enter the formula: DUE_VAL1—PAY_VAL1.
      Click OK.
   b. Second Derive Node: Do likewise for UNDR_PAY2 and its formula.

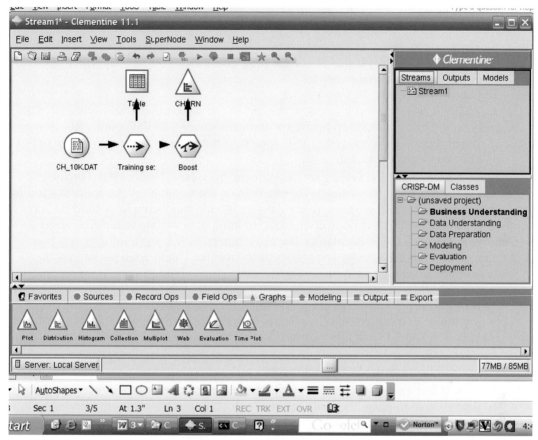

FIGURE F.1   SPSS Clementine 11.1 version interface.

    c. Third Derive Node: Do likewise for UNDR_PAY3 and its formula.

        The Derive node is a very powerful data generator. The built-in CLEM expression language is rich in functions and operations you can use to derive new variables to test as predictors.

7. Link a Type node into the stream after the third Derive node. (*Note:* Data typing was done in the Variable File node, but Clementine models still require an explicit Type node from which the algorithms get the target variable. The provision of data typing in the input node is a recent addition; previous versions did not do that.) The target variable specification does not pass from the input node for some reason.

    a. Scroll down the field list in the node and set the role of the CHURN variable to Out. That will specify that CHURN is the target variable (the output of the model). The terminology here is derived from the artificial intelligence world, where a data input role is an "in" and a data output role is an "out"; most data mining tools refer to the output variable as the "Target."

8. Link a Neural Net node to the stream to the right of the Type node (from the Modeling palette).

   a. The name assigned to the node should be CHURN, picking up on the name of the Target variable. If this is not the case, check that you nave specified the role of the CHURN variable as "In."

   b. Execute the Neural Net node.

   c. You will see a feedback graph at the top of the screen that shows the current accuracy of classification during the iterative training of the neural net.

   d. When the model is trained, it will appear in the Models tab screen in the upper-right corner of the palette.

   Now, we will build the Testing set data stream. You should evaluate the predictive power of the model on the testing set rather than on the training set. Actually, this testing data set will be the third partition of the data set used in this example. Behind the scenes, the Clementine Neural Net node divided the training data set into two pieces, sized by settings in the node (50:50 is default). The algorithm trains the model during one pass (iteration) through the data set and then tests the results. The internal testing results are used to optimize settings for the number of neurons in the hidden layer, learning rate, and momentum settings of the final model. The Clementine neural net is a highly automated adaptive algorithm that is very easy to use to create a very good model, even with its default settings. You can modify the settings to optimize performance with your data set.

   Your screen should look like Figure F.2 now. Notice the trained model icon in the upper-right box in the interface.

9. Double-click the Training Set Sample node.

   a. Select Copy node (or just press Ctrl-C).

   b. Press Ctrl-V to paste the node into the modeling palette below the first sample node (standard Windows keystrokes).

   c. Double-click the pasted node, make sure the Settings tab is open, and change the radio button to Discard Sample. This operation will pass the other half of the incoming records to become the Testing data set.

   d. Click the Annotations tab and change the name to Testing Set.

   e. Connect the Testing set node to the variable file input node.

10. Add the derived nodes like you did in the training stream. (Note that you can click on the first one, hold the Shift key, and click on the third one. All three nodes will be highlighted. You can copy and paste them as a group.)

11. Copy the Type node (like you did the copy operation before).

    Paste the second Type node in the palette and connect it to the Testing Set node.

12. Now drag and drop the trained model icon from the Models tab (upper-right box in the interface) and connect it to the Type node.

13. Finally, connect an Evaluation node (from the Graphics palette) to the neural net model icon. Your screen should look like Figure F.3.

    a. Double-click the Evaluation node and make sure the Plot tab is open.

    b. In the middle of the screen, you will see a box labeled Percentiles, with a drop-down arrow in it. Click on the arrow to view the different kinds of bins (quantiles)

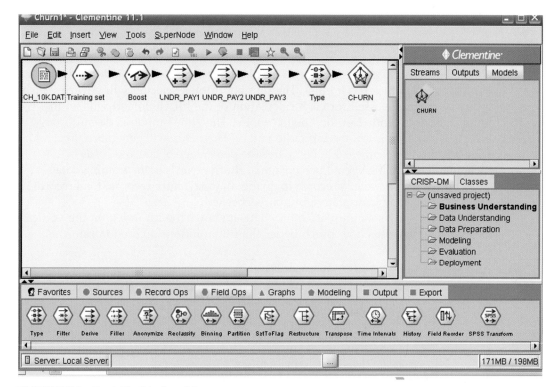

FIGURE F.2　Partial build of model stream.

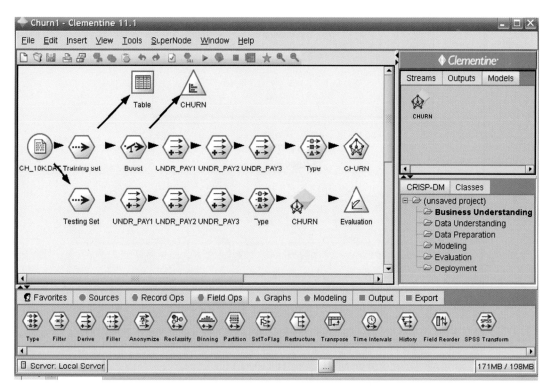

FIGURE F.3　Completed model.

into which the records are assorted. The default value is Percentiles (100 bins). Other choices are Deciles (10 bins), Vingtiles (20 bins), Quintiles (5 bins), Quartiles (4 bins), or 1000-tiles (1000 bins). The number of bins selected will govern which group of records is used for each calculation of the lift value and will set the number of dots plotted in the "point" mode. The "line" mode is default.

c. Next to Chart Type, click the Lift radio button. This setup will generate a lift chart with 100 bins (percentiles). Lift compares the percentage of churn records in each quantile with the overall percentage of churn records in the training data set = (# churns in a quantile/# records in quantile)/(total churn records/total records).

d. Make sure that the Cumulative button is checked.

Figure F.4 shows a cumulative lift chart. It shows the lift index on the vertical axis, which expresses how much better than the random prediction rate was

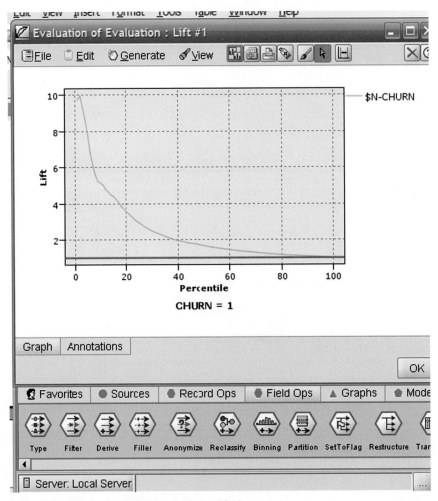

FIGURE F.4   Evaluation results showing cumulative lift chart.

produced by the model. For example, in the first two deciles (up to the 20th percentile), the model performed about 3.8 times better than random (red line).

e. Now edit the Evaluation node and uncheck the Cumulative button.

f. Re-execute the Evaluation node. The resulting graph, shown in Figure F.5, is an incremental lift chart.

g. The incremental chart in Figure F.5 shows the lift in each percentile (no accumulation is done here). Notice that the lift line descends below the random line (in red) at about the 18th percentile. This means that all of the benefit of the model (compared to random expectation) is achieved in the first 18% of the records (note that the lift chart is created from a list sorted on the predicted probability in descending order). A CRM manager could be guided by this model

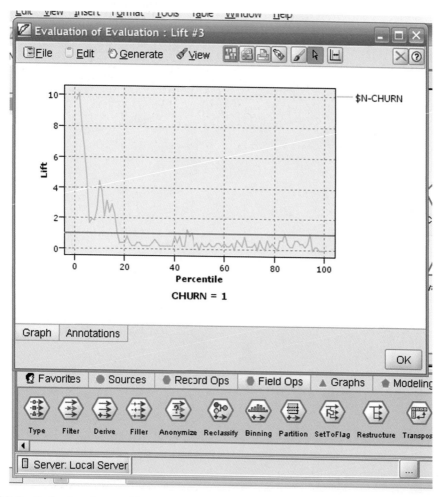

FIGURE F.5   Evaluation showing lift in each Percentile.

to send interdiction offers to only the top 18% of the customers in this sorted scored list, and expect to contact more than three times the number of high-probability churners than normal (compare the cumulative lift chart with the incremental lift chart to see this). The response rate from the retention campaign would then be driven by the effectiveness of the interdiction offers (incentives to stay with the company).

This churn model is an example of a customer response model. You can analyze virtually any customer response this way (cross-sell, up-sell, customer acquisition).

Text Mining: Automobile Brand Review Using *STATISTICA* Data Miner and Text Miner

Sachin Lahoti and Kiron Mathew
Edited by Gary Miner, Ph.D.

INTRODUCTION

Data mining has played a promising role in the extraction of implicit, previously unknown, and potentially useful information from databases (structured data). However, it is estimated that around 80% of the data in an organization is in unstructured form. Data trapped in "text form" (unstructured data) also express a vast and rich source of such information. Text mining is all about analyzing text for extracting information from unstructured data.

TEXT MINING

The purpose of this *STATISTICA* Text Mining and Document Retrieval tutorial is to provide powerful tools to process unstructured (textual) information, extract meaningful numeric indices from the text, and, thus, make the information contained in the text accessible to the various data mining (statistical and machine learning) algorithms available in the *STATISTICA* system. Information can be extracted to derive summaries for the words contained in the documents or to compute summaries for the documents based on the words contained in them. Hence, you can analyze words, clusters of words used in documents, etc., or you could analyze documents and determine similarities between them or how they are related to other variables of interest in the data mining project.

STATISTICA Text Mining and Document Retrieval is a text-mining tool for indexing text in various languages, i.e., for meaningfully representing the number of times that terms occur in the input documents. The program includes numerous options for stemming words (terms), for handling synonym lists and phrases, and for summarizing the results of the indexing using various indices and statistical techniques. Flexible options are available for finalizing a list of terms that can be "deployed," to quickly score ("numericize") new input texts. Efficient methods for searching indexed documents are also supported.

Input Documents

The software accepts as input documents in a variety of formats, including MS Word® document files and rich text files (RTF), PDF (Acrobat Reader®), PS (PostScript®), htm and html (Web pages or URL addresses), XML, and text files. You can also specify a variable in a *STATISTICA* input spreadsheet containing the actual text itself.

Selecting Input Documents

Input documents can be selected in a variety of ways. File names and directories (references to input documents) can be stored in a variable in an input spreadsheet, or you can "crawl" through directories and subdirectory structures to retrieve files of particular types. In addition, various methods for accessing Web pages and for "crawling" the Web (retrieving all Web pages linked to a particular document specified as the root; e.g., you could retrieve all documents referenced or linked to the StatSoft home page at www.statsoft. com). Web crawling can be performed to a user-defined depth; e.g., you can request to retrieve all web sites linked to pages that are referenced from a particular root URL, pages that are referenced in those pages, and so on.

Stop Lists, Synonyms, and Phrases

Various options are available for specifying lists of words (terms) that are to be excluded from the indexing of the input documents or pairs of terms that are to be treated as synonyms (i.e., counted as the same word). It can be specified to treat specific phrases (e.g., *Eiffel Tower*) as single terms and entries in the index. These lists can be edited and saved for future and repetitive use, so the system can be customized to specific terminologies for different domains.

Stemming and Support for Different Languages

Stemming refers to the reduction of words to their roots so that, for example, different grammatical forms or declinations of verbs are identified and indexed (counted) as the same word. For example, stemming will ensure that both *travel* and *traveled* will be recognized by the program as the same word. The software includes stemming algorithms for most European languages including English, French, German, Italian, and Spanish.

Indexing of Input Documents: Scalability of *STATISTICA* Text Mining and Document Retrieval

The indexing of the input documents is extremely fast and efficient, and based on relational database components built into the program. The contents of this database can be saved for further updating in future sessions, or for "deployment," i.e., to score input documents using only previously selected key terms.

Results, Summaries, and Transformations

The Text Mining Results dialog contains numerous options for summarizing the frequency counts of different words and terms. You can also combine terms or phrases (to count them as a single term or phrase), or clear only some of the terms in the analyses.

Options are available for reviewing word/term frequencies or document frequencies, as well as transformations of those frequencies better suited for subsequent analyses (e.g., inverse document frequencies). The Results dialog also contains options for performing singular value decomposition on the documents-by-terms frequency matrix (or transformations of frequencies) to extract dominant "dimensions" into which terms and documents can be mapped.

The scores and coefficients for the extracted dimensions can also be saved for subsequent processing of new documents to map those documents into the same space. Because of the integrated architecture of the *STATISTICA* system, all results spreadsheets can be used as input data for subsequent analyses or graphs. Hence, it is easy to apply any of the large number of analytic algorithms available in the software to the outputs generated by the Text Mining and Document Retrieval module, for example, to apply cluster analysis methods or any of the methods for predictive data mining to include textual information in those projects.

Let's look at some examples to see how interesting information can be extracted from unstructured data (or text corpus) using the text mining approach. In the process of the following illustrations, the different features available within the *STATISTICA* Text Mining and Document Retrieval module will be demonstrated. At the end of this tutorial, you will also understand how the outputs from the Text Mining module can be integrated into the Data Mining module to analyze the numerated text data.

CAR REVIEW EXAMPLE

The sample data file, 4Cars.sta, contains car reviews written by automobile owners. Car reviews related to four popular brands were extracted from the web sites (www.carreview.com), and one of the brands was renamed to conceal certain findings. The attributes of this file are detailed in Table G.1.

Each row (or case or instance) contains opinion (Summary, Strengths, or Weaknesses) filled by car owners about the car they own, along with other information, the rating for the car, the overall price they paid, and the car type. The purpose of this analysis is to see whether we can extract some information from textual corpus via the nascent concept of text mining. Follow these steps to start:

- Open *STATISTICA* by choosing Start, Programs, and then *STATISTICA*.
- Close the Welcome to *STATISTICA* dialog, the Data Miner workspace, and the spreadsheet.
- From the Files menu, select Open. Open the 4Cars.sta file from the Examples/Data sets folder, as shown in Figure G.1. (*Note*: In most default installations of *STATISTICA*, you will find the sample data files in the Examples/Data sets file.)
- Next, select Text Mining & Document Retrieval from the Statistics, Text & Document Mining, Web Crawling submenu to display the Text Mining startup panel (see Figure G.2).

This dialog contains nine tabs: Quick, Advanced, Filters, Characters, Index, Synonyms & Phrases, Delimiters, Project, and Defaults. Use the options on this dialog to specify the documents to be analyzed; the words, terms, and phrases that are to be included or ignored; and the database (internal, and used only for this module) where the indexed terms are to be stored. You can also select an existing database and thus "score" new documents using the terms selected (and saved) in that database. If you would like to learn more about these options, refer to the *STATISTICA* electronic manual.

- To start, first extract the word frequencies from the text contents from the first variable, Summary.
- From the Quick tab, select the From Variable option under the Retrieve Text Contents section.
- Next, click the Text Variable button to display the Select a Variable Containing Texts dialog and then select the variable Summary (which is the variable containing the text body; see Figure G.3).
- Click OK to return to the Quick tab.

TABLE G.1   Attributes of Car Review Example

Unstructured Data	Structured Data
1. Summary	4. Overall Rating
2. Strengths	5. Price paid
3. Weaknesses	6. Car type

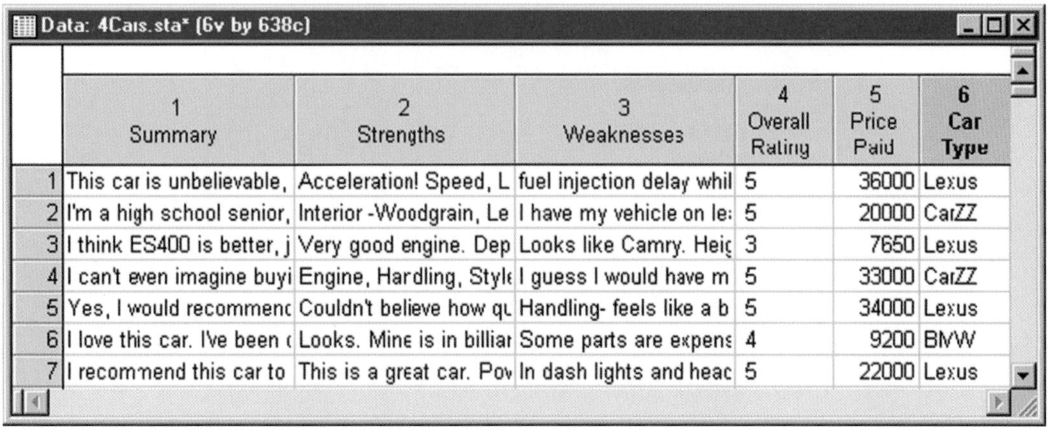

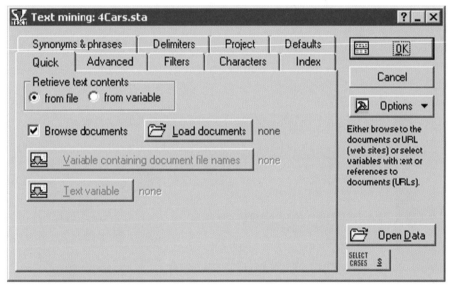

	1 Summary	2 Strengths	3 Weaknesses	4 Overall Rating	5 Price Paid	6 Car Type
1	This car is unbelievable,	Acceleration! Speed, L	fuel injection delay whil	5	36000	Lexus
2	I'm a high school senior,	Interior -Woodgrain, Le	I have my vehicle on lea	5	20000	CarZZ
3	I think ES400 is better, j	Very good engine. Dep	Looks like Camry. Heig	3	7650	Lexus
4	I can't even imagine buyi	Engine, Hardling, Style	I guess I would have m	5	33000	CarZZ
5	Yes, I would recommend	Couldn't believe how qu	Handling- feels like a b	5	34000	Lexus
6	I love this car. I've been	Looks. Mine is in billiar	Some parts are expens	4	9200	BMW
7	I recommend this car to	This is a great car. Pow	In dash lights and head	5	22000	Lexus

Data: 4Cars.sta* (6v by 638c)

FIGURE G.1   Text Mining data spreadsheet.

FIGURE G.2   Text Mining Quick tab.

- Now click on the Index tab and click on the Edit Stop-Word File button. Drag down to the bottom of the stop word list and add the word *car* to the list, as shown in Figure G.4.
- Click the OK [Save] button to add the word *car* to the existing stop word file (i.e., the words and terms contained in that stop list will be excluded from the indexing that occurs during the processing of the documents).
- Next, click the OK button on the Text Mining: 4Cars.sta dialog to begin the processing of the documents. After a few seconds (or minutes, depending on the speed of your computer hardware), the TM Results: 4Cars.sta dialog will be displayed, as shown in Figure G.5.

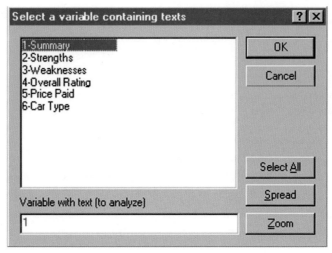

FIGURE G.3   Select Text Variable dialog.

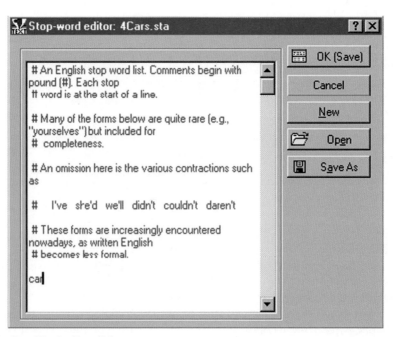

FIGURE G.4   Stop-Word editor dialog.

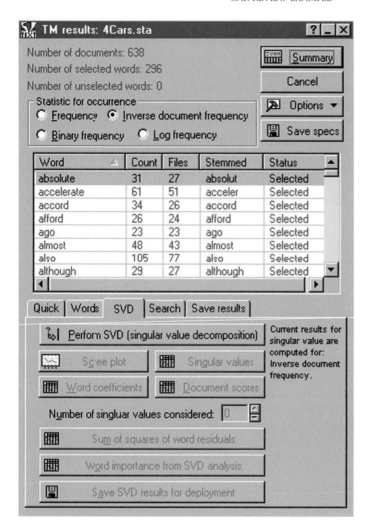

FIGURE G.5 *STATISTICA* Text Miner Results dialog.

The TM Results dialog gives a brief summary (number of documents, selected words and unselected words) and displays the words extracted by the Text Mining module.

• Click on the Count header to sort the words by frequency. You will see that the word *drive* appeared with the highest frequency, followed by *great*, etc.

The options available in the TM Results dialog are as follows:

Frequency: Select this option button to analyze and report the simple word frequencies.

Binary Frequency: Select this option button to analyze and report binary indicators instead of word frequencies. Specifically, this option will simply enumerate whether or not a term is used in a document. The resulting documents-by-words matrix will contain only 1s and 0s, to indicate the presence or absence of the respective word. As the other transformations of simple word frequencies, this transformation will dampen the effect of the raw frequency counts on subsequent computations and analyses.

Inverse Document Frequency: Select this option button to analyze and report inverse document frequencies. One issue that you may want to consider more carefully, and reflect in the indices used in further analyses, is the relative document frequencies (df) of different words. For example, a term such as *guess* may occur frequently in all documents, while another term such as *software* may occur in only a few. The reason is that one might make *guesses* in various contexts, regardless of the specific topic, while *software* is a more semantically focused term that is only likely to occur in documents that deal with computer software. A common and very useful transformation that reflects both the specificity of words (document frequencies) as well as the overall frequency of their occurrences (word frequencies) is the so-called inverse document frequency. This option includes both the dampening of the simple word frequencies via the log function and includes a weighting factor that evaluates to 0 if the word occurs in all documents and to the maximum value when a word occurs in only a single document. It can easily be seen how this transformation will create indices that both reflect the relative frequencies-of-occurrences of words, as well as their semantic specificities over the documents included in the analysis.

Log Frequency: Select this option button to analyze and report logs of the raw word frequencies. A common transformation of the raw word frequency counts is to "dampen" the raw frequencies and see how they will affect the results of subsequent computations.

Summary: Click the Summary button to compute the summary of word occurrence in document (same results as the option by the longer name on the Quick tab). Specifically, the results spreadsheet will contain a row for each input document and a column for each word or term. The entries in the cells of the results dialog depend on the option selection in the Statistic for Occurrence Group box on this dialog. The summary spreadsheet can quickly be turned into an input spreadsheet for subsequent analyses (use the options on the Save Results tab to write the respective word statistics to another file or database).

Let's get back to the example:

- In this case, select the option Inverse Document Frequency (it's the most efficient and frequently used option to represent the word counts) from the Statistic for Occurrence group box. Also, select the SVD tab (see Figure G.6).

At this point, there are different features/techniques available from different tabs (Quick, Words, SVD, Search, and Save Results) from which the "numericized" words from the TM results dialog can be saved into a standalone spreadsheet for further analysis. (Refer to the *STATISTICA* electronic manual to learn more about the features available within these tabs.)

Let us next perform the "singular value decomposition."

The SVD tab of the Text Mining Results dialog gives options to perform singular value decomposition on the document-by-words matrix, based on the selected words only, and with the word frequencies or transformed word frequencies as currently selected in the Statistic for Occurrence group box on the Text Miner Results dialog. Note that the results will be available for only one particular type of matrix (transformation of the word frequencies),

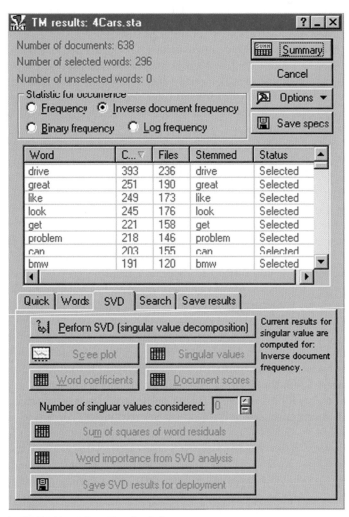

FIGURE G.6 *STATISTICA* Text Miner Results dialog.

and once you change your selection on the Text Miner Results dialog, any previously computed results for singular value decomposition will be discarded. In addition, when you save SVD results for deployment, only the singular value decomposition results for the specified word frequencies or their transformations will be saved.

Singular value decomposition is an analytic tool for feature extraction that can be used to determine a few underlying "dimensions" that account for most of the common contents or "meaning" of the documents and words that were extracted.

For computational details of the statistics and results available on the SVD tab of the Text Mining Results dialog, you can use the following options:

Perform SVD (Singular Value Decomposition): Click this button to start computing the singular value decomposition of the documents-by-words matrix (word occurrences or their transformations as currently selected on the Text Mining Results dialog). When the computations are completed, the various options available for reviewing the coefficients and document scores, and for saving the results for deployment, will become available (not dimmed).

Scree Plot: Click this button to create a scree plot of the singular values extracted from the word occurrence matrix. This plot is useful for determining the number of singular values that are useful and informative, and that should be retained for subsequent analyses. Usually, the number of "informative" dimensions to retain for subsequent analysis is determined by locating the "elbow" in this plot, to the right of which you presumably find on the factorial "scree" due to random noise.

Singular Values: Click this button to display a results spreadsheet with the singular values.

Word Coefficients: Click this button to display a results spreadsheet with the word coefficients. You can use the standard spreadsheet options to turn these results into an input spreadsheet in order to, for example, create 2D scatterplots for selected dimensions. Such scatterplots, when they contain labeled points, can be very useful for exploring the meaning of the dimensions into which the words and documents are mapped, i.e., to understand the semantic space for the extracted words or terms and documents. See also the section on latent semantic indexing via singular value decomposition in the Introductory Overview.

Document Scores: Click this button to display a results spreadsheet with the document scores; like the Word coefficients, these can be plotted in 2D or 3D scatterplots to aid in the interpretation of the semantic space defined by the extracted words and documents in the analysis.

Sum of Squares of Word Residuals: Click this button to display the word residuals from the singular value decomposition. As described in Singular Value Decomposition in *STATISTICA* Text Mining and Document Retrieval, these values are related to the extent to which each word is represented well by the semantic space defined by the dimensions extracted via singular value decomposition.

Word Importance from SVD Analysis: Click this button to display the word importance values computed from the singular value decomposition. As described in Singular Value Decomposition in *STATISTICA* Text Mining and Document Retrieval, the reported values (indices) are proportional to and can be interpreted as the extent to which the individual words are represented or reproduced by the dimensions extracted via singular value decomposition and, hence, how important the words are for defining the semantic space extracted by this technique.

Save SVD Results for Deployment: Click this button to save the SVD results for deployment, i.e., to "score" new documents. Specifically, this option will save the current SVD results for the currently selected words; this information will be saved in the current database, which can then be used in subsequent analyses to automatically index and score new documents. Thus, this option is essential for many applications of text mining (as, for example, discussed in the Introductory Overview), for example, to implement automatic mail filters or text routing systems.

- Next, perform the singular value decomposition by clicking on the Perform SVD (Singular Value Decomposition) button available from the SVD tab. Once this computation is over, you can see all the other options available within this tab being enabled.

At this point, you can extract different details of the statistics and results that can be used for further analysis.

- Click on the Scree Plot button to view the graph shown in Figure G.7.

The plot in this figure is useful for determining the number of singular values that are useful and informative, and that should be retained for subsequent analyses. It helps to visually determine the number of components that explains the variance among the inputs. We can tell by looking at the graph that the first component explains slightly more than 18% of the total variance for 295 words that were used as inputs, followed by the second component, which explains 7%, etc. So 25% of the variance present within the inputs is explained by the top two components. Usually, the number of "informative" dimensions to retain for subsequent analysis is determined by locating the "elbow" in this plot. The scree test involves finding the place where the smooth decrease of singular values appears to level off to the right of the plot. To the right of this point, presumably, you find only *SVD scree* (*scree* is the geological term referring to the debris that collects on the lower part of a rocky slope). Thus, no more than the number of components to the left of this point will be useful for analysis.

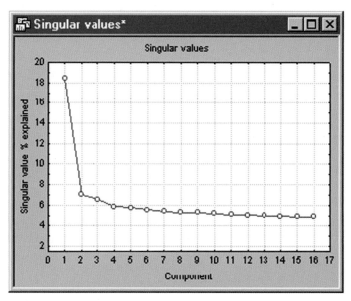

FIGURE G.7 Scree Plot screen of singular values.

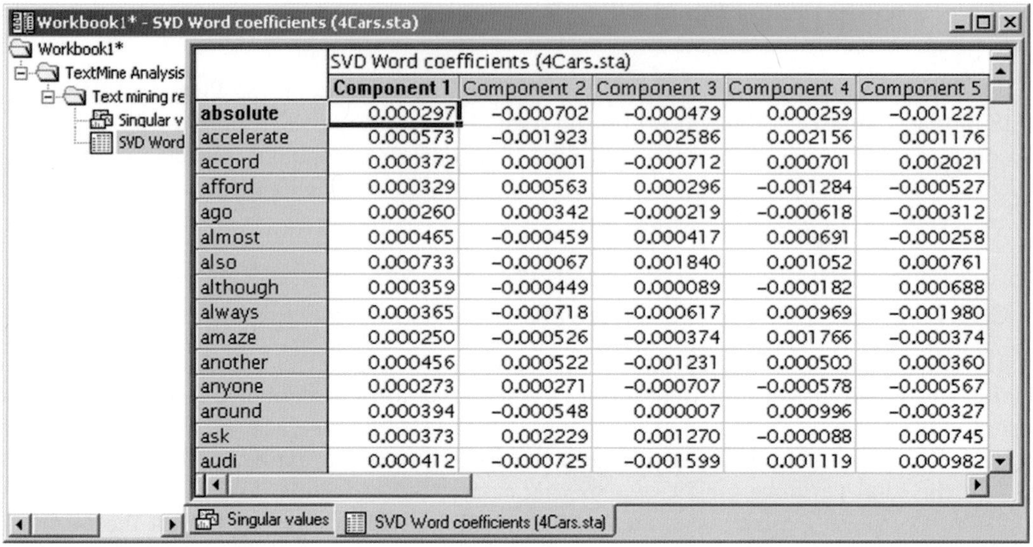

Workbook1* - SVD Word coefficients (4Cars.sta)

	Component 1	Component 2	Component 3	Component 4	Component 5
absolute	0.000297	−0.000702	−0.000479	0.000259	−0.001227
accelerate	0.000573	−0.001923	0.002586	0.002156	0.001176
accord	0.000372	0.000001	−0.000712	0.000701	0.002021
afford	0.000329	0.000563	0.000296	−0.001284	−0.000527
ago	0.000260	0.000342	−0.000219	−0.000618	−0.000312
almost	0.000465	−0.000459	0.000417	0.000691	−0.000258
also	0.000733	−0.000067	0.001840	0.001052	0.000761
although	0.000359	−0.000449	0.000089	−0.000182	0.000688
always	0.000365	−0.000718	−0.000617	0.000969	−0.001980
amaze	0.000250	−0.000526	−0.000374	0.001766	−0.000374
another	0.000456	0.000522	−0.001231	0.000500	0.000360
anyone	0.000273	0.000271	−0.000707	−0.000578	−0.000567
around	0.000394	−0.000548	0.000007	0.000996	−0.000327
ask	0.000373	0.002229	0.001270	−0.000088	0.000745
audi	0.000412	−0.000725	−0.001599	0.001119	0.000982

FIGURE G.8   SVD Word Coefficients spreadsheet.

- Now click on the Word coefficient button to view the results spreadsheet holding the word coefficient based on the results from the SVD (see Figure G.8).

As explained earlier, we can now draw scatterplots to explore the meaning of the dimensions into which the words and documents are mapped, i.e., to understand the semantic space for the extracted words or terms or documents. Let's now use the top two components to draw a scatterplot to plot the important words picked by the components.

- Right-click on the SVD Word Coefficients spreadsheet header (appearing within the left pane of the workbook) and select the option Use as Active Input.
- Next select Scatterplots from the menu option Graphs. Click on the Variables button within the 2D Scatterplot dialog.
- Select Component 1 as the X-variable and Component 2 as the Y-variable (see Figure G.9).
- Click OK on the Select Variables for Scatterplot dialog and then the 2D Scatterplot dialog to view the 2D scatterplot graph shown in Figure G.10.
- Next, click on the Brushing toolbar button 🔍 from the Graphs toolbar (by default displayed as the third layer of menu options within the *STATISTICA* application) to display the Brushing 2D dialog (see Figure G.11).

The Brushing 2D dialog contains tools for identifying points or groups of points on both 2D and 3D graphs to be marked, labeled, or temporarily turned off (i.e., removed from the graph and from considerations for fit lines applied, etc.) When brushing is activated, the mouse pointer turns by default into a "gun-sight style" cross-hair 🔍.

The pointer can be used to select/highlight either individual points (select the Simple option button under Selection Brush on the Brushing dialog) or groups of points

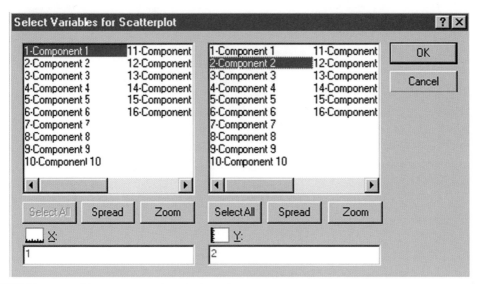

FIGURE G.9   Variable-selection window for SVD Components.

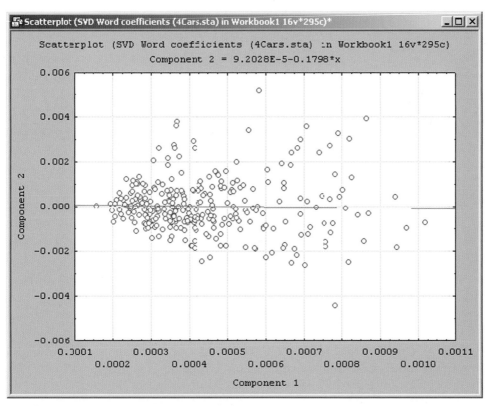

FIGURE G.10   Scatterplot of SVD Word Coefficients.

FIGURE G.11   Brushing dialog in *STATISTICA*.

(Lasso or Box). Other options such as the Slice X, Y, and Z and the Cube can be used to define areas on a 2D or 3D plot or volumes on a 3D plot. The areas or volumes defined by the Lasso, Box, Slice, and Cube options can be animated to move over the extent of the plot (or a matrix of plots in some cases) to explore the spatial distribution of values.

With the points highlighted, clicking the Update button on the Brushing dialog causes the action specified (labeling, marking, turning off) to be executed. Actions taken can be reversed by clicking the Reset All button. The Quit button closes the dialog, leaving the actions already applied intact.

- Next, check the Auto Apply option (displayed below the Reset All button).
- Select the Toggle tab and then select the Toggle Label option and select the Simple option from the Selection Brush section.

Figure G.12 shows how the Brushing 2D dialog should look.

FIGURE G.12   Brushing 2D dialog.

• Next, use the "gun-sight style" mouse pointer 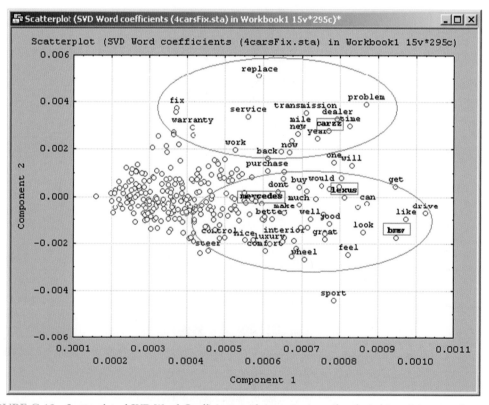 and click on each point to view the actual words that are represented by points. As you start clicking on the points, you will see the words being displayed.

Figure G.13 shows some of the words that were used in the reviews along with the brands that were picked for this analysis.

You can look at such graphs to visualize the "semantic" (of or related to meaning in language) spaces of related words. Words appearing close to one another are related to one another. You can see from the graph in Figure G.13 that the words within the second ellipse contain positive words (*comfort*, *better*, *quality*, *good*, *great*, etc.) when compared to the first ellipse, which contains negative words (*problem*, *replace*, *fix*, etc.). We can say that the reviewers used positive words to describe brands BMW, Lexus, and Mercedes, and more of negative words were used to describe *CarZZ*. This gives you a clear picture of how the reviewers described their experience with the brand they were using. You can also use other pairs of components to draw scatterplots and further understand/drill down into the other dimensions.

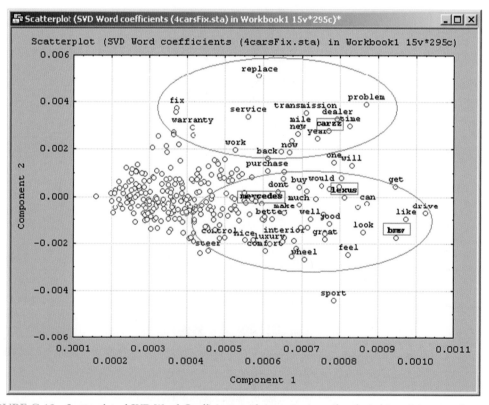

FIGURE G.13   Scatterplot of SVD Word Coefficients with important words selected to print on graph.

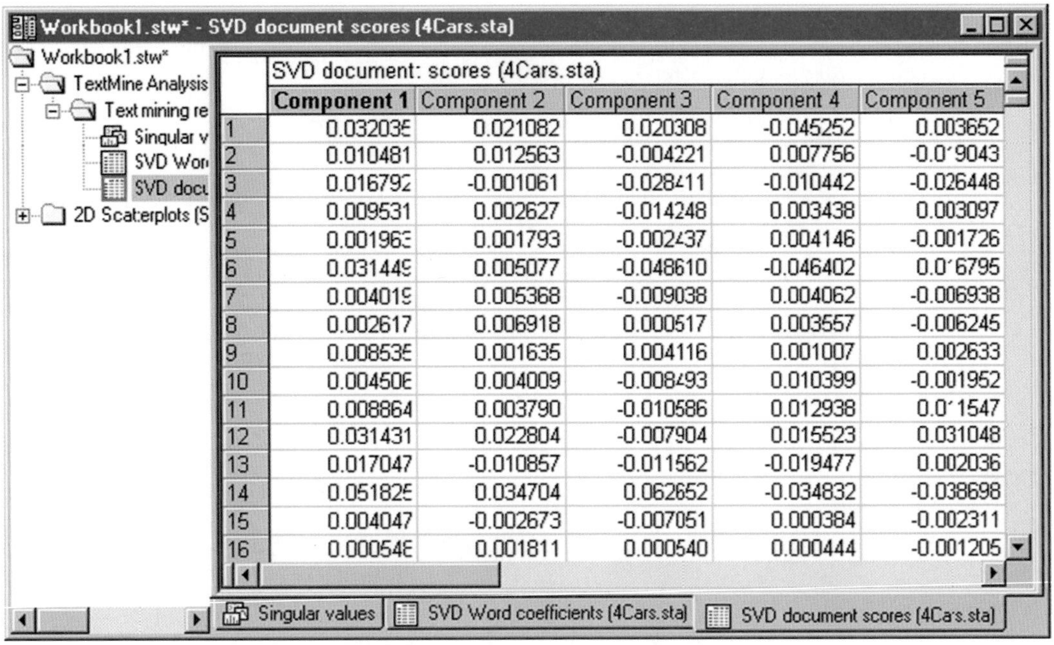

SVD document: scores (4Cars.sta)

	Component 1	Component 2	Component 3	Component 4	Component 5
1	0.032035	0.021082	0.020308	-0.045252	0.003652
2	0.010481	0.012563	-0.004221	0.007756	-0.019043
3	0.016792	-0.001061	-0.028411	-0.010442	-0.026448
4	0.009531	0.002627	-0.014248	0.003438	0.003097
5	0.001963	0.001793	-0.002437	0.004146	-0.001726
6	0.031449	0.005077	-0.048610	-0.046402	0.016795
7	0.004019	0.005368	-0.009038	0.004062	-0.006938
8	0.002617	0.006918	0.000517	0.003557	-0.006245
9	0.008535	0.001635	0.004116	0.001007	0.002633
10	0.004506	0.004009	-0.008493	0.010399	-0.001952
11	0.008864	0.003790	-0.010586	0.012938	0.011547
12	0.031431	0.022804	-0.007904	0.015523	0.031048
13	0.017047	-0.010857	-0.011562	-0.019477	0.002036
14	0.051825	0.034704	0.062652	-0.034832	-0.038698
15	0.004047	-0.002673	-0.007051	0.000384	-0.002311
16	0.000548	0.001811	0.000540	0.000444	-0.001205

FIGURE G.14 SVD Document Scores.

- Next, click on Document Scores option to view the spreadsheet, shown in Figure G.14, that holds document scores.

You can now use the document score results (displayed as components) within the spreadsheet to draw scatterplots as illustrated in the previous section. This time you will see the documents' IDs instead of the words, and the resulting scatterplot can be used to visualize the documents that are related (Document IDs falling close are related to each other). You can also try clustering techniques using the top components (that explains a good percentage of variance) as inputs to identify clusters or groups of related reviews or documents. You can then drill down into these cluster results using other tools (such as Feature Selection, Classification Trees, etc.), to further explore/understand the differentiating factors of these clusters.

You can also use the other options such as the Sum of Squares of Word Residuals, Word Importance from the SVD Analysis, etc. to view other results computed from the Singular Value Decomposition (SVD) analysis.

The following section will illustrate how the results from the Text Mining module can be written back into a spreadsheet or an input file to make it available for other analytic techniques/tools.

- Click on the Save Results tab, as shown in Figure G.15, to view the options available within this tab.

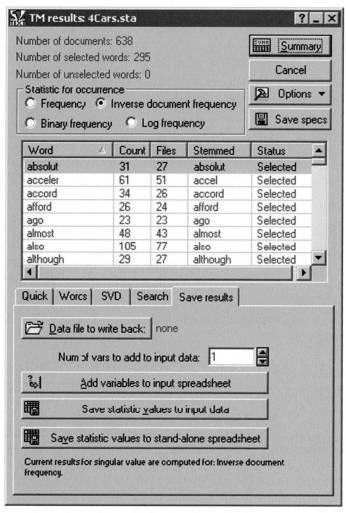

FIGURE G.15   Results of Singular Value Decomposition (SVD) analysis.

Saving Results into Input Spreadsheet

Select the Save Results tab of the Text Mining Results dialog to access options to save the current results, i.e., the word frequencies or transformed word frequencies (see the documentation for the Statistic for Occurrence option on the Text Mining Results dialog) for the documents-by-word matrix as well as the singular values, if singular value decomposition was performed (see the documentation for the SVD tab). Specifically, with the options available on this tab, you can write these results back into the input data spreadsheet or database [see also In-Place Database Processing (IDP)]; you can also use the Save Statistic Values to Stand-Alone Spreadsheet option (see below) to create a new spreadsheet with these results, along with selected variables from the input file. Thus, these options are extremely useful for joining the computed results with other information available in the input spreadsheet from which they were computed. This is a common requirement and operation in many applications related to data mining projects, where the input data consist of both structured information as well as unstructured textual information, both of which are to be included in the analyses.

- Next, click within the Num of Vars to Add to Input Data field and increase the value from 1 to 311 (to insert word frequencies of 295 words and 16 components from the SVD analysis).
- Next, click on the Add Variables to Input Spreadsheet option to create 311 new variables within the input spreadsheet.

Now you will see 311 new variables created to the right of the existing variables, as shown in Figure G.16.

- Next click on Save Statistics Values to Input Data on the TM Results: 4Cars.sta dialog to view the Assign Statistics to Variables, to Save Them to the Input Data dialog, shown in Figure G.17.
- Select all the words appearing within the Statistics pane and then select new variables from 7 to 318 (use the Shift key to perform this operation).
- Click on the Assign button to assign the word frequencies and the SVD Scores to the new variables.

Figure G.18 shows how the Assign Statistics to Variables, to Save Them to Input Data dialog should look now.

- Next, click OK to write the results extracted from the Text Mining module to the input spreadsheet.

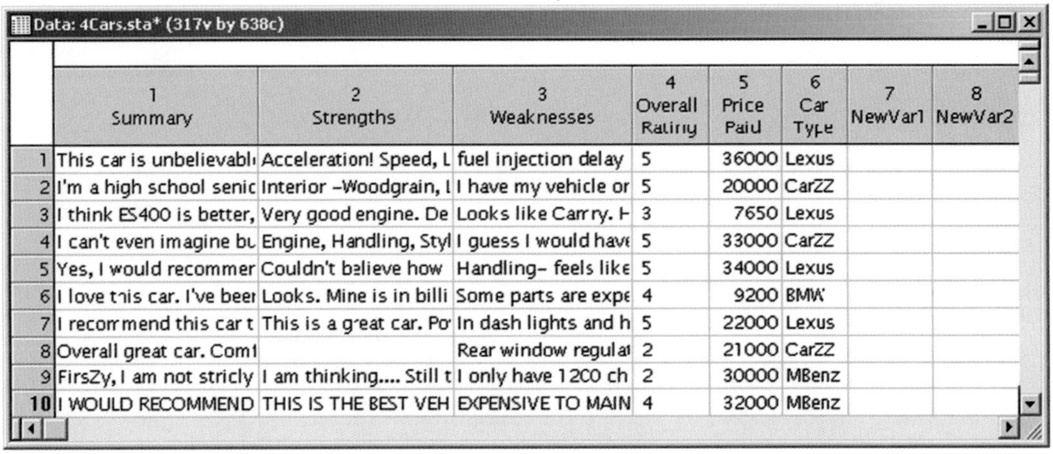

	1 Summary	2 Strengths	3 Weaknesses	4 Overall Rating	5 Price Paid	6 Car Type	7 NewVar1	8 NewVar2
1	This car is unbelievabl	Acceleration! Speed, L	fuel injection delay	5	36000	Lexus		
2	I'm a high school senic	Interior –Woodgrain, I	I have my vehicle or	5	20000	CarZZ		
3	I think ES400 is better,	Very good engine. De	Looks like Carrry. H	3	7650	Lexus		
4	I can't even imagine bu	Engine, Handling, Styl	I guess I would have	5	33000	CarZZ		
5	Yes, I would recommer	Couldn't believe how	Handling– feels like	5	34000	Lexus		
6	I love this car. I've beer	Looks. Mine is in billi	Some parts are expe	4	9200	BMW		
7	I recommend this car t	This is a great car. Po	In dash lights and h	5	22000	Lexus		
8	Overall great car. Com1		Rear window regulat	2	21000	CarZZ		
9	FirsZy, I am not stricly	I am thinking.... Still t	I only have 1200 ch	2	30000	MBenz		
10	I WOULD RECOMMEND	THIS IS THE BEST VEH	EXPENSIVE TO MAIN	4	32000	MBenz		

Data: 4Cars.sta* (317v by 638c)

FIGURE G.16   New Variables created to the right of existing Variables 1–6.

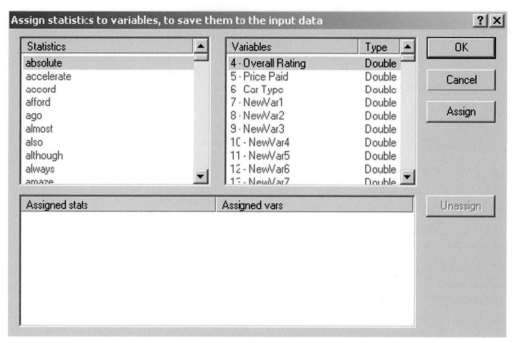

FIGURE G.17 Assign Statistics to Variables in *STATISTICA* Text Miner.

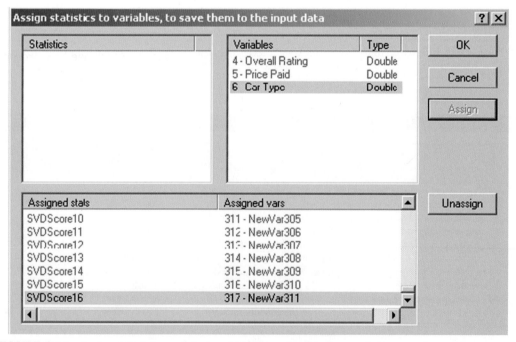

FIGURE G.18 Assigned Statistics and Assigned Variables placed in lower window.

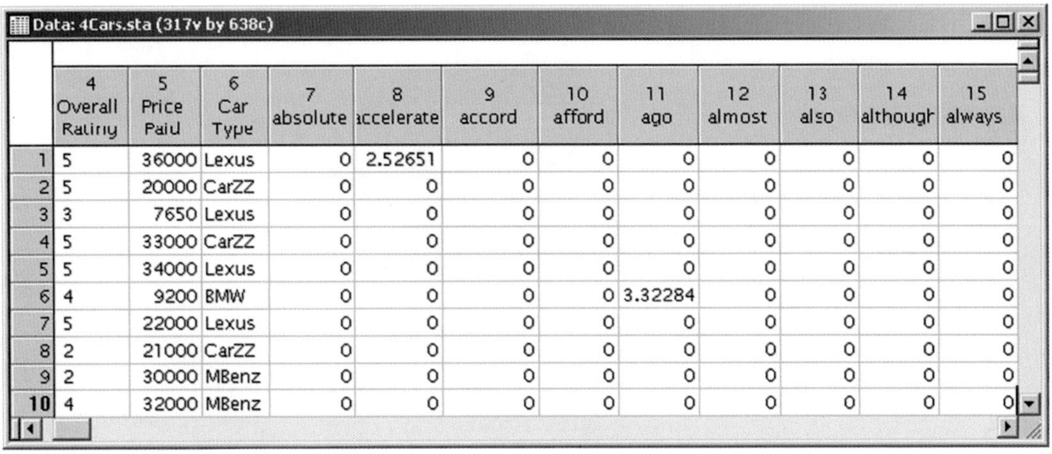

Data: 4Cars.sta (317v by 638c)												
	4 Overall Rating	5 Price Paid	6 Car Type	7 absolute	8 accelerate	9 accord	10 afford	11 ago	12 almost	13 also	14 although	15 always
1	5	36000	Lexus	0	2.52651	0	0	0	0	0	0	0
2	5	20000	CarZZ	0	0	0	0	0	0	0	0	0
3	3	7650	Lexus	0	0	0	0	0	0	0	0	0
4	5	33000	CarZZ	0	0	0	0	0	0	0	0	0
5	5	34000	Lexus	0	0	0	0	0	0	0	0	0
6	4	9200	BMW	0	0	0	0	3.32284	0	0	0	0
7	5	22000	Lexus	0	0	0	0	0	0	0	0	0
8	2	21000	CarZZ	0	0	0	0	0	0	0	0	0
9	2	30000	MBenz	0	0	0	0	0	0	0	0	0
10	4	32000	MBenz	0	0	0	0	0	0	0	0	0

FIGURE G.19   Enlarged data spreadsheet following assigning of Variables and Statistics; frequencies of specific words are placed in Variable columns 7, 8, 9, ...

Once you click OK, you will see the extracted word frequencies and the SVD Components being inserted into the spreadsheet. The spreadsheet holding the results is displayed in Figure G.19.

You now have the file ready for trying numerous analytic techniques available within the *STATISTICA* toolset. Let's next see how we can drill down further into specific words of interest. Recall that the first ellipse in the scatterplot contained negative words, and *transmission* was one word that appeared within this ellipse. Let us say you need to find the words that are related to *transmission*. There are several ways in which we can proceed (you can use Feature Selection tool, classification trees, correlations matrices, etc.). In this case, let's use the Feature Selection tool to identify the Best Predictors for the word *transmission*.

- Select Feature Selection and Variable Screening from the Statistics-Data Mining menu.
- Click on the Variables button and select Transmission as Dependent; Continuous: and all the other extracted words as Predictor; Continuous: as shown in Figure G.20.
- Click OK on the Select Dependent Variables and Predictors dialog and the Feature Selection and Variable Screening dialog to view the FSL Results dialog, as shown in Figure G.21.

Let's look at some options available in the FSL Results dialog.

Display k best predictors: At this point, you can request the best k predictors; for regression-type problems (for continuous dependent variables), the k predictors with the largest F values will be chosen; for classification-type problems, the k predictors with the largest chi-square values will be chosen.

Display best predictors with $p<$: Select this option button to display the list of best predictors for which the p value is less than the value specified in the adjacent edit field. The list of predictors will be sorted in ascending order by p.

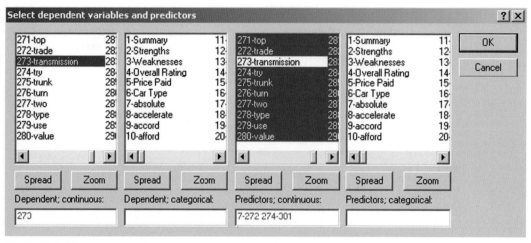

FIGURE G.20   Variable selection screen in interactive feature selection module.

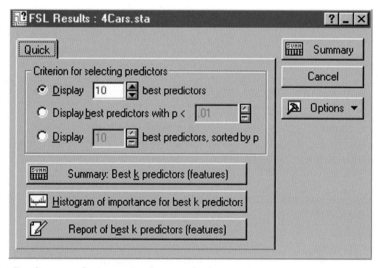

FIGURE G.21   Results screen for interactive feature selection.

Display _k_ best predictors sorted by _p_: Select this option button to be able to select the _k_ best predictors based on the probability (_p_) criterion.

- In this case, leave the option to the default setting and click on Histogram of Importance for Best _k_ Predictors button to view the graph shown in Figure G.22.

This graph displays the top 10 important words that are related to the word _transmission_. From this histogram, you can tell that when the word _transmission_ was mentioned,

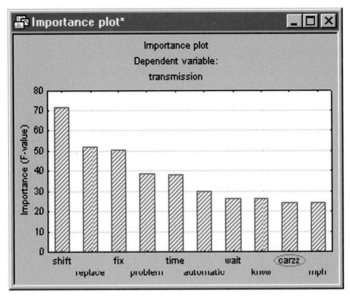

FIGURE G.22   Importance plot.

reviewers also used words like *shift, replace, fix, problem, time,* etc. The Feature Selection tool also identified CarZZ as one of the important predictors. We can also use other techniques such as classification trees to see whether we can extract similar useful information using the prepared file containing inverse document frequencies.

INTERACTIVE TREES (C&RT, CHAID)

The *STATISTICA* Interactive Trees (C&RT or CART, CHAID) module builds ("grows") classification and regression trees as well as CHAID trees based on automatic (algorithmic) methods, user-defined rules and criteria specified via a highly interactive graphical user interface (brushing tools), or combinations of both. The purpose of the module is to provide a highly interactive environment for building classification or regression trees (via classic C&RT methods or CHAID) to enable users to try various predictors and split criteria in combination with almost all functionality for automatic tree building provided in the General Classification and Regression Trees (GC&RT) and General CHAID Models (GCHAID) modules of *STATISTICA*. You can select the variables to use for each split (branch) from a list of suggested variables, determine how and where to split a variable, interactively grow the tree branch by branch or level by level, grow the entire tree automatically, delete ("prune back") individual branches of trees, and more. All of these options are provided in an efficient graphical user interface, where you can "brush" the current tree, i.e., select a specific node to grow a branch, delete a branch, etc. We will next use Interactive C&RT

to classify the car type and use the extracted words as predictors to see whether the tree will expose any hidden information useful for this study.

- Select Interactive Trees (C&RT and CHAID) from the Statistics-Data Mining menu.
- Click OK to perform a Classification Analysis using the C&RT model.
- Click on the Variables button within the Quick tab to view the Select Dependent Vars, Categorical, and Continuous Predictors.

Remember that we are trying to find the words that can classify (discriminate between) the car type (structured information), and there will be many reviews that mention about the brand name within the text corpus. Therefore, we need to eliminate the words that refer to the brands from the predictor list because if these words are included, they will turn out to be the best classifier/discriminator for the car type. For instance, say we include words like *Mercedes Benz, MZ, BMW, Lexus, CarZZ, ES 300*, etc. as predictors to classify the car types (CarZZ, BMW, Mercedes, and Lexus). It is quite obvious that the tree algorithm will pick these words that will discriminate one car type against the other, which will not reveal any information that is relevant for our study.

- Select Car Type as Dependent: and all the extracted words from variable 7 to variable 301 as Continuous pred: (as discussed let us exclude the words *accord, audi, benz, bmw, carzz, class, es300, honda, infiniti, lexus, mb, mercedes, series*, and *slk* that directly associate to brand names; use the Ctrl key for this purpose).
- Click OK on the Select Dependent Vars, Categorical and Continuous Predictors: dialog and the ITrees C&RT Extended Options: 4Cars.sta dialog to view the ITrees C&RT Results: 4Cars.sta dialog, as shown in Figure G.23.

You now have several options to extract the tree according to your choice (refer to the *STATISTICA* electronic manual to further study the different options available within this dialog). First, let us grow the tree to its full size and remove one level to resize the tree for easy visual interpretation.

- Click on the Grow Tree button to build the tree and next click on the Remove 1 Level button to reduce the tree by one level.
- Click on the Tree Graph button under the Review Tree: section.

You should now see the tree graph shown in Figure G.24 for car type.

The tree solutions are relatively simple and straightforward for interpretation. As you can see from the graph in Figure G.24, the C&RT algorithm had distinguished eight decision outcomes (contained in eight terminal nodes highlighted in red) built on seven if-then conditions to classify the car type. Terminal nodes, or terminal leaves as they are sometimes called, are points on the tree beyond which no further decisions are made. The tree starts with the top decision node (also called the root node) with all the 638 cases (reviews in our case) predominated by the car type CarZZ category. CarZZ had the highest frequency of reviews among the four car types, as indicated in the histogram. The legend identifying which bars in the node histograms correspond to the four categories is located in the top-left corner of the graph.

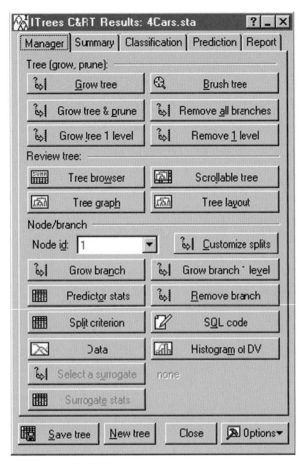

FIGURE G.23   Interactive Trees results dialog.

Recall that the purpose of this analysis is to learn how you can discriminate between the different car types based on the extracted inverse document frequency of words that are used as predictors. The interpretation of this tree is straightforward. The root node is split on the inverse document frequency of the word *top* forming two new nodes, one with car type BMW as the predominant category among 38 cases falling into a terminal node. This simply means that whenever the importance of the word *top* is high (> or *higher* inverse document frequencies describing the importance of its occurrence), reviewers were mostly mentioning BMW. Similarly, when the interactions of "word importance" (represented by inverse document frequencies) for *value and sport, price and use* were high, they were referring to BMW. You can also tell by looking at this tree that when the words *value, price, great, transmission*, etc., were used, the reviewers were also mentioning CarZZ. Therefore, we can tell that CarZZ was the brand that had the

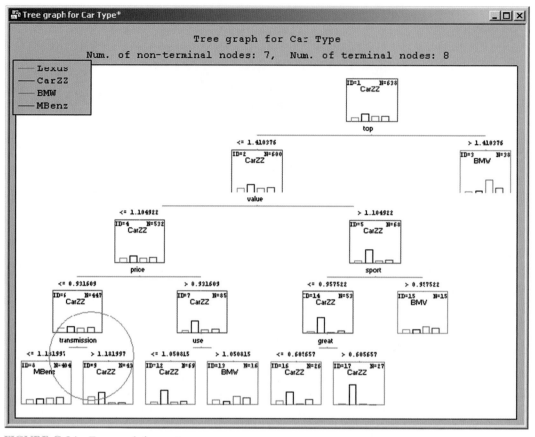

FIGURE G.24   Tree graph for car type.

transmission problem. This gives us further corroborating evidence for results identified by the Feature Selection tool.

Another possible approach to make use of this rich textual information would be to recode a new indicator variable for comparative study. For instance, we can create a new indicator variable derived from the negative connotation words (*complaint, disappoint, fix, noise, problem, repair, replace,* etc.) and use crosstabs or interaction plots for comparing which brand accumulated the most number of negative connotation words. Let's first create a new variable named Negative Connotations for this purpose.

- Make the spreadsheet 4Cars.sta active (minimize the Result workbook that is open). Then select the Add Variables option from Insert menu.
- Type in **SVDScore16** within the After: text box. Name the new variable Negative Connotations.
- Click OK to add a new variable named Negative Connotations to the spreadsheet.

Next, we will use the Recode function to create a binary indicator to determine whether or not the reviews contain any negative connotations (*complaint, disappoint, fix, noise, problem, repair, replace*, etc.).

- Select column 320, named Negative Connotation. Then select the Recode option from the Vars menu list.
- Type the "if" conditions to derive the new variable from the variables holding the negative connotation word. (i.e., if the inverse document frequency of variables *complaint, disappoint, fix, noise, problem, replace*, and *repair* is greater than 0, then flag the case or review as 1, else 0). Note that you can enter up to 256 conditions within this dialog.

Figure G.25 shows how the Recode Values of Variable dialog will look after you enter all the conditions.

- Click OK to perform the recode operation.
- Save the file 4Cars.sta for future requirements.

You will now see the binary values within the Negative Connotation variable, indicating whether or not the reviews contained a negative connotation word. Before we perform the comparative study, we have to make sure that there is an equal number of cases for each car

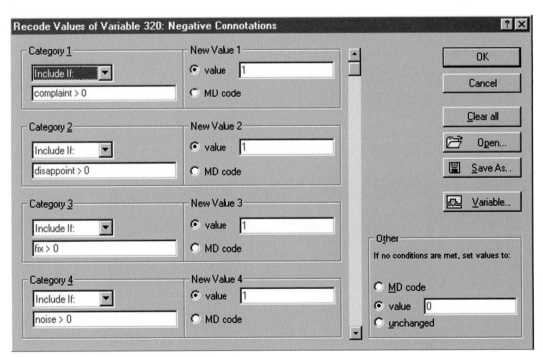

FIGURE G.25   Recode Values of Variables dialog.

type so that the study does not bias itself for a particular car type. The first question then would be to decide the number of cases that has to be extracted for each car type. We will next find the frequency of the four categories of car types so that we can identify the car type or brand with the lowest frequency. That number of cases for each car type will then be extracted.

- Select column 6, Car Type, in your 4Cars.sta spreadsheet. Then select the Basic Statistics/ Tables option from the Statistics menu.
- Select the Frequency Tables option within the Basic Statistics and Tables: dialog and click OK.
- The car type should be already selected within the Variables: Selection dialog. Next, click on the Summary button to view the Frequency Table: Car Type, as shown in Figure G.26.

The results table shows that Lexus had the least number of reviews when compared to the other car types. Therefore, we will extract 119 cases for each car type for comparative study.

- Minimize the active workbook holding the frequency table results.
- Select the Subset/Random Sampling option from the Data menu. Click on the Options tab and then select Calculate Based on Approximate *N*.
- Next, click on the Stratified Sampling tab. Click on the Strata Variables button and select the variable Car Type. Click OK to make the selection.
- Next, click on the Codes button. Click on the All button and then click OK. You will now see all the categories of car types under the Stratification Groups column.
- Enter **119** against each strata under the Approximate *N* column to extract a sample set with 119 brands for each category. You should be looking at a dialog like the one shown in Figure G.27.
- Click OK to view the new stratified sample spreadsheet that holds an equal number of cases for each car type.

Data: Frequency table: Car Type (4Cars)*

Frequency table: Car Type (4Cars.sta)

Category	Count	Cumulative Count	Percent	Cumulative Percent
Lexus	119	119	18.65204	18.6520
CarZZ	218	337	34.16928	52.8213
BMW	148	485	23.19749	76.0188
MBenz	153	638	23.98119	100.0000
Missing	0	638	0.00000	100.0000

FIGURE G.26 Frequency table.

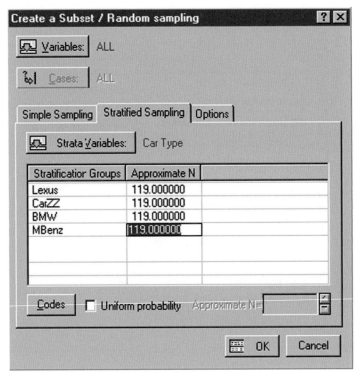

FIGURE G.27   Create Subset–Random Sampling dialog.

If you now check the frequency for the variable Car Types, you will have a balanced spreadsheet with an equal number of categories. We will use this spreadsheet to draw an interaction plot to perform a comparative study.

- Select Basic Statistics/Tables from the Statistics menu to open the dialog shown in Figure G.28.
- Then select the Tables and Banners option from the Basic Statistics and Tables dialog and click OK.
- Next, click on the Specify Tables (Select Variables) button on the Crosstabulation Tables dialog.
- Select the variable Car Type in the List1: section and Negative Connotations in the List2: section, as shown in the Select Up to Six Lists of Grouping Variables: dialog in Figure G.29.
- Click OK on the Variable Selection and Crosstabulation Tables dialogs to view the Crosstabulation Tables Results: dialog. Select the Advanced tab to view the options shown in Figure G.30.
- Click on the Interaction Plots of Frequencies button to view the Interaction Plot: Car Type X Negative Connotations dialog shown in Figure G.31.

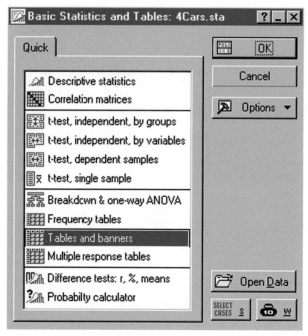

FIGURE G.28   Basic Statistics and Tables module dialog.

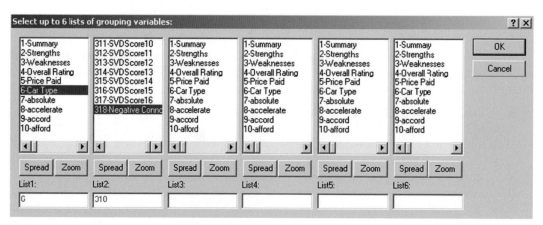

FIGURE G.29   Variable Selection in Tables and Banners.

From this graph, we can tell that Mercedes Benz accumulated the greatest number of reviews containing negative connotations (around 50 or so), followed by CarZZ, BMW, and Lexus (category 1 representing the reviews having negative connotations). We have identified by this simple approach that Lexus had the fewest number of negative

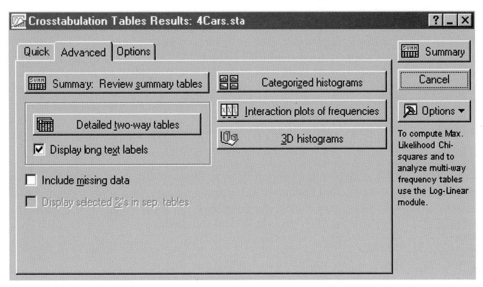

FIGURE G.30   Crosstabulation Tables Results dialog.

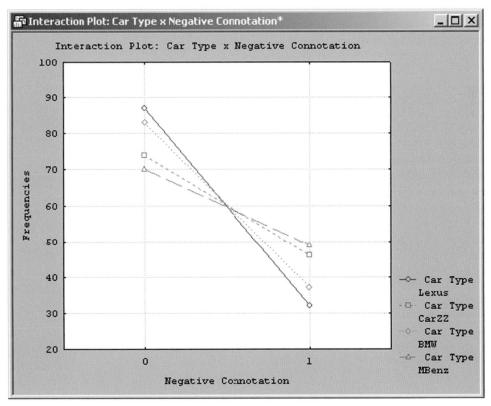

FIGURE G.31   Interaction Plot.

connotation words when compared to the other car types. If we had more information about the state, city, manufacturing unit for each car/brand, etc., we could have extracted useful information that could identify the places/units that elicited the greatest number of complaints.

OTHER APPLICATIONS OF TEXT MINING

Unstructured text is very common and, in fact, may represent the majority of information available to a particular research or data mining project. The selection of tools or techniques available with *STATISTICA*, along with the Text Mining module, can help organizations to solve a variety of problems. A few to mention are the following:

1. Extracting information reflecting customers/employees/public—opinions, needs, and interest (e.g., visualizing semantic spaces using 2D, 3D plots);
2. Filtering unwanted documents/emails (using stop list, include lists, etc.);
3. Predicting customer satisfaction levels (e.g., negative connotations);
4. Clustering similar words/documents. (e.g., reviews, research papers, survey data, etc.);
5. Classifying or organizing documents (e.g., electronic documents about general information can be classified into different subgroups);
6. Predicting/routing new documents, etc. (The rules for clustering or classifying or predicting can be used to score new documents.)

CONCLUSION

This simple tutorial is intended to help you understand how the *STATISTICA* Text Miner module, along with numerous *STATISTICA* Data Miner tools and techniques, can be used for finding solutions to problems that require knowledge of language and computing technology. More importantly, extraction of useful insights or information from unstructured data could be used as input for decision-making purposes.

Predictive Process Control: QC-Data Mining Using *STATISTICA* Data Miner and QC-Miner

Sachin Lahoti and Kiron Mathew
Edited by Gary Miner, Ph.D.

Notes to users:

1. The data set on the DVD that goes with tutorial is titled ProcessControl.sta.
2. The Data Miner Workspace illustrated in this tutorial uses an older NN module icon; you will need to replace the NN icon with the current NEW SANN–Neural Networks Icon, obtained from the Node Browser, if you want to re-run the Data Miner Workspace.

PREDICTIVE PROCESS CONTROL USING *STATISTICA* AND *STATISTICA* QC-MINER

In today's competitive world, remarkable progress has been made in the control of many different kinds of processes. Predictive Process Control (PPC) is an approach to identify variations of controllable parameters to stabilize processes within a manufacturing unit to maintain/enhance quality. This tutorial will use an example to illustrate how **advanced predictive models can be used to predict malfunctions within processes, in fact *even before such situations occur*.** The earlier these malfunctions are detected, the less time and money are wasted processing defective products. Furthermore, because of the high cost of modern equipment, if a scheme can quickly detect malfunctions, it will result in considerable savings due to higher equipment utilization.

It's often observed that quality can be improved by reducing the variability in process and raw materials. Since variability can be described only in statistical terms, statistical methods play a vital role in quality improvement efforts. The case study explains a complete set of tools and techniques and supporting systems for process control, which would be an invaluable resource for any technical manager, production engineer, or technician in any manufacturing enterprise.

CASE STUDY: PREDICTIVE PROCESS CONTROL

This case study will be illustrated using a real-world example to demonstrate the possible application of predictive data mining in the field of process control. The proposed analyses workflow used in this case integrates advanced predictive models that will be trained, tested, and automatically compared to find the best model for deployment. Various design approaches used in this specific example to tackle the problem will provide you some useful insight into how predictive models can be used to detect quality problems ahead of time, thus helping floor engineers to adjust parameter settings even before quality starts deteriorating.

Understanding Manufacturing Processes

Manufacturing processes are inherently complex; as a result, process development is often a tedious and experimental task. Process parameters settings, such as temperature, pressure, speed, etc., are typically chosen by costly trial-and-error prototyping, with the result that solutions are often suboptimal. Producing high-quality products within such a suboptimal environment is not easy. It's often noticed that too little attention is paid to achieve all dimensions of an optimal process: economy, productivity, and quality. Every manufacturer should realize that all three aspects could be accomplished by focusing on just one dimension—quality—because quality helps in increasing productivity and reducing cost.

STATISTICA QC Data Miner is a powerful software solution designed for manufacturing enterprises to help achieve an optimum level of quality. A wide array of advanced analytic tools and techniques helps to monitor processes and not only identifies but also anticipates problems related to quality control, providing improvement with unmatched sensitivity and effectiveness. It combines the most powerful tools for QC and SPC with data mining technology. *STATISTICA* QC Miner integrates the complete functionality of *STATISTICA* software for quality control and improvement with *STATISTICA* Data Miner software for uncovering hidden trends, explaining known patterns, and predicting the future.

Data File: ProcessControl.sta

The data file to be used in the example, ProcessControl.sta, consists of data collected from a soda pop-manufacturing unit. The file contains 37 variables and 2,838 cases. Most of the variables in this data file hold information about different parameter settings and readings measured between intervals along the production process. The dependent variable PC Volume is the final quality measure for a batch/cart of pop.

Variable Information

- **Predictors (also called *independent variables*)**: VolumeCO2, Pressure_CO2, Temp_CO2, Filler_Speed, Bowl_Setpoint, Alcohol_Release, etc.
- **Outcome or target variable (or variable of interest, also called the *dependent variable*)**: PC Volume (Quality decline if PC Volume > 0.15, referred to as "spike" in the following sections)
- **New dependent variable (also called a *derived dependent variable*)**: High_PC (1 if PC Volume > 0.15, else 0)

Problem Definition

Based on expert feedback, if the PC Volume measure goes above 0.15, the quality of the pop starts deteriorating. It's often noticed that a significant aspect of these problems may be caused both by controllable and noncontrollable factors. In this case, the analyses will be focused to find variation in the controllable parameter setting that deteriorates pop quality.

Graphical methods can be used for both data analyses and the presentation of results. Figure H.1 shows a visual presentation of PC Volume distribution. (PC Volume stands for Process Control Volume.)

Such simplified quality control charts can be used to visually observe the distribution of PC Volume falling above and below standard cutoff levels (in this case PC Volume was categorized as high/low based on a cutoff level of 0.15). The gravity of the problem can be visually detected by looking at such a graph.

Design Approaches

Three main design approaches were used to tackle this problem, with tasks defined to predict variation in factors/parameters that adversely affect the quality of soda pop:

1. **Static analyses:** The most usual approach is Static, i.e., given the data "as is," build a model that will learn the patterns of predictors corresponding to the predicted/outcome variable.
2. **Dynamic analyses:** In the Dynamic approach, we will lag (move one step backward) the dependent variable and then try to predict, to see how well the model performs on unseen data. When we lag the dependent variable, each case/row will hold the outcome observation of the following case, with the task now defined to build a model to predict the lagged outcome (in our case, predict possible deterioration in quality ahead of time).
3. **Transformation of change:** In most of the continuous processes, outcome observations (in our case, observations of PC Volume) follow certain patterns, and these patterns continue for some time; in short these observations are not random occurrences. Our

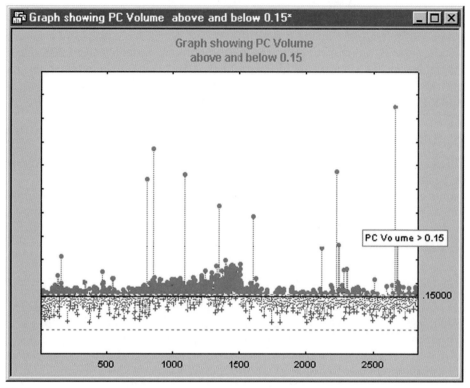

FIGURE H.1   Quality control charts.

third approach will focus on understanding causes that trigger changes in PC Volume observations, i.e., pattern of changes from (see Figure H.2)

    a. Normal PC_Volume level (PC_Vol < 0.15) to Spike Occurrence (PC_Vol > 0.15)
    b. Spike Occurrence (PC_Vol > 0.15) to Spike Continuing (PC_Vol continuing at 0.15)
    c. Spike Continuing (PC_Vol at 0.15) to Back to Normal (PC_Vol < 0.15)

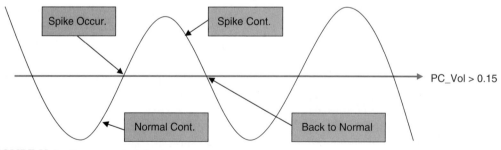

FIGURE H.2   Process control charts.

Next, we will move to data analyses to see how different tools and techniques can be used to possibly reveal hidden causes of quality problems, which could provide invaluable insights to engineers.

DATA ANALYSES WITH *STATISTICA*

The following sections will follow the order in which the analyses were performed, from data preparation to the final deployment of the model.

First, we will start with the Static approach.

Split Input Data into the Training and Testing Sample

During real-time model building, you often need to keep aside a test set of independent instances that have played no role in building the predictive model. Once the training is over, the test set can be used to predict the outcome variable (PC_Volume in our case) using inputs that went in as predictors and then cross-validate the result with the original outcome contained in the variable PC_Volume. The Split Input Data into Training and Testing Sample node available within the Data Miner Workspace will split the data set not marked for deployment into two input data sets: one marked for deployment (Testing) and the other one marked not for deployment (Training). In this case, 34% of the cases (out of 2,838 cases) were kept aside for testing.

Stratified Random Sampling

Blindly accepting the proportion of two categories of PC_Volume (1 and 0) to build a model will produce unlikely predictive results because the model may not capture the patterns present in the small proportion of the category (see Figure H.3).

In Figure H.3, you easily can see that there is a high disproportion between the two categories of High_PC (1 and 0). Hence, while model building, you need to extract data sets with equal proportion of outcome categories (PC_Volume > 0.15 denoted by 1 and PC_Volume < 0.15 denoted by 0) to clearly differentiate the characteristics or patterns underlying the variation of parameter/measurements. This can be achieved by extracting an equal number of observations falling above and below the PC_Volume cutoff level of 0.15. The Stratified Random Sampling tool available within the Node Browser will help you to achieve this task.

Feature Selection and Root Cause Analyses

As an exploratory step, you can use the Feature Selection and Root Cause Analyses tool to identify the Best Predictors (in this case, different parameter settings/measurements) that clearly discriminate between High/Low PC_Volume levels (dependent categorical), which can help to identify factors that cause variance in PC_Volume.

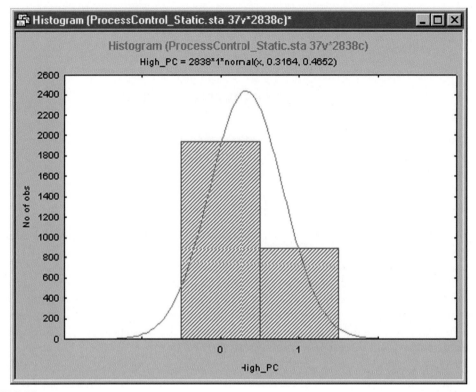

FIGURE H.3   Histogram for High_PC.

The Feature Selection tool also helps to short-list standalone predictors, in cases where there are thousands of predictors that are directly associated with quality. Having such a tool can help engineers to identify important factors and use them for model-building purposes. These tools also lend themselves to easy interpretation because the results (ranked order of predictors) are visually depicted on an importance plot.

Let's now analyze the results from the Feature Selection tool to identify the best predictor (factors causing variation) for dependent variable High_PC (see Figure H.4).

The Feature Selection tool identified Bowl_Setpoint as the most important predictor for High_PC, followed by Pressure_H1, Filler_Speed, etc. For cases in which there are many predictors, you can now select the best predictors for model building based on the importance they hold to explain the variation of the dependent variable. In this case, we will use the top 16 predictors out of the 36 variables for model building.

Different Models Used for Prediction

Now that we have the right proportion of observations from both the categories (PC_Volume < 0.15 and PC_Volume > 0.15) and have selected the right predictors, we will

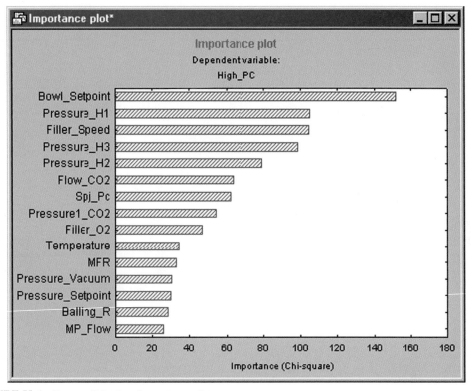

FIGURE H.4   Feature Selection: Importance plot.

next build different predictive models and then use comparative tools to automatically select the one that best predicts the dependent variable (variable of interest). We will try a few of these predictive models for this example:

1. SANN – *STATISTICA* Automated Neural
2. Support Vector Machine with Deployment
3. MARSplines for Classification with Deployment
4. Standard Classification CHAID with Deployment
5. Stochastic Gradient Boosting with Deployment

The Data Miner Workspace in Figure H.5 shows the analyses workflow after the preceding models were inserted.

Next, you run these models to see downstream spreadsheets and workbooks that hold the results. After you have all the models ready, the analyst can easily detect the best models using different comparative tools available within the Data Miner Workspace, such as gains chart, lift charts, goodness of fit of multiple inputs, cross-tabulation tables, etc. In this case, we will use lift charts to find the model that performs the best.

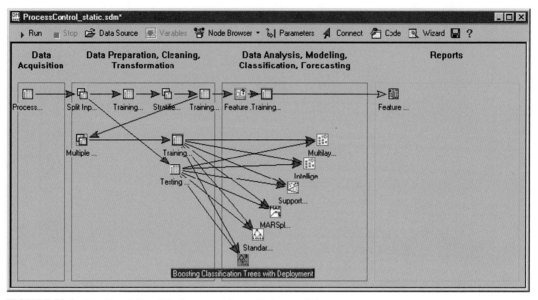

FIGURE H.5   The Data Miner Workspace with predictive models.

Compute Overlaid Lift Charts from All Models: Static Analyses

This node will search through the entire workspace for deployment information computed for the participating classification models. If more than one model was used to compute predictions, then overlaid lift or gains charts will be generated for each category of the dependent variable (see Figures H.6 and H.7).

Specifically, the chart summarizes the utility that you can expect by using the respective predictive models, as compared to using Model information only. Where you'd like to be in a lift chart is near the upper-right corner: the place where you have the maximum gain of predictive accuracy by using all your data. Any model that can fall close to this point will be your best choice. You can see from the graphs in Figures H.6 and H.7 that at most of the percentile levels the gain or lift value for the MARSplines model (shaded in green) has the highest gain in predictive accuracy when compared to the other models. Therefore, the MARSplines model works the best for this data set.

Let's try to interpret the graph for category 0 (High_PC < 0.15). If you consider the top two deciles, you would end up with a sample that had almost 1.325 times the number of category 0 (High_PC < 0.15) when compared to the baseline model. In other words, the relative gain or lift value by using MARSplines with the deployment model is approximately 1.325 for predicting category 0. When we take the top two deciles for category 1 (High_PC > 0.15), we can see that the CHAID model outperformed the MARSplines model with a lift value of 2.2 when compared to the baseline model. Now deciding the number of top deciles to target using the best model entirely depends on expertise and heuristics.

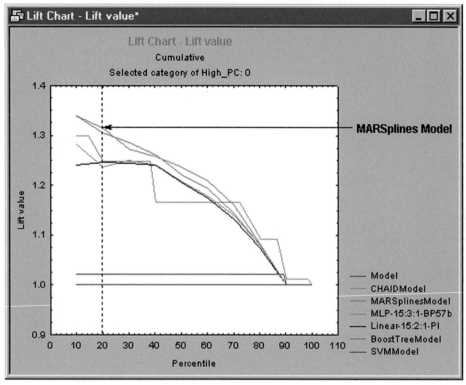

FIGURE H.6   Lift Chart: High_PC: 0.

Classification Trees: CHAID

Classification trees are used to predict membership of cases or objects into classes of a categorical dependent variable from their measurements on one or more predictor variables. Classification tree analysis has traditionally been one of the main techniques used in data mining. The Classification Trees module in *STATISTICA* Data Miner is a full-featured implementation of techniques for computing binary classification trees based on univariate splits for categorical predictor variables, ordered predictor variables (measured on at least an ordinal scale), or a mix of both types of predictors. It also has options for computing classification trees based on linear combination splits for interval scale predictor variables.

The flexibility of classification trees makes it a very attractive analysis option, but this is not to say that its use is recommended to the exclusion of other methods. As an exploratory technique, classification trees are, in the opinion of many researchers, unsurpassed. Classification trees readily lend themselves to being displayed graphically, helping to make them easier to interpret than they would be if only a strict numerical interpretation were possible.

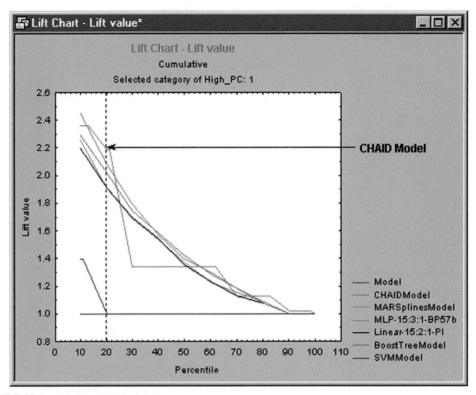

FIGURE H.7   Lift Chart: High_PC: 1.

Let's next review the results generated from the CHAID algorithm (see Figure H.8).

Interpreting these trees is quite straightforward. The CHAID algorithm identified parameter settings for Pressure_H1, Bowl_Setpoint, MP_Flow, and Flow_CO2 to explain the interactions that led to PC_Volume observations falling above and below the cutoff level of 0.15.

The rules generated by these trees (also available from tree structure table) can help engineers to pinpoint the interaction effect of parameter setting that causes fluctuation in PC_Volume. As you can see from the graph in Figure H.8, the CHAID algorithm has distinguished 11 decision outcomes (contained in 11 terminal nodes highlighted in red) built on 11 if-then conditions to predict the category of PC_Volume (1 if PC_Volume > 0.15 and 0 if PC_Volume < 0.15). By following the path from the root node (ID = 1) to terminal node (ID = 15), we can derive a rule for category 1 (PC_Volume > 0.15). We can say by looking at the classification tree that if Pressure_H1 is greater than 13.40 and Flow_CO2 is greater than 547.00, then there were 195 observations recorded, out of which most of the observations fall into the PC_Volume > 0.15 category. Similarly, we can analyze the other branches and draw further conclusions. The legend that identifies which bars in the node histograms

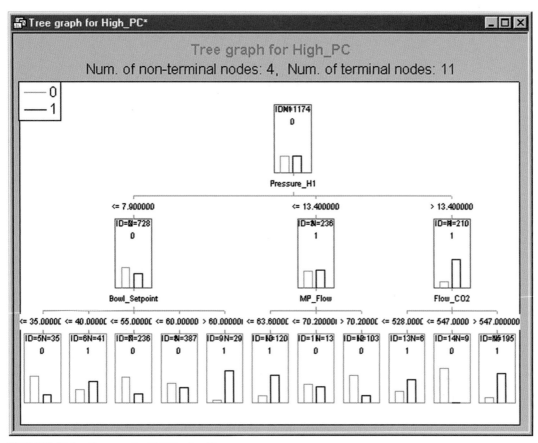

FIGURE H.8   Tree graph: CHAID algorithm.

correspond to the two categories of PC_Volume is located in the top-left corner of the graph.

Similar analysis steps were followed to predict the lagged predicted variable (dynamic analysis).

Compute Overlaid Lift/Gain Charts from All Models: Dynamic Analyses

The overlaid lift/gain charts in Figures H.9 and H.10 will give you a clear idea about the model that performed the best for the category of interest (in this case, PC_Volume > 0.15).

We can clearly say from the these charts that, even for predicting the lagged dependent variable, MARSplines performed the best as compared to the other models. We may now want to further drill down to see the percentage of predictive accuracy the model achieved on the test data.

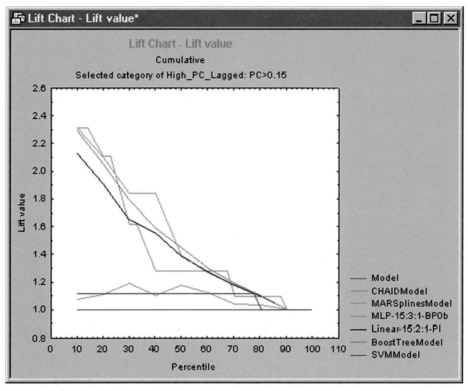

FIGURE H.9   Lift Chart: High_PC_Lagged: PC > 0.15.

Cross-Tabulation Matrix

After the predicted values for each model are generated using the Compute Best Pre-dicted Classification from All Models node, we can now run cross-tabs on the predicted and observed variables to find the accuracy rate of each model. The Cross-Tabulation node creates summary cross-tabulation tables. If categorical predictors and categorical dependent variables are selected, then two-way tables for these two lists will automatically be con-structed. Let's now try to interpret the results using cross-tabs to find the predictive accu-racy of MARSplines (see Figure H.11).

The overall hit-ratio (or overall correct prediction percentage) for the cross-validation set (or test set) using the MARSplines model for dynamic prediction is 77.40% [or (556+191)/ 965]. You can also see that the predictive accuracy for the category of interest (PC > 0.15)is approximately 64.31%, or the model will be able to predict the spikes (PC > 0.15) ahead of time with 64.31% of accuracy. Let us next check how the other models performed on the same data set.

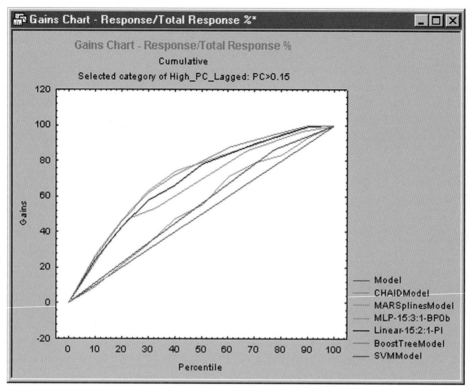

FIGURE H.10 Gains Chart: High_PC_Lagged: PC > 0.15.

Data: 2-Way Summary Table: Observed Frequencies (Final Prediction for High_...

2-Way Summary Table: Observed Frequencies (Final Prediction for Hig
Marked cells have counts > 10

High_PC_Lagged	PMML_CMARSplines29Pred for High_PC_Lagged PC<0.15	**PMML_CMARSplines29Pred for High_PC_Lagged PC>0.15**	Row Totals
PC<0.15	556	112	668
Row %	83.23%	16.77%	
PC>0.15	106	191	297
Row %	35.69%	64.31%	
Totals	662	303	965

FIGURE H.11 Two-way cross-tabulation table: MARSplines.

Comparative Evaluation of Models: Dynamic Analyses

Note to readers: Lines No. 1 and No. 2 of Table H.1, the neural networks, need to be redone with the new SANN module. Replace the NN icon in the Data Miner Workspace with the new SANN, and see what you can do; this will be a good exercise in learning.

Based on the overall accuracy, MARSplines performs the best as compared to the other models, but you may also notice that Support Vector Machines predicts the correct positives (category of interest or spikes) with 86.53% accuracy. *Which model would you choose?* If you look at the third column for Support Vector Machines, you can see that the number of false positives (or misclassified PC_Volume < 0.15) is 73.65%. For instance, if an alarm system is built in the factory, and 73.65% of the time a spike is predicted (or an alarm goes on), it could be a misclassified category of 0 (PC_Volume < 0.15). This explains why MARSplines would be the best model that could be deployed in this scenario. Another approach to reduce the false alarm would be to target only the observations that were predicted with certain levels of confidence. Table H.2 shows the hit ratio for percentile of cases (predicted using MARSplines) sorted by the confidence level of prediction.

Gains Analyses by Deciles: Dynamic Analyses

You can infer from Table H.2 that if we target more cases (sorted by confidence level), then the percentage of correct positives starts to fall, whereas the false positives tends to rise. Now taking necessary actions based on the predictions of these models totally depends on expertise and heuristics to decide the percentile of cases that have to be targeted or the confidence level to be maintained.

Next, we will discuss the third approach, transformation of change, to view from another angle the possible interactions of parameter settings that trigger changes in quality patterns.

TABLE H.1   Comparative Evaluation of Models—Dynamic Analyses

Predictive Models	Overall Accuracy	Correct Positives PC Vol > 0.15	False Positives PC Vol > 0.15
1. Multilayer Perceptron	57.20%	51.52%	40.27%
2. Intelligent Problem Solver	72.53%	63.64%	23.50%
3. Support Vector Machines	44.87%	86.53%	73.65%
4. MARSplines for classification	77.40%	64.31%	16.77%
5. Standard Classification CHAID	75.95%	47.81%	11.53%
6. Stochastic Gradient Boosting	76.68%	68.69%	19.76%

TABLE H.2   Gains Analyses by Deciles—Dynamic Analyses

Percentile	Percentile:N	Gain:N	Correct Positives	False Positives
10	96	40	100	10.71
20	192	76	96.05	11.21
30	288	104	86.54	11.96
40	384	131	80.92	11.46
50	480	152	76.32	11.59
60	576	179	72.63	12.09
70	672	203	71.43	12.15
80	768	236	68.64	13.16
90	864	265	66.42	14.36
100	966	297	64.31	16.77

Transformation of Change

A new variable was derived/calculated for the transformation of change analysis, which explains patterns in the observations of quality. The following steps explain how the new variable was derived from the dependent variable High_PC that had two categories (0 if PC_Volume < 0.15 and 1 if PC_Volume > 0.15):

1. If the PC_Volume is less than 0.15 and the following observation doesn't show any change, then that particular observation is categorized as NorCont (normal condition continuing, or 0 categories were observed after another).
2. Next, if there is a change in the pattern of NorCont or when PC_Volume goes above the cutoff level of 0.15, the particular observation that showed the spike is categorized as SpkOccr (spike occurrence or category 1 was observed after a 0 category).
3. If the PC_Volume level of the following observations are still above the cutoff level after the spike occurs, the cases are categorized as SpkCont (spike continuing or category 1 followed by another 1).
4. Next, if the SpkCont pattern changes and the process goes back to normality (PC_Volume < 0.15), then those cases/observations that showed this trend are categorized as Bck2Nor (back to normal or category 0 was observed after category 1).

Now that we have the newly defined variable, different analyses can be run to study the pattern of parameters that trigger these changes. Let's next try to understand the factor that triggers spikes (PC_Volume > 0.15). In this scenario, we first extract equal numbers of cases using the Stratified Random Sampling feature from both the category NorCont (normal continuing) and SpkOccr (spike occurrences) and then use the Feature Selection tool to select the important predictor for model building.

Feature Selection and Root Cause Analyses

The Feature Selection tool identifies Bowl_Setpoint as the most important predictor for the derived dependent variable (with two categories NorCont and SpkOccr) followed by Flow_CO2, Balling, etc. (see Figure H.12). Next, we use Interactive Trees C&RT to understand the interaction effect of the different parameter settings that led to the occurrence of spikes.

Interactive Trees: C&RT

The C&RT algorithm identifies interactions of parameter settings MP_Flow, Filler_Speed, Balling, and Filler_O2 to explain the variations in dependent variable TransChng-NorCont-SpkOccr (see Figure H.13). Rules derived from these trees can help engineers to understand the causes that trigger changes in pattern from NorCont to SpkOccr. By following the path from Node ID = 1 to Node ID = 7, we can derive a rule for category SpkOccr. We can say that if MP_Flow is < 67.55 and if Filler_Speed is < 2003.50 and if Filler O2 is > 0.038, then 77 observations were reported, from which the majority of the observations fall into the SpkOccr category. Now follow the path from Node ID = 1 to Node ID = 25 to derive a rule

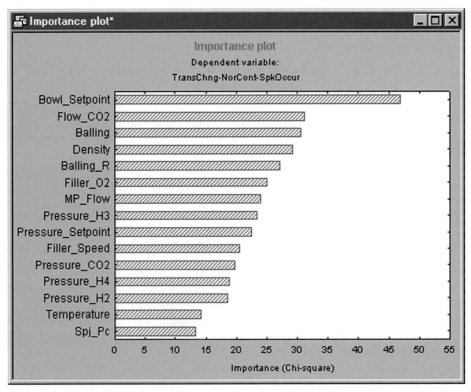

FIGURE H.12   Feature Selection: Importance plot.

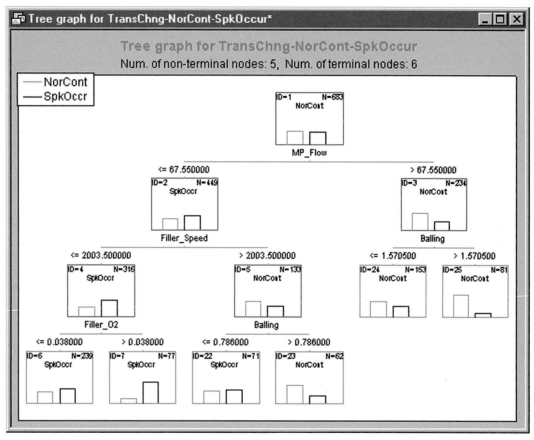

FIGURE H.13   Interactive Decision Tree: C&RT.

for NorCont. We can say that if the parameter setting for MP_Flow is maintained above 67.55 and the Balling is above 1.57, then we can maintain PC_Volume lower than 0.15. Such rules provide invaluable insights for engineers to understand the causes that affect quality, design process flows and determine optimal parameter settings, control the processes by tuning the parameter settings to maintain quality, etc.

CONCLUSION

Predictive process control, as explained here, involves the capability to monitor and control a continuous process in real time. This allows the conditions of the process to be adjusted quickly and responsively, and avoids the delay associated with monitoring only the final product. The tools and techniques demonstrated in this case should have provided you with a solid understanding of how data mining tools can be used to control and adjust parameters to maintain or enhance final quality. Engineers can also analyze these results, gaining information to take action, e.g., to make key decisions.

Three Short Tutorials Showing the Use of Data Mining and Particularly C&RT to Predict and Display Possible Structural Relationships among Data

Linda A. Miner, Ph.D.

The following tutorials show parts of three analyses in which similar data mining techniques were used to discover possible patterns in data:

1. The first example sought reliable predictors for length of stay in a medical facility—e.g., administrative concerns.
2. The second example sought to hone a questionnaire and display the relationships of the variables to each other in a clinical psychology instrument.
3. The third example used data mining to help with the validation of a questionnaire for assessing leadership training success for both business administrators and educators.

Business Administration in a Medical Industry: Determining Possible Predictors for Days with Hospice Service for Patients with Dementia

Linda A. Miner, Ph.D., James Ross, MD, and Karen James, RN, BSN, CHPN

Medicare has a set of guidelines for hospices for admitting patients with dementias to their care. The ideal number of days with the service is 6 months or less, but prognostication is difficult for the noncancer patient. The following example is part of a project that we did to determine what variables might predict length of stay and particularly which might predict a stay of ≤ 180 days.

Data were gathered for 6 years from a large hospice on patients with dementia, many of whom, had Alzheimer's disease. There were 449 cases in the data set. The following tutorial provides the steps we used while attempting to find predictors that would accurately separate the patients into the 180 days or less or the greater than 180 days group.

First, the 449 cases were separated randomly into two groups—50/50 using a random selection. We did this by first opening a Data Mining Workspace, inserting the data set and selecting classification and discrimination under the node browser, and finally, selecting the first option, Split Data into Training and Testing Sets (see Figures I.1 and I.2).

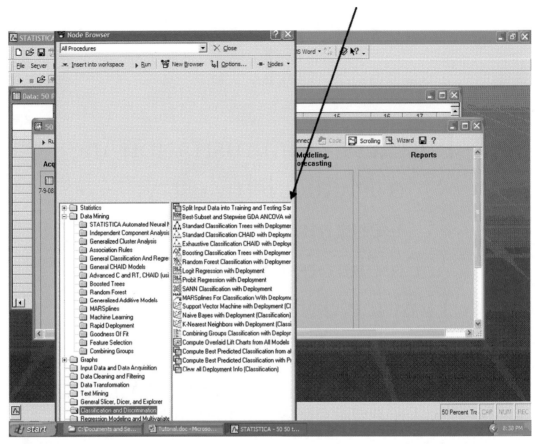

FIGURE I.1   Select Split Input Data into Training and Testing Sets from the node browser in the Data Miner Workspace.

FIGURE I.2   Split Input Data node in the Data Preparation, Cleaning, Transformation panel of the Data Miner Workspace.

Then we right-clicked on the node to edit the parameters (see Figure I.3).

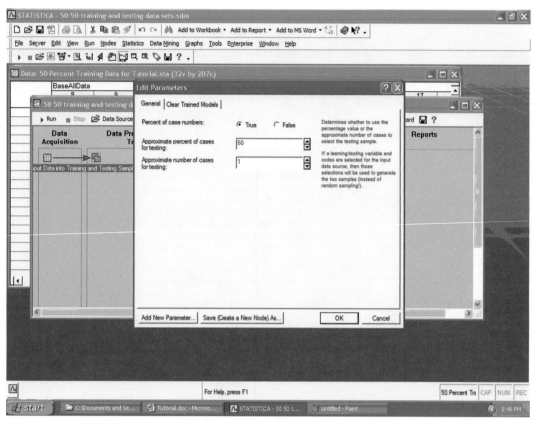

FIGURE I.3   Edit Parameters dialog.

The default is a 50/50 random split. We used that. After we clicked OK, the two data sets were formed, as shown in Figure I.4.

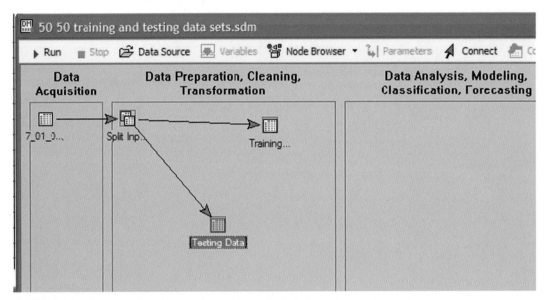

FIGURE I.4   Train and Test data sets formed from the Split Input Data node.

The first group was called the Training data set and second was the Testing or Holdout data set. Training data were analyzed in an exploratory manner, seeking the variables that seemed most predictive. Then these variables were applied to the testing set. This tutorial involves only the Training data set.

The dependent variable was the length of stay (variables 31 and 32) with the hospice and was used in two forms: a discrete variable of ≤180 days versus >180 days with the service until death, or the actual number of days with the service until death (see Figure I.5).

Data: 50 Percent Training Data for Tutorial.sta (32v by 207c)

	26 Referral Group	27 First Admit	28 Last Status Date	29 Age at Admission	30 Age at "last status"	31 Days from First Admit until Death	32 180 days
1		38475	39030	72.260274		556	>180
2		37929	38004	81.2465753		76	<=180
3		39148	39151	92.1260274		4	<=180
4		38394	39092	84.4821918		699	>180
5		39334	39345	76.5835616		12	<=180
6		38301	38402	84.939726		102	<=180
7		38169	38500	93.0986301		332	>180
8		37574	37664	76.1780822		91	<=180
9		37518	38098	72.5945205		581	>180
10		38453	38463	93.3424658		11	<=180
11		38028	38040	82.739726		13	<=180
12		38268	38353	93.430137		86	<=180
13		38852	38953	93.8136986		102	<=180
14		38217	38225	94.4657534		9	<=180
15		38525	38540	72.9835616		16	<=180
16		38846	38926	84.8575342		81	<=180
17		38797	38810	82.9972603		14	<=180
18		37524	37531	92.8356164		8	<=180
19		37763	37765	86.1616438		3	<=180

FIGURE I.5   Data set for the 50 Percent Training.

The independent variables were gender, BMI, PPS, FAST, Coronary, COPD, Delirium, Pressure, Aspiration, Pneumonia, UTI, Septicemia, Fever Recurrent, Use of Foley, Location at Referral, Marital Status, and Age (Variables 7–18, 20–22, and 29). The variable Location at Referral was not the domiciles of the patients, but rather where they were located at the time they were referred. For example, someone might have been hospitalized even though living at home. The Location at Referral would then be listed as hospital. BMI and PPS were both continuous variables, as was Age. The rest were categorical variables. The variables included those mandated by Medicare for providing services to Alzheimer's patients.

Working first with the training data, we did a stepwise multiple regression using the number of days (variable 31) as the dependent variable (continuous variable) and variables 7–18, 20–22, and 29 as the independent variables (see Figure I.6). We used multiple regression rather than feature selection because it was a more powerful procedure than feature

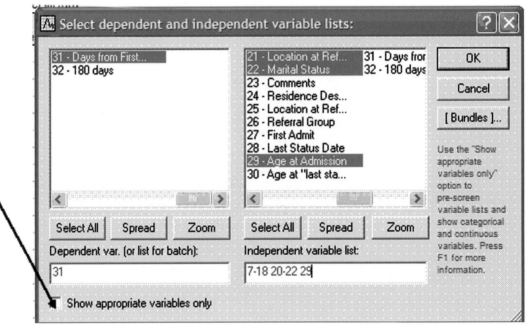

FIGURE I.6	Variables selected.

selection although the latter procedure was used later. Given the relatively small data available for this tutorial, we still went ahead and used it so we could get a beginning answer as to which variables might be important for pre-modeling the data. We left the appropriate variables box unchecked and told the program to continue with the current selection (see Figure I.7).

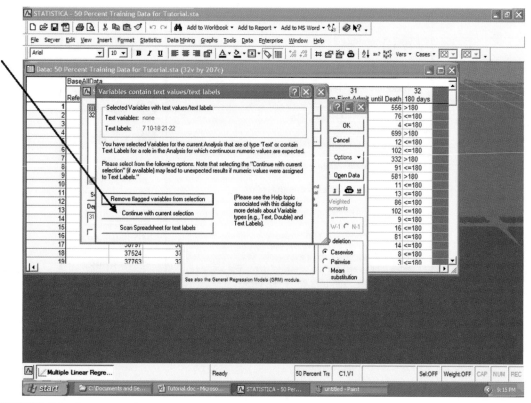

FIGURE I.7   Continue with current selection.

Under the Advanced tab, we selected a forward stepwise model, as shown in Figure I.8, and then clicked OK.

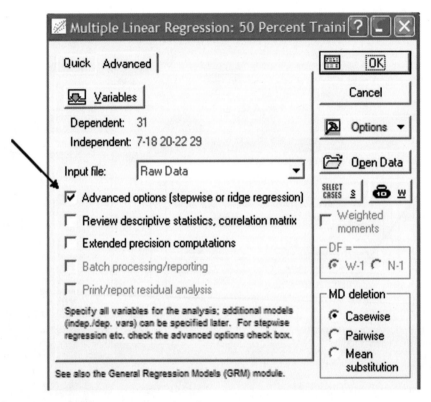

FIGURE I.8   Forward stepwise regression selected.

The result in Figure I.9 emerged.

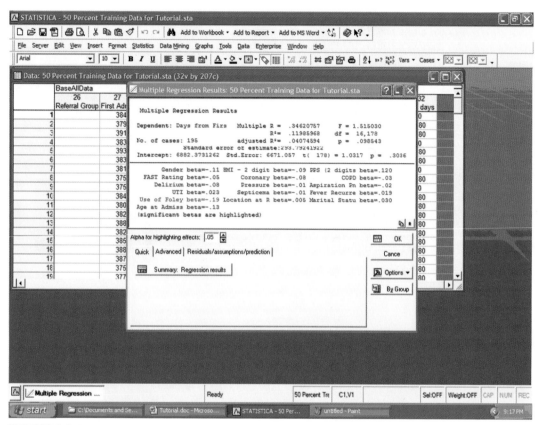

FIGURE I.9   Results dialog from Multiple Regression Analysis.

The summary is shown in Figure I.10.

N=195	Beta	Std.Err. of Beta	B	Std.Err. of B	t(178)	p-level
Intercept			6882.373	6671.057	1.03168	0.303623
Gender	-0.106453	0.075637	-71.593	50.868	-1.40742	0.161048
BMI - 2 digits	-0.094948	0.075722	-5.844	4.660	-1.25390	0.211521
PPS (2 digits)	0.119715	0.086710	4.130	2.991	1.38064	0.169119
FAST Rating	-0.054265	0.082364	-10.385	15.762	-0.65885	0.510847
Coronary	-0.084253	0.074279	-54.618	48.152	-1.13428	0.258201
COPD	-0.026160	0.073986	-26.394	74.646	-0.35358	0.724070
Delirium	-0.075708	0.074492	-51.543	50.715	-1.01633	0.310853
Pressure	-0.009155	0.076611	-6.663	55.759	-0.11949	0.905019
Aspiration Pneumonia	-0.017893	0.073810	-14.838	61.207	-0.24242	0.808730
UTI	0.023284	0.076906	14.478	47.820	0.30275	0.762431
Septicemia	-0.006604	0.075859	-7.921	90.989	-0.08705	0.930727
Fever Recurrent	0.018871	0.075216	19.506	77.746	0.25089	0.802187
Use of Foley	-0.190707	0.079579	-173.682	72.475	-2.39646	0.017590
Location at Referral	0.004994	0.074940	2.272	34.085	0.06664	0.946940
Marital Status	0.029784	0.075326	9.697	24.524	0.39541	0.693015
Age at Admission	-0.126555	0.074791	-5.353	3.163	-1.69211	0.092374

Regression Summary for Dependent Variable: Days from First Admit until Death (50 Percent Training Data for Tutorial.sta
R= .34620757 R²= .11985968 Adjusted R²= .04074594
F(16,178)=1.5150 p<.09854 Std.Error of estimate: 293.79

FIGURE I.10   Summary spreadsheet from Regression Analysis.

The variable highlighted in red, Use of Foley, was the one deemed probably most important to the model and would most certainly be selected for the further data mining analysis. Because we mixed continuous and categorical variables, we did a feature selection just to check this result. We opened a Data Mining Workspace, copied in the data, and selected the variables of interest. Under the Node Browser, we selected Feature Selection and Root Cause Analysis. We inserted it into the workspace (see Figure I.11).

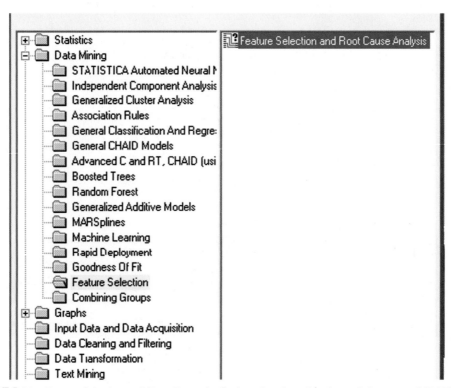

FIGURE I.11   Feature Selection and Root Cause Analysis node selected in the node browser of *STATISTICA*.

We right-clicked on the node and chose Edit the Parameters, as shown in Figure I.12.

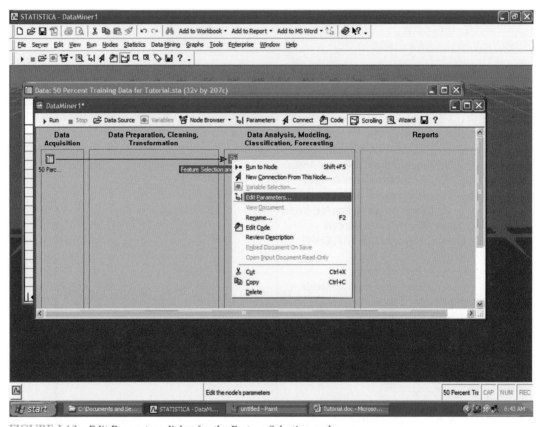

FIGURE I.12   Edit Parameters dialog for the Feature Selection node.

The only thing we changed in the Edit Parameters dialog shown in Figure I.13 was All Results rather than Minimal.

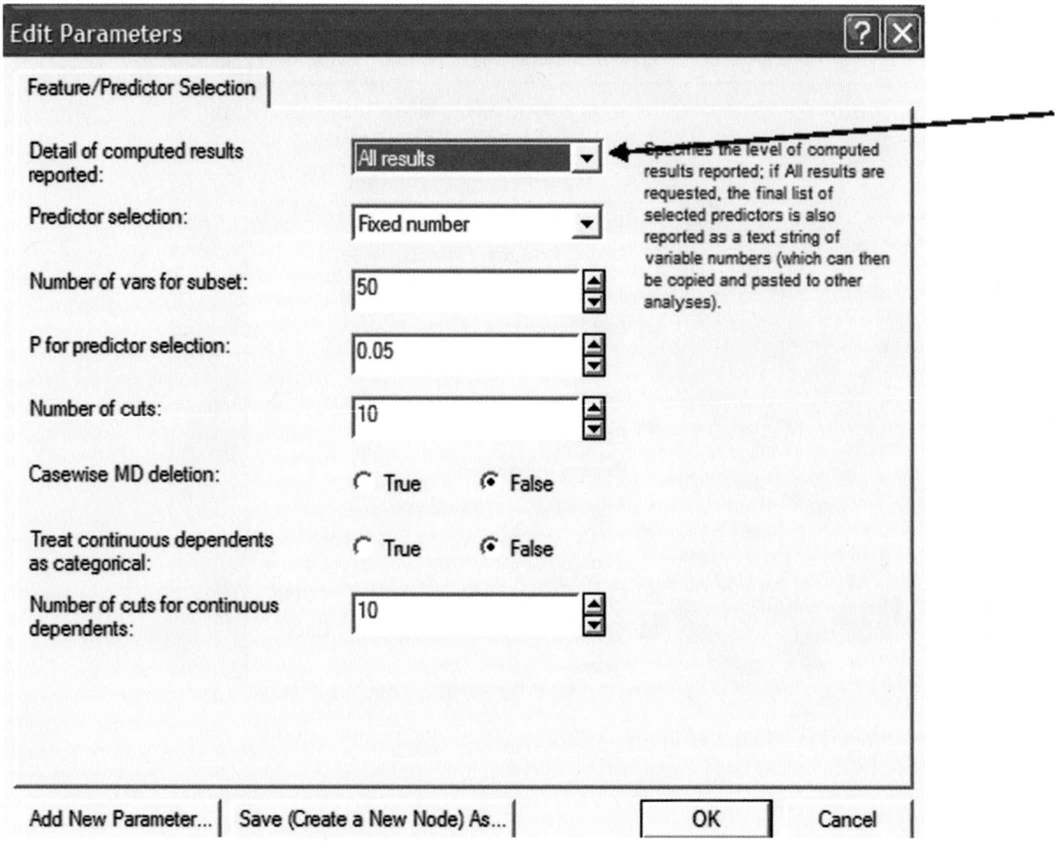

FIGURE I.13 All Results selected and other parameters left at default values.

We then right-clicked and selected Run to Node, as shown in Figure I.14.

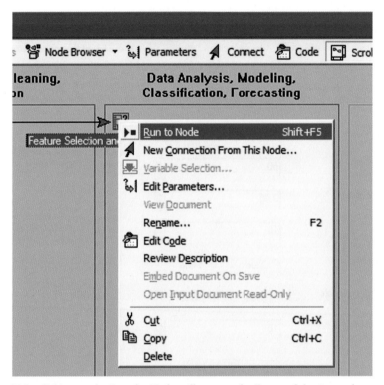

FIGURE I.14   Right-clicking on the Run the Node will execute the Feature Selection node.

To make sure we could open the output later, we also selected Embed Document on Save, as shown in Figure I.15.

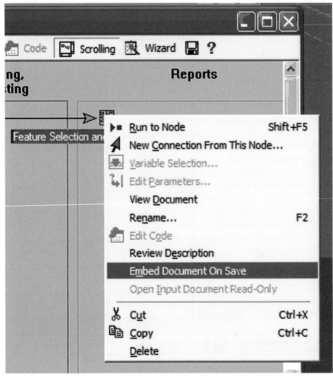

FIGURE I.15 Select Embed Document on Save to keep the results of Feature Selection saved in the Data Miner Workspace project model file.

Double-clicking the output showed the list in Figure I.16.

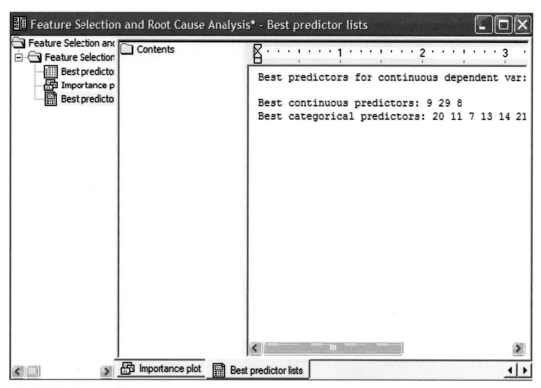

FIGURE I.16   List of Best Predictors obtained from Feature Selection.

We clicked on the importance plot to see a view of the relative importance of the independent variables to the dependent variable, as shown in Figure I.17.

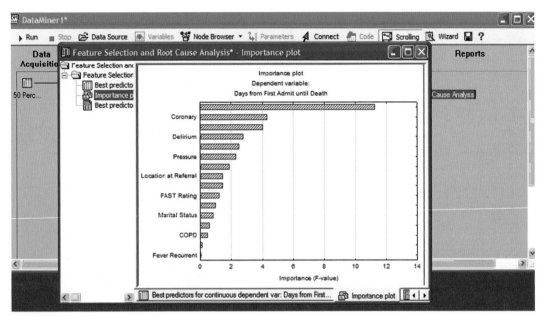

FIGURE I.17   Importance Plot from Feature Selection.

The top bar was Use of Foley and, once again, seemed to be an important variable from both techniques of exploring the data—at least for the training data. The listing of variables and p-values is shown in Figure I.18. However, it is important to note that this procedure is not a hypothesis test and should not be viewed as such. It is only a pattern-seeking procedure.

	Best predictors for continuous dependent var: Days from First Admit until Death	
	F-value	p-value
Use of Foley	11.27134	0.000940
Coronary	4.33031	0.038694
Gender	4.00465	0.046694
Delirium	2.76289	0.098017
PPS (2 digits)	2.48301	0.045015
Pressure	2.30410	0.130589
Age at Admission	1.88462	0.056131
Location at Referral	1.45114	0.236723
Septicemia	1.42929	0.233281
FAST Rating	1.20925	0.306158
Aspiration Pneumonia	0.98382	0.322438
Marital Status	0.83748	0.502801
BMI - 2 digits	0.59540	0.780980
COPD	0.46401	0.496534
UTI	0.10123	0.750688
Fever Recurrent	0.05358	0.817188

FIGURE I.18   Best Predictors from Feature Selection in descending order of F-value.

We also did a feature selection using the discrete ($\leq$180 days versus >180 days) variable 32. Figure I.19 shows the importance plot from that analysis in which seven of the variables were identified by the procedure.

We decided to concentrate on the discrete variable as the dependent variable for the subsequent analyses.

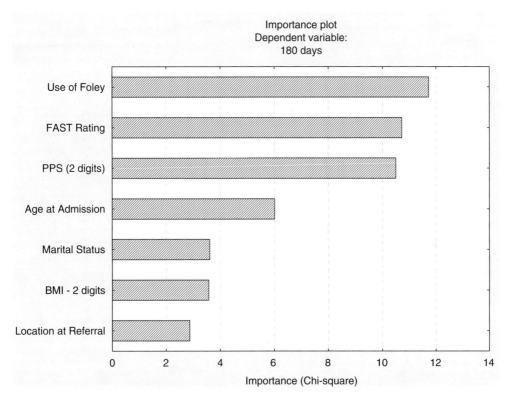

FIGURE I.19   Importance Plot using only variable 32 as the target (e.g. dependent) variable.

Next, we ran a Data Mining Recipe using 180 days (the discrete variable) as the dependent variable and all the variables (7–18, 20–22, and 29) as above for the independent variable (see Figure I.20).

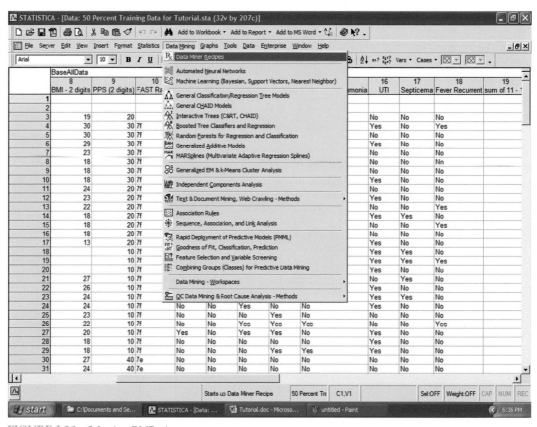

FIGURE I.20   Selecting DMRecipe.

Next, we selected New in the Data Miner Recipes dialog, as shown in Figure I.21.

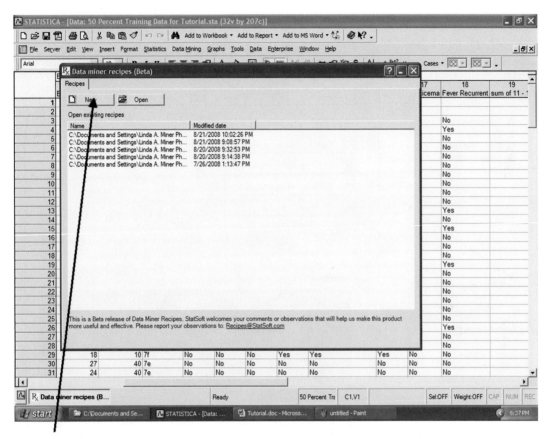

FIGURE I.21   New button in DMRecipe must be clicked to start a new project.

Then, when the new screen emerged, we clicked to place a check in the Configure All Steps checkbox, as shown in Figure I.22.

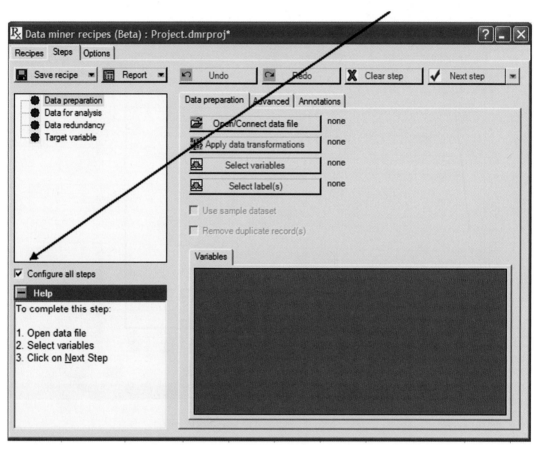

FIGURE I.22   Selecting Configure All Steps.

Next, we connected the data file, as shown in Figure I.23.

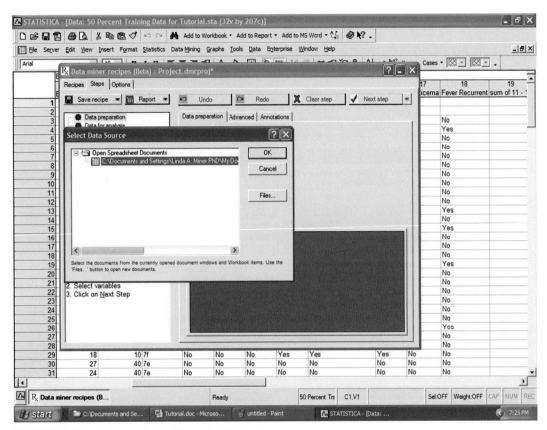

FIGURE I.23   Selection of data file to the DMRecipe.

Then we selected the variables in the Select Variables dialog, as shown in Figure I.24.

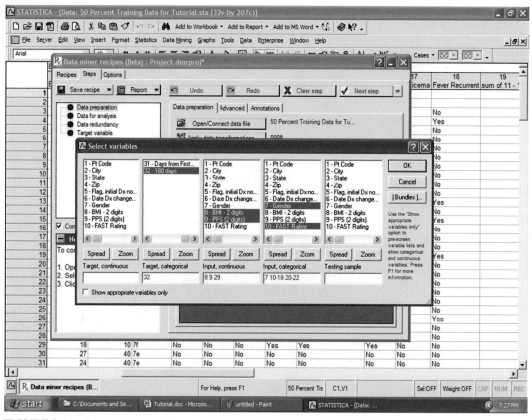

FIGURE I.24   Selection of variables.

After we clicked OK, the variables showed up in a box (see Figure I.25). The target is the dependent variable.

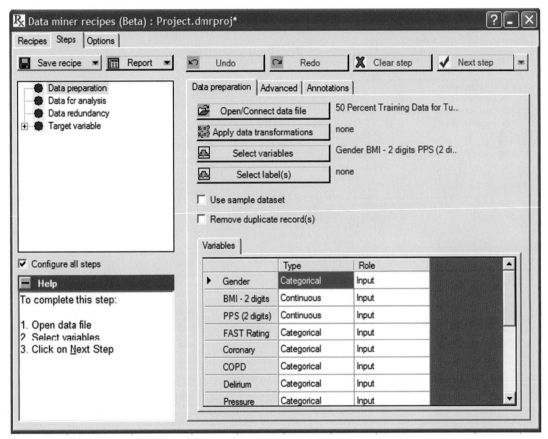

FIGURE I.25   After selecting variables and clicking OK on variable section dialog, the variables show on the DMRecipe screen.

We next clicked on Target Variable (see Figure I.26) and then on 180 Days (see Figure I.27).

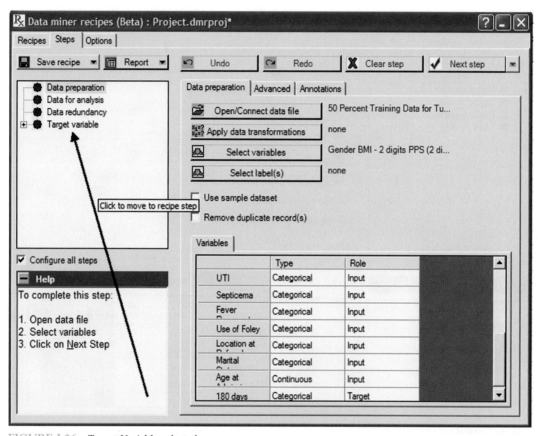

FIGURE I.26   Target Variable selected.

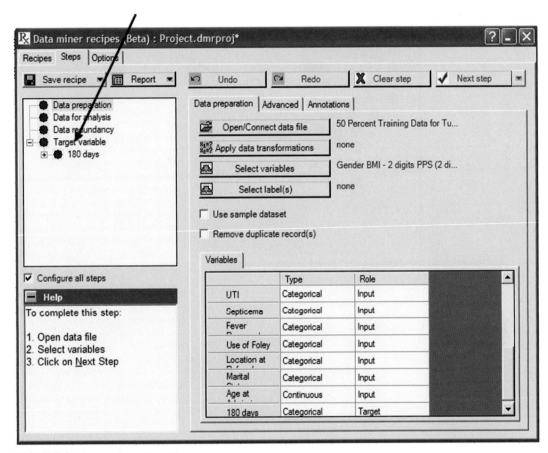

FIGURE I.27   180 days selected.

We then clicked on Model Building to select the models we wanted the recipe to compute (see Figure I.28).

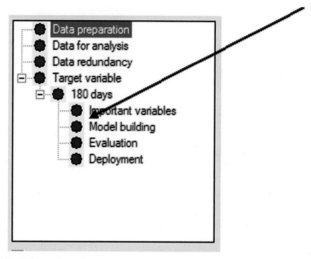

FIGURE I.28   Model Building selected.

We were mainly interested in C&RT so that we could visualize the relationships, but we wanted to know if other models could result in predictions that were more accurate. We chose C&RT, boosted trees, and SVM (see Figure I.29).

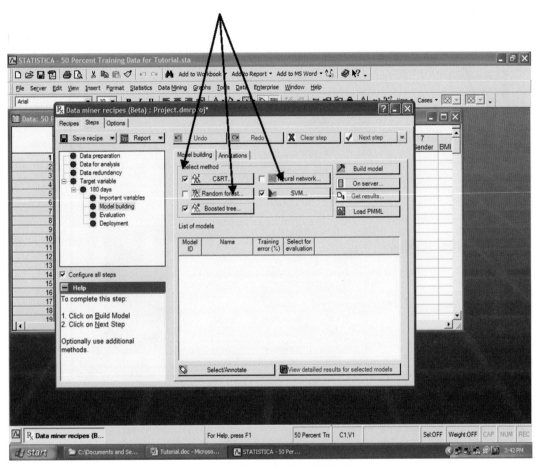

FIGURE I.29   C&RT, Boosted Trees, and SVM (Support Vector Machines) selected.

Finally, we clicked to remove the check mark from Configure All Steps, and then under the Next Step tab, we chose Run to Completion (see Figure I.30).

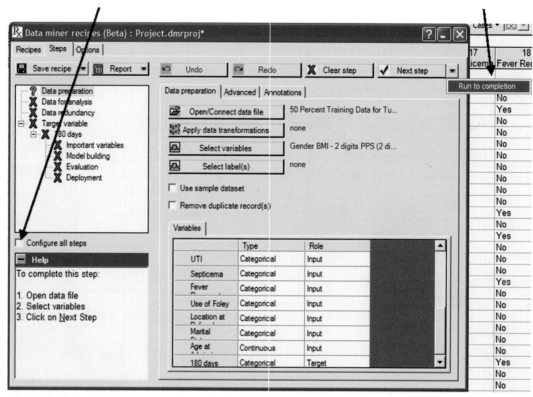

FIGURE I.30   Unselection of Configure All Steps.

After the program had run, we opened the summary report, as shown in Figure I.31.

The error rates were lowest for the SVM, so it might produce the best model in the end. However, the C&RT provided decision trees and was the second best prediction model. We opened it in the Data Miner Recipe (DMR) to reveal the output.

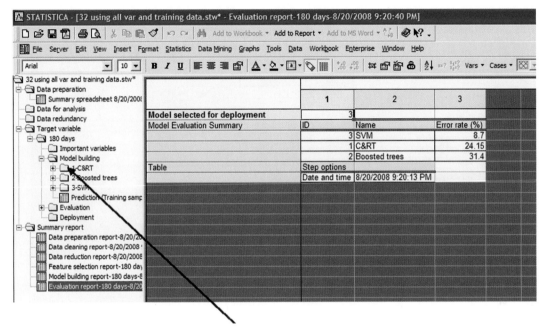

FIGURE I.31   Results workbook of DMRecipe with the Evaluation Report (e.g. Summary Report) selected so that the summary appears in the right-side window. SVM had the lowest error rate of 8.7%. The Accuracy Rate of the models is (100 − error rate), thus 91% for SVM.

The prediction accuracies were displayed in the cross-tabulation. The >180 days was accurately predicted about 76% of the time and the ≤180 correctly predicted about 75% (see Figure I.32).

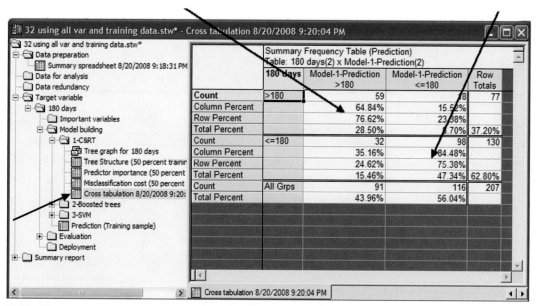

FIGURE I.32   Prediction accuracies displayed in the cross tabulation.

Next, we examined the decision trees (see Figure I.33).

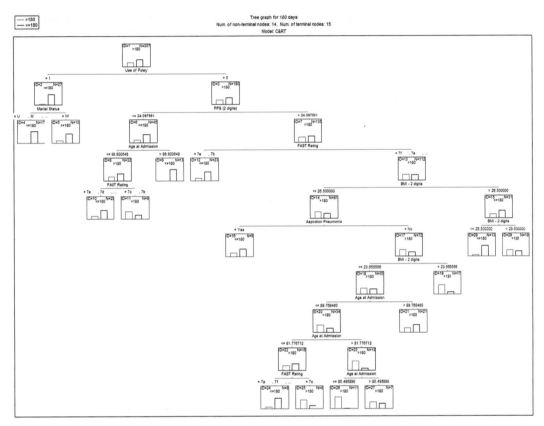

FIGURE I.33   Decision trees result from C&RT in *STATISTICA* Data Miner.

The decision tree in Figure I.34 shows fewer rows, so it was easier to view.

Again, the use of the Foley and then Age at Admission and some form of functioning (PPS and FAST) showed up, as well as Marital Status.

Additional analyses were done for these data. Additional C&RT analyses were completed. After all the training analyses were completed, the most promising variables were selected, new analyses were run on the training data, and then the same analyses were completed for the testing data. The outputs were compared.

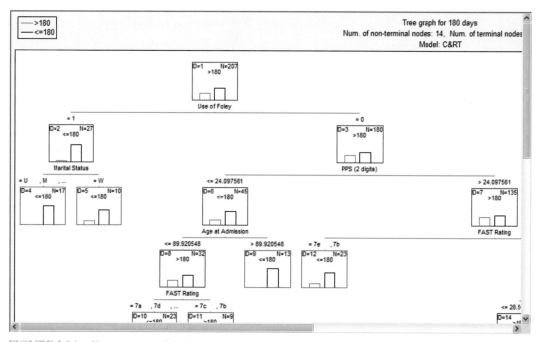

FIGURE I.34   Close-up view of decision trees.

Clinical Psychology: Making Decisions about Best Therapy for a Client:
Using Data Mining to Explore the Structure of a Depression Instrument

David P. Armentrout, Ph.D. and
Linda A. Miner, Ph.D.

The original data set for this example had 359 cases. The intent of the instrument was to measure various components of depression as an aid to practitioners as they organize the therapy of a client. It was important that the instrument did, in fact, measure depression and that it did so reliably. There were 164 questions in the original survey.

To determine the structure, we followed these procedures. A factor analysis first grouped the questions into meaningful groups. We thought that people suffering from depression would appreciate a smaller survey rather than a longer one. Feature selection indicated which of the individual questions were most important to each factor grouping. We reduced the number of questions in the survey by eliminating the questions that contributed the least to each factor. By finding the questions that most associated with their meaningful groups, we shortened the survey while retaining valuable information. The resulting data set produced the one that we used for this tutorial, which illustrates part of our procedures. We wanted to focus on the structure of the relationships of variables, not predicting the amount of depression. In fact, in predicting who was depressed and who was not, we found that simply asking the question "Are you depressed, yes or no?" was quite good at predicting. Using the entire data set (variable numbers given below for the training data provided), we selected mean with error plots to view this relationship. We used four depression instruments—the Zung, the PHQ-9, the Beck Depression Inventory, and the CES-D (variables 199, 201, 202, and 203 in the training data)—as the dependent variables and question 161 as the independent variable (see Figure J.1).

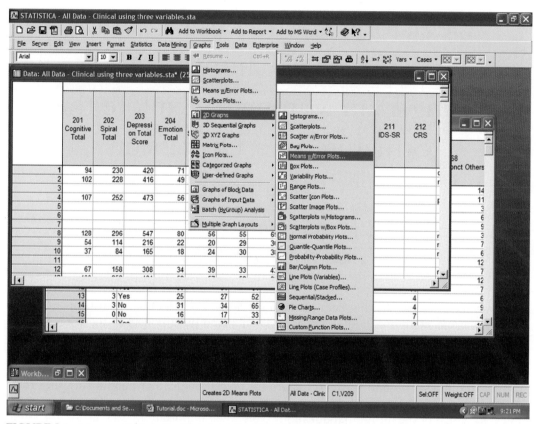

FIGURE J.1   Means with Error Bar Plots from the *STATISTICA* graphs → 2D Graphs menu.

We left the defaults as they were (see Figure J.2).

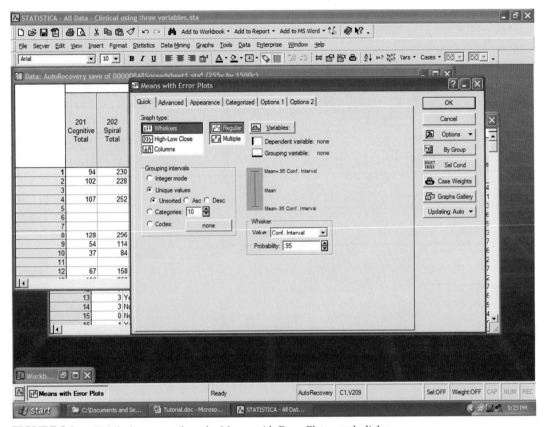

FIGURE J.2   All defaults accepted on the Means with Error Plots graph dialog.

We selected the variables, as shown in Figure J.3, and clicked OK.

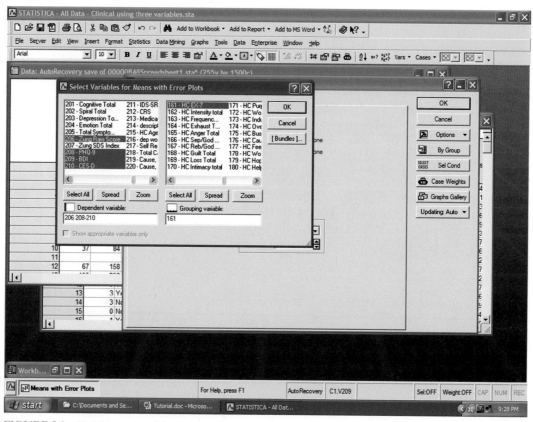

FIGURE J.3   Variables selected from data set.

The resulting four graphs showed us the huge differences in scores between the yeses and the nos (see Figure J.4).

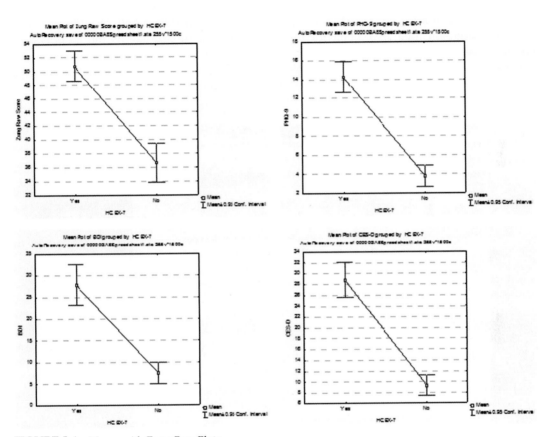

FIGURE J.4   Means with Error Bars Plots.

We wanted to determine the pathways that might exist in the data among the subscales of the depression inventory. For this task, we decided to run a series of data mining recipes on the training data (provided), which would give us C&RT decision trees and the prediction accuracies as well. We also wanted to decide exactly which procedures would be most informative, and therefore, we once again separated the data into training and testing groups by using the random split module.

The first data mining recipe set the yes/no question as the dependent variable and variables 182, emotion total; 190, behavior total; and 196, cognitive total, as the independent variables (see Figure J.5).

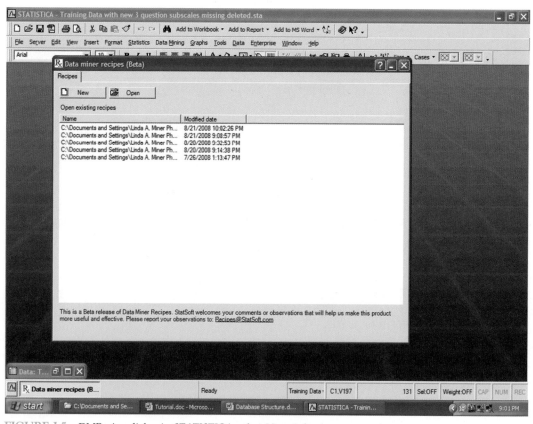

FIGURE J.5   DMRecipe dialog in *STATISTICA*; select New to begin a new project.

We selected a new recipe, as shown in Figure J.6.

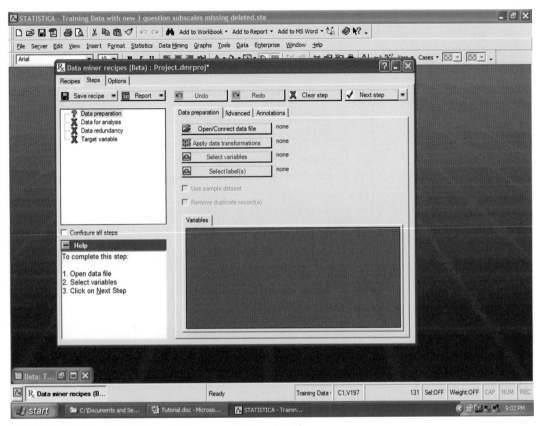

FIGURE J.6   New DMRecipe window; no variables selected.

Next, we clicked Configure All Steps, as shown in Figure J.7.

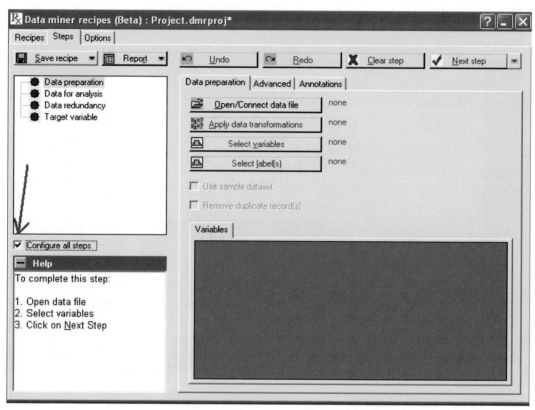

FIGURE J.7   Configure All Steps selected.

Next, we placed the data set into the recipe, as shown in Figure J.8.

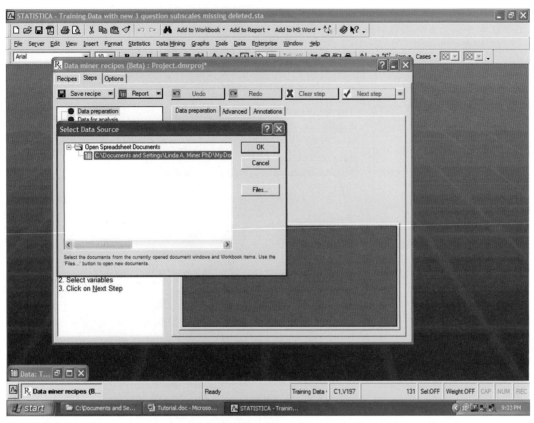

FIGURE J.8   Selecting data set; this window obtained by clicking on the Open/Connect Data Source.

Next, we selected variables (see Figures J.9 and J.10).

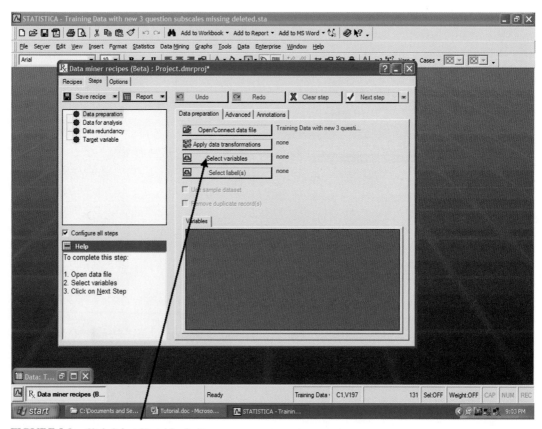

FIGURE J.9   Click Select Variables button.

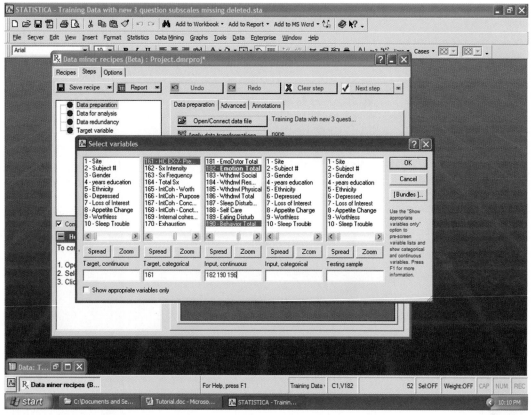

FIGURE J.10   Variables selected.

We clicked OK. The variables were then displayed (see Figure J.11).

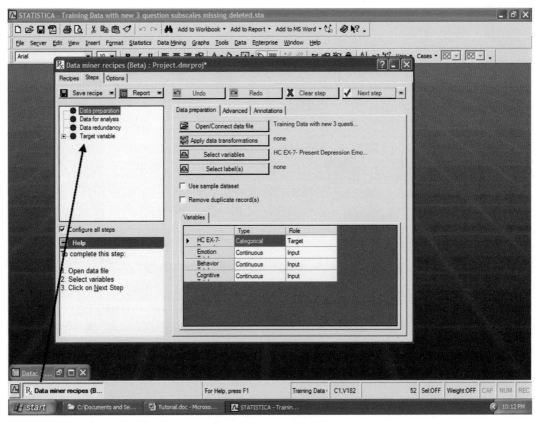

FIGURE J.11   Target Variable "+" sign clicked to open Tree Outline.

We clicked on Target Variable, then HC EX-7 (the dependent variable 161), and finally on Model Building (see Figure J.12).

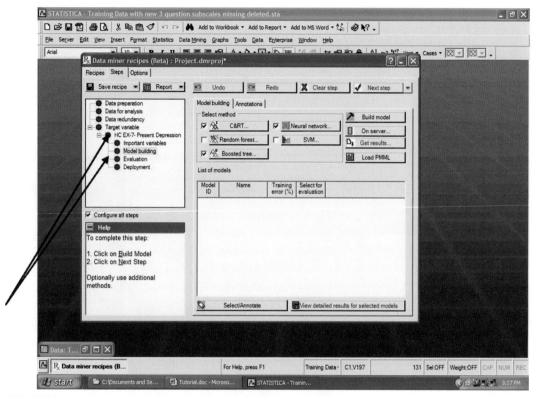

FIGURE J.12   Next, click on HC EX-7 in outline and then Model Building.

We unchecked everything except for C&RT, as shown in Figure J.13.

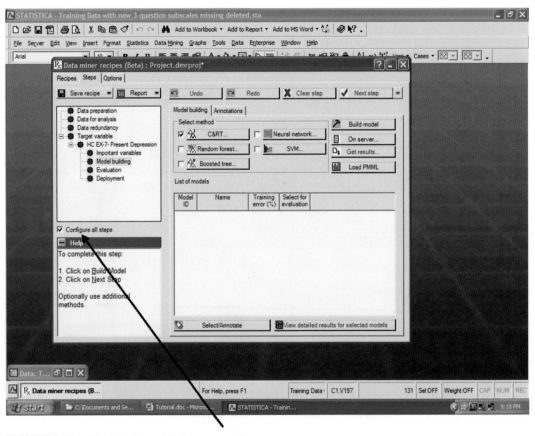

FIGURE J.13   Click Configure All Steps to unselect.

We removed the check mark from Configure All Steps. and then clicked Run to Completion (under the Next Step tab), as shown in Figure J.14.

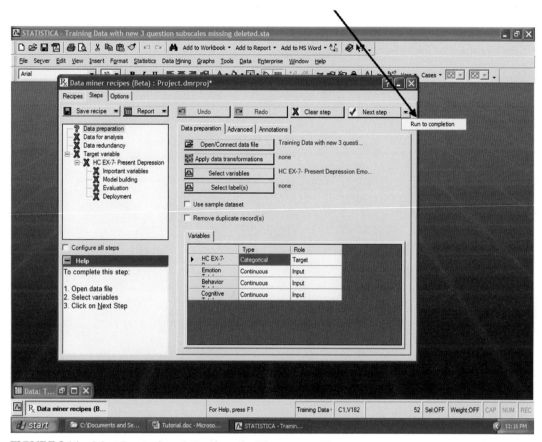

FIGURE J.14   Select Run to Completion from the "down arrow" button.

Then we let it run its course. Figure J.15 shows the status bar.

FIGURE J.15   DMRecipe processing.

The first screen of the results let us know there was an error rate of 13.08%.

We enlarged the screen and clicked on Evaluation and Crosstabulation, which then showed us the accuracies for prediction of the decision trees (see Figure J.16).

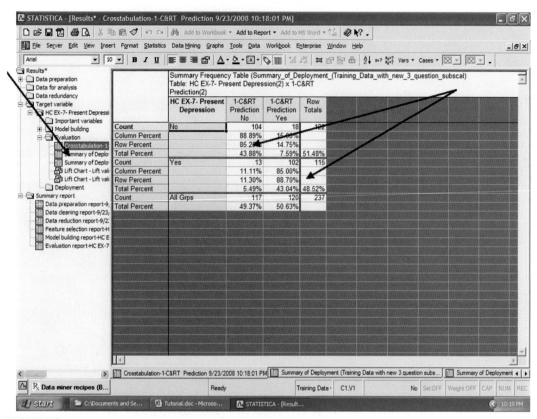

FIGURE J.16   DMRecipe results workbook with Crosstabulations selected.

About 85% of the nos were predicted, and almost 89% of the yeses were correctly predicted.

We viewed the predictor importance spreadsheet, shown in Figure J.17, to find that the behavior total seemed to be the most important.

Next, we looked at the decision tree to see how the variables might have been related (see Figures J.18 and J.19). It is found under Model Building and then 1–C&RT.

We will explain the first three levels. The behavior total was the most important and formed the trunk of the tree. There were 237 cases in the trunk with a few more nos (in

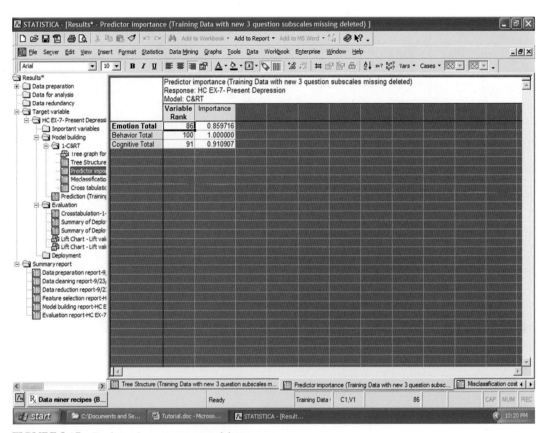

FIGURE J.17   Predictor Importance spreadsheet.

red) than yeses (in black). The behavior score was split at 40.75. One hundred twenty-two cases had behavior scores of 40.75 or less. Of those, most were nos. One hundred fifteen cases scored more than 40.75, and of them most were yeses.

Looking at those who scored 40.75 or less on behavior, we saw that the next split was on the cognitive total. For those that scored less than or equal to 40.5, most were nos. For those that scored above 40.5 on cognitive, all were yeses. But there were only five more captured this way.

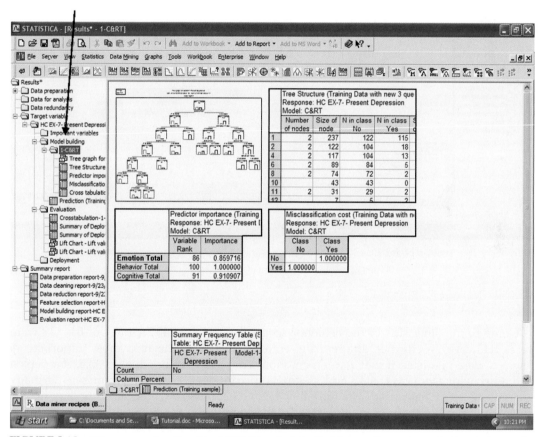

FIGURE J.18   When a superheading, like 1-CR&T is selected, all of the results are shown in the window to the right. Each result can be shown separately by clicking that selection under the 1-CR&T heading.

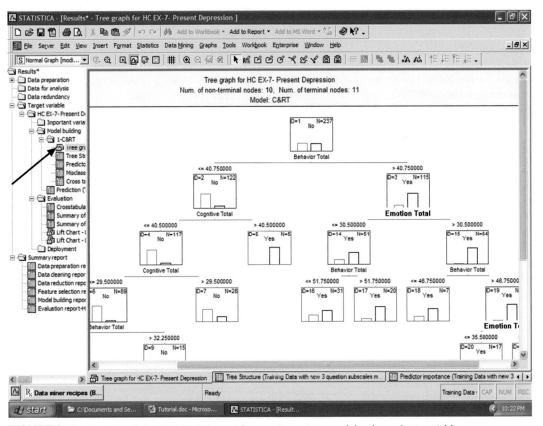

FIGURE J.19   Tree Graph for HC EX-7 Present Depression category of the dependant variable.

Going to the other side of the behavior scores, greater than 40.75, most of the 118 were yeses. However, to further differentiate, emotion total was the next most important. A few nos were predicted if they had an emotion score less than or equal to 30.5. If a person had a behavior score over 40.75 and an emotion score over 30.5, that person was virtually guaranteed to be depressed. On the other hand, if a person scored 40.75 or less on the behavior scale, less than or equal to 40.5 on the cognitive scale, he or she was most likely not depressed.

Education–Leadership Training for Business and Education Using C&RT to Predict and Display Possible Structural Relationships

Greg S. Robinson, Ph.D., Linda A. Miner, Ph.D., and Mary A. Millikin, Ph.D.

The following tutorial was part of our efforts at examining leadership patterns among business students at a university while exploring relationships between concepts measured by several instruments. The first instrument, the Collaborative Leader Profile (Robinson, 2004) measured learning, adaptability and open collaboration, necessary for differentiated leadership (scales: v126–130; total v131). The second instrument (Millikin and Miner, 1995) measured risk-taking behavior (v134) and social desirability (v135), and the third instrument was the State Trait Anxiety Scale (v133) (Spielberger, 1983). More differentiated people tend to have less anxiety in their interpersonal relationships (Thorberg and Lyvers, 2005), so we would expect to see an inverse relationship between those scores. It was difficult to know how fear of intimacy might relate to risk-taking behavior, but we were interested in investigating the relationship. Differentiation was measured by The Differentiation of Self Inventory (DSI; v132) (Skrowron and Friedlander, 1998).

The variables comprising the Collaborative Leader were 111–126. To find the reliability of the questions, we did a Cronbach Alpha. First, we selected Multivariate Exploratory Techniques from the Statistics menu in *STATISTICA* and then chose Reliability/Item Analysis, as shown in Figure K.1.

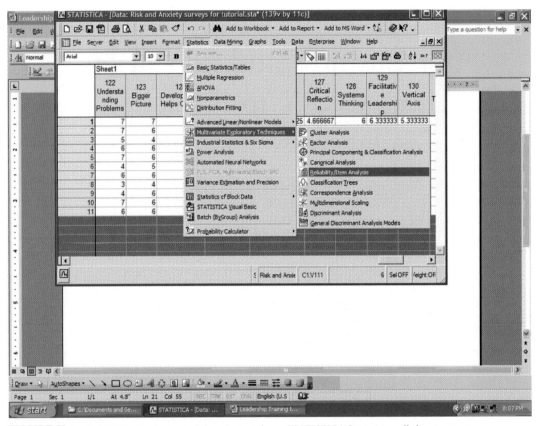

FIGURE K.1   Reliability Analysis module selection from *STATISTICA* Statistics pull-down menu.

We put in the variables and clicked OK on dialog boxes, until the Result Dialog Box popped up, showing the Cronbach statistic; this is illustrated in the following few figures (see Figure K.2–K.6).

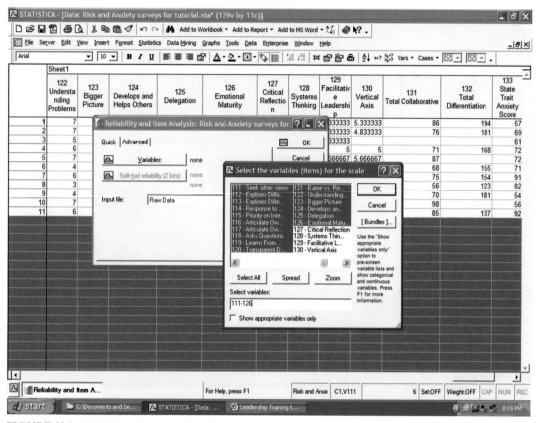

FIGURE K.2   Variable selection.

After clicking OK twice, we found the results shown in Figure K.3.

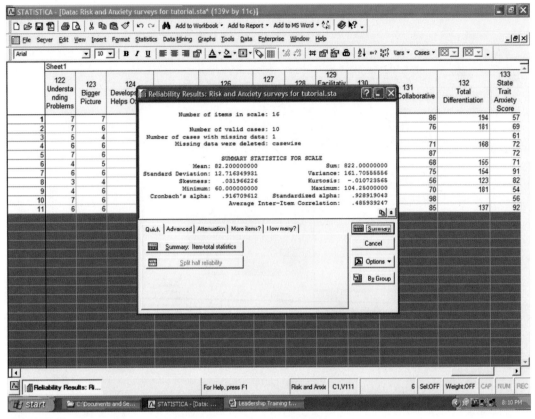

FIGURE K.3   Reliability Results dialog after running the module.

We found that the Cronbach Alpha was 0.9167 and the standardized alpha was 0.9289. This was good reliability, although we wondered if it was too high and perhaps indicated too much colinearity. By clicking the Advanced tab, we could find other options, as you can see in Figure K.4.

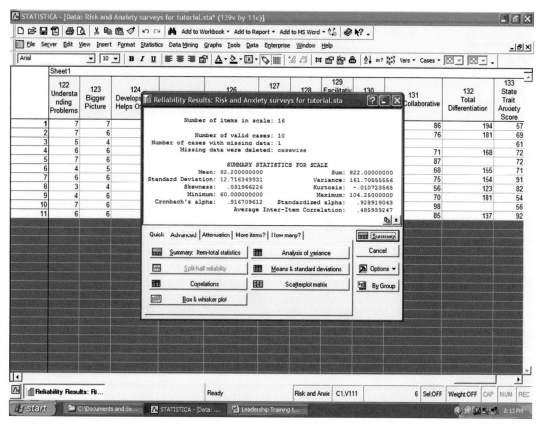

FIGURE K.4   Advanced tab selections in the Reliability Analysis Results windows.

Clicking Summary, Item-Total statistics gave us the outcome shown in Figure K.5.

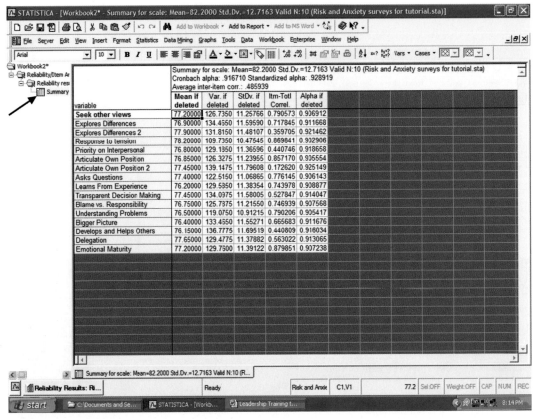

FIGURE K.5   Reliability Item Analysis summary spreadsheet.

The Correlations button let us see how the items were correlated one to another (see Figure K.6).

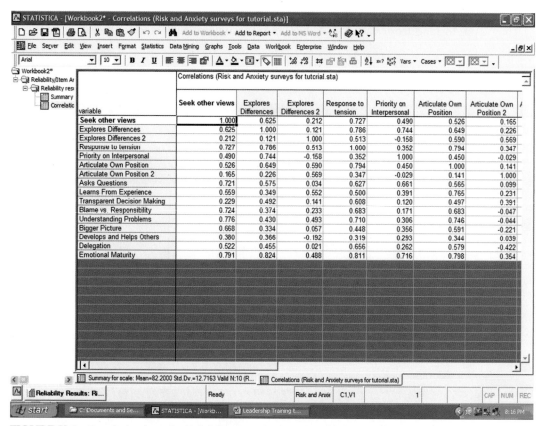

FIGURE K.6   Correlation from the Reliability Item Analysis Results dialog.

We next selected Factor Analysis, as shown in Figure K.7, to see if the clusters were what was meant as the instrument was written (i.e., variables 127–130):

Emotional Maturity   Critical Reflection   Systems Thinking   Facilitative Leadership   Vertical Axis

They were formed from variables 111–125, so these were subjected to the factor analysis.

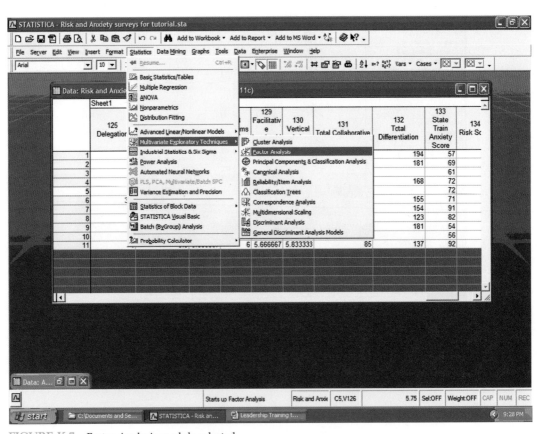

FIGURE K.7   Factor Analysis module selected.

We selected our variables, as shown in Figure K.8.

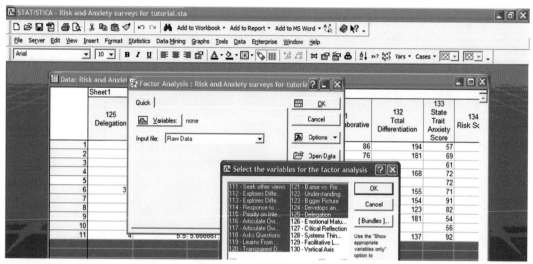

FIGURE K.8   Variables used for Factor Analysis.

Then we clicked OK twice to reach the window shown in Figure K.9.

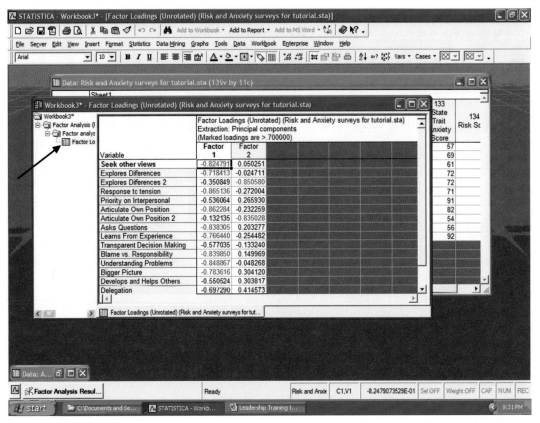

FIGURE K.9   Factor Loadings from Factor Analysis; Unrotated format.

There were two factors identified using the raw data default and unrotated (see Figure K.10). Choosing varimax rotation made no real difference.

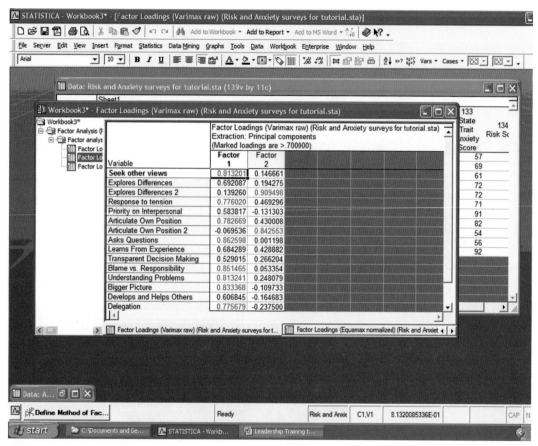

FIGURE K.10   Factor Loadings using Varimax Raw rotation of factors.

Still there were only two factors. These were not the planned subscales. Numbers do not always follow theory. However, there were so few cases that it would be hard to know what the true structure of the instrument was from these numbers. We persevered for the sake of the tutorial.

We decided to use the individual questions in a DMR, especially looking at the C&RT to see how those questions might interrelate with the risk survey on the anxiety variable.

First, we opened a new Data Mining Recipe and our data (see Figure K.11).

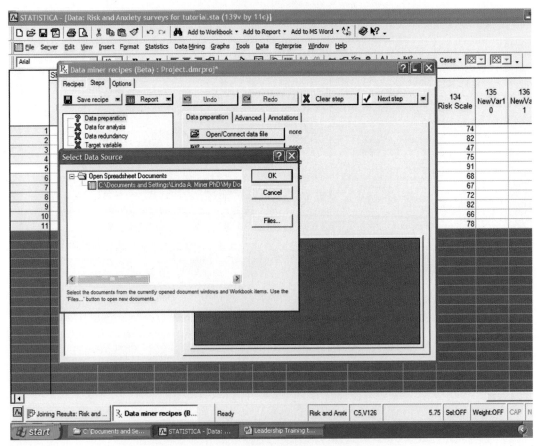

FIGURE K.11   Select Data Source window from DMRecipe.

We selected our variables, as shown in Figure K.12.

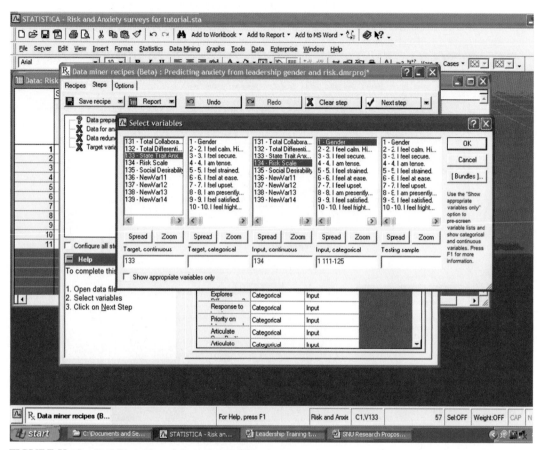

FIGURE K.12   Variables selected from the DMR analysis.

We selected Configure All Steps and then clicked on Target Variable so that we would be able to select our procedures (see Figure K.13).

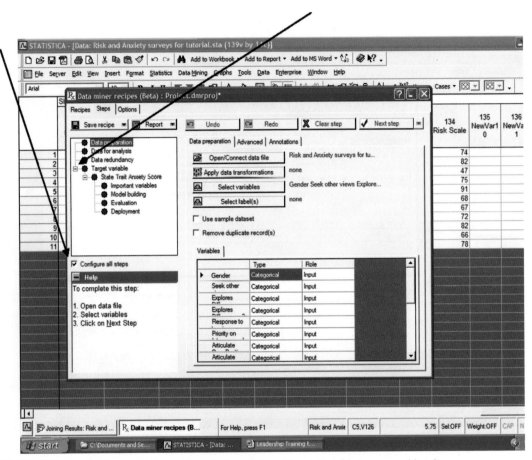

FIGURE K.13   Configure All Steps selected, then click on Target Variable to open subheadings.

Then we selected C&RT and Boosted Trees, as shown in Figure K.14.

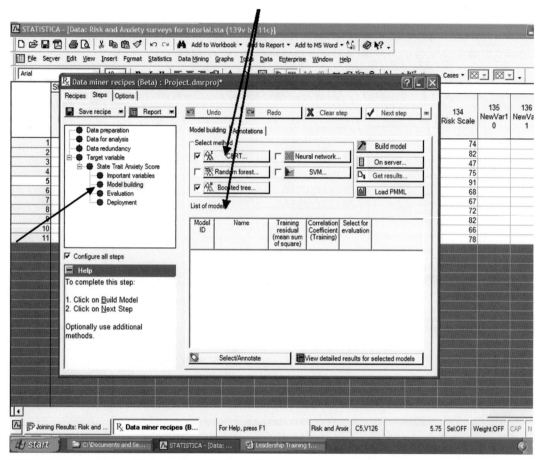

FIGURE K.14   After highlighting Model Building, C&RT and Boosted Trees are selected as the only two models to evaluate the data.

We click to remove the check mark from Configure All Steps and then clicked on Run to Completion, as shown in Figure K.15.

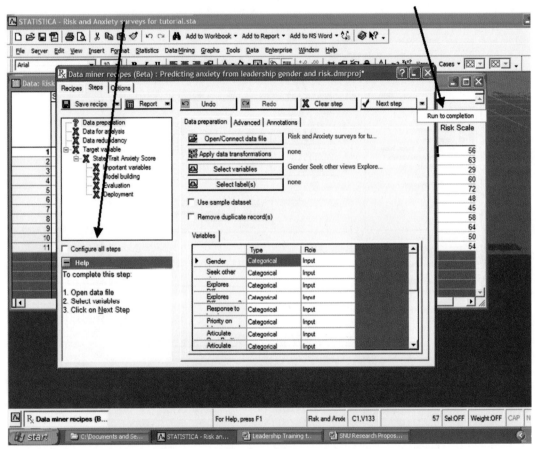

FIGURE K.15   Deselect Configure All Steps and then click Run to Completion; get Run to Completion by clicking the "down arrow".

We then let the program run. Figure K.16 shows the work in progress.

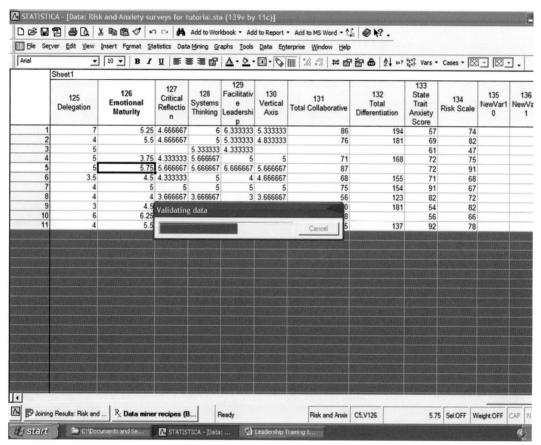

FIGURE K.16　DMRecipe processing.

We found that the boosted trees algorithm would not run because there were not enough cases (see Figure K.17). That was not surprising because there were only 11 cases and one had missing data.

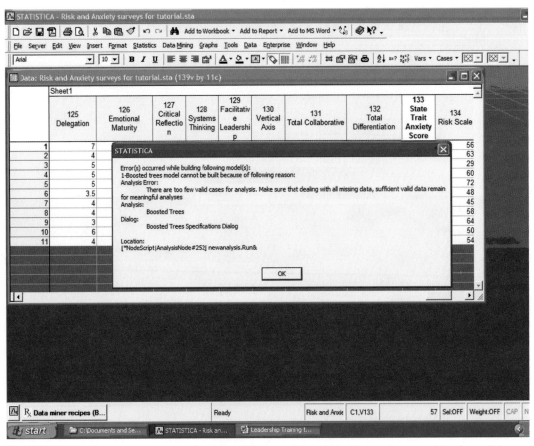

FIGURE K.17   Error message stating the Boosted Trees model could not be run because of too few valid cases.

When we opened the C&RT, there was only one node and nothing was revealed, as you can see in Figure K.18.

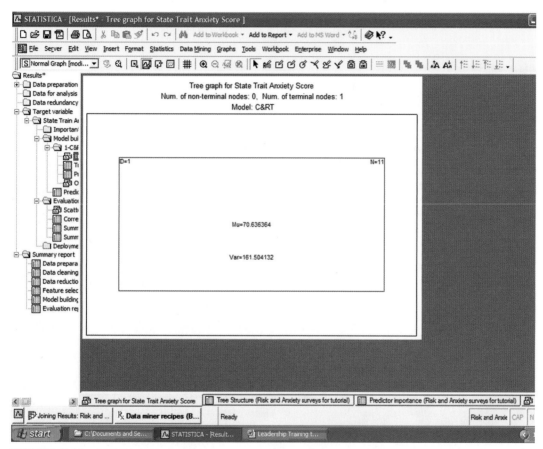

FIGURE K.18   C&RT did run, but there was only one tree node.

We decided to go a different route. But, again, we have only 11 cases and so could not expect profound findings.

We opened another DMR. We decided to predict the Total Collaboration score (131) from Total Differentiation (132), State Trait (133) and the Risk Scale (134), as shown in Figure K.19.

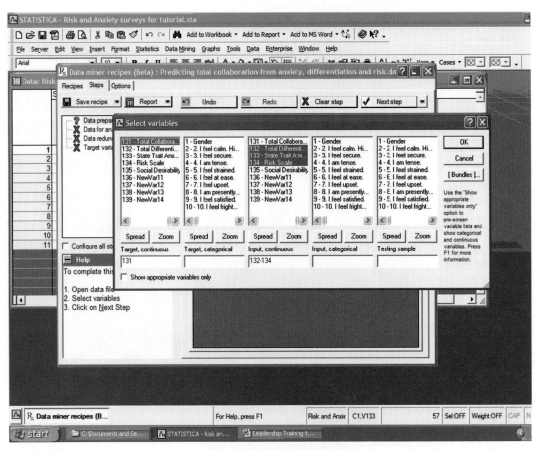

FIGURE K.19   Variables selected for a second DMRecipe model.

We checked Configure All Steps and then clicked to remove the check mark from Neural Networks (see Figure K.20).

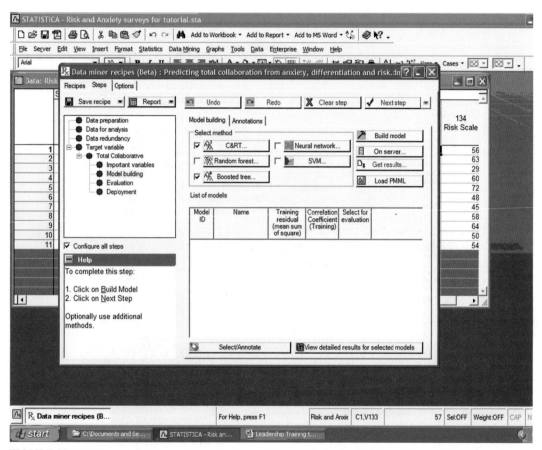

FIGURE K.20   The default Neural Networks is deselected so that only C&RT and Boosted Trees would run.

Then we unchecked Configure All Steps and ran all steps by choosing Run to Completion, as shown in Figure K.21.

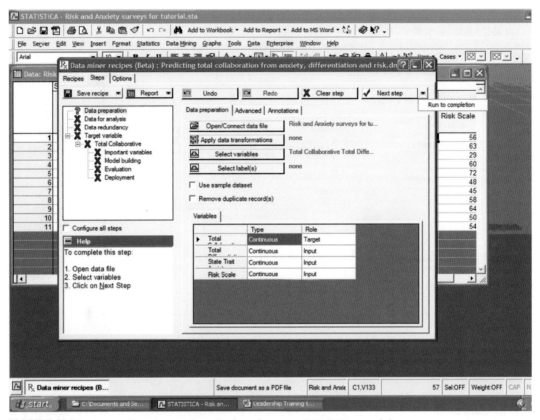

FIGURE K.21   Run to Completion selected to run this second DMRecipe model.

The boosted trees algorithm would not run once again (see Figure K.22). Unfortunately, there was not a prediction with this analysis either.

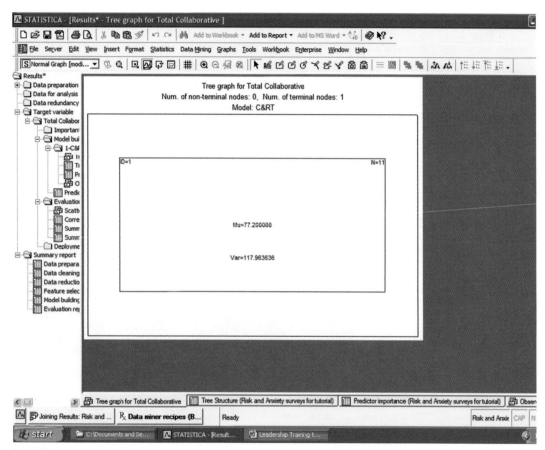

FIGURE K.22   Again, the C&RT would not produce useful trees, and Boosted Trees would not run.

Undaunted, we decided to try one more approach. This time we predicted risk from the four subscales of the leadership instrument, as shown in Figure K.23.

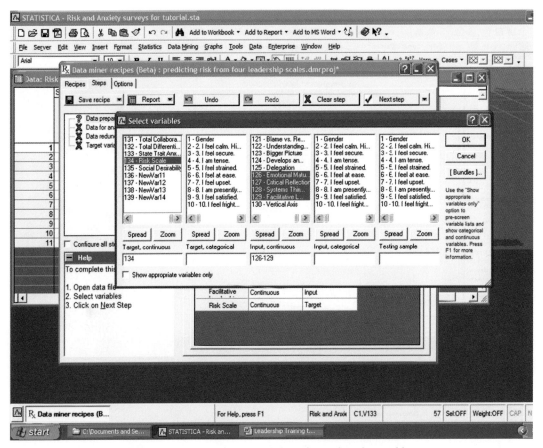

FIGURE K.23   Third DMRecipe model predicting Risk from four sub-scale variables.

We left the defaults, as shown in Figure K.24, and ran the program.

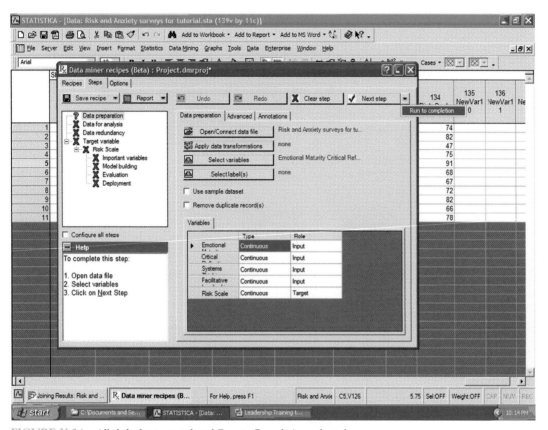

FIGURE K.24   All defaults accepted and Run to Completion selected.

Boosted trees still would not work. The C&RT had a higher correlation than neural networks, but neither was significant (see Figure K.25).

C&RT gave us a decision tree to think about, however, as you can see in Figure K.26.

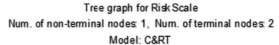

Variable	Correlations (Prediction) Marked correlations are significant at p < .05000 N=11 (Casewise deletion of missing data)			
	Risk Scale			
1-C&RT Prediction	0.38			
3-Neural network Prediction	0.20			

FIGURE K.25   Neural Networks and C&RT ran but both had low scores on Risk Scale; Boosted Trees still would not work.

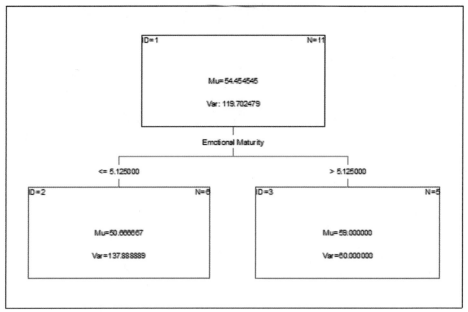

Tree graph for Risk Scale

Num. of non-terminal nodes: 1, Num. of terminal nodes: 2

Model: C&RT

FIGURE K.26   C&RT did give a decision tree split in this third DMRecipe model.

The variable Emotional Maturity separated the group into two groups on the risk scale. Those with higher scores on Emotional Maturity scored higher on the risk scale.

In one last desperate attempt, we ran a fourth DMR predicting total differentiation (v132) from all the questions in the Leadership survey (see Figure K.27).

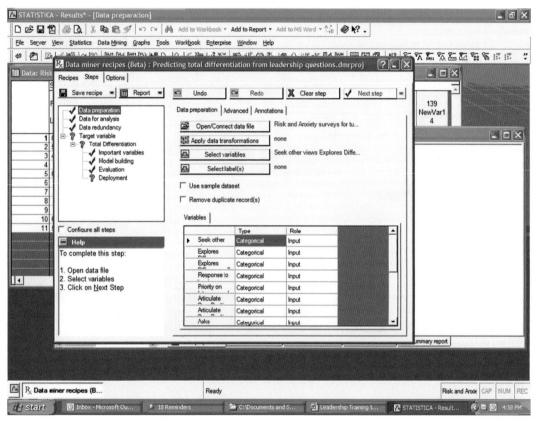

FIGURE K.27   Fourth DMRecipe model using var132 (Total Differentiation) predicted from all the questions in the Leadership survey.

We left all the defaults, as shown in Figure K.28.

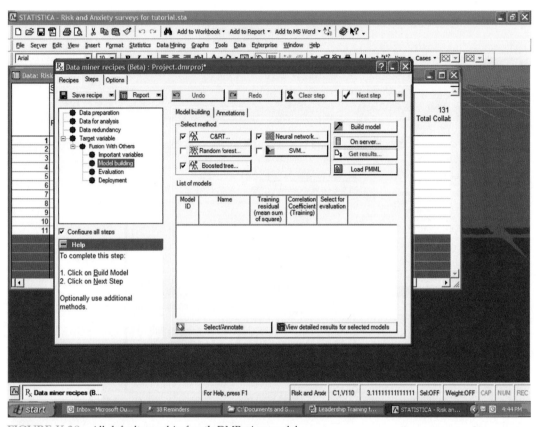

FIGURE K.28   All defaults used in fourth DMRecipe model.

Not surprisingly, boosted trees and neural networks would not run. C&RT, however, indicated that Transparent Decision Making might have been important, as you can see in Figure K.29.

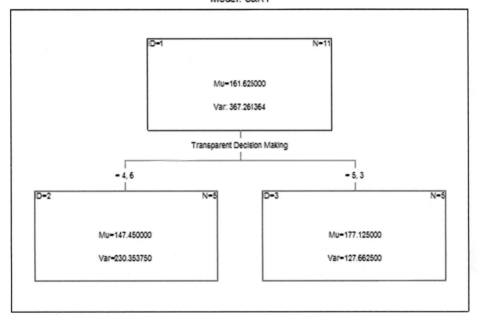

FIGURE K.29   C&RT ran in fourth DMRecipe model; Boosted Trees and Neural Networks would not run.

We also tried an interactive C&RT to see if the leadership questions might separate males and females (see Figure K.30).

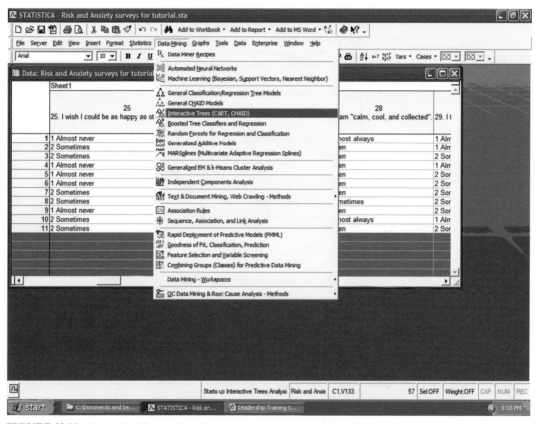

FIGURE K.30   Interactive Trees selected.

Next, we selected the variables, as shown in Figure K.31.

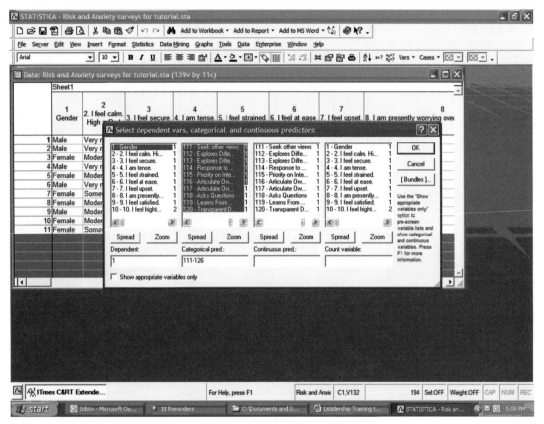

FIGURE K.31   Variables selected for Interactive Trees model.

Then we reduced the default minimums in an effort to see structure (see Figure K.32).

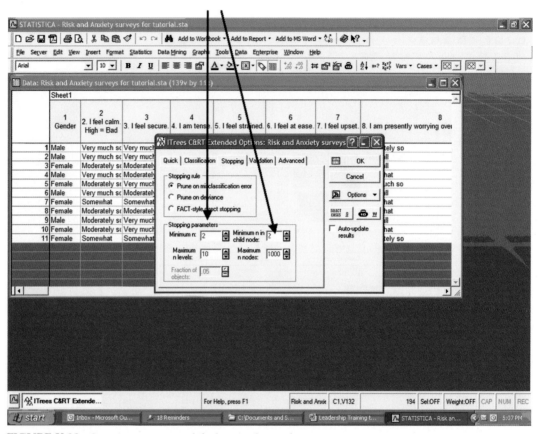

FIGURE K.32   Stopping parameters defaults were lowered.

We then selected the option for cross-validation, as shown in Figure K.33, and then clicked OK.

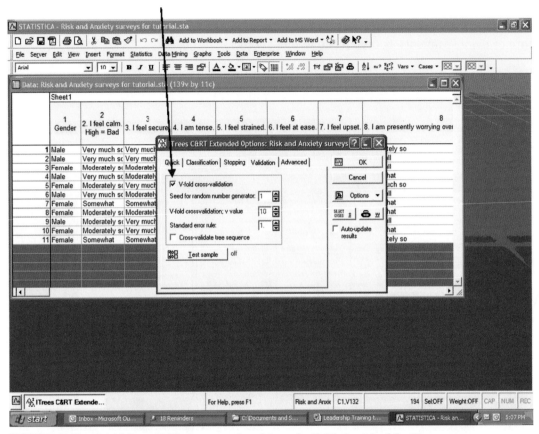

FIGURE K.33   V-fold Cross-Validation was selected from the Validation tab of the ITrees menu.

We clicked the Tree Graph (see Figure K.34) to view the decision trees (see Figure K.35).

Again, Transparent Decision Making seemed possibly important in separating the males and the females. Higher scores on transparency were related to males. Relating to earlier findings, perhaps these males were more self-differentiated and were transparent leaders. These were ideas that could lead us to additional hypotheses and encouraged us to continue gathering data.

We were certain of one thing from all of the above: data mining couldn't make up for a lack of data. We proceeded to go after more data if we were going to investigate the structure of the instruments as they relate to one another.

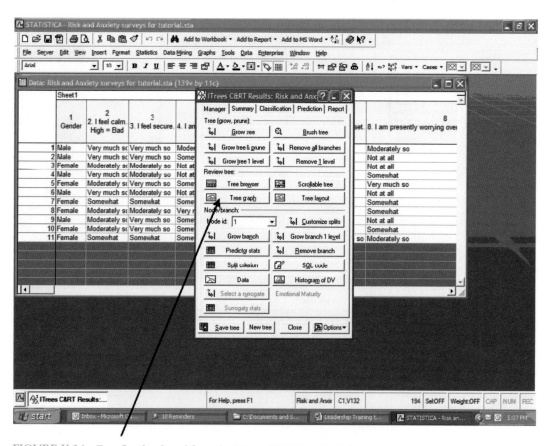

FIGURE K.34   Tree Graph selected from the ITrees C&RT Results dialog.

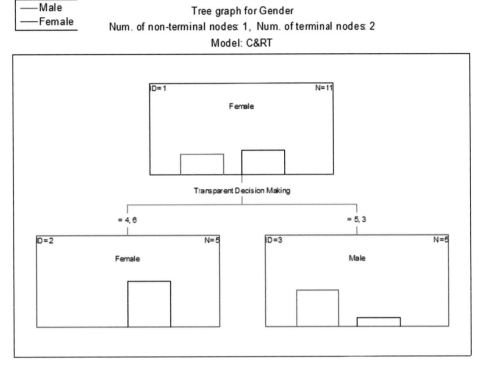

FIGURE K.35   Tree Graph for gender from ITrees model.

References

Millikin, M. A., & Miner, L. A. (1995). *A Risk Inventory*. Tulsa, OK: Right-Brain, Ink.

Robinson, G. (2004). *The Collaborative Leadership Profile*. Prior, OK: Challengequest.

Skrowron, E. A., & Friedlander, M. L. (1998). The differentiation of self inventory: Development and initial validation. *Journal of Counseling Psychology, 45*(3), 235–246.

Spielberger, C. D. (1983). *State-Train Anxiety Inventory*. Redwood City, CA: Mind Garden.

Thorberg, F. A., & Lyvers, M. (2005). Attachment, fear of intimacy and differentiation of self among clients in substance disorder treatment facilities. *Humanities and Social Science Papers*. Queensland, Australia: Bond University.

L

Dentistry: Facial Pain Study Based on 84 Predictor Variables (Both Categorical and Continuous)

Charles G. Widmer, DDS, MS
Edited by Gary Miner, Ph.D.

Charlie Widmer, using DMRecipe for the first time and following the DMRecipe tutorial earlier in this book, achieved the DMRecipe results shown in Figure L.1 in less than 30 minutes.

Looking at the summary of the spreadsheet (e.g., data) that was used by this automatic DMRecipe where all defaults were used (see Figure L.2), we see that all *predictor variables* no. 1–84 were used against *target variable* (also called a *dependent variable*) no. 85. Variable 85 is a facial pain variable, but because it has so many categories, it is considered continuous.

The default DMRecipe uses only three DM algorithms: neural networks, boosted trees, and CART. When we utilized these, the boosted trees algorithm gave the best model, with only a 30% error rate, which translates into 70% accuracy rate in predicting facial pain based on these 84 predictor variables.

This example has a lot of predictor variables for only 120 cases, so in this tutorial we will try to reduce the number of variables used as predictors and see whether we can get a more accurate model by tweaking various parameters.

To start, we need to explore our variables and decide which ones may be the most important by doing a feature selection. For the feature selection, we will use the Data Miner Workspace because the types of results obtained in the workspace are more useful, in our example, than the feature selection in an interactive module, or even the feature selection that happens automatically—behind the scenes in the DMRecipe format. However, after we decide which of the 84 variables we will use, we will complete most of the rest of this

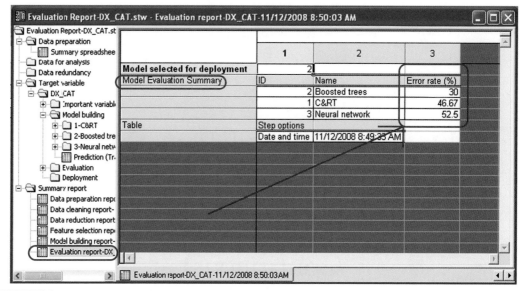

FIGURE L.1   Results of a DMRecipe project obtained in 30 minutes by a first time user of DMRecipe.

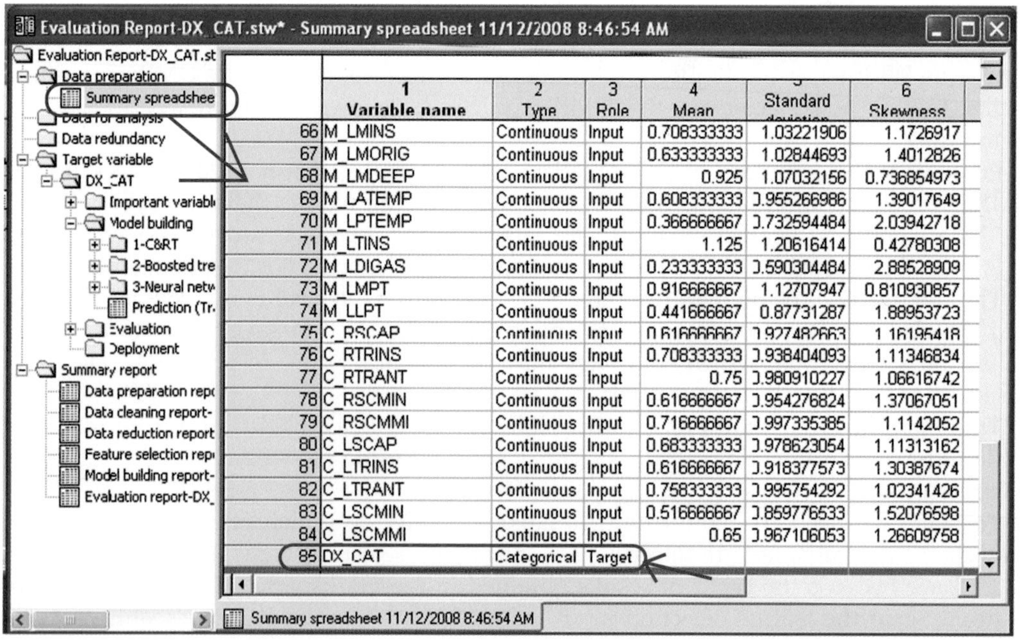

FIGURE L.2   Summary spreadsheet of 84 predictor variables and 1 target variable produced in the results of DMRecipe project.

tutorial using the DMRecipe format (i.e., Data Miner Recipe, where you can do data mining with just a few clicks of the mouse).

The dependent variable here, Variable 86, consists of several categorical levels of facial pain following various dental procedures; Variables 1–84 are various measures of dental procedures done and other measures of the patient, setting, or procedures.

Our overall goal is to use the predictor variables to determine whether we can make a predictive model as to what will cause facial pain following a dental procedure and thus find ways to minimize this pain, from our model, rather than just give pain-killing medications following the procedures.

Figure L.3 shows our Data Miner Workspace, with the dental data set embedded into the workspace and two feature selection icons put in: one for a chi-square feature selection (variables ordered on descending value of chi-square, using a chi-square iterative method of determining which variables are most important) and the other a p-value feature selection, where the variables will be output to a table showing them in ever increasing p-values (where the variables at the top of the table will have the lowest p-values, which is what we want).

Chi-square feature selection, putting all 84 predictor variables into the computations and asking for the top 25 to be displayed, is shown in the importance plot in Figure L.4.

In Figure L.5, the feature selection setting for the parameters is *Select only those with p-values less than 0.05.*

Note that all 84 predictor variables are significant at $p < 0.05$. However, using all of these variables probably will reduce the accuracy of any data mining methods we use, so our goal will be to reduce them in data mining analysis.

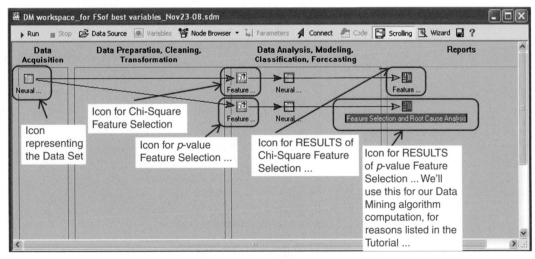

FIGURE L.3   Data Miner Workspace using Feature Selection.

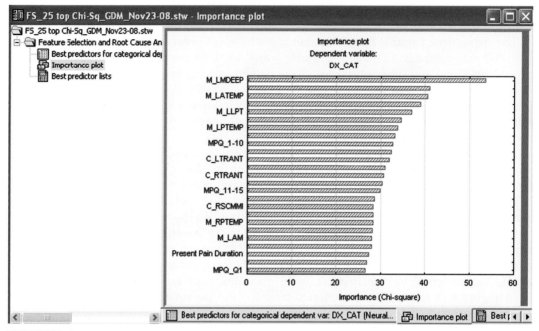

FIGURE L.4 Importance plot.

	Best predictors for categorica dependent var: DX_CAT (Neural N	
	Chi-square	p-value
C_LTRINS	28.26998	0.005050
SITE14	14.71154	0.005338
M_LAM	27.94872	0.005628
TMJ_RPL	27.89474	0.005732
C_LSCMIN	20.59050	0.008319
C_RTRINS	26.23327	0.009947
C_LSCMMI	26.15873	0.010192
SITE11	12.78724	0.012364
SITE13	12.74700	0.012580
C_RSCMIN	25.45788	0.012796
MPQ_Q18	30.86190	0.014013
Sex	12.47136	0.014170
M_RMORIG	24.37852	0.018059
C_RSCAP	24.20047	0.019101
SITE7	11.37636	0.022645
M_RMDEEP	22.94996	0.028154
TMJ_LPL	22.03261	0.037156
MPQ_Q6	21.91565	0.038476
M_LMINS	21.60337	0.042213
M_RAM	21.56080	0.042747
M_RMPT	21.13859	0.048387

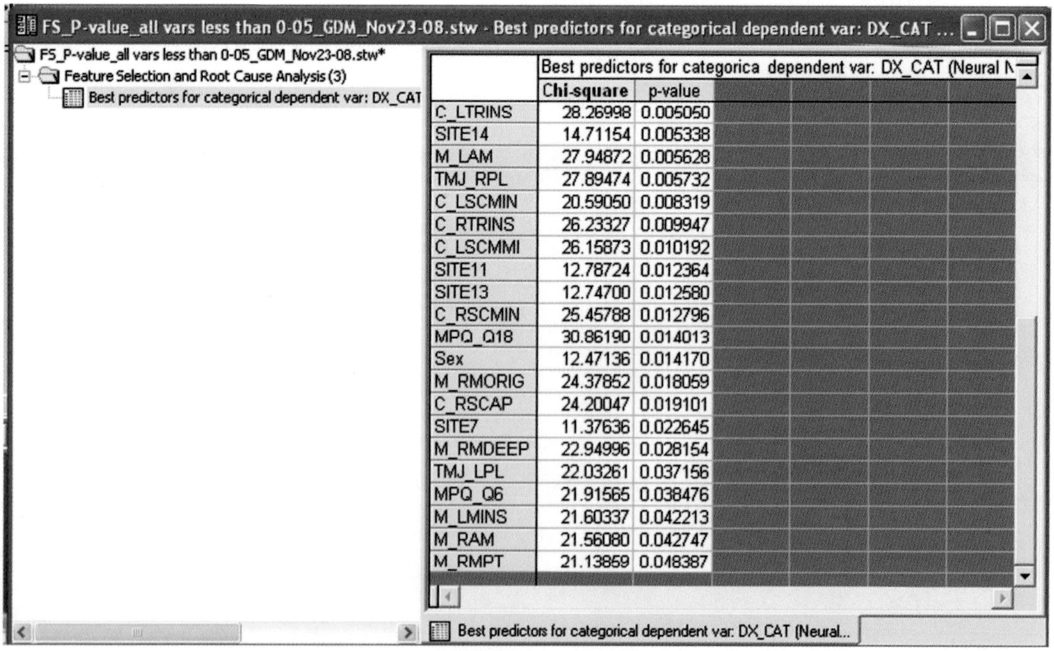

FIGURE L.5 Best predictor variables listed in order of ascending *p*-value.

We might, in this case, go with the chi-square selection, or we could use a p-value where there appears to be a break or cutoff in the preceding values.

A cutoff does appear in the p-value feature selection at about

C_RTRANT variable with a $p = 0.002290$, and
C_LTRINS variable with a $p = 0.005050$.

There are five variables between these two, so we could probably make the cutoff point anywhere among these five. For our example, let's take the cutoff at Variable M_RLPT (p-value 0.003076), as shown in Figure L.6.

The number of variables from the top down to M_RLPT is 16, as illustrated in Figure L.7.

We note that only the best predictor table was given in the results of the p-value feature selection; the reason is that we forgot to select all results in the Parameters dialog, which we can open by right-clicking on the FS icon in the Data Miner Workspace. So we'll go back and select All Results, and then rerun the p-value FS, getting the results shown in Figure L.8.

Note that the order of variables from the top of the graph to the bottom is based on the p-value, but the level of the variable (e.g., how far the bar extends to the right) is based on the chi-square. Thus, the chi-squares and the p-values do not run in the same order, causing the up and down nature of the graph.

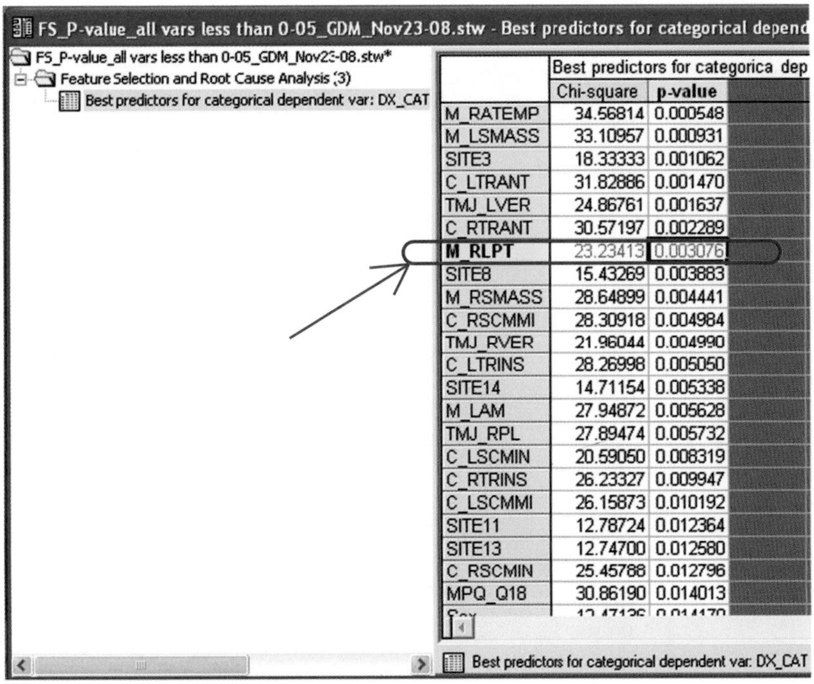

	Best predictors for categorica dep	
	Chi-square	p-value
M_RATEMP	34.56814	0.000548
M_LSMASS	33.10957	0.000931
SITE3	18.33333	0.001062
C_LTRANT	31.82886	0.001470
TMJ_LVER	24.86761	0.001637
C_RTRANT	30.57197	0.002289
M_RLPT	23.23413	0.003076
SITE8	15.43269	0.003883
M_RSMASS	28.64899	0.004441
C_RSCMMI	28.30918	0.004984
TMJ_RVER	21.96044	0.004990
C_LTRINS	28.26998	0.005050
SITE14	14.71154	0.005338
M_LAM	27.94872	0.005628
TMJ_RPL	27.89474	0.005732
C_LSCMIN	20.59050	0.008319
C_RTRINS	26.23327	0.009947
C_LSCMMI	26.15873	0.010192
SITE11	12.78724	0.012364
SITE13	12.74700	0.012580
C_RSCMIN	25.45788	0.012796
MPQ_Q18	30.86190	0.014013

FIGURE L.6 Using predictor variable M-RLPT as the cutoff point; e.g., all variables above this in the table including M-RLPT were used for further statistical analysis.

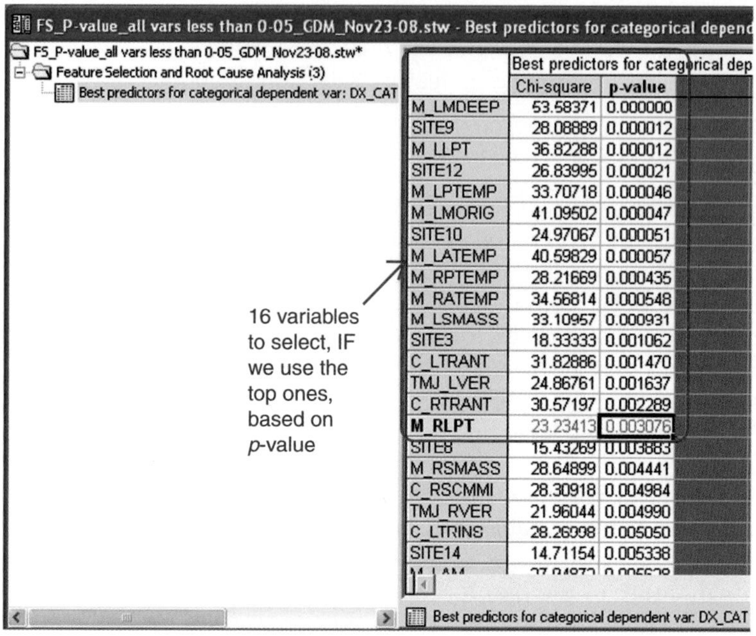

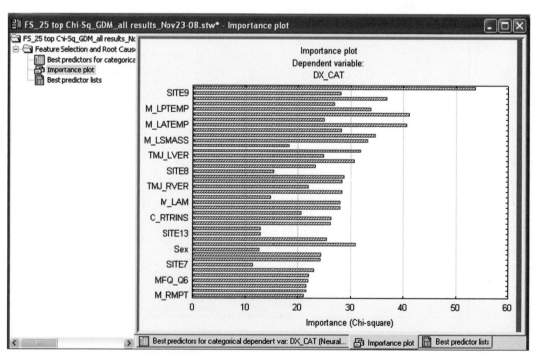

FIGURE L.7   Total number of variables from top down M-RLPT is 16.

FIGURE L.8   Importance plot based on *p*-value. Note that the *p*-values and chi-square values do not run in the same order; if they did, the plot would show a continuously descending curve based on chi-square importance.

We can do two things at this point, in deciding on what variables to select for further analysis:

1. We can take the top 16 variables, as seen in the *p*-value chart in Figure L.8, but we'll have to get all of these Var numbers from the results listing and probably select them one by one because they are in the results listing in order of presentation on the graph. Some are categorical and some are continuous, and both numeric types are in separate lists. But this may be the best thing to do.
2. Alternatively, we can take the top 16 variables from the chi-square results with the thinking that these are approximately the same variables; e.g., there is a lot of overlap between the chi-square method and the *p*-value method of feature selection. This approach is easy because we can just copy and paste the list into the Variables Selection dialog when selecting variables.

For this tutorial, we will take the *p*-value top 16, even though it is going to take a little bit more work to select the variables one by one, while looking at the *p*-value results listing, and then selecting each of those variables in the Variable dialog boxes.

Now we need to bring up the Data Miner Recipe dialog by selecting Data Miner Recipes from the Data Mining menu, as shown in Figure L.9.

Selecting this option brings up the DMRecipe dialog, as shown in Figure L.10. As you can see, we have placed the *p*-value results spreadsheet to the right on our screen so that we can see both at the same time; we'll need this listing to select the variables in the DMRecipe dialogs.

After we highlight the name of data file shown in Figure L.10 and then click the New button, a new DMRecipe dialog will appear, as shown in Figure L.11.

Now click on the Open/Connect Data File button, as shown in Figure L.12.

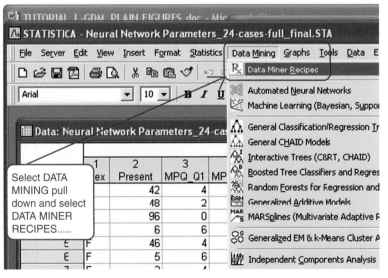

FIGURE L.9 DMRecipe Selection.

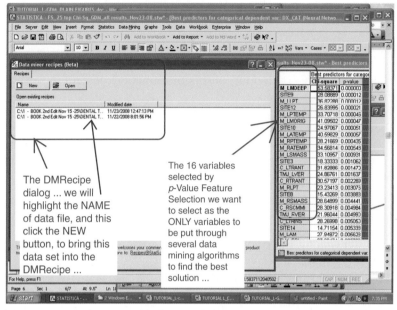

FIGURE L.10   Selecting previously saved DMRecipe data file.

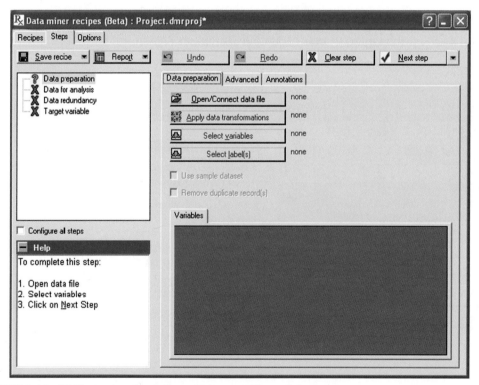

FIGURE L.11   DMRecipe interface before selecting variables and data set.

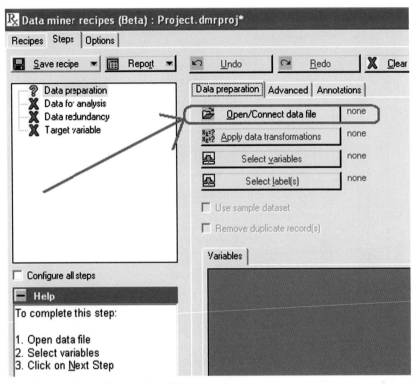

FIGURE L.12   Click on Open/Connect Data File to get data set.

Click OK in the Select Data Source dialog, shown in Figure L.13, and the data file will be connected to the DMRecipe format, as shown in Figure L.14.

Next, click on the Select Variables button, as shown in Figure L.15.

The Select Variables dialog appears, as shown in Figure L.16.

Select the 16 variables, as shown in Figure L.17.

Not all of the input-continuous variables are shown in the dialog in Figure L.17; they are hidden in the white input box, but here is the entire listing:

Input continuous: 52 58–59 63–64 67–70 74 77 82
Input categorical: 31 37–38 40

Now click OK to select these variables and close the Select Variables dialog. You then return to the DMRecipes dialog shown in Figure L.18.

Next, click on the plus (+) sign next to Target Variable in the tree hierarchy pane on the left of the dialog, as shown in Figure L.19.

The tree expands as shown in Figure L.20.

Now click to place a check in the Configure All Steps checkbox, and the tree will expand to look like that shown in Figure L.21.

Click on Model Building, as shown in Figure L.22.

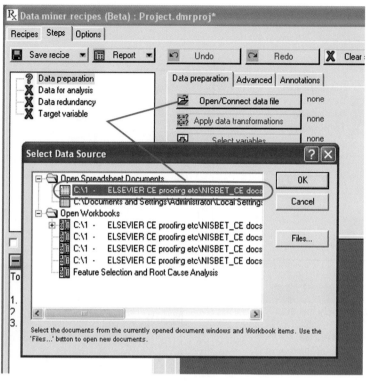

FIGURE L.13   Select Data Source window.

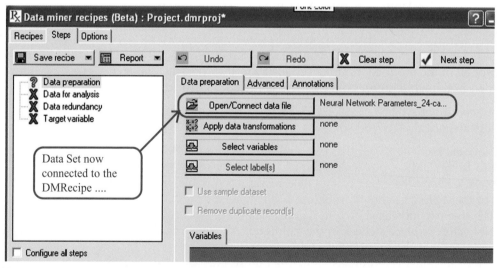

FIGURE L.14   Data set connected to DMRecipe.

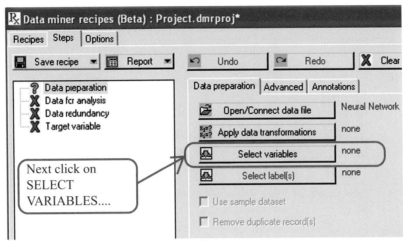

FIGURE L.15 Select Variables button.

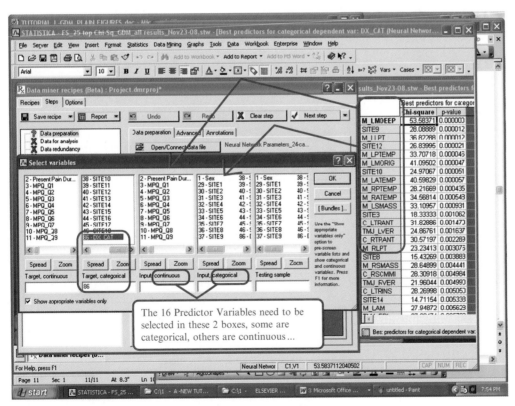

FIGURE L.16 How to select categorical and continuous variables.

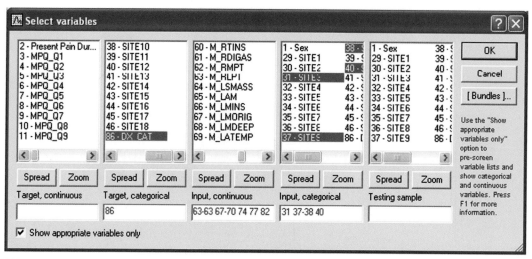

FIGURE L.17   Variables selected are highlighted.

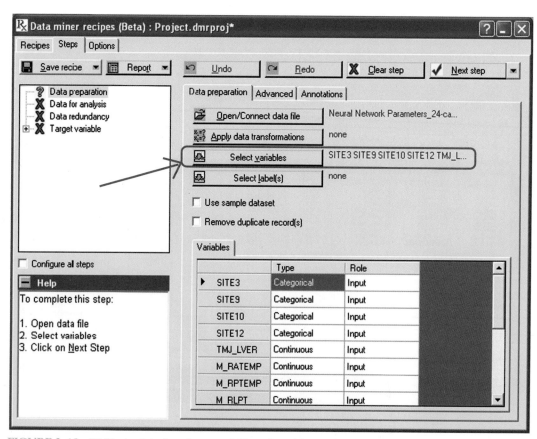

FIGURE L.18   DMRecipe interface shows variables selected for analysis.

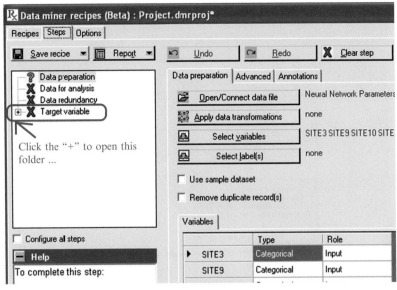

FIGURE L.19   Open left window flow tree by clicking on "+".

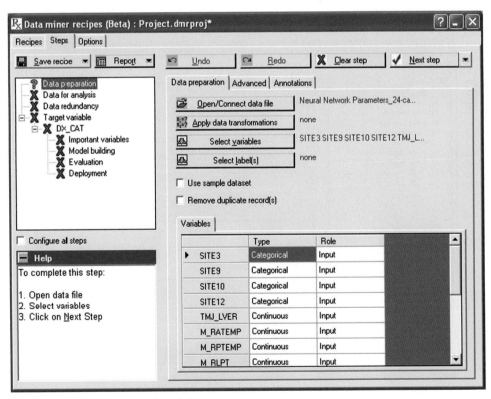

FIGURE L.20   Left window tree flow fully open.

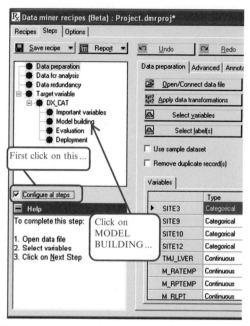

FIGURE L.21   Steps needed to set parameters of model.

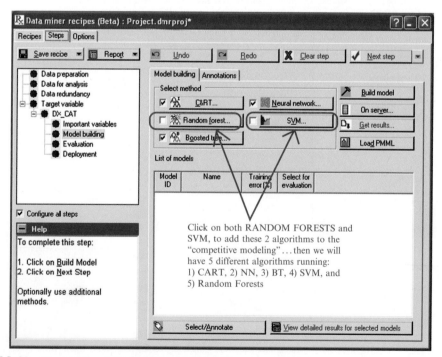

FIGURE L.22   Random Forests and SVM are not defaults.

Click to check all the algorithms shown in the Model Building tab on the right, as shown in Figure L.23. Next, click to remove the check from the Configure All Steps checkbox, thus turning it off.

Now click on the down arrow in the upper-right corner of the DMRecipe dialog, and select Run to Completion, as shown in Figure L.24. The entire DMRecipe will run, while you go off and have lunch or take a walk.

As the DMRecipe is running, you will see status bars like those shown in Figures L.25 through L.27.

In our particular case, the DMRecipe did all the computations in less than 2 minutes. Figure L.28 shows the summary workbook, with the evaluation report comparing the error rates of the five algorithms.

Keep in mind that

$$\text{Accuracy Rates} = (1 - \text{Error Rate})$$

Thus, the accuracy rates of the SVM is the best, e.g., about 77%. Boosted trees has the next best accuracy rate, almost the same as SVM, 75%. But when we look at neural networks,

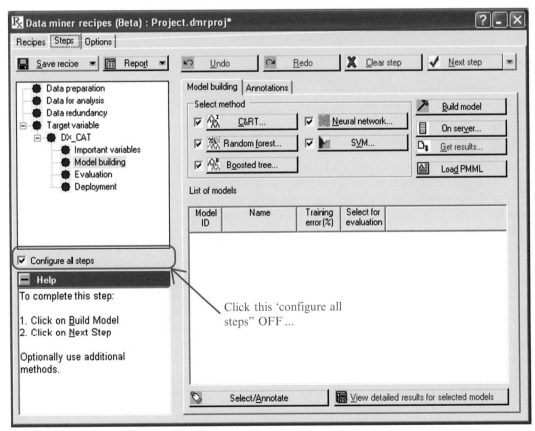

FIGURE L.23   After model parameters are set, click the Configure All Steps off.

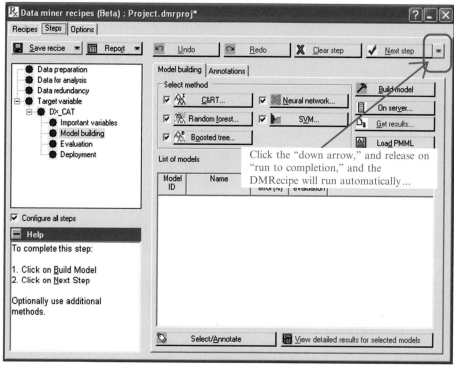

FIGURE L.24   Click Run to Completion to have the model computed.

random forests, and CART (or C&RT), we see that the accuracy rates in these algorithms has fallen to the 52–60% range. Therefore, we will probably want to concentrate any predictive models on SVM and maybe even make a hybrid voted model combining SVM and boosted trees.

However, we may also decide from this analysis that the 77% accuracy rate is not enough, and we will try other things with our data. Or we may gather more data and/or some more variables because we may have learned something from this analysis that suggests that important variables are missing from our data set.

However, now we will put these same data through an interactive SVM method, where we can tweak some of the parameters more thoroughly and also get a V-fold cross-validation measure to compare with the test sample accuracy rates. So let's do this interactive SVM model right now. To start, open the Data Mining menu and select Machine Learning, as shown in Figure L.29.

Selecting Machine Learning opens the dialog shown in Figure L.30.

Because it defaults to Support Vector Machine, click OK to select this option and open the dialog box shown in Figure L.31.

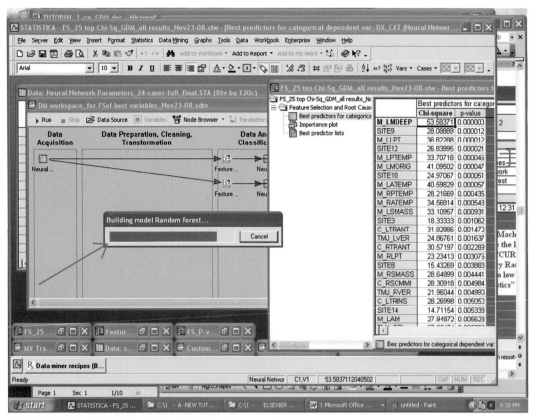

FIGURE L.25   DMRecipe model computing.

Now click on the Variables button to open the dialog shown in Figure L.32. Here, we'll select the variables, copying our predictor variables, as selected for the DMRecipe earlier, e.g., just the 16 top *p*-value feature selection variables.

Then click OK to return to the Support Vector Machines dialog, as shown in Figure L.33.

Now let's click on the Cross-Validation tab to bring it to the front, as shown in Figure L.34.

Next, we will click to put a check in the Apply *V*-Fold Cross-Validation checkbox and leave the default setting of V = 10. This means 10 separate bootstrapping random samples of the data will be taken, and analysis will be done with each of these, to compare; this is like doing 10 separate experiments in 10 separate labs around the world to see whether all labs can reach the same conclusion. We will use this *V*-fold cross-validation accuracy score, if needed; e.g., if the training accuracy and testing accuracy scores are not almost

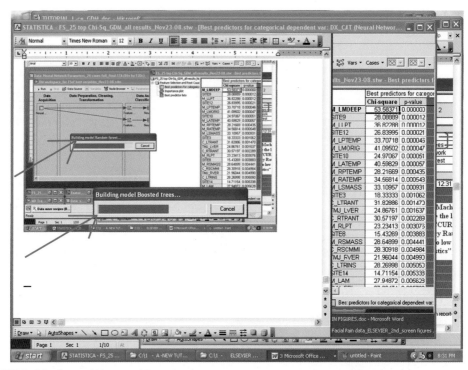

FIGURE L.26 Boosted Trees and Basic Statistics computing.

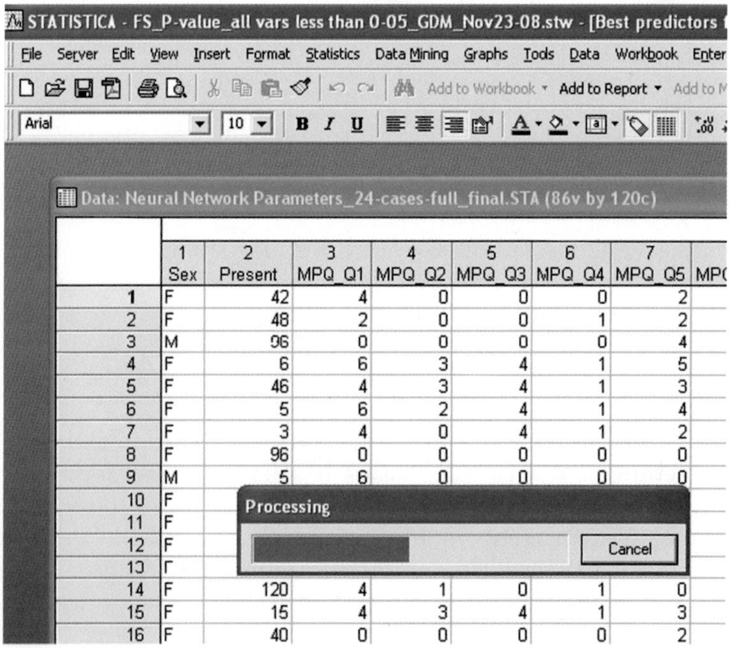

FIGURE L.27 Final processing of models.

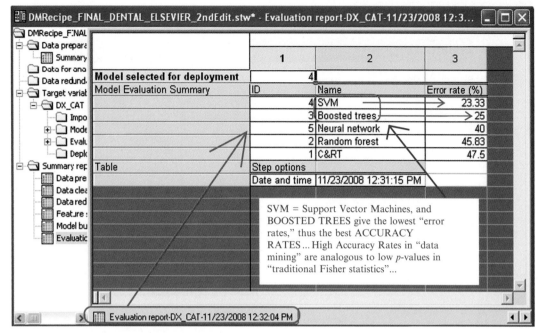

FIGURE L.28 DMRecipe results workbook.

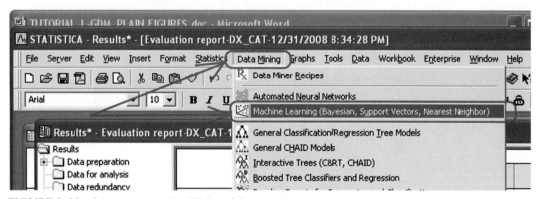

FIGURE L.29 Selecting interactive SVM module.

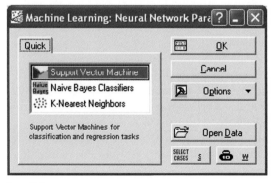

FIGURE L.30 SVM selection dialog.

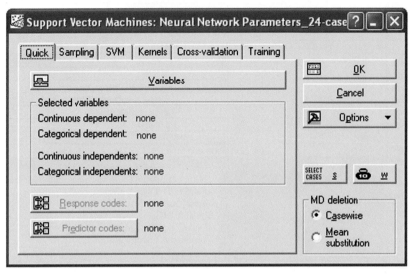

FIGURE L.31   SVM quick-tab dialog.

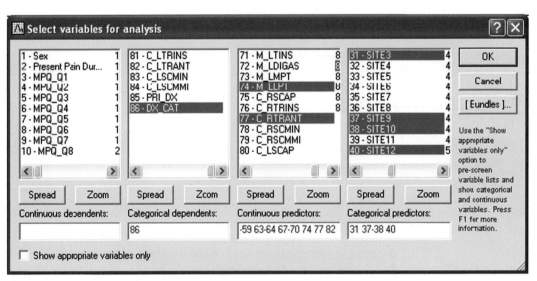

FIGURE L.32   Variable selection.

identical, then we will compare the testing sample accuracy with the *V*-fold cross-validation, and if both of these are about the same, then maybe we have a pretty good model.

Next, we will looking at the Kernels tab (see Figure L.35), where we will leave the default
settings.

The RBF kernel setting generally does best with most data sets, in general; however, if we
don't like the model we get from using it, we can come back and select a Linear,

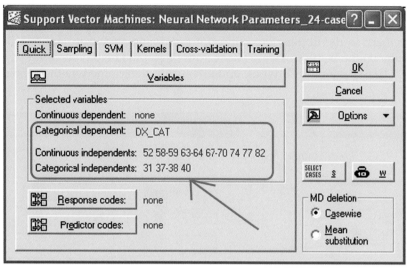

FIGURE L.33   When variables are selected, their names appear on the SVM dialog.

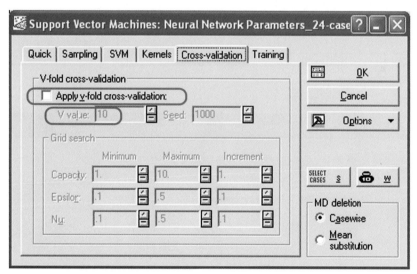

FIGURE L.34   Cross-Validation tab.

Polynomial, or a Sigmoid kernel type if we suspect our data have a curve that is more closely related to the others. The RBF kernel does not need a particular curve type; i.e., it can work with data accurately regardless of whether data follow a linear or very nonlinear curve; that is why the RBF usually will work on all or most data sets.

For the SVM tab (see Figure L.36), we can leave the default settings.

If we don't like the results we get, we can come back to this tab on the dialog and select Type 2 Classification SVM. To find the difference between Type 2 and Type 1, click the question (?) mark in the upper-right corner of the dialog shown in Figure L.36. This will take you to the online help of *STATISTICA*; from the online help, you can read about Type 1 and Type 2 Classification SVB types, and attempt to determine whether one or the other

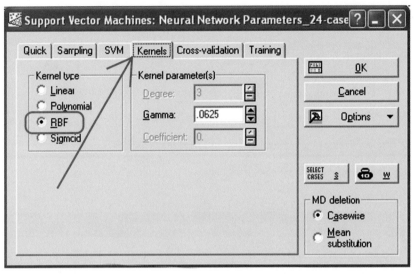

FIGURE L.35   Kernals tab.

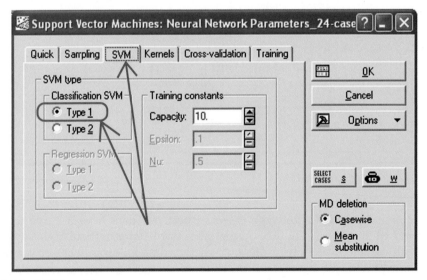

FIGURE L.36   SVM type tab.

will be better for your data. For practical purposes, it is faster just to try each and see which gives you a more accurate model.

Now let's look at the Sampling tab (see Figure L.37), where we can decide what proportion of the data we want as the training sample, and thus the remainder will be the testing sample.

We generally like to use about two-thirds as the training sample, so we'll change this amount to 66%, as shown in Figure L.38.

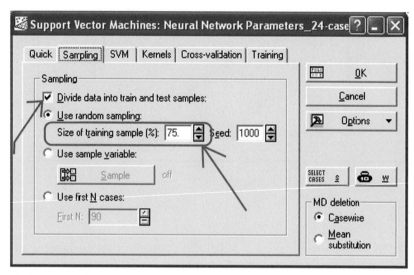

FIGURE L.37   Sampling tab.

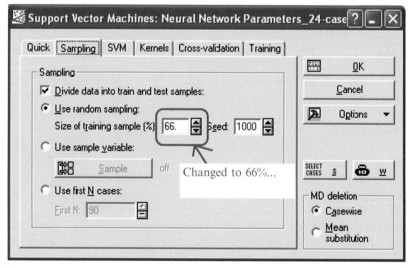

FIGURE L.38   Sampling tab with defaults changed.

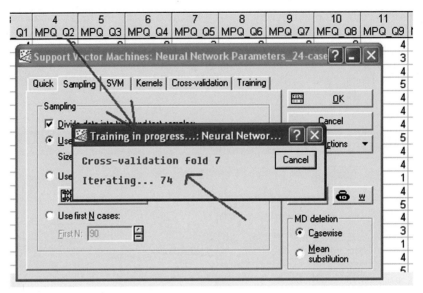

FIGURE L.39   Training of SVM model In Progress.

Then click OK in the Support Vector Machines dialog and let the computations begin.

During this process, you will see a message like the one in Figure L.39, but you'll have to watch closely, as this data set took only about 11 seconds to run.

The results are shown in Figure L.40.

We ideally want the Train accuracy, Overall accuracy, and Test accuracy to be almost identical or very close to one another. As you can see, even though the Train and Overall are about the same, the Test accuracy is very low, only about 39%. When this happens, we can look at the *V*-fold cross-validation to see if it is about the same as the test accuracy; in the example, it is, e.g., about 48% versus 39%.

If all of our accuracy values were in the same range, say all around 75%, then we could say that we have a model that will predict new cases accurately 75% of the time. But since we didn't get this, we have to say that this interactive model of SVM is not doing the job we'd really like to see.

We tried all of the other kernel types and classification types, and also reset the training sample to 75% to get about five additional models. Type 2 classification brought the *V*-fold and Test accuracies up closer to 50%, but all of these models, no matter what curve type (e.g., Kernel type) was used, were about the same. So we have to conclude that we need to examine these variables more closely.

As the editor of this tutorial, I do not know the meaning of the predictor variables at this point, so I would have to consult with Dr. Charles Widmer, the professor of Orthodontics,

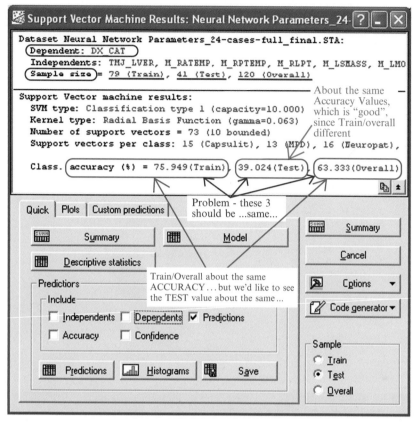

FIGURE L.40   Interactive SVM results.

to understand them better. Possibly we could select a better set of predictor variables and get a good model. However, feature selection is very powerful, and I suspect we did find the most important variables. But they still may be highly correlated.

We could try one more thing: reduce the number of predictor variables put into the model and try this with both the SVM interactive approach and also DMRecipe. Let's take a quick look.

We'll just select the top categorical and continuous variables from the chi-square feature selection, as shown in Figures L.41 and L.42.

Figure L.43 shows the results for the Interactive SVM algorithm method, using 75% Train, the RFB kernel, and V-fold cross-validation, but you can see that they are not much better than achieved previously.

Now let's try the DMRecipe again, using just the six predictor variables shown in Figure L.44.

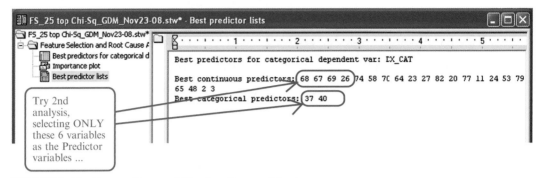

FIGURE L.41   Best predictor variables from Feature Selection.

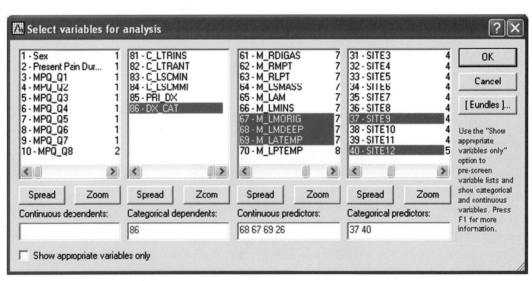

FIGURE L.42   Variables selected.

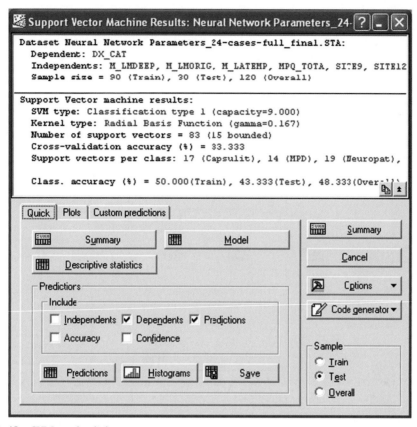

Dataset Neural Network Parameters_24-cases-full_final.STA:
  Dependent: DX_CAT
  Independents: M_LMDEEP, M_LMORIG, M_LATEMP, MPQ_TOTA, SITE9, SITE12
  Sample size = 90 (Train), 30 (Test), 120 (Overall)

Support Vector machine results:
  SVM type: Classification type 1 (capacity=9.000)
  Kernel type: Radial Basis Function (gamma=0.167)
  Number of support vectors = 83 (15 bounded)
  Cross-validation accuracy (%) = 33.333
  Support vectors per class: 17 (Capsulit), 14 (MPD), 19 (Neuropat),

  Class. accuracy (%) = 50.000(Train), 43.333(Test), 48.333(Over-'

FIGURE L.43   SVM results dialog.

Again, we don't see any improvement with fewer variables, as you can tell in Figure L.45. We probably could have expected this outcome, as the feature selection importance plots did not have a distinctive pivot point in the curve they made; instead, there was just a steady decline in the levels of Variables 1–84.

Many data sets will show a distinct change in direction in this importance plot curve, and this usually is the place to cut off use of additional variables. Generally, those variables with the high chi-square values (or low *p*-values, if using the *p*-value feature selection) will give very good models with high accuracy scores.

With this data set, we may need to go back to the drawing board and find the variable or variables missing that are really critical to this dependant variable of facial pain.

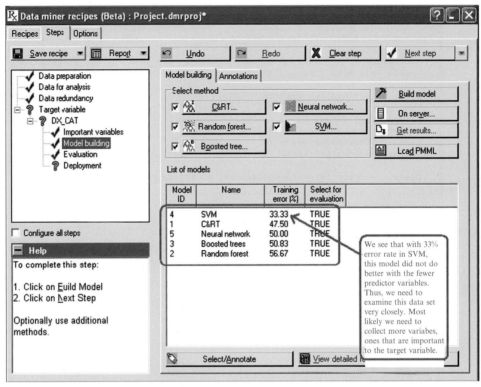

FIGURE L.44   DMRecipe results dialog.

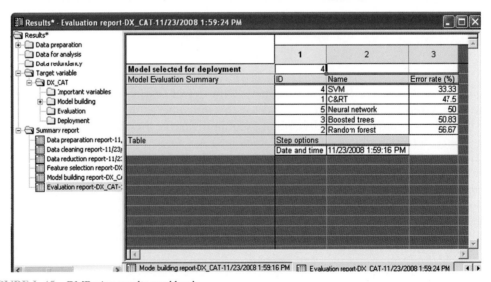

FIGURE L.45   DMRecipe results workbook.

M

Profit Analysis of the German Credit Data Using SAS-EM Version 5.3

Chamont Wang, Ph.D.
Edited by Gary Miner, Ph.D.

INTRODUCTION

In this tutorial, we will approach the German credit data from a cost/profit perspective. Specifically, we assume that a correct decision of the bank would result in 35% of the profit at the end of a specific period, say 3–5 years. Here, a *correct decision* means that the bank predicts that a customer's credit is in good standing (and hence would obtain the loan), and the customer indeed has good credit. On the other hand, if the model or the manager makes a false prediction that the customer's credit is in good standing, yet the opposite is

TABLE M.1   Profit Matrix

	Good Customer (predicted)	Bad Customer (predicted)
Good Customer (observed)	+0.35	0
Bad Customer (observed)	−1.00	0

true, then the bank will result in a unit loss. This concludes the first column of the profit matrix shown in Table M.1.

In the second column of the matrix, the bank predicted that the customer's credit is not in good standing and declined the loan. Hence, there is no gain or loss in the decision.

Note that the data have 70% credit-worthy (good) customers and 30% not-credit-worthy (bad) customers. A manager who doesn't have any model and who gives everybody the loan would result in the following negative profit per customer:

$$(700^*0.35 - 300^*1.00)/1000 = -55/1000 = -0.055 \text{ unit loss}$$

This number (–0.055 unit loss) may seem small. But if the average of the loan is $20,000 for this population ($n = 1000$), then the total loss will be

$$(-0.055 \text{ unit loss}) * (\$20,000 \text{ per unit per customer}) * (1,000 \text{ customers}) = -\$1,100,000$$

a whopping $1,100,000 loss. On the other hand, say a model produces the classification matrix shown in Table M.2.

In this case, the total profit would be

$$\text{Profit} = \text{True\_Positive}^*\$20,000^*0.35 - \text{False\_Positive}^*\$20,000$$
$$= 608^*\$20,000^*0.35 - 192^*\$20,000 = \$416,000$$

The difference of model versus no-model is

$$\$416,000 - (-\$1,100,000) = \$1,516,000,$$

or about $1.5 million of profit. The goal of this tutorial is to build statistical models to maximize the profit.

TABLE M.2   Classification Matrix

	Good (predicted)	Bad (predicted)	Row Total
Good (observed)	608 customers (76%, True_Positive)	46 customers	700 customers
Bad (observed)	192 customers (24%, False_Positive)	154 customers	300 customers
Column total & percentages	800 customers (100%)	200 customers	1,000 customers

MODELING STRATEGY

Assume that the data are already cleaned. Then the following steps would help maximize the profit:

1. **Try Different Tools:** SAS-EM provides a variety of data mining tools, including Decision Tree, Regression, Neural Network, Stochastic Gradient Boosting, Support Vector Machine, Ensemble model, and countless variations of these tools. In this tutorial, we will use mainly the default settings of some of these models.
2. **Use Variable Selection:** The original data set has 20 predictors. Some of the predictors may not be as important as others, and the exclusion of these variables may improve the model performance.
3. **Bundle the Variables:** Some of the predictors may be correlated to each other. In SAS-EM, the grouping of these predictors via two different techniques (Variable Clustering node and Principle Components node) often improves the model performance.
4. **Employ Binning, Filtering, and Variable Transformation**
5. **Tune Parameters**
6. **Change Nominal Predictors to Ordinal and Interval Variables:** This subsection discusses a very powerful node is SAS-EM: the Replacement Node. The node can be very handy and very useful in many studies where input variables are intrinsicially in **Ordinal** scale but are coded in **Nominal** scale. This is exactly what happens with the German credit data, and this is the reason the Neural Network failed in the section "A Primer of SAS-EM Predictive Modeling".
7. **Mega Models**
8. **Use Different Cutoff Values:** Given the study population, the model will produce the probabilities of all customers with regard to their credit standing. If the probability of a specific customer is above the cutoff (a.k.a. threshold), then the customer will be placed in the category of good customers; otherwise, the customer loan application will be denied. By adjusting different cutoff values, we may be able to increase the total profit. In our experience, this technique is one of the most important in the maximization of the profit.
9. **Incorporate Decision Rules of Complicated Models:** Machine learning techniques such as Neural Network and Gradient Boosting are often criticized for being black-box models; that is, it is "impossible to figure out how an individual input is affecting the predicted outcome" (Ayres, 2007, p. 143). This is not true. Given any Neural Network, you can plot its response surface and calculate its marginal effects (Wang and Liu, 2008). For Boosted Trees, you can also calculate Interaction Effects (Friedman and Popescu, 2005) and draw Partial Dependence Plots for the understanding and the interpretation of the model (Friedman, 2002). In SAS-EM, a special technique is to build a Decision Tree after a Neural Network (or other complicated model) to extract decision rules that can be very helpful for managers or other decision makers in real-world applications.

Due to the allocated space of (and time constraints of writing) this tutorial, we will skip steps 2, 3, 4, 5, 6 and 8 presented here. Furthermore, in step 6, we will use the default in SAS-EM, which gives cutoff values at 5% increments. For finer resolution at 1% or 0.5% increments, you need to write SAS codes to accomplish the task.

SAS-EM 5.3 INTERFACE

The diagram in Figure M.1 shows the key components of the SAS-EM 5.3 interface.

- **Toolbars:** Three rows of tools can be activated by clicking or can be dragged to the workspace. Move the cursor to a specific tool, and a small window will pop up giving a brief description of the tool functionality.

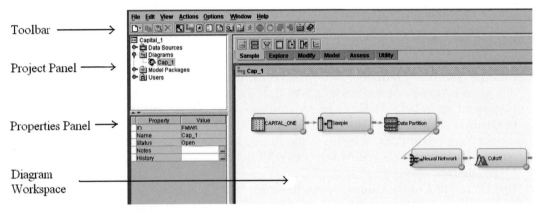

FIGURE M.1   SAS-EM interface.

- **Project Panel:** To manage and view data sources, diagrams, and results.
- **Properties Panel:** To view and edit the settings of data sources, diagrams, and nodes.
- **Diagram Workshop:** To graphically build, edit, run, and save process flow diagrams.

A PRIMER OF SAS-EM PREDICTIVE MODELING

This section provides information on the construction of the SAS-EM process.

The construction of the process flow shown in Figure M.2 is sufficient for small and medium-sized data sets. For large data sets, a Sample node can be added with little effort.

1. Creating a Project:

Select **File** → **New** → **Project** from the main menu. Specify the project name (Profit Analysis) in the name field of the pop-up window, as shown in Figure M.3.

credit_scoring
Data Partition
Replacement
Regression
Neural Network
Decision Tree
Model
Comparison

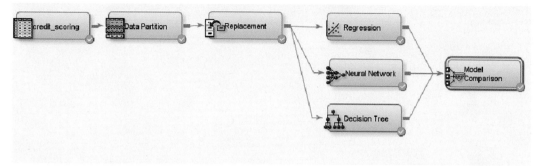

FIGURE M.2 SAS-EM process flow.

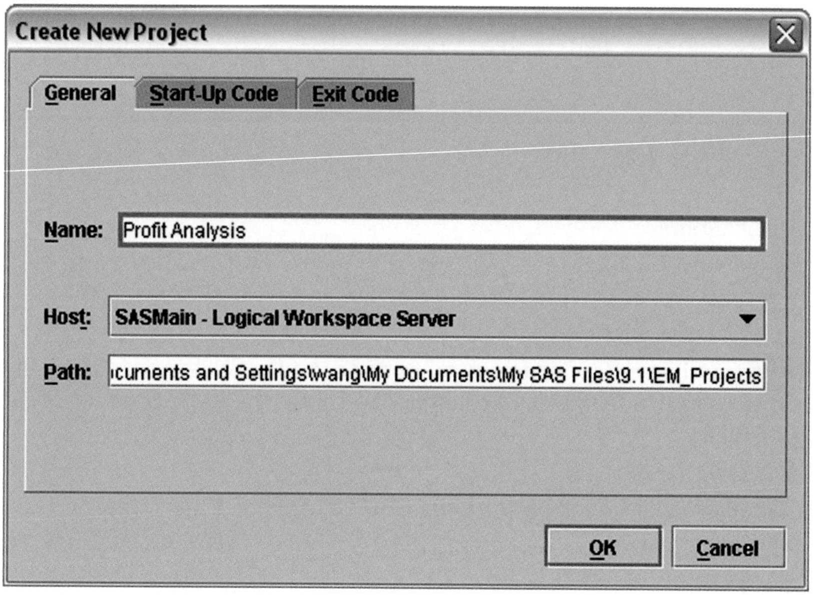

Create New Project
General
Start-Up Code
Exit Code
Name:
Profit Analysis
Host:
SASMain - Logical Workspace Server
Path:
cuments and Settings\wang\My Documents\My SAS Files\9.1\EM_Projects
OK
Cancel

FIGURE M.3 Create New Project dialog.

2. Creating a Data Source:

   Select **File** → **New** → **Data Source** to open the Data Source Wizard, as shown in Figure M.4.

   Click on the **Next** button to browse the folder of the Credit_scoring data, which resides in the Sasuser library (see **Appendix** on how to Credit_scoring import Excel data to the Sasuser library) (see Figure M.5).

   After you locate the data, click on **OK**, as shown at the bottom of Figure M.6, and then click on **Next** four times and then on **Finish** to import the data.

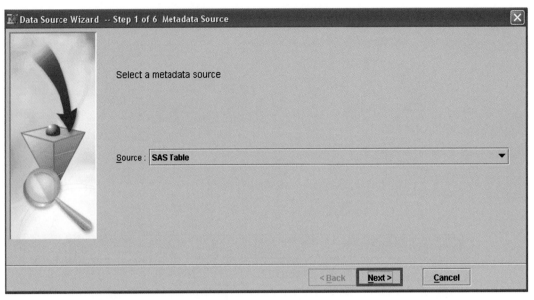

FIGURE M.4   SAS-EM Data Source Wizard.

3. Creating a Diagram:

Select **File** → **New** → **Diagram**. Then type the name, Profit Analysis, in the Name field of the pop-up window shown in Figure M.7 and then click on **OK**.

4. Creating the Process Flow:

From the **Project Panel** (upper-left corner), drag the Profit_Analysis icon directly under Data Sources to the **Diagram Workspace** to create the **Data Sources** node, as shown in Figure M.8.

5. Editing Variables:

In the **Workspace**, right-click on the **Data Source node** and then click on **Edit Variables**, as shown in Figure M.9.

In the pop-up window, identify the target variable, Credibility. Change the model **Role** from **Input** to **Target**, and change the **Level** from **Nominal** to **Binary**, as shown in Figure M.10. Then click on **OK**:

6. Replacement Node:

The node can be very handy and very useful in many studies where the predictors are intrinsically **Ordinal** variables but are coded in **Nominal** scale.

A click of the **Input Data** node would produce the window shown in Figure M.11. This window shows that most predictors are Nominal variables.

To change the nominal predictors into ordinal variables, click on the **Modify** tab and then drag the **Replacement** node to the Workspace.

Right-click the **Replacement** node and select **Run**.

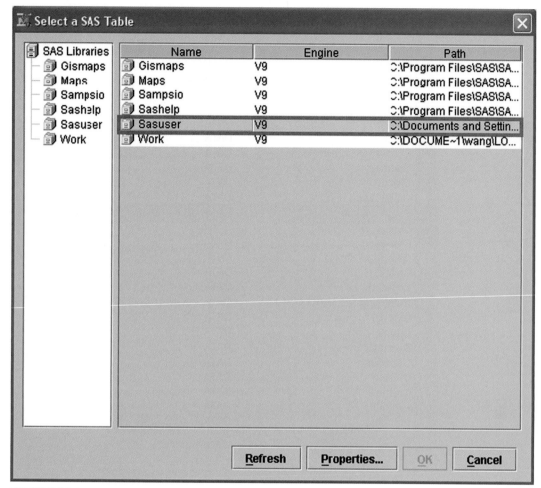

FIGURE M.5   Select an SAS Table dialog.

Click on the **Replacement** node to activate its **Property Panel** and then click on the
ellipses button 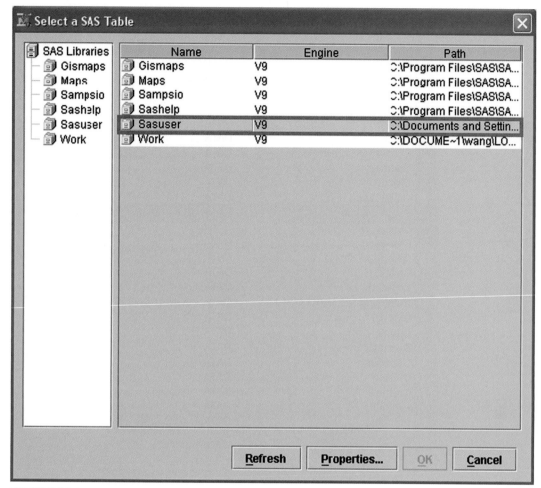 at the right of the Class Variables **Replacement Editor** (see
Figure M.12).

In the pop-up window, change the **Level** of the selected Input Variables as shown in
Figure M.13. Once you are done, select **OK**.

The conversion of Nominal variables to Ordinal scale requires a lot of subject-matter
judgment and sometimes can be controversial. Readers of this tutorial are urged
to examine the conversion and use his/her own numbers when deemed necessary.

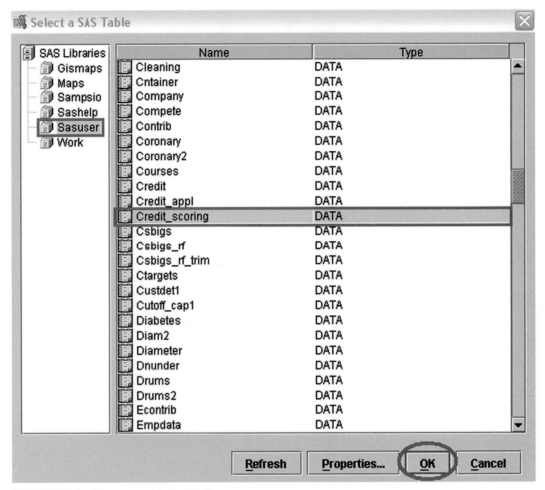

FIGURE M.6   Selecting a table in SAS library.

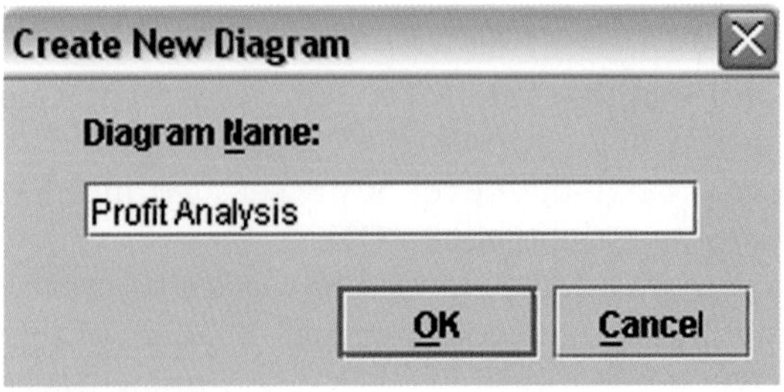

FIGURE M.7   Create New Diagram window.

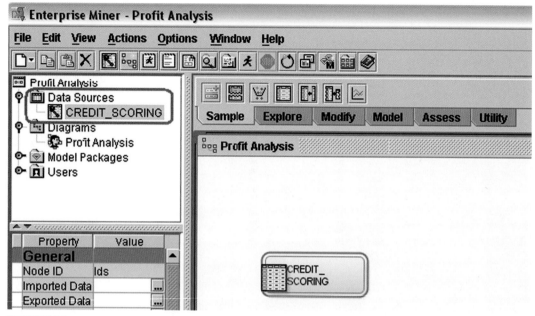

FIGURE M.8   Profit Analysis in SAS-EM.

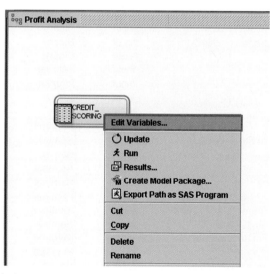

FIGURE M.9   Profit Analysis pop-up menu.

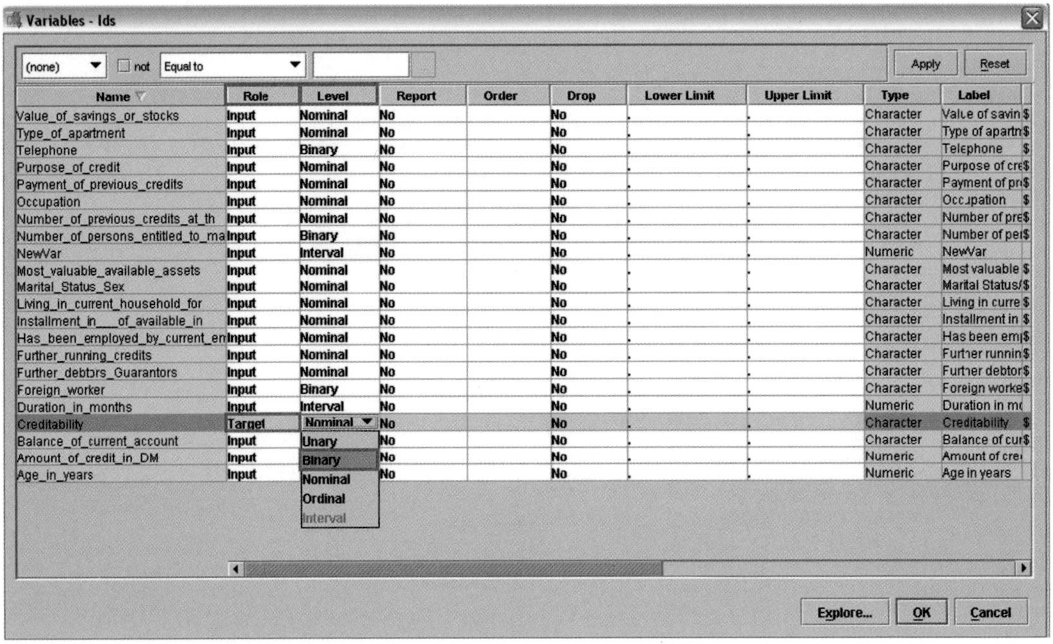

Variables - Ids

(none) | not | Equal to | Apply | Reset

Name	Role	Level	Report	Order	Drop	Lower Limit	Upper Limit	Type	Label
Value_of_savings_or_stocks	Input	Nominal	No		No	.	.	Character	Value of savin $
Type_of_apartment	Input	Nominal	No		No	.	.	Character	Type of apartn $
Telephone	Input	Binary	No		No	.	.	Character	Telephone $
Purpose_of_credit	Input	Nominal	No		No	.	.	Character	Purpose of cre $
Payment_of_previous_credits	Input	Nominal	No		No	.	.	Character	Payment of pr $
Occupation	Input	Nominal	No		No	.	.	Character	Occupation $
Number_of_previous_credits_at_th	Input	Nominal	No		No	.	.	Character	Number of pre $
Number_of_persons_entitled_to_ma	Input	Binary	No		No	.	.	Character	Number of pe $
NewVar	Input	Interval	No		No	.	.	Numeric	NewVar
Most_valuable_available_assets	Input	Nominal	No		No	.	.	Character	Most valuable $
Marital_Status_Sex	Input	Nominal	No		No	.	.	Character	Marital Status/ $
Living_in_current_household_for	Input	Nominal	No		No	.	.	Character	Living in curre $
Installment_in___of_available_in	Input	Nominal	No		No	.	.	Character	Installment in $
Has_been_employed_by_current_em	Input	Nominal	No		No	.	.	Character	Has been em $
Further_running_credits	Input	Nominal	No		No	.	.	Character	Further runnin $
Further_debtors_Guarantors	Input	Nominal	No		No	.	.	Character	Further debtor $
Foreign_worker	Input	Binary	No		No	.	.	Character	Foreign worke $
Duration_in_months	Input	Interval	No		No	.	.	Numeric	Duration in m
Creditability	Target	Nominal	No		No	.	.	Character	Creditability $
Balance_of_current_account	Input	Unary	No		No	.	.	Character	Balance of cur $
Amount_of_credit_in_DM	Input	Binary	No		No	.	.	Numeric	Amount of cre
Age_in_years	Input	Nominal	No		No	.	.	Numeric	Age in years

FIGURE M.10 SAS-EM variable listing.

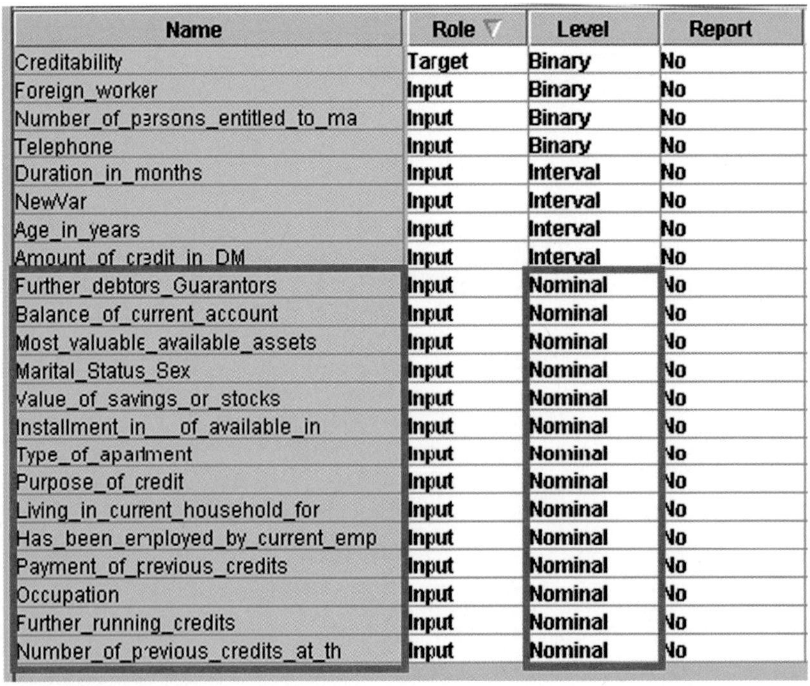

Name	Role	Level	Report
Creditability	Target	Binary	No
Foreign_worker	Input	Binary	No
Number_of_persons_entitled_to_ma	Input	Binary	No
Telephone	Input	Binary	No
Duration_in_months	Input	Interval	No
NewVar	Input	Interval	No
Age_in_years	Input	Interval	No
Amount_of_credit_in_DM	Input	Interval	No
Further_debtors_Guarantors	Input	Nominal	No
Balance_of_current_account	Input	Nominal	No
Most_valuable_available_assets	Input	Nominal	No
Marital_Status_Sex	Input	Nominal	No
Value_of_savings_or_stocks	Input	Nominal	No
Installment_in___of_available_in	Input	Nominal	No
Type_of_apartment	Input	Nominal	No
Purpose_of_credit	Input	Nominal	No
Living_in_current_household_for	Input	Nominal	No
Has_been_employed_by_current_emp	Input	Nominal	No
Payment_of_previous_credits	Input	Nominal	No
Occupation	Input	Nominal	No
Further_running_credits	Input	Nominal	No
Number_of_previous_credits_at_th	Input	Nominal	No

FIGURE M.11 Data Input node selection produces this window showing variables.

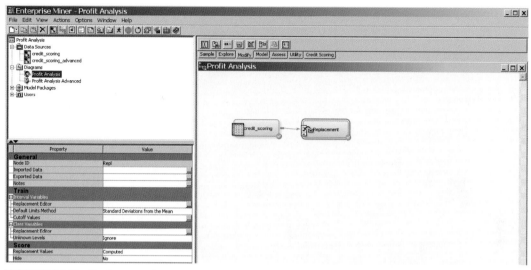

FIGURE M.12   Replacement button.

Variable	Level	Frequency				Replace...
Balance_of_current_account	>=200 DM	394	C	...	.	1
Balance_of_current_account	no running account	274	C	...	.	0
Balance_of_current_account	no balance	269	C	...	.	3
Balance_of_current_account	<= 200 DM	63	C	...	.	2
Balance_of_current_account	_UNKNOWN_	.	C		.	_DEFA...
Creditability	good	700	C	...	.	
Creditability	bad	300	C	...	.	
Creditability	_UNKNOWN_	.	C		.	_DEFA...
Foreign_worker	yes	963	C	...	.	0
Foreign_worker	no	37	C	...	.	1
Foreign_worker	_UNKNOWN_		C		.	_DEFA...
Further_debtors_Guarantors	none	907	C	...	.	2
Further_debtors_Guarantors	guarantor	52	C	...	.	0
Further_debtors_Guarantors	co-applicant	41	C	...	.	1
Further_debtors_Guarantors	_UNKNOWN_		C		.	_DEFA...
Further_running_credits	no further running credits	814	C	...	.	1
Further_running_credits	at other banks	139	C	...	.	0
Further_running_credits	at department store or mail ...	47	C	...	.	0
Further_running_credits	_UNKNOWN_	.	C		.	_DEFA...
Has_been_employed_by_current_emp	between 1 and 4 years	339	C	...	.	2
Has_been_employed_by_current_emp	greater than 7 years	253	C	...	.	4
Has_been_employed_by_current_emp	between 4 and 7 years	174	C	...	.	3
Has_been_employed_by_current_emp	less than 1 year	172	C	l...	.	1
Has_been_employed_by_current_emp	unemployed	62	C	...	.	0

FIGURE M.13   Replacement Editor in SAS-EM.

Installment_in___of_available_in	less than 20	476	C	l...	.	1
Installment_in___of_available_in	between 25 and 35	231	C	...	.	3
Installment_in___of_available_in	between 20 and 25	157	C	...	.	2
Installment_in___of_available_in	greater than 35	136	C	...	.	4
Installment_in___of_available_in	UNKNOWN	.	C		.	DEFA...
Living_in_current_household_for	greater than 7 years	413	C	...	.	3
Living_in_current_household_for	between 1 and 4 years	308	C	...	.	1
Living_in_current_household_for	between 4 and 7 years	149	C	...	.	2
Living_in_current_household_for	less than 1 year	130	C	l...	.	0
Living_in_current_household_for	_UNKNOWN_	.	C		.	_DEFA...
Marital_Status_Sex	male:single	548	C	...	.	
Marital_Status_Sex	female:divorced/living apart/...	310	C	f...	.	
Marital_Status_Sex	male:married/widowed	92	C	...	.	
Marital_Status_Sex	male:divorced/living apart	50	C	...	.	
Marital_Status_Sex	UNKNOWN	.	C		.	DEFA...
Most_valuable_available_assets	savings contract with a build..	332	C	...	.	3
Most_valuable_available_assets	no assets	282	C	...	.	0
Most_valuable_available_assets	car/other	232	C	...	.	1
Most_valuable_available_assets	ownership of house or land	154	C	...	.	2
Number_of_previous_credits_at_th	one	633	C	...	.	0
Number_of_previous_credits_at_th	2 or 3	333	C	...	.	1
Number_of_previous_credits_at_th	4 or 5	28	C	...	.	2
Number_of_previous_credits_at_th	6 or more	6	C	...	.	3
Number_of_previous_credits_at_th	_UNKNOWN_	.	C		.	DEFA...
Occupation	skilled worker/skilled emplo..	630	C	...	.	2
Occupation	unskilled with permanant re...	200	C	...	.	1
Occupation	executive/self-employed/hig...	148	C	...	.	3
Occupation	unemployed/unskilled with ...	22	C	...	.	0
Occupation	UNKNOWN	.	C		.	DEFA...
Payment_of_previous_credits	no previous credits or paid ...	530	C	...	.	1
Payment_of_previous_credits	paid back previous credits a...	293	C	...	.	2
Payment_of_previous_credits	no problems with current cr...	88	C	...	.	2
Payment_of_previous_credits	problematic running accounts	49	C	...	.	0
Payment_of_previous_credits	hesistant payment of previo...	40	C	...	.	0
Purpose_of_credit	items of furniture	280	C	i...	.	4
Purpose_of_credit	other	234	C	...	.	1
Purpose_of_credit	used car	181	C	...	.	7
Purpose_of_credit	new car	103	C	...	.	8
Purpose_of_credit	retraining	97	C	...	.	9
Purpose_of_credit	repair	50	C	...	.	5
Purpose_of_credit	household appliances	22	C	...	.	3
Purpose_of_credit	business	12	C	...	.	10
Purpose_of_credit	radio/television	12	C	...	.	2
Purpose_of_credit	vacation	9	C	...	.	6
Purpose_of_credit	_UNKNOWN_	.	C		.	_DEFA...
Telephone	no	596	C	...	.	0
Telephone	yes	404	C	...	.	1
Telephone	_UNKNOWN_	.	C		.	DEFA...
Type_of_apartment	rented	714	C	...	.	1
Type_of_apartment	free apartment	179	C	f...	.	2
Type_of_apartment	owner	107	C	...	.	3
Type_of_apartment	UNKNOWN	.	C		.	DEFA...
Value_of_savings_or_stocks	no savings	603	C	...	.	0
Value_of_savings_or_stocks	greater than 1000 DM	183	C	...	.	4
Value_of_savings_or_stocks	less than 100 DM	103	C	l...	.	1
Value_of_savings_or_stocks	between 100 and 500 DM	63	C	...	.	2
Value_of_savings_or_stocks	between 500 and 1000 DM	48	C	...	.	3
Value_of_savings_or_stocks	_UNKNOWN_		C			_DEFA

FIGURE M.13—Cont'd

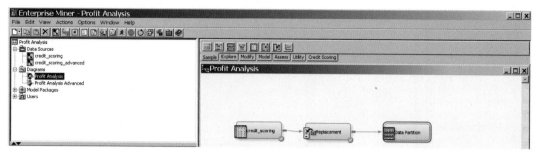

FIGURE M.14   Data Partition icon in SAS-EM.

7. Data Partition Node:

Drag the **Data Partition** icon (in the third icon under the Sample tab) to the **Diagram Workspace**. Connect the **Replacement** node to the **Data Partition** node, as shown in Figure M.14.

In SAS-EM 5.3, the default setting is 40%–30%–30% for the partition of the original data into Training, Validation, and Test data sets.

8. Regression node:

To use this node, click on the **Model** tab (in the third row of toolbars) to activate the **Regression** icon [icon]. Drag the icon to the Workspace and then connect the Data Partition node and the **Neural Network** node (Figure M.15).

Right-click on the **Regression** node and select **Run**. Click on **Yes** in the pop-up window. Wait for the next pop-up window and then click on **Results**. The next pop-up window contains a lot of information which is useful in many other studies. In this case, we will skip these results and go straight to the profit calculation.

Almost all data mining packages are confused by the subtle difference between a *misclassification matrix* and a *decision matrix*. SAS-EM is a rare exception. Note that the first matrix does not allow non-zero entries on the diagonal line, while the second matrix is able to accommodate different kinds of cost-profit considerations. The difference may seem small, but the consequence is enormous. The following steps show how to accomplish this with SAS-EM.

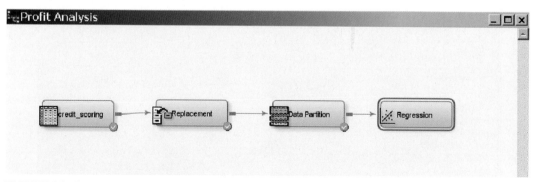

FIGURE M.15   Completed process flow model.

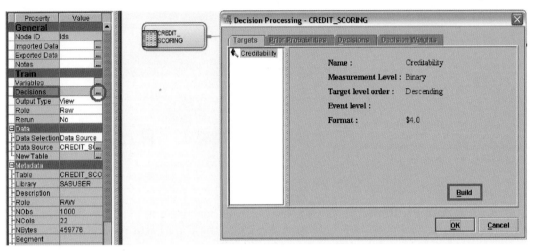

FIGURE M.16   Click on Decisions selection.

9. Decision weights in Data Source node:

Click on the **Data Source** node to activate its **Property Panel**. Then click on the ellipses button 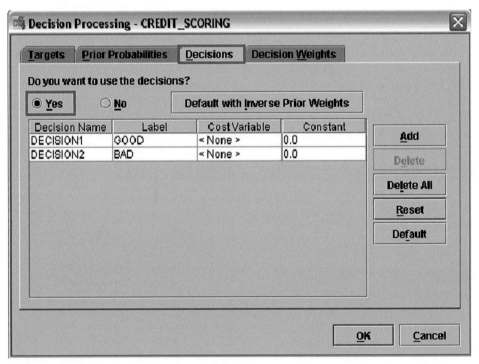 to the right of **Decisions**. On the **Decision Processing** pop-up window, click on **Build** to activate the Decisions menu (see Figure M.16).

Click on **Decisions** in the menu and select **Yes**, as shown in Figure M.17.

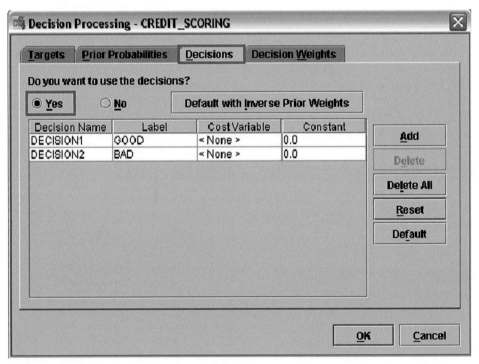

FIGURE M.17   Decision Processing window in SAS-EM.

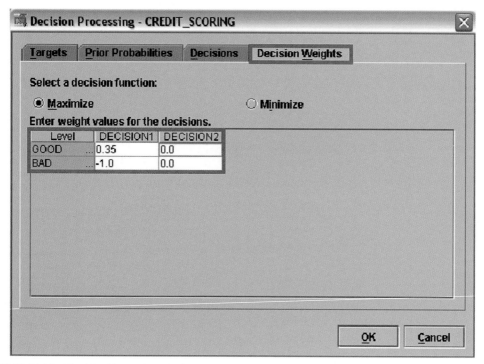

FIGURE M.18   Decision Processing window.

Click on **Decision Weights** in the menu and enter *weight values* for the decision, as shown in Figure M.18.

10. Comparison of the Profits:

Build the process flow to compare the profits of different models. To do so, first click on the **Model** tab to activate a number of predictive modeling tools.

Then drag **Neural Network** and **Decision Tree** icons to the workspace (see Figure M.19).

Next click on the **Assess** tab to activate a number of new tools. This tab contains a special icon called **Model Comparison**. Drag a **Model Comparison** icon to the Workspace and connect the Regression, Tree, and Neural Network nodes to it (see Figure M.20).

Run the **Model Comparison** node and select **Results** in the popup window.

In the pop-up Results window, go to the lower-left corner and click on the arrow of **Cumulative Lift** to open a drop-down window (see Figure M.21).

In the drop-down window, scroll downward to locate the **Expected Profit** for Cumulative Profit (the Total Profit gives non-Cumulative Profit and is useful for other applications) (see Figure M.22).

At the bottom of the **Expected Profit** window, click on the **Neural Network** model. Go to the sub-window with Data Role = TEST. Hold the key at any of the blue boxes to see the Expected Profit Mean (see Figure M.23):

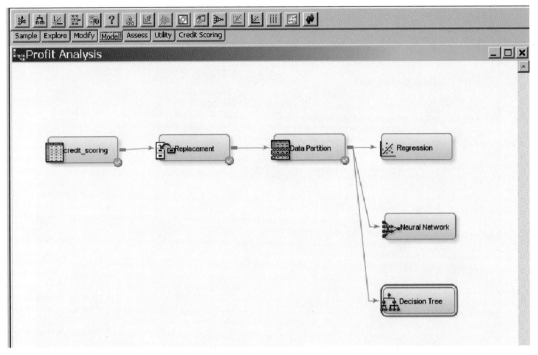

FIGURE M.19   Model process flow in SAS-EM.

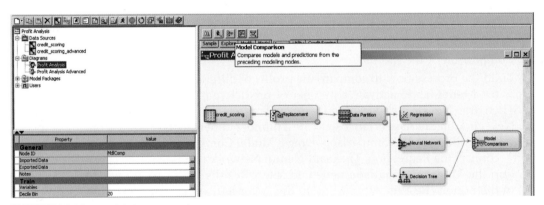

FIGURE M.20   Model Comparison button.

The 0.34 Expected Profit in the top 10% of the data for the Neural Network curve corresponds very closely to the number entered into the profit matrix earlier. Thus, the Neural Network was able to achieve 100% accuracy in the top 10% of the customers. Table M.3 compares the Total Profit of the Tree and Dmine Regression at different cutoff values (Total Profit = Mean Profit*Cutoff*Population Size).

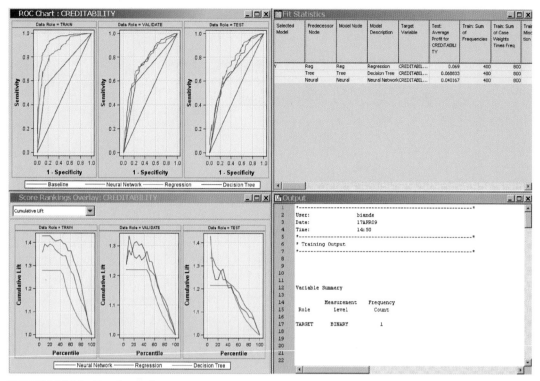

FIGURE M.21   Results window.

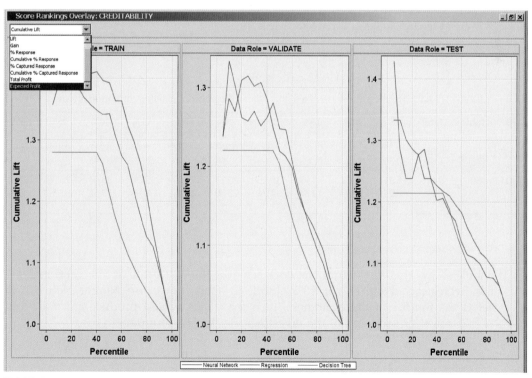

FIGURE M.22   Selecting Expected Profit.

III. TUTORIALS—STEP-BY-STEP CASE STUDIES

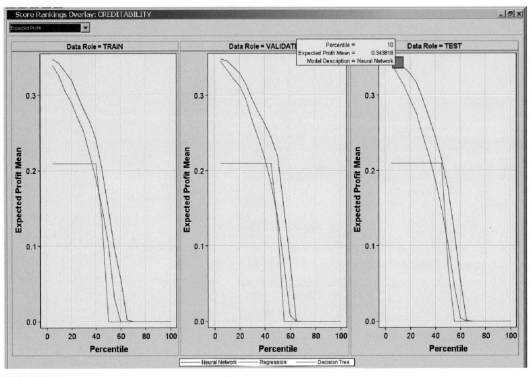

FIGURE M.23   Expected Profit results graph for Dmine Regression.

TABLE M.3   Total Profit of the Tree and Dmine Regression at Different Cutoff Values

N=300 (Holdout Data)		5%	10%	15%	20%	25%	30%	35%	40%	45%	50%
Mean Profit	Reg	0.35	0.341	0.333	0.321	0.303	0.287	0.260	0.217	0.188	0.137
	Tree	0.35	0.35	0.303	0.296	0.296	0.296	0.296	0.296	0.296	0.265
Total Profit	Reg	5.25	10.23	14.99	19.26	22.73	25.83	27.30	26.04	25.38	20.55
	Tree	5.25	10.5	13.64	17.76	22.20	26.64	31.08	35.52	39.96	39.81

The table shows that the Neural Network achieves the best profit at 5% cutoff and the Regression at the 5% or 10% cutoff. In short, if we use Neural Network to select the top 5% of the customers, then the model would produce a Total Profit of 5.25 units for each unit of the investment in the Holdout data (n = 300).

Discussions

a. Assume that we have a new population of 1,000 customers with an average loan of $20,000. The Neural Network model would select the top 5% of the customers and result in a total profit of quite a bit of money indeed.

$$0.35^*0.05^*1000^*\$20,000 = \$350,000$$

ADVANCED TECHNIQUES OF PREDICTIVE MODELING

1. Gradient Boosting and Ensemble Models

   SAS-EM 5.3 has several advanced techniques for building predictive models for classification and regression and many of which also incorporate profit based model selection and assessment. A new technique available in this version is Stochastic Gradient Boosting based on published work (Friedman, 2002) which has shown great promise in many applications. A Support Vector Machine procedure is also available; however, its status is experimental and should not be used for production model development.

   Consequently it would be desirable to compare the performance of Gradient Boosting with other advanced techniques in this specific study. We will again start with the Credit Scoring data.

2. This time, we will use another advanced feature of SAS EM 5.3— the Advanced Advisor in the Datasource Wizard. This selection, as shown in Figure M.24, will execute a

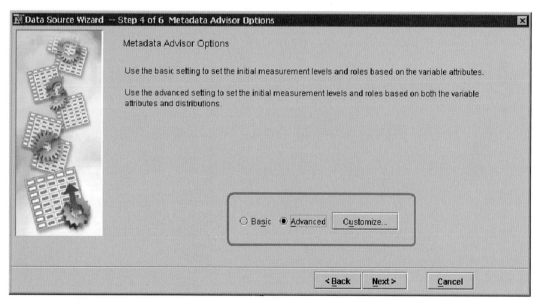

FIGURE M.24   Advanced Advisor feature in the Datasource Wizard.

process that will scan the data for variables that should be ordinal, nominal, or interval, and for variables that should be rejected due to having too many or too few class levels. The user may control distribution thresholds for these assignments.

Using this method, we find that the variables have been assigned a new set of level and role attributes. Again, we have set the Creditability variable as the dependent target variable (see Figure M.25).

3. To complete the analysis, we now click on the **Model** tab to activate the predictive modeling tools. Then drag the **Gradient Boosting** icon to the Workspace.

4. In addition, we will add an **Ensemble** node to the diagram. Ensembles work by combining the predictions from multiple models into a single prediction that often produces superior results to any of the constituent models. In this case, the Ensemble node will average the probabilities of the input models and will be compared to the input models (see Figure M.26).

5. We will again highlight the use of the **Replacement** node. In this case, we want to replace the values of the BALANCE_OF_CURRENT_ACCOUNT with simple values for high, none, low, and missing. Cleaning data is a frequent activity for most data miners. Select the **Replacement** node on your diagram.

6. Change the default interval replacement method to **None**.

7. Select the class variables replacement editor and make the changes shown in Figure M.27.

8. Now that we have cleared that task, select the **Model Compare** node and run the path. Once the path as completed, we will select a champion model.

9. Open the results of the **Model Compare** node and select the output listing. In this case, the Gradient Boosting model has posted the best Test Average Profit value, followed by the Decision Tree, and has been automatically selected as the champion

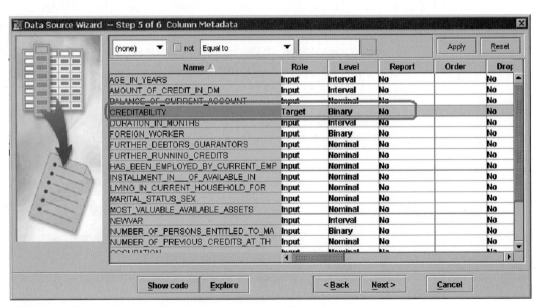

FIGURE M.25   Creditability set as dependent target variable.

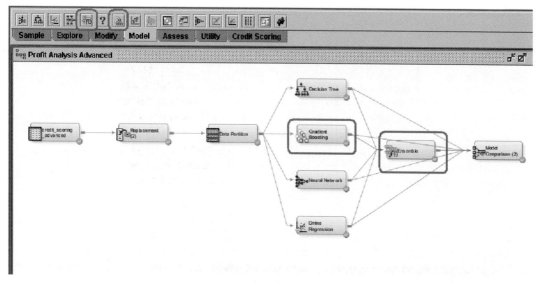

FIGURE M.26   Total profit from decision tree.

Variable	Level	Frequency	Type	Char Raw value	Num Raw Va	Replacemen.
BALANCE_OF_CURRENT_ACCOUNT	>=200 DM	394	C	>=200 DM	.	high
BALANCE_OF_CURRENT_ACCOUNT	no running acc...	274	C	no running account	.	none
BALANCE_OF_CURRENT_ACCOUNT	no balance	269	C	no balance	.	none
BALANCE_OF_CURRENT_ACCOUNT	<= 200 DM	63	C	<= 200 DM	.	low
BALANCE_OF_CURRENT_ACCOUNT	_UNKNOWN_		C		.	_DEFAULT_
CREDITABILITY	good	700	C	good	.	
CREDITABILITY	bad	300	C	bad	.	
CREDITABILITY	_UNKNOWN_		C		.	_DEFAULT_
FOREIGN_WORKER	yes	963	C	yes	.	
FOREIGN_WORKER	no	37	C	no	.	
FOREIGN_WORKER	_UNKNOWN_		C		.	_DEFAULT_
FURTHER_DEBTORS_GUARANTORS	none	907	C	none	.	
FURTHER_DEBTORS_GUARANTORS	guarantor	52	C	guarantor	.	
FURTHER_DEBTORS_GUARANTORS	co-applicant	41	C	co-applicant	.	
FURTHER_DEBTORS_GUARANTORS	_UNKNOWN_		C		.	_DEFAULT_
FURTHER_RUNNING_CREDITS	no further runni...	814	C	no further running cr...	.	
FURTHER_RUNNING_CREDITS	at other banks	139	C	at other banks	.	
FURTHER_RUNNING_CREDITS	at department...	47	C	at department store ...	.	
FURTHER_RUNNING_CREDITS	_UNKNOWN_		C		.	_DEFAULT_
HAS_BEEN_EMPLOYED_BY_CURRENT_EMP	between 1 and ...	339	C	between 1 and 4 yea...	.	
HAS_BEEN_EMPLOYED_BY_CURRENT_EMP	greater than 7 y...	253	C	greater than 7 years	.	
HAS_BEEN_EMPLOYED_BY_CURRENT_EMP	between 4 and ...	174	C	between 4 and 7 yea...	.	
HAS_BEEN_EMPLOYED_BY_CURRENT_EMP	less than 1 year	172	C	less than 1 year	.	
HAS_BEEN_EMPLOYED_BY_CURRENT_EMP	unemployed	62	C	unemployed	.	
HAS_BEEN_EMPLOYED_BY_CURRENT_EMP	_UNKNOWN_		C		.	_DEFAULT_

Reset   OK   Cancel

FIGURE M.27   Replacement Editor.

model (see Figure M.28). In terms of Average Square Error (ASE), the Ensemble model is the champion followed by the Gradient Boosting model. Why are these orderings different? The profit matrix that we entered unevenly weights the distribution of classification matrix; thus, models that are not monotonically related will produce

Fit Statistics
Model selection based on _TAPROF_

Selected Model	Model Node	Test: Average Profit for CREDITABILITY	Train: Average Squared Error	Train: Misclassification Rate	Valid: Average Squared Error	Valid: Misclassification Rate
Y	Boost	0.069333	0.14640	0.2175	0.17604	0.28667
	Tree	0.068833	0.18074	0.3000	0.18892	0.30000
	Ensmbl	0.067000	0.11441	0.1350	0.16880	0.24667
	Neural	0.063500	0.09335	0.1600	0.19300	0.26333
	DmineReg	0.056333	0.11617	0.1675	0.20793	0.27333

FIGURE M.28

different orders in those measures. Alternatively, you may think that these three models have identified different features within the overall pattern detection.

The ability of Ensembles to outperform their constituent models on classification tasks (Elder, 2003) is a very interesting effect. If we look at the expected profit curve we find a different story. In this case, the Neural Network is the champion with a consistent expected profit of 0.35 in the 5% percentile over the train, validation, and test data sets (see Figure M.29).

10. We can also look at the more conventional measures such as ROC, shown in Figure M.30. If we were selecting models based on the tradeoff between specificity and sensitivity, the Ensemble model is consistently superior with the Neural Network is a perhaps insignificantly close and second. The Decision Tree and Gradient boosting models are ranked lower on this measure. In fact, the sharp line shape of the Decision Tree indicates that a shallow tree was created which produces a harsh distribution of probabilities.

11. We may also examine the lift charts where we find the Neural Network is now consistently the highest ranked model, followed by the Gradient Boosting. The Decision Tree is again ranked lower due to its highly pruned structure (see Figure M.31).

12. To better understand these results, we can look at the distributions of scores. In the results of the model compare node, select the menu item **View** → **Assessment** → **Score Distribution Plots**. First look at the plot of the Gradient Boosting model and find a good separation of true and false events (see Figure M.32).

Now, we look at the Ensemble model and the Neural Network models. The Ensemble model shows a similar distribution with more separation between the two cases which can be related to its slightly better score on ASE. The Neural model, on the other hand, shows a very different score distribution that produces more low probabilities and more overlap between the models (see Figure M.33). Remember that the Ensemble is a mixture of the constituent models including both the Gradient Boosting model and the Neural Network model.

13. Now that we have established that the Gradient Boosting model is our champion, we can look at its results in more detail. Open the results of the **Gradient Boosting** node

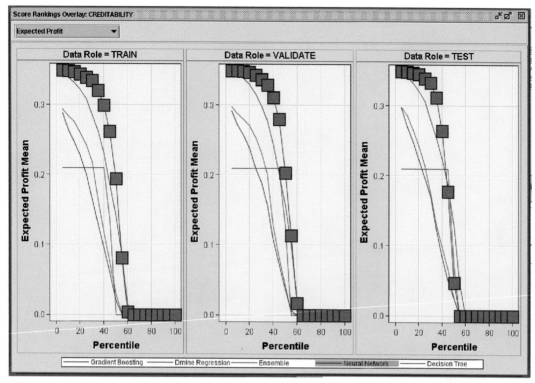

FIGURE M.29   NN expected profit.

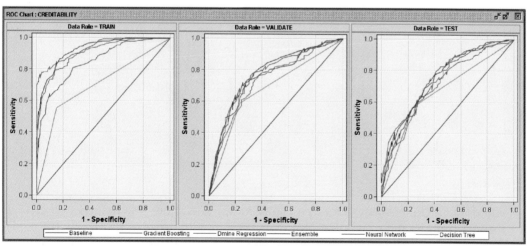

FIGURE M.30

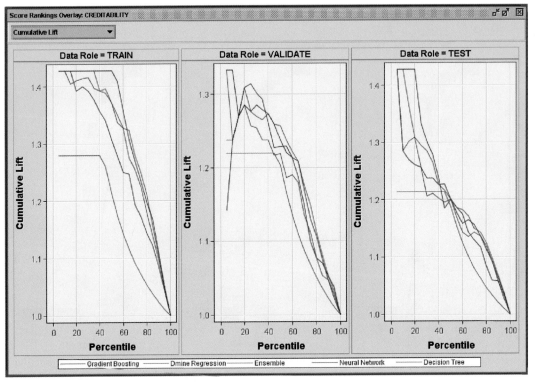

FIGURE M.31

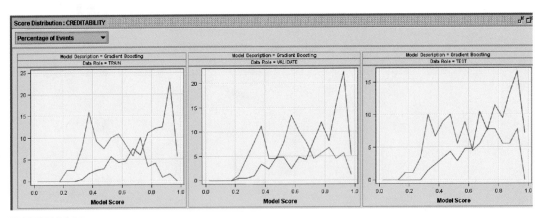

FIGURE M.32

and examine the Subseries Plot (see Figure M.34). This shows the reduction of error as the model grows more complex. However, the ASE plot does not show why the model was selected at iteration 39. Since we entered a profit matrix, Enterprise Miner also shows the evolution of profit. The Gradient Boosting model was selected to maximize profit which helps us develop a better campaign.

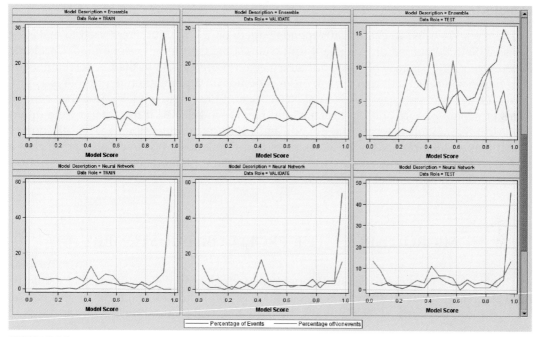

FIGURE M.33

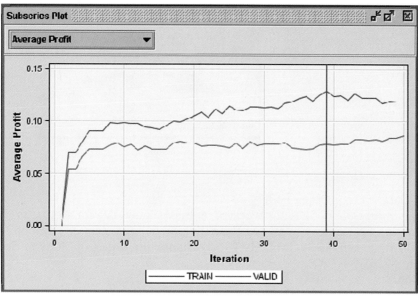

FIGURE M.34

Conclusion

The model selected as champion depends largely on the measurement used to make the decision. Selection based on Average Profit chooses the Gradient Boosting model, yielding an average profit per case of 0.069, but an expected profit of 0.35 in the 5% percentile. The data miner will select and report rank order measures at a population depth that is appropriate for the business case. The Neural Network and Ensemble models were very close in terms of overall performance and each would have been selected if the criteria were different. All three models detected the pattern and produced usable models. The selection of model is often determined by business rules and regulations; however, also producing a best model from modern techniques is valuable for setting bounds on the expected performance of the chosen model.

MICRO-TARGET THE PROFITABLE CUSTOMERS

This sub-section presents detailed steps on how to identify the customers that would be most profitable. Recall that Decision Tree is the best and hence we will focus on this model. The red block in Figure M.35 highlights the parts that will be used for micro-targeting.

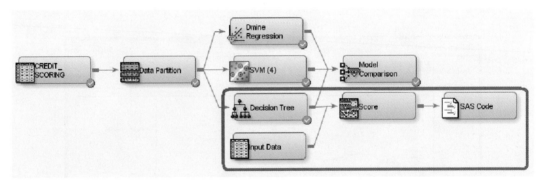

FIGURE M.35

1. Scoring Data via Input Data node:
   Drag the Input Data node to the workspace. Import new data for scoring (see details in the section "A Primer of SAS-EM Predictive Modeling," item 2). Click on the **Input Data** node to activate its Property Panel. Change the **Role** from **Raw** to **Score** (see Figure M.36).
2. Score Node:
   Add the Score node to the Decision Tree node. **Run** the Score node (see Figure M.37).
3. SAS Code Node (Figure M.38):

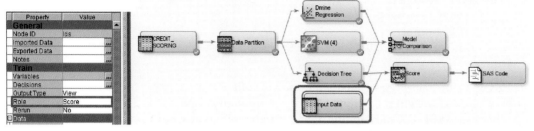

FIGURE M.36

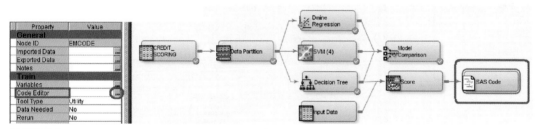

FIGURE M.37

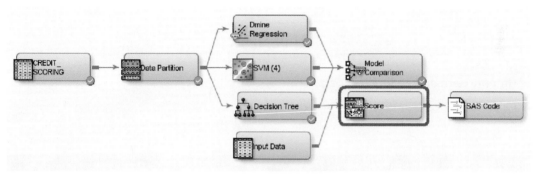

FIGURE M.38

```
Data Customers;
        Set &EM_Import_Score;
        Customer_ID = _N_;
Run;

PROC Sort data = Customers;
        By descending P_CreditabilityGood;
Run;
```

Data good_customers;
        Set Customers;
        Obsnum = _N_;
        If Obsnum > **0.05*1000** THEN delete;
Run;

PROC Print data = good_customers noobs split = '*';
        VAR obsnum Customer_ID P_CreditabilityGood;
        LABEL P_CreditabilityGood= 'Predicted*Good*Credit';
        TITLE "Credit Worthy Applicants";
Run;

Proc print;

APPENDIX

(Import German Credit Excel data to the SASuser library)

1. Open base SAS. Select **File ➔ Import Data**.

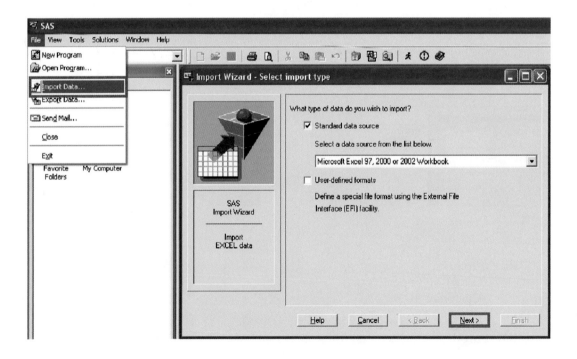

2. Click on Next and then Browse the location of the data:

3. Select the SASUSER library.

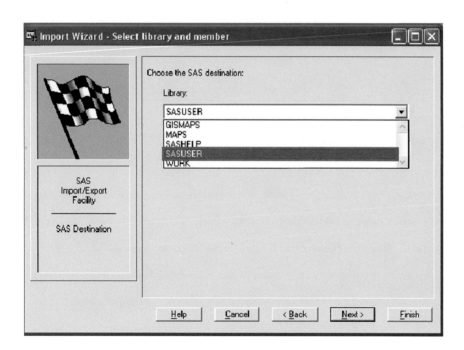

4. Type the name of the file, click on **Next** and then in the next window click on Finish to complete the data import.

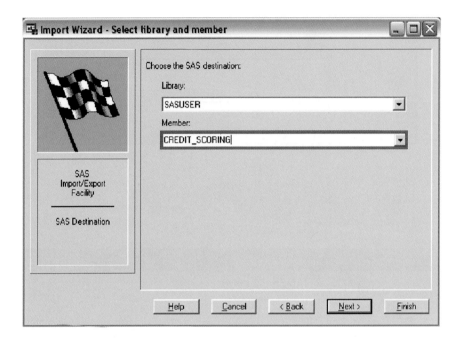

References

Adnan, A. & Bastos, E. (2005). A comparative estimation of machine learning methods on QSAR data sets. *SUGI-30 Proceedings*.

Ayres, I. (2007). *Super Crunchers: Why Thinking-by-Numbers Is the New Way to Be Smart*. New York, NY: Bantam Books.

Elder, J. F., IV. (2003). The generalization paradox of ensembles. *Journal of Computational and Graphical Statistics*, *12*, No. 4, 853–864.

Foster, D. P., & Stine, R. A. (2004). Variable selection in data mining: Building a predictive model for bankruptcy. *JASA*, *99*(466), 303–313.

Friedman, J. H. (2002). Greedy function approximation: A gradient boosting machine. *Annuals of Statistics*, *29*, 1189–1232.

Friedman, J. H., & Popescu, B. E. (2005). *Uncovering interaction effects*. Presented in the Second International Salford Systems Data Mining Conference. New York, NY, March 29–30, 2005.

Wang, C., & Liu, B. (2008). Data mining for large datasets and hotspot detection in an urban development project. *Journal of Data Science*, *6*, 389–414.

Yu, J. S., Ongarello, S., Fiedler, R., Chen, X. W., Toffolo, G., Cobelli, C., et al. (2005). Ovarian cancer identification based on dimensionality reduction for high-throughput mass spectrometry data. *Bioinformatics*, *21*(10), 2200–2209.

Predicting Self-Reported Health Status Using Artificial Neural Networks

Nephi Walton, Stacey Knight, MStat, and Mollie R. Poynton, Ph.D., APRN, BC

Edited by Gary Miner, Ph.D.

BACKGROUND

Self-reported health status in the form of the question "How would you say your health in general is?" has been shown to be an excellent predictor of mortality, health care utilization, and disability (Mossey and Shapiro, 1982; Idler et al., 1992; Idler and Kasl, 1995; Idler and Benyamini, 1997; Idler et al., 2004; Maciejewski et al., 2005). While the strength of effect varies, the predictive power of self-reported health status has been found in different countries, racial/ethnic groups, age groups, and patient populations (Mossey and Shapiro, 1982; Idler and Kasl, 1991; Miilunpalo et al., 1997; Leinonen et al., 1998; Mansson and Merlo, 2001; Mackenbach et al., 2002; Bath, 2003; Larson et al., 2008). There appears to be an intrinsic prediction power of self-reported health status on health outcome above that explained by other factors including gender, age, social and economic resources, and medical condition (Idler and Kasl, 1991; Burstrom and Fredlund, 2001; Fiscella and Franks, 2000; Fan et al., 2002; Idler et al., 2004). While the relationship between health outcome and self-reported health status has been greatly studied, few studies have examined factors that might contribute to or predict self-reported health status (Wan, 1976; Manderbacka et al., 1998; Fiscella and Franks, 2000; Mansson and Merlo, 2001; Mackenbach et al., 2002; Franks et al., 2003; Idler et al., 2004; Jylha et al., 2006). None of these studies have been comprehensive, with most focusing on specific factors such as social support, economic resources, biomarkers, or risk factors. Furthermore, all of these studies have used standard statistical methods (e.g., Cox-regression and logistic regression) for determining association with the factors and health status, and none of these studies have used data mining techniques designed for the purpose of prediction.

In this tutorial we will use neural networks to derive a model that will accurately predict self-reported health status. The National Health and Nutrition Examination Survey (NHANES) 2003–2004 data were used for this tutorial (CDC, 2008). These data encompass a wide range of factors that span many aspects of health, providing for a more comprehensive prediction model. Neural networks allow us to detect complex interactions and patterns in the data, and increase the classification accuracy. A classification algorithm for self-reported health status should give health care providers a better understanding of the underlying factors associated with self-reported health status and in turn provide better knowledge of factors associated with mortality and morbidity.

DATA

The NHANES 2003–2004 data set of more than 4,000 variables was collected by the Centers for Disease Control and Prevention (CDC) National Center for Health Statistics (NCHS). Participants undergo an interview that consists of questions about demographics, personal health and medical conditions, social support and resources, nutritional and dietary intake, and risk behaviors. A selected number of individuals receive a physical examination. Specimens collected from the examination are analyzed, and measures of various biomarkers and toxins are recorded. The class variable of interest is the response to the survey question "How would you say your health in general is—excellent, very good, good, fair, or poor?" The participants' answers were categorized into two groups: good as defined by a response of excellent, very good, or good versus poor as a defined by a response of fair or poor. The variable in our data set for this question is "Status." For the modeling, we chose to limit the age range to individuals over the age of 39 because certain data of interest were not collected on subjects less than 40 years of age. Subjects were restricted to be those with both a physical examination and laboratory data because these factors were deemed to be important for the analysis.

Preprocessing and Filtering

There are more than 4,000 variables in this data set, so a filtering process was necessary. The first step was to select factors based on prior literature and domain expert opinion that could be indicators of self-reported health status. This approach was chosen based on an understanding of the study question and to eliminate all but the most relevant and interesting variables. Some numeric variables were categorized based on standards found in the literature, but the majority were left as continuous variables. Variables with >10% missing data were removed, resulting in a list of 105 variables.

The variables were further filtered using a tier process. First, each variable's association with self-reported health status was determined using a chi-square test for nominal or ordinal variables and logistic regression for continuous variables. Variables with a p-value less than 0.20 were used in the next stage of filtering. This threshold was set high to allow for the detection of variables that had an interaction effect, but whose main effects would only be marginal. Before the next stage of selection, we imputed all of the missing values in the data set using a model-based algorithm that uses the values of other variables of an instance to compute the value for the missing variable. This filtering stage resulted in the selection of 85 variables. This subset of variables is the data set used in this tutorial. We start the first part of the tutorial by selecting the most important subset of these variables using Weka.

Note: When doing this tutorial, please take just a small part of the data set and redo with this smaller set of data; using the entire data set will take the following Weka procedures several hours (e.g., 4–6 hours or more) to run. To make this process go faster, reduce the size of the data. Alternatively, read through the Weka part and then take the results part of the data into the *STATISTICA* Automated Neural Networks, which should run in just a few minutes.

Part 1: Using a Wrapper Approach in Weka to Determine the Most Appropriate Variables for Your Neural Network Model

1. Open the Weka Explorer, as shown in Figure N.1.

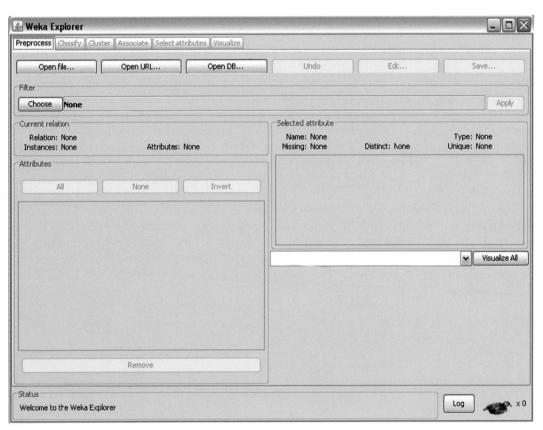

FIGURE N.1   Weka Explorer interface window.

2. Make sure the Preprocess tab is selected and then click the Open File ... button (see Figure N.2).
3. When the Open file dialog appears, select CSV Data Files from the Files of Type drop-down box.
4. Select and open the file Tutorial_NHANES_Data.csv.

FIGURE N.2   Open data file window.

5. Once the file has been selected, click on the Select Attributes tab at the top of the explorer (see Figure N.3). Under Attribute Selection Mode, notice that the default variable we are predicting is set to Status. Weka defaults this variable to the last variable in your data set. You could change this variable by clicking on it and selecting another variable from the pop-up list. Since Status is our target variable, we will not change it this time.
6. Click on the Choose button under Attribute Evaluator.

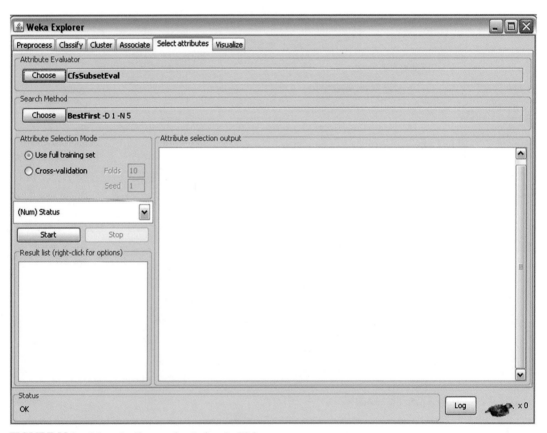

FIGURE N.3   Select Attributes tab window in Weka.

7. Select the WrapperSubsetEval method, as shown in Figure N.4.

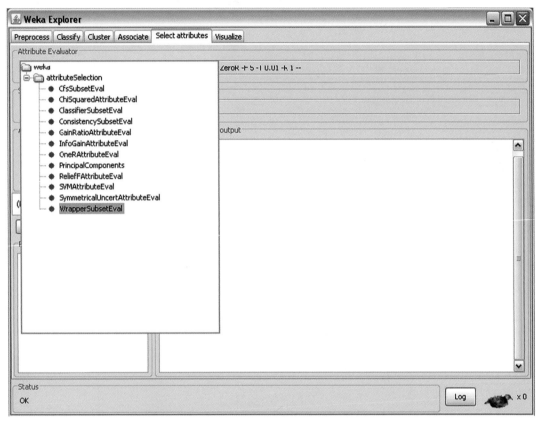

WrapperSubsetEval method selected from the Attribute Evaluator menu in Weka.

8. Double-click WrapperSubsetEval under Attribute Evaluator; this should open up the dialog box shown in Figure N.5.

9. In the dialog box where it says Classifier, click on the Choose button.

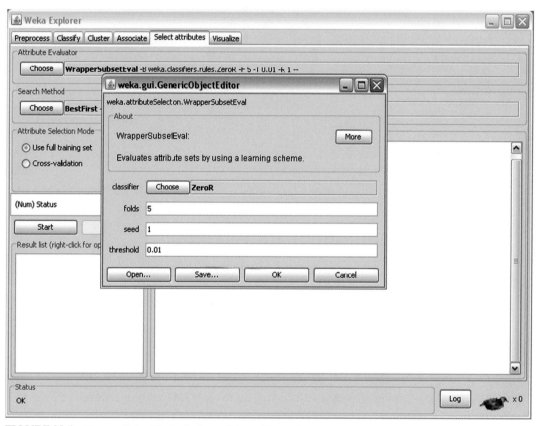

FIGURE N.5   WrapperSubsetEval window; click on Choose button.

10. This pops up a list of folders and classifiers, as shown in Figure N.6. Open the Functions folder and select MultiLayerPerceptron.
11. Leave the rest of the values at defaults and click OK
12. Click on the Choose button under Search Method.

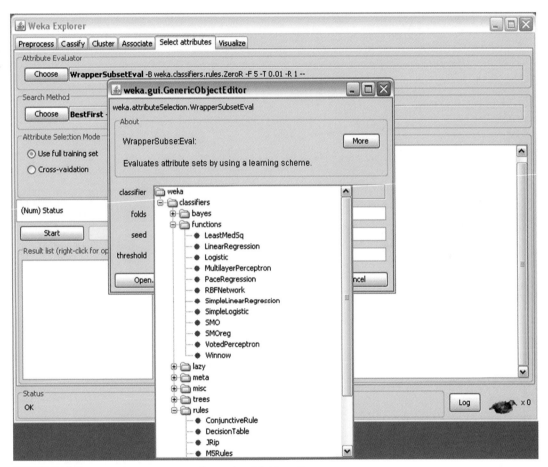

FIGURE N.6   Classifiers pop-up window; select MultlayerPerceptron.

13. Select the GeneticSearch method, as shown in Figure N.7.
14. Under the Attribute Selection Mode in the Explorer, select the Cross-validation radio button and leave the values at their defaults.
15. Click the Start button to start the process; this may take several hours to run depending on your CPU speed.
16. Once the wrapper is finished running, it will display a list of selected variables in the output window. We will use these selected variables in our neural network model. The results of this method, shown here, could vary slightly with each run.

Resulting Variables from the Weka Wrapper Subset Selection

Age
Education level
Income
Problem with balance
Chest pain ever
Taking antacid
Need dental work

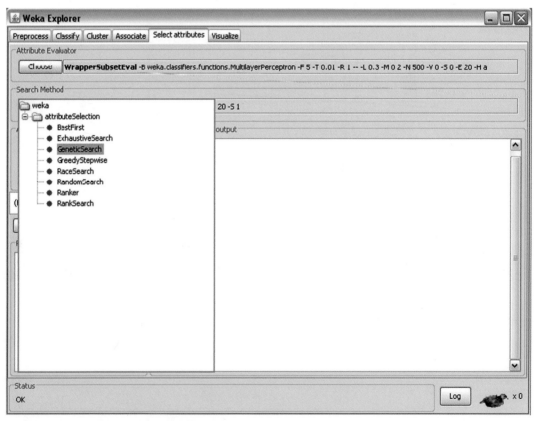

FIGURE N.7   GeneticSearch method selected in search window.

Part 2: Taking the Results from the Wrapper Approach in Weka into *STATISTICA* Data Miner to do Neural Network Analyses

Now following this Wrapper Subset selection in Weka, we will take these variables into *STATISTICA* and use the Automatic Neural Networks to do a classification and find a model.

1. First, we need to open our data file in *STATISTICA*. We have saved an identical copy of the CSV file used in Weka as an Excel.xls file for ease of importing into *STATISTICA*. Select Open from the File menu of *STATISTICA*, as shown in Figure N.8.

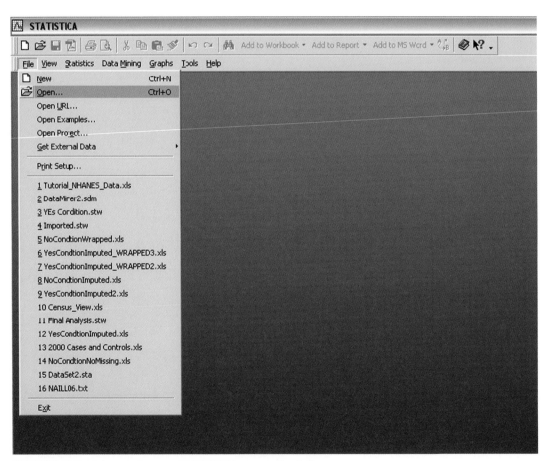

FIGURE N.8   Selecting Open in *STATISTICA* to open Excel file.

2. In the Open dialog, select Excel Files (*.xls) from the Files of Type drop-down box (see Figure N.9). Now find the file Tutorial_NHANES_data.xls and open it by double-clicking it or selecting the file and clicking the Open button.

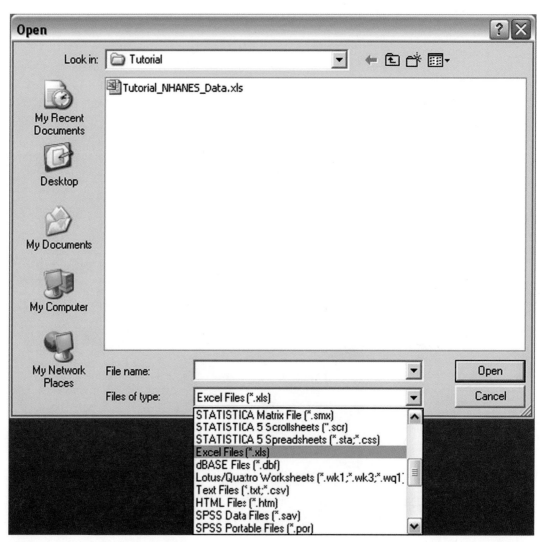

FIGURE N.9   Selecting an Excel File to open in *STATISTICA*.

3. Once you open the file, the dialog shown in Figure N.10 will appear. Click on Import All Sheets to a Workbook. This selection will load all the sheets in the Excel spreadsheet into *STATISTICA*. This Excel file has only one sheet in it. You could selectively import any sheet from an Excel file by clicking on the Import Selected Sheet to a Spreadsheet button.

4. Another dialog appears with some options for importing your data (see Figure N.11). Check the box next to Get Variable Names from First Row. This selection will take the variable names from the first row of the Excel file and use them as names for all your variables in *STATISTICA*. Otherwise, each variable would be assigned a number in sequential order in *STATISTICA*. Now click OK, and the data file will be opened in *STATISTICA*.

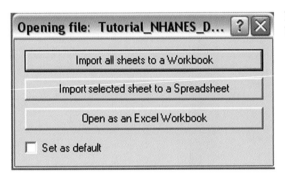

FIGURE N.10   Opening an Excel File in *STATISTICA*; select Import All Sheets to a Workbook.

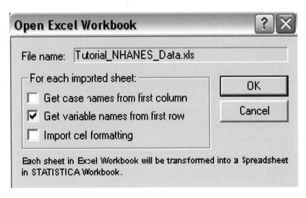

FIGURE N.11   Our Excel data sheet has variable names listed in the first row; thus, select Get Variable Names From First Row.

5. To start Automated Neural Networks, select it from the Data Mining menu, as shown in Figure N.12.
6. In the resulting dialog, select Classification as the type of neural network because we are classifying subjects into either poor or good health status (see Figure N.13).
7. Click OK.

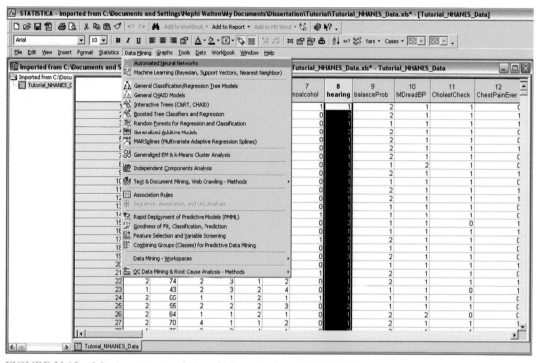

FIGURE N.12   Selecting Automated Neural Networks from *STATISTICA*.

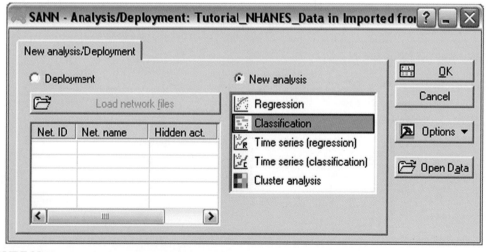

FIGURE N.13   *STATISTICA* Automated Neural Networks new analysis dialog.

8. A data selection dialog appears, as shown in Figure N.14. Click on the Variables button, and a variable selection dialog will appear.

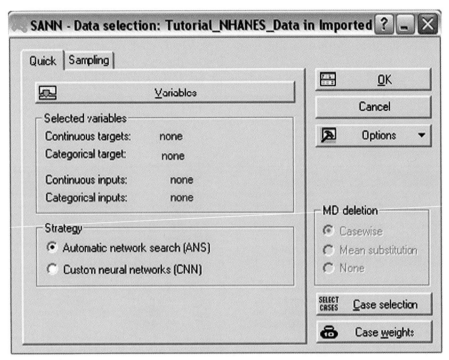

FIGURE N.14   Variable selection dialog in *STATISTICA* Automated Neural Networks.

9. On the variable selection dialog, first select the output variable as Status, as shown in Figure N.15. You can do this by clicking on it from the list or typing in variable number 85 in the box below the list.

10. Now you have to select your input variables. Input variables can have categorical values such as education level or continuous values such as lab values. Income has been selected as one of our variables, and this could be continuous or categorical. Since this study categorized income into groups of income levels, it is categorical in this study. All the variables selected in Weka happen to be categorical, so we will select those variables under the categorical inputs heading. Again you can select these by clicking on them or typing their assigned numbers in the box below the list. To select multiple variables from the list, click on the first one and then hold down Ctrl while you click on the other variables you want to include.

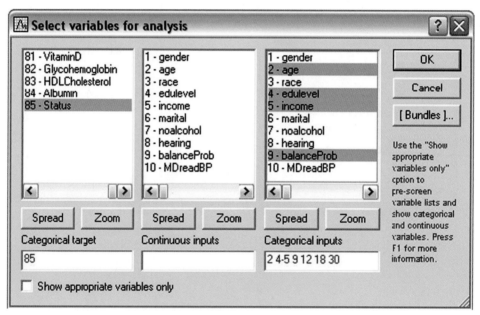

FIGURE N.15   Variables selected for analysis are highlighted above.

11. Now click on the Sampling tab at the top of the dialog box (see Figure N.16). We will keep the default of using 80% of the data set for training and 20% for testing. You can change this however you would like.
12. Click OK and you will go to the next dialog.

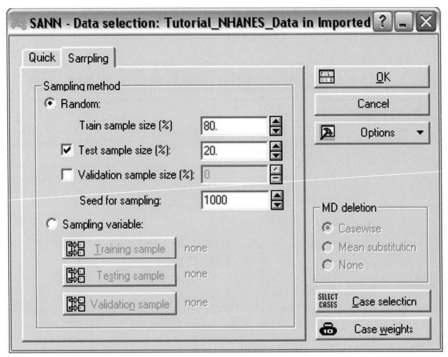

FIGURE N.16   Sampling tab in *STATISTICA* Automated Neural Networks (SANN).

13. On this dialog, shown in Figure N.17, you can select the types of networks to use and how many networks you want to test and how many to retain. *STATISTICA* will retain the best performing networks. For this tutorial, we will uncheck Radial Basis Function (RBF) networks, since our wrapper was optimized for multilayer perceptron networks. We will leave the number of networks to test at 20 and leave the number of networks to retain at 5.

14. Click on Train, and *STATISTICA* will start testing neural networks. This process may take several minutes to an hour, depending on your CPU speed.

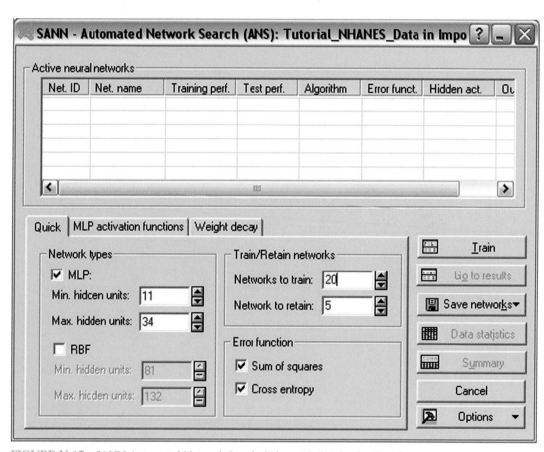

FIGURE N.17   SANN Automated Network Search dialog with Quick tab selected.

15. Once the testing is complete, *STATISTICA* displays the five top performing networks (see Figure N.18). Results may vary each time this process is run, but you should achieve accuracies around 70%. You can see in the results the details of each network and its performance on both the training and testing data set.

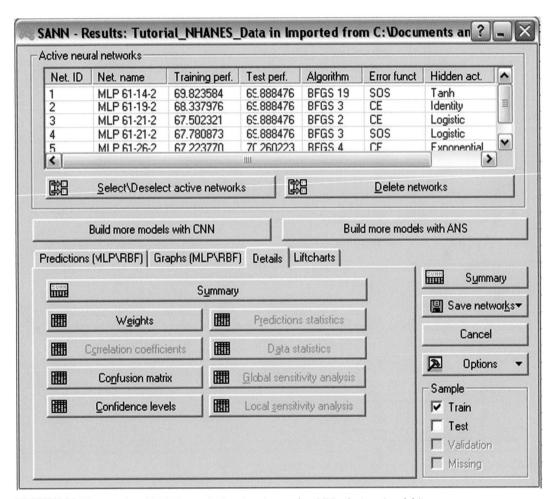

FIGURE N.18   Results of SANN search showing the top five NN solutions (models).

16. To compare ROC curves of the five different networks, click on the Liftcharts tab (see Figure N.19).

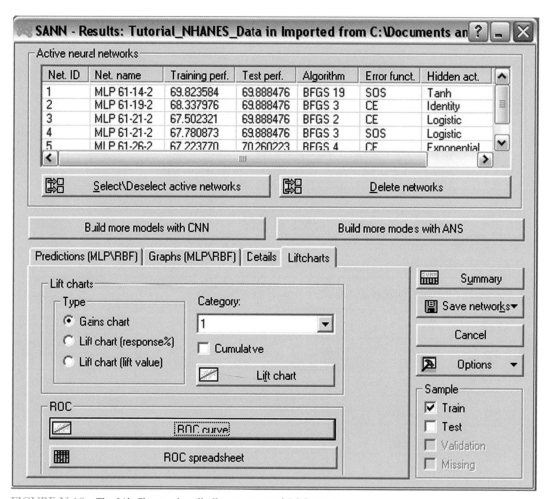

FIGURE N.19   The Lift Charts tab will allow outputs of ROC curves.

17. Click the ROC curve button to display ROC curves for each of the retained networks (see Figure N.20).

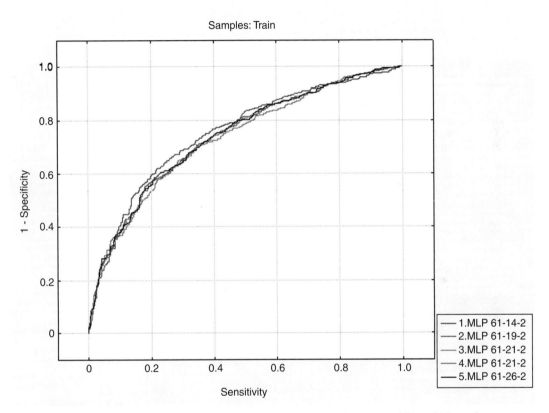

FIGURE N.20   ROC curves for each of the five retained *STATISTICA* Automated Neural Networks.

18. You can also look at the actual output of the network and the confidence for each prediction. Click on the Predictions (MLP\RBF) tab, as shown in Figure N.21. Then check the boxes to include Targets, Output, and Confidence. On the right side, deselect Train and select Test to look at the testing data instead of the training data. After you have checked the items you want to look at, click on the Predictions button. A spreadsheet will appear showing all the values you checked off.

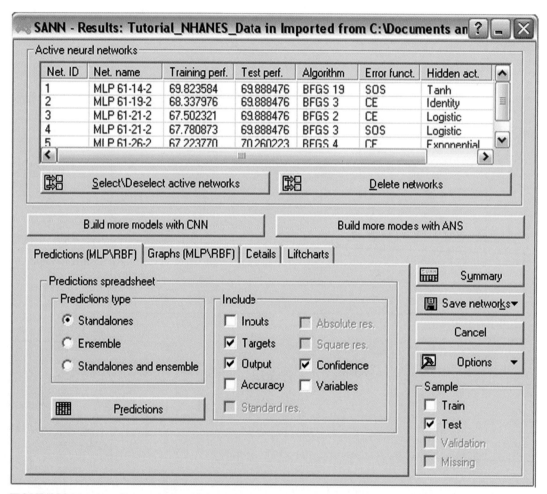

FIGURE N.21   Actual output and confidence for each prediction can be obtained from the Prediction tab; select type of prediction, what output you want, and select either Train or Test Sample. Click OK and a spreadsheet will appear showing all the values checked.

References

Bath, A. (2003). Differences between older men and women in the self-rated health-mortality relationship. *The Gerontologist*, 43(3), 387–395; discussion 372–375.

Burstrom, B., & Fredlund, P. (2001). Self rated health: Is it as good a predictor of subsequent mortality among adults in lower as well as in higher social classes? *Journal of Epidemiology and Community Health*, 55(11), 836–840.

Centers for Disease Control and Prevention (CDC). (2008). National Center for Health Statistics (NCHS). National Health and Nutrition Examination Survey Data. March 30, 2008; Available from http://www.cdc.gov/nchs/about/major/nhanes/datalink.htm

Fan, V. S., et al. (2002). Validation of case-mix measures derived from self-reports of diagnoses and health. *Journal of Clinical Epidemiology*, 55(4), 371–380.

Fiscella, K., & Franks, P. (2000). Individual income, income inequality, health, and mortality: What are the relationships? *Health Services Research*, 35(1 Pt 2), 307–318.

Franks, P., Gold, M. R., & Fiscella, K. (2003). Sociodemographics, self-rated health, and mortality in the US. *Social Science & Medicine (1982)*, 56(12), 2505–2514.

Idler, E., et al. (2004). In sickness but not in health: Self-ratings, identity, and mortality. *Journal of Health and Social Behavior*, 45(3), 336–356.

Idler, E. L., & Benyamini, Y. (1997). Self-rated health and mortality: A review of twenty-seven community studies. *Journal of Health and Social Behavior*, 38(1), 21–37.

Idler, E. L., & Kasl, S. (1991). Health perceptions and survival: Do global evaluations of health status really predict mortality? *Journal of Gerontology*, 46(2), S55–65.

Idler, E. L., & Kasl, S. V. (1995). Self-ratings of health: Do they also predict change in functional ability? *The Journals of Gerontology. Series B, Psychological Sciences and Social Sciences*, 50(6), S344–353.

Idler, E. L., Russell, L. B., & Davis, D. (2000). Survival, functional limitations, and self-rated health in the NHANES I Epidemiologic Follow-up Study, 1992. First National Health and Nutrition Examination Survey. *American Journal of Epidemiology*, 152(9), 874–883.

Jylha, M., Volpato, S., & Guralnik, J. M. (2006). Self-rated health showed a graded association with frequently used biomarkers in a large population sample. *Journal of Clinical Epidemiology*, 59(5), 465–471.

Larson, C. O., et al. (2008). Validity of the SF-12 for use in a low-income African American community-based research initiative (REACH 2010). *Preventing Chronic Disease*, 5(2), A44.

Leinonen, R., Heikkinen, E., & Jylha, M. (1998). Self-rated health and self-assessed change in health in elderly men and women—A five-year longitudinal study. *Social Science & Medicine (1982)*, 46(4–5), 591–597.

Maciejewski, M. L., et al. (2005). The performance of administrative and self-reported measures for risk adjustment of Veterans Affairs expenditures. *Health Services Research*, 40(3), 887–904.

Mackenbach, J. P., et al. (2002). Self-assessed health and mortality: Could psychosocial factors explain the association? *International Journal of Epidemiology*, 31(6), 1162–1168.

Manderbacka, K., Lahelma, E. & Martikainen, P. (1998). Examining the continuity of self-rated health. *International Journal of Epidemiology*, 27(2), 208–213.

Mansson, N. O., & Merlo, J. (2001). The relation between self-rated health, socioeconomic status, body mass index and disability pension among middle-aged men. *European Journal of Epidemiology*, 17(1), 65–69.

Miilunpalo, S., et al. (1997). Self-rated health status as a health measure: The predictive value of self-reported health status on the use of physician services and on mortality in the working-age population. *Journal of Clinical Epidemiology*, 50(5), 517–528.

Mossey, J. M., & Shapiro, E. (1982). Self-rated health: A predictor of mortality among the elderly. *American Journal of Public Health*, 72(8), 800–808.

Wan, T. T. (1976). Predicting self-assessed health status: A multivariate approach. *Health Services Research*, 11(4), 464–477.

MEASURING TRUE COMPLEXITY, THE "RIGHT MODEL FOR THE RIGHT USE," TOP MISTAKES, AND THE FUTURE OF ANALYTICS

In many ways, the purpose of this book is distilled in the following chapters of Part IV. The information in the previous chapters and the practical experience provided by the tutorials are but a prelude to the "symphony" of data mining the authors offer in these chapters. To a large extent, this book was written backwards from the way it was viewed initially. Indeed, we could have written only the chapters in Part IV, and we would have fulfilled our desire to share the combined experience of over 100 years of analytical work. We hope that you will enjoy these chapters and profit from them in multiple and unanticipated ways.

Model Complexity (and How Ensembles Help)

PREAMBLE

How do you choose the next President of the United States (or any elected official)? Different people and groups of people have different ideas about who would be the best candidate to put into office. Totalitarian systems install the strongest person into office (or the one with the most military power). This system leads almost always to repressive and unilateral governments from the perspective of one person. Demographic systems form more representative governments by giving all or part of the people a vote for the winner. In the U.S. system, all eligible people have an equal vote, but the "super-voters" representing each state in the Electoral College can reverse the popular vote. This foray into civics is an introduction to the philosophy behind creating models that come as close as possible to representing the significant "voices" in a data set (aspects of the target signal), and selecting the best model to reflect them all. No single model can do it. We must let groups of equal "super-voters" (like the Electoral College) cast their votes and add them up to decide the "winning" predictions for any given case in the data set. We do this by creating an *ensemble* of models, each perhaps using a different mathematical algorithm to predict the outcomes.

These algorithms "look" at the data in slightly different ways, just like the different states of the union view a presidential candidate. And, the surprise of this chapter is that ensembles are actually less complex in behavior than single models; that is, they are less flexible in their adjustment to arbitrary changes in the training data, and thus can generalize to new data more accurately.

Ensemble models—built by methods such as *bagging* (Breiman, 1996), *boosting* (Freund and Shapire, 1996), and *Bayesian model averaging*—appear dauntingly complex, yet tend to strongly outperform their component models on new data. This is a statistical paradox, as the result appears to violate Occam's Razor—the widespread belief that "the simpler of competing alternatives is preferred." We argue, however, that complexity has traditionally been measured incorrectly. Instead of counting parameters (as with regression) to assess the complexity of a modeling process (a "black box"), we need to measure the flexibility of the modeling process—e.g., according to Generalized Degrees of Freedom, GDF (Ye, 1998). By measuring a model's function rather than its form, the role of Occam's Razor is restored. We'll demonstrate this on a two-dimensional decision tree problem, where an ensemble of several trees is shown to actually have less GDF complexity than any of the single trees contained within it.

MODEL ENSEMBLES

A wide variety of competing methods is available for inducing models, and their relative strengths are of keen interest. Clearly, results can depend strongly on the details of the problems addressed, as shown in Figure 18.1 (from Elder and Lee, 1997), which plots the relative out-of-sample error of five algorithms for six public-domain problems. Every algorithm scored best or next-to-best on at least two of the six data sets. Michie et al. (1994) built a decision tree from a larger such study (23 algorithms on 22 data sets) to forecast the best algorithm to use given a data set's properties. Though the study was skewed toward trees—they were nine of the algorithms studied, and several selected data sets exhibited sharp thresholds—it did reveal some useful lessons for algorithm selection (Elder, 1996a).

Still, a method for improving accuracy more powerful than tailoring the algorithm has been discovered: bundling models into ensembles. Figure 18.2 reveals the out-of-sample accuracy of the models of Figure 18.1 when they are combined four different ways, including averaging, voting, and "advisor perceptrons" (Elder and Lee, 1997).

Building an ensemble consists of two steps: (1) constructing varied models and (2) combining their estimates. One may generate component models by varying case weights, data values, guidance parameters, variable subsets, or partitions of the input space. Combination can be done by voting, but is primarily accomplished through weights, with gating and advisor perceptrons as special cases. For example, Bayesian model averaging sums estimates of possible models, weighted by their posterior evidence. Bagging (*bootstrap aggregating*; Breiman, 1996) bootstraps the training data set (usually to build varied decision trees) and takes the majority vote or the average of their estimates. Boosting (Freund and Shapire, 1996)

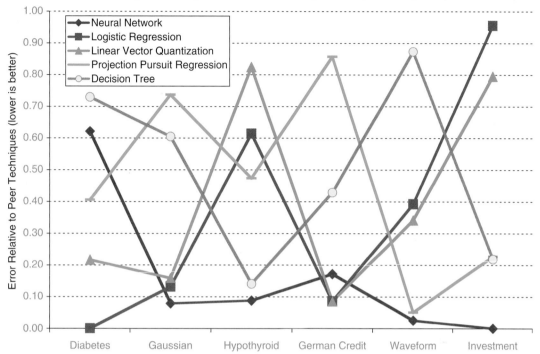

FIGURE 18.1 Relative out-of-sample error of five algorithms on six public-domain problems (from Elder and Lee, 1997).

and ARCing (Breiman, 1996) iteratively build models by varying case weights (up-weighting cases with large current errors and down-weighting those accurately estimated) and employs the weighted sum of the estimates of the sequence of models.

The Group Method of Data Handling (GMDH; Ivakhenko, 1968) and its descendent, Polynomial Networks (Barron et al., 1984; Elder and Brown, 2000), can be thought of as early ensemble techniques. They build multiple layers of moderate-order polynomials, fit by linear regression, where variety arises from different variable sets being employed by each node. Their combination is nonlinear since the outputs of interior nodes are inputs to polynomial nodes in subsequent layers. Network construction is stopped by a simple cross-validation test (GMDH) or a complexity penalty. Another popular method, stacking (Wolpert, 1992), employs neural networks as components (whose variety can stem from simply using different guidance parameters, such as initialization weights), combined in a linear regression trained on leave-one-out estimates from the networks.

Lastly, model fusion (Elder, 1996b) achieves variety by averaging estimates of models built from very different algorithms (as in Figures 18.1 and 18.2). Their different basis functions and structures often lead to their fitting the data well in different regions, as suggested by the two-dimensional surface plots of Figure 18.3 for five different algorithms.

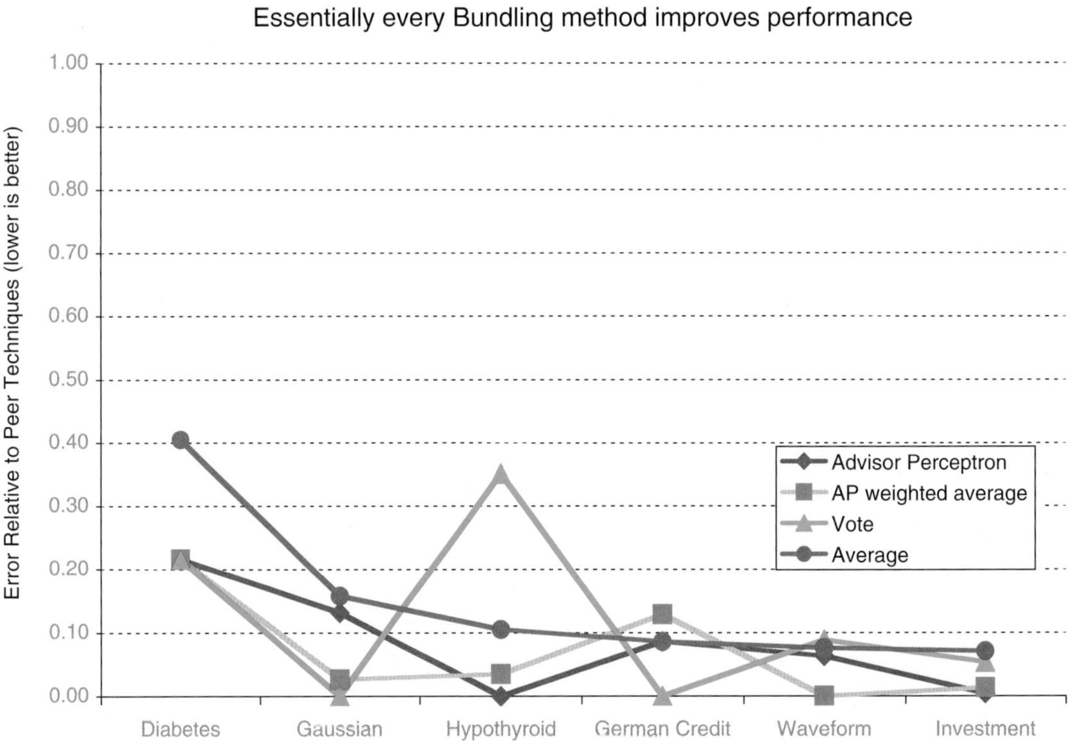

FIGURE 18.2   Relative out-of-sample error of four ensemble methods on the problems of Figure 18.1 (from Elder and Lee, 1997).

Figure 18.4 reveals the out-of-sample results of so fusing up to five different types of models on a credit scoring application. The combinations are ordered by the number of models involved, and Figure 18.5 highlights the finding that the mean error reduces with increasing degree of combination. Note that the final model with all five components does better than the best of the single models.

COMPLEXITY

One criticism of ensembles is that interpretation of the model is now even less possible. For example, decision trees have properties so attractive that, second to linear regression (LR), they are the modeling method most widely employed, despite having the worst accuracy of the major algorithms. Bundling trees into an ensemble makes them competitive on this crucial property, though at a serious loss in interpretability. To quantify this loss, note that an ensemble of trees can itself be represented as a tree, as it produces a piecewise constant response surface. But the tree equivalent to an ensemble can have vastly more nodes than the component trees; for example, a bag of M "stumps" (single-split binary trees) can require up to 2^M leaves to be represented by a single tree.

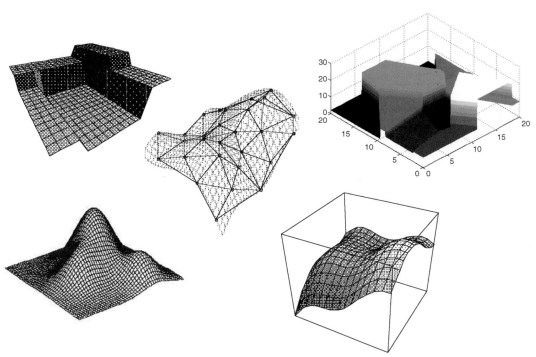

FIGURE 18.3 Estimation surfaces of five modeling algorithms. Clockwise from top left: decision tree, nearest neighbor, polynomial network, kernel; center: Delaunay planes (Elder, 1993).

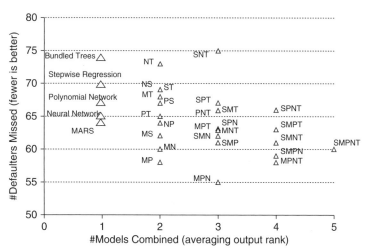

FIGURE 18.4 Out-of-sample errors on a credit scoring application when fusing one to five different types of models into ensembles.

IV. TRUE COMPLEXITY, THE "RIGHT MODEL," TOP MISTAKES, AND THE FUTURE

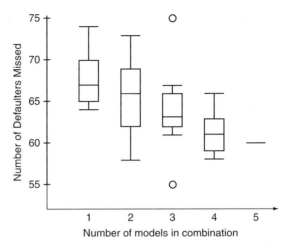

FIGURE 18.5   Box plot for Figure 18.4; median (and mean) error decreased as degree of combination increased.

Indeed, bumping (Tibshirani and Knight, 1999a) was designed to get some of the benefit of bagging without requiring multiple models, in order to retain some interpretability. It builds competing models from bootstrapped data sets and keeps only the one with least error on the original data. This typically outperforms, on new data, a model built simply on the original data, likely due to a bumped model being robust enough to do well on two related but different data sets. But the accuracy increase is less than with ensembles.

Another criticism of ensembles—more serious to those for whom an incremental increase in accuracy is worth a multiplied decrease in interpretability—is that surely their increased complexity will lead to overfit and, thus, inaccuracy on new data. In fact, not observing ensemble overfit in practical applications has helped throw into doubt, for many, the Occam's Razor axiom that generalization is hurt by complexity. [This and other critiques of the axiom are argued in an award-winning paper by Domingues (1998).]

But are ensembles truly complex? They appear so, but do they *act* so? The key question is how we should measure complexity. For LR, you can merely count terms, yet this is known to fail for nonlinear models. It is possible for a single parameter in a nonlinear method to have the influence of less than a single linear parameter, or greater than several—e.g., three effective degrees of freedom for each parameter in Multivariate Adaptive Regression Splines (Friedman, 1991; Owen, 1991). The under-linear case can occur with, say, a neural network that hasn't trained long enough to pull all its weights into play. The over-linear case is more widely known. For example, Friedman and Silverman (1989) note: "[The results of Hastie and Tibshirani (1985)], together with those of (Hinkley, 1969, 1970) and (Feder, 1975), indicate that the number of degrees of freedom associated with nonlinear least squares regression can be considerably more than the number of parameters involved in the fit."

The number of parameters and their degree of optimization is not all that contributes to a model's complexity or its potential for overfit. The model form alone doesn't reveal the *extent of the search for structure*. For example, the winning model for the 2001 Knowledge Discovery and Data Mining (KDD) Cup employed only three variables. But the data had

140,000 candidate variables, constrained by only 2,000 cases. Given a large enough ratio of unique candidate variables to cases, searches are bound to find some variables that look explanatory even when there is no true relationship. As Hjorth (1989) warned: "... the evaluation of a selected model can not be based on that model alone, but requires information about the class of models and the selection procedure." We thus need to employ model selection metrics that include the effect of model selection!

There is a growing realization that complexity should be measured not just for a model, but for an entire modeling *procedure*, and that it is closely related to that procedure's *flexibility*. For example, the recent Covariance Inflation Criterion (Tibshirani and Knight, 1999b) fits a model and saves the estimates, then randomly shuffles the output variable, reruns the modeling procedure, and measures the covariance between the new and old estimates. The greater the change (adaptation to randomness, or flexibility), the greater the complexity penalty needed to restrain the model from overfit. Somewhat more simply, Generalized Degrees of Freedom, GDF (Ye, 1998), randomly perturbs (adds noise to) the output variable, reruns the modeling procedure, and measures the changes to the estimates. Again, the more a modeling procedure adapts to match the added noise, the more flexible (and therefore more complex) its model is deemed to be.

The key step in both—a randomized loop around a modeling procedure—is reminiscent of the Regression Analysis Tool (Faraway, 1991), which measured, through resampling, the robustness of results from multistep automated modeling. Whereas at that time sufficient resamples of a 2-second procedure took 2 days, increases in computing power have made such empirical measures much more practical.

GENERALIZED DEGREES OF FREEDOM

For LR, the degrees of freedom, K, equal the number of terms, though this does not extrapolate to nonlinear regression. But there exists another definition that does:

$$K = \text{trace(Hat Matrix)} = \Sigma \; \delta Y\_hat / \delta Y \tag{1}$$

where

$$\delta Y = Ye - Y, \text{ and } \delta Y\_hat = Ye\_hat - Y\_hat \tag{2}$$

$$Y\_hat = f(Y, X) \text{ for model } f(), \text{ output } Y, \text{ and input vectors, } X; Ye\_hat = f(Ye, X) \tag{3}$$

$$Ye = Y + N(0, \sigma_\varepsilon).[1] \tag{4}$$

GDF is thus defined to be the sum of the sensitivity of each fitted value, Y_hat_i, to perturbations in its corresponding output, Y_i. (Similarly, the effective degrees of freedom of a spline model is estimated by the trace of the projection matrix, S: $Y\_hat = SY$.) Ye (1998)

---

[1] We enjoyed naming the perturbed output, (Y + error) after GDF's inventor, Ye.

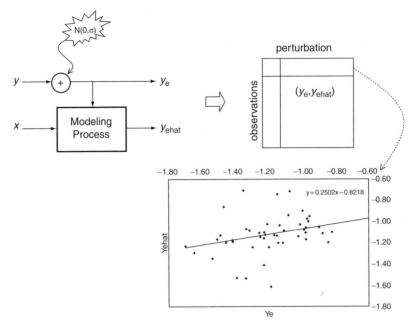

FIGURE 18.6   Diagram of GDF computation process.

suggests generating a table of perturbation sensitivities, then employing a "horizontal" method of calculating GDF, as diagrammed in Figure 18.6.

Fit an LR to δY_hat_i versus δY_i using the row of data corresponding to case i; then add together the slopes, m_i. (Since Y_i and Y_hat_i are constant, the LR simplifies to be of Ye_hat_i versus Ye_i.) This estimate appears more robust than that obtained by the "vertical" method of averaging the value obtained for each column of data (i.e., the GDF for each model or perturbation data set).

EXAMPLES: DECISION TREE SURFACE WITH NOISE

We take as a starting point for our tests the two-dimensional piecewise constant surface used to introduce GDF (Ye, 1998), shown in Figure 18.7.

It is generated by (and so can be perfectly fit by) a decision tree with five terminal (leaf) nodes (i.e., four splits), whose smallest structural change is 0.5. Figure 18.8 illustrates the "surface" after Gaussian noise $N(0, 0.5)$ has been added, and Figure 18.9 shows 100 random samples of that space. These tree + noise data are the (X,Y) data set employed for the experiments. For GDF perturbations, we employed 50 replications, where each added to Y Gaussian noise, $N(0, 0.25)$, having half the standard deviation of the noise already in the training data (a rule of thumb for perturbation magnitude).

Figure 18.10 shows the GDF versus K (number of parameters) sequence for LR models, single trees, and ensembles of five trees (and two more sequences described later).

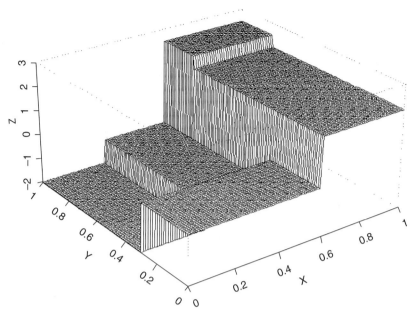

FIGURE 18.7   (Noiseless version of) two-dimensional tree surface used in experiments (after Ye, 1998).

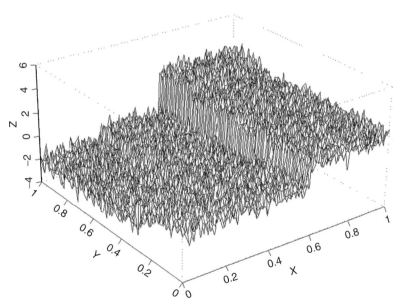

FIGURE 18.8   Tree surface of Figure 18.7 after adding N(0, 0.5) noise.

IV. TRUE COMPLEXITY, THE "RIGHT MODEL," TOP MISTAKES, AND THE FUTURE

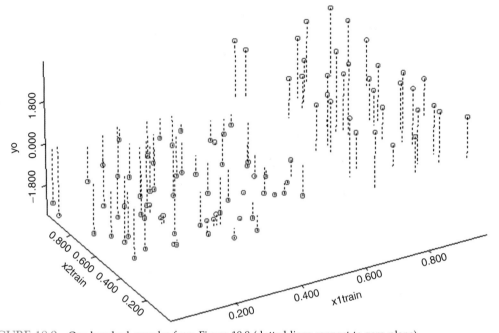

FIGURE 18.9   One hundred samples from Figure 18.8 (dotted lines connect to zero plane).

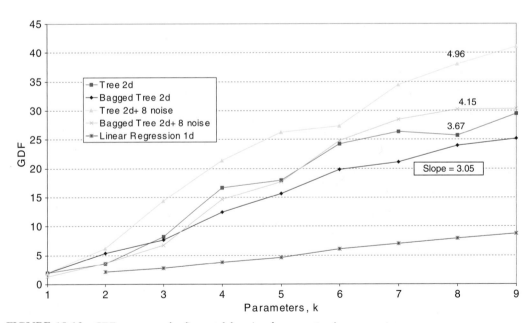

FIGURE 18.10   GDF sequences for five models using from one to nine parameters.

Confirming theory, note that the GDF for the LR models closely matches the number of terms, K. For decision trees of different sizes, K (i.e., maximum number of split thresholds), the GDF grew at about 3.67 times the rate of K. Bagging (bootstrap sampling the data sets and averaging the outputs) five trees together, the rate of complexity growth is 3.05. Surprisingly, perhaps, the bagged trees of a given size, K, are about a fifth simpler, by GDF, than each of their components!

Figure 18.11 illustrates two of the surfaces in the sequence of bagged trees. Bagging five trees limited to four leaf nodes (three splits) each produces the estimation surface of Figure 18.11a. Allowing eight leaves (seven splits) produces that of Figure 18.11b. The bag of more complex trees creates a surface with finer detail (most of which here does not relate to actual structure in the underlying data-generating function, as the tree is more complex than needed). For both bags, the surface has gentler stairsteps than those of a lone tree, revealing how bagging trees can especially improve their generalization on smooth functions.

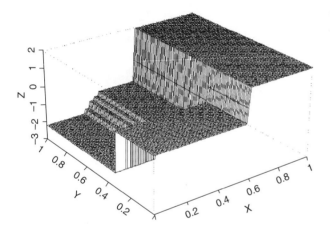

FIGURE 18.11A Surface of bag of five trees using three splits.

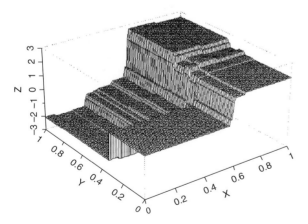

FIGURE 18.11B Surface of bag of five trees using seven splits.

IV. TRUE COMPLEXITY, THE "RIGHT MODEL," TOP MISTAKES, AND THE FUTURE

Expanding the experiment (after Ye, 1998), we appended eight random candidate input variables to X, to introduce *selection noise*, and reran the sequence of individual and bagged trees. Figures 18.12a and 18.12b illustrate two of the resulting bagged surfaces (projected onto the space of the two real inputs), again for component trees with three and seven splits, respectively. The structure in the data is clear enough for the under-complex model to avoid using the random inputs, but the over-complex model picks some up. The GDF progression for the individual and bagged trees with 10 candidate inputs is also shown in Figure 18.10. Note that the complexity slope for the bag (4.15) is again less than that for its components (4.96). Note also that the complexity for each 10-input experiment is greater than its corresponding 2-input one. Thus, even though you cannot tell—by looking

FIGURE 18.12A   Surface of bag of five trees using three splits with eight noise inputs.

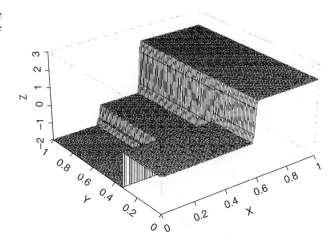

FIGURE 18.12B   Surface of bag of five trees using seven splits with eight noise inputs (projected onto plane of two real inputs).

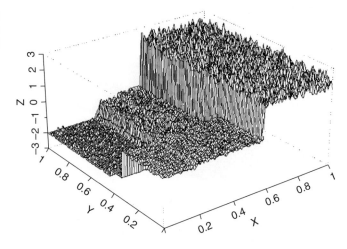

at a final model using only the real inputs X_1 and X_2—that random variables were considered, the chance for overfit was greater, and this is appropriately reflected in the GDF measure of complexity.

SUMMARY AND DISCUSSION

Bundling competing models into ensembles almost always improves generalization—and using different algorithms is an effective way to obtain the requisite diversity of components. Ensembles *appear* to increase complexity, as they have many more parameters than their components, so their ability to generalize better seems to violate the preference for simplicity embodied by Occam's Razor. Yet, if we employ GDF—an empirical measure of the *flexibility* of a modeling *process*—to measure complexity, we find that ensembles can be simpler than their components. We argue that when complexity is thereby more properly measured, Occam's Razor is restored.

Under GDF, the more a modeling process can match an arbitrary change made to its output, the more complex it is. It agrees with linear theory but can also fairly compare very different, multistage modeling processes. In our tree experiments, GDF increased in the presence of distracting input variables, and with parameter power (trees versus LR). It is expected to also increase with search thoroughness, and to decrease with use of model priors, with parameter shrinkage, and when the structure in the data is more clear relative to the noise. Additional observations (constraints) may affect GDF either way.

Lastly, case-wise (horizontal) computation of GDF has an interesting byproduct: an identification of the complexity contribution of each case. Figure 18.13 illustrates these contributions for two of the single-tree models of Figure 18.10 (having three and seven splits, respectively). The under-fit tree results of Figure 18.13a reveal only a few observations to be complex; that is, to lead to changes in the model's estimates when perturbed by random noise. (Contrastingly, the complexity is more diffuse for the results of the overfit tree, in

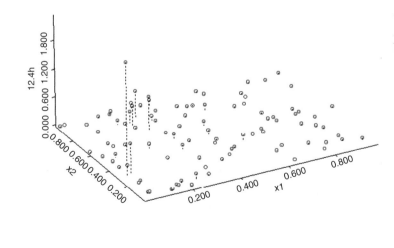

FIGURE 18.13A Complexity contribution of each sample for bag of five trees using three splits.

FIGURE 18.13B Complexity contribution of each sample for bag of five trees using seven splits.

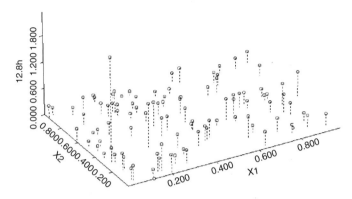

Figure 18.13b.) A future modeling algorithm could recursively seek such *complexity contribution outliers* and focus its attention on the local model structure necessary to reduce them, without increasing model detail in regions that are stable.

POSTSCRIPT

Complex predictive systems arm us with some powerful techniques for building models that represent the major elements of the target signal dynamics among all cases in the data set. But we must be careful not to go overboard in our search for complex solutions to what we expect is a complex problem. Sometimes, we can capture the major dynamics in the target signal with a relatively simple program—a solution is said to be is more *elegant*. Chapter 19 presents the other half of the story of the search for the elegant solution. Just like models can be overtrained, complexity can be overapplied. We need a modeling "stopping function," like that used to prevent overtraining of a neural net. Chapter 19 provides a philosophical approach to implementing stopping functions in our modeling design.

References

Barron, R. L., Mucciardi, A. N., Cook, A. N., Craig, J. N., & Barron, A. R. (1984). Adaptive learning networks: Development and application in the United States of algorithms related to GMDH. In S. J. Farlow (Ed.), *Self-Organizing Methods in Modeling: GMDH Type Algorithms* (pp. 25–65). New York: Marcel Dekker.

Breiman, L. (1996). Bagging predictors. *Machine Learning, 26*(2), 123–140.

Domingues, P. (1998). *Occam's two razors: The sharp and the blunt. Proceedings of the 5th International Conference on Knowledge Discovery and Data Mining* (37–43). New York: AAAI Press.

Elder, J. F., IV (1993). *Efficient optimization through response surface modeling: A GROPE algorithm.* Charlottesville: Dissertation, School of Engineering and Applied Science, University of Virginia.

Elder, J. F., IV (1996a). A review of *Machine Learning, Neural and Statistical Classification* (Michie, Spiegelhalter & Taylor, Eds., 1994), *Journal of the American Statistical Association* 91(433), 436–437.

Elder, J. F., IV (1996b). Heuristic search for model structure: The benefits of restraining greed. In D. Fisher & M. Lenz (Eds.), *Learning from Data: Artificial Intelligence and Statistics*, Chapter 13, New York: Springer-Verlag.

Elder, J. F., IV, & Brown, D. E. (2000). Induction and polynomial networks. In M. D. Fraser (Ed.), *Network models for control and processing* (pp. 143–198). Portland: Intellect.

Elder, J. F., IV, & Lee, S. S. (1997). *Bundling heterogeneous classifiers with advisor perceptrons.* University of Idaho Technical Report, October, 14.

Faraway, J. J. (1991). On the cost of data analysis. Technical Report. UNC Chapel Hill: Dept. Statistics.

Feder, P. I. (1975). The log likelihood ratio in segmented regression. *The Annals of Statistics, 3,* 84–97.

Freund, Y., & Schapire, R. E. (1996). Experiments with a new boosting algorithm. *Machine Learning: Proceedings of the 13th International Conference,* July, 148–156.

Friedman, J. H. (1991). Multivariate Adaptive Regression Splines. *Annals of Statistics, 19,* 1–67.

Friedman, J. H., & Silverman, B. W. (1989). Flexible parsimonious smoothing and additive modeling. *Technometrics, 31*(1), 3–21.

Hastie, T., & Tibshirani, R. (1985). Discussion of "Projection Pursuit" by P. Huber. *The Annals of Statistics, 13,* 502–508.

Hinkley, D. V. (1969). Inference about the intersection in two-phase regression. *Biometrika, 56,* 495–504.

Hinkley, D. V. (1970). Inference in two-phase regression. *Journal of the American Statistical Association, 66,* 736–743.

Hjorth, U. (1989). On model selection in the computer age. *Journal of Statistical Planning and Inference, 23,* 101–115.

Ivakhenko, A. G. (1968). The group method of data handling—A rival of the method of stochastic approximation. *Soviet Automatic Control, 3,* 43–71.

Michie, D., Spiegelhalter, D. J., & Taylor, C. C. (1994). *Machine Learning, Neural and Statistical Classification.* New York: Ellis Horwood.

Owen, A. (1991). Discussion of "Multivariate Adaptive Regression Splines" by J. H. Friedman. *Annals of Statistics, 19,* 82–90.

Tibshirani, R., & Knight, K. (1999a). The covariance inflation criterion for adaptive model selection. *Journal of the Royal Statistical Society, Series B-Statistical Methodology, 61*(pt. 3), 529–546.

Tibshirani, R., & Knight, K. (1999b). Model search and inference by bootstrap "bumping." *Journal of Computational and Graphical Statistics, 8,* 671–686.

Wolpert, D. (1992). Stacked generalization. *Neural Networks, 5,* 241–259.

Ye, J. (1998). On measuring and correcting the effects of data mining and model selection. *Journal of the American Statistical Association, 93*(441), 120–131.

IV. TRUE COMPLEXITY, THE "RIGHT MODEL," TOP MISTAKES, AND THE FUTURE

The Right Model for the Right Purpose: When Less Is Good Enough

PREAMBLE

The problem with seeking solutions to problems in the world is not that the world is just more complicated than we think, but it is more complicated than we *can* think. Consequently, our solutions are bound to be less complicated than the problems demand. Most often, our response is to oversimplify, and one of our strategies is to follow what we think are common perceptions to simplify our tasks. For example, many critics of early efforts of man flying in airplanes rationalized that if God had meant man to fly, he would have given us wings. Today, we understand that our technological capabilities are every bit as much an enabler to do things as are our bodily appendages. The error of those critics long ago is not in their understanding that men had never flown in recorded history, but in their presumption that man *could not* fly. Not only is nature more complicated than we think it is, but our ability to apply human technology to understand it and harness it is far greater than many can imagine.

The way in which we deal with most problems in the world follows that pathway of thinking, in one way or another. Often, we are victims of our own narrow perceptions. And we substitute those narrow perceptions for the much more complicated reality of the situations. Rather than take the time to study and understand (at least in principle) all aspects of a situation, we jump to a solution to a problem following a very narrow path through a decision landscape constrained by our assumptions and presuppositions. Even if the assumption is valid, we ignore effects of many other influences for the sake of generating a solution in the required time frame. Generating acceptable solutions in a given time frame is commendable, but our usual way of doing it is not!

One of the common perceptions in data mining is that *more is better*. This is expressed in the belief that

1. More is better.
2. Efficiency or sufficiency must be selected. (This is a false dichotomy, discussed next.)

MORE IS NOT NECESSARILY BETTER: LESSONS FROM NATURE AND ENGINEERING

Efficiency is usually defined in terms that involve maximizing output and minimizing input. The best solution is often defined in terms of the most efficient solution: This idea is often referred to as the *Efficiency Paradigm*. This paradigm assumes that the goal of efficiency is to maximize output while minimizing input.

In statistical analysis, the Efficiency Paradigm is expressed to define an *efficient solution* as one that has a relatively small variance. One solution (e.g., an estimator) can be considered more efficient than another if the covariance matrix of the second minus the covariance matrix of the first is composed largely of positive numbers. When all the elements of the resultant matrix are positive, it is called a *positive semi-definite matrix*. The most efficient solution is one that is closest to a positive semi-definite matrix.

The statistical definition can be useful if the Efficiency Paradigm is correct in the context of the solution. But what if it isn't? Many examples in the real world appear to violate this paradigm. For example, ecological succession occurs on a previously forested area when a highly efficient grass community (defined in terms of productivity per gram of biomass) is replaced by a less-efficient shrub community, which in turn is replaced by an even less efficient forest community. The Efficiency Paradigm might still apply to the mature forest if efficiency is defined in terms of accumulation of biomass over time rather than in terms of productivity rate. In that case, sufficient productivity occurs to permit crown closure and shading out of competing species, including highly productive grasses.

This approach to defining efficiency in terms of sufficiency is at the core of the current debate on the definition of sustainable agriculture (Falvey, 2004). Voices supporting sustainable agriculture maintain that we should avoid thinking in terms of the false dichotomy of efficiency and sufficiency, and embrace both. Rather than make sufficiency dependent on efficiency, we should turn the concept around and let efficiency be defined in terms of sufficient solutions, rather than solutions of maximum productivity. In this natural context, we might call this the *Sufficiency Paradigm*.

We might even replace the Efficiency Paradigm with the Sufficiency Paradigm in business also. This paradigm shift in business should be reflected not only in the business goals of a company, but also in the business processes followed to achieve them. In the context of the theme of this book, we can prescribe this paradigm shift in the IT departments as a technical specific to enable and promote the development of the business organism.

If we follow the Sufficiency Paradigm in analytical data mart design and data mining solution development, it will change the way we create data mining models. It will lead us to accept solutions that are good enough (sufficient) to build the synergies and products of the business organism, which will maximize productivity *within* the constraints of our goals to promote long-term stability and growth.

Under the Sufficiency Paradigm, the best data mining solutions will not be defined solely in terms of maximizing financial productivity. Rather, they will be defined in terms of how well they work together with other business processes to enhance the cohesive action throughout the entire profit chain. This cohesive action permits the company to be proactive and adaptive to change, rather than reactive and hampered by it. This set of features is intrinsically *organic* rather than *mechanical*.

EMBRACE CHANGE RATHER THAN FLEE FROM IT

In his book *Bionomics: Economy as Ecosystem*, Rothschild (1991) maintains that mechanistic organizations fear change because it happens faster than their rather rigid business processes can respond to it. Peters (1987) recommends that companies design business processes to take advantage of change to evolve new market niches. Peters also recommends that information should flow freely to encourage (and spread the contagion of) innovation. This free flow will permit even bad news to travel fast, and even encourage it to do so. Mistakes become like pain that the whole body feels, not just at the receptor site. Sharing of this information by a foot in the business organism prevents the other foot from making the same mistake. The business organism learns from mistakes and can be driven by these mistakes to evolve into a more successful state.

DECISION MAKING BREEDS TRUE IN THE BUSINESS ORGANISM

Often, data mining results may drive decision-making activities to design actions in remote parts of the organization. But these decisions may be difficult or impossible to implement. For example, it may be very difficult (or impossible under constraints of time and budget) to re-create the Customer Analytic Record (CAR) in the production database.

One reason for this is that it may be very difficult to access all data sources used to create the CAR. Another reason could be that no business processes exist in production operations to do the necessary data preparation to create the CAR. This disconnect between modeling and production operations represents a gap in the information flow pathway. This is why many data mining models just sit on the shelf in IT and are not implemented in production.

The business organism must have a decision flow pathway between analysis and production (sites of action) that is properly designed to transmit information quickly and efficiently. This is the *Digital Nervous System* (Gates, 1999).

Muscles in the Business Organism

Another very important system in the pathway leading to business action is represented by the business processes (muscles) that are properly trained to turn the decision information into action. These business processes at the site of action must be developed *before* they can receive the decision information and act on it. *Therefore, the data mining solution must be reverse-engineered from the point of action.* That means you must model the model development process from the back end in the business unit, rather than from the front end in IT. Implementation requirements must be designed for each step in the decision information pathway, starting with the action site, and proceeding toward model development. Inputs for each step must be coupled with necessary information transforms to generate the precise nature of the outputs of that step required as inputs of the next step in the process.

The overall design of the IT network systems and business process for each step along the way *is an expression of building the solution model from the* **top down (following Plato)**. The detailed design of the business and analytical processes at each step *is an expression of building the solution model from the* **bottom up (following Aristotle).** Along the way, compromises and assumptions must be made to jump over problems that would otherwise prevent information flow. *If we try to build the solution model solely from the top down or from the bottom up, we will reach a point at which we discover that we don't know enough or understand enough to link the steps in the process.* System engineers quantify these links in simple ways and call them *transfer functions.* This is the way complex systems are modeled in the real world.

What Is a Complex System?

A *complex system* is an organization of interconnected and interacting components, whose behaviour is not obvious from the properties of the individual parts. We introduced these systems-level properties in Chapter 1 as *emergent properties*. These emergent properties are not obvious and may not even exist in the set of properties of the individual parts. The example of the rain forest given in Chapter 1 was used to illustrate how these emergent properties can be among the primary aspects of the complex system that permit it to exist and function as a system.

The concept of the business organism can be viewed in the context of a complex system. Like the forest system which changes over time, the business organism can adapt to changing business conditions. Therefore, to facilitate data mining solutions in such an adaptive business organism, the entire decision pathway must be designed with such solutions in mind. This can be done very effectively in an exploration data mart, dependent on an enterprise data warehouse. The concept of the *business ecosystem* as a part of the *corporate information factory* was spawned by Inmon et al. (1998) based on articles in *DM-Review* by Imhoff and Sousa (1997). In the first article, Imhoff and Sousa presented the concept of the business ecosystem driven by a brain, composed partly of memory (the relational data warehouse) and partly of an analytical system served by information stored in denormalized form—the analytical data mart (Figure 19.1). This approach to data warehousing support for analytical modeling was refined by Inmon et al. (1998).

This data mart, along with others designed for other reporting purposes, could be expressed either in logical format or physical format. A physical data mart is hosted on a separate system with database schemas designed to serve specialized purposes. The logical data mart is hosted on the data warehouse system, and is implemented in the form of database views or just composed of a group of denormalized tables containing aggregate data suitable for creating the CAR.

Often, the easiest place to begin is to build a logical data mart by designing tables, such as householding tables, with summary data aggregated at the account, individual, and household levels. Other tables containing demographic data (for example) can be added to the logical data mart by joining other kinds of data with keys in common with the householding tables. This logical structure is well suited as a data source for data mining.

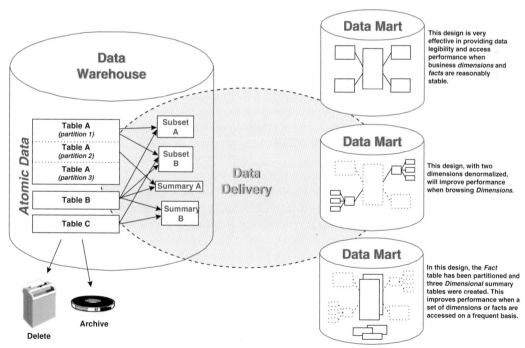

FIGURE 19.1   The corporate information factory.

But a properly designed data source is only one element in the decision chain. The other elements include the following:

- The IT network structure in a company is the "digital nervous system," through which data mining results are communicated to the site of business action.
- The business processes (muscles) properly conditioned move the service and product function of the company (the bones) to generate profit.
- The right actions are produced.
- Results of the actions (e.g., customer responses) generate more or less value for the company.
- Results are fed back to the business brain and provide a basis for learning how to build better models (the *virtuous cycle* of Berry and Linoff, 1997);

This learned response is characteristic of an adaptive organism rather than a static machine. In the system composing the business organism, the decisions designed in IT breed true, as they pass through the time-steps in the process from one part of the system to another. That is, decisions remain unchanged in nature as they pass through various functions in the business organism through time. The reason they can remain unchanged is that they are designed right up-front, so they fit the decision-response pathway through each process from perception in the brain to the site of action.

THE 80:20 RULE IN ACTION

In 1906, an Italian economist Vilfredo Pareto observed that about 20% of the people owned about 80% of the wealth. This principle was picked up by quality management pioneer Joseph Juran in the late 1940s, in which he cast his concept of the *vital few* and the *trivial many* (Juran, 1951). Juran generalized this principle in quality management to postulate that 20% of something is always responsible for 80% of the results. He adopted Pareto's principle to explain this, and named it *Pareto's Principle*, or *Pareto's Law* (AKA the "80:20 Rule").

The creation of sufficient data mining modeling solutions may follow the 80:20 Rule also. Certainly, the structure of the modeling process follows this rule, in that about 80% of the modeling project time is spent in data preparation, and only about 20% is spent in training and testing the model. Based on the 80:20 Rule, we might expect to achieve a sufficient modeling solution in many cases, with only 20% of the effort we could spend modeling to create the solution with maximum predictability. There is some support for believing this in the concept of *agile modeling*.

AGILE MODELING: AN EXAMPLE OF HOW TO CRAFT SUFFICIENT SOLUTIONS

One of the most insightful approaches to modeling comes from the environment of Extreme Programming (XP) software development. The premise of XP is to deliver the software the customer needs when it is needed. Niceties, enhancements, and other bells and whistles have to take the back seat to utility and timeliness. The approach of XP was extended by Ambler (2002) to cover the modeling of the entire software development process, referred to as *agile modeling*. One of the most important utility functions in agile modeling is the feedback loop to the stakeholders. Stakeholders are brought into the development process at key points in the project to validate the current state of the potential utility *in their perception*.

Ambler cites six propositions of agile modeling that pertain very closely to the development of data mining models.

1. **Just barely good enough (JBGE) is actually the most effective policy:** JBGE is analogous to the inflection point on a curved response graph. The JBGE point on Figure 19.2 is the most reasonable position for effort to end. Assuming that additional effort could be spent on creating other JBGE models for other purposes, the optimum benefit across the entire modeling operations would restrict modeling efforts to the JBGE levels of effort.
2. **JBGE does not imply poor quality:** The JBGE level of model production may not produce the highest accuracy, but it is sufficient to get the job done for which the model was commissioned. The stakeholders of the model are the best judges of the utility of the model, not the modeler.
3. **JBGE depends on the situation:** What is good enough for one situation may not be good enough for another situation. The classic example is discussed in Chapter 17, where a 90% accurate model, good enough for most situations, is certainly not good enough for a fraud model.
4. **The JBGE model evolves over time:** The initial model can be refined and/or updated over time. As needs and conditions change, the model can change. The characteristics of the model can adapt to new conditions. In this way, a model is like a biological species, which can respond to changing environmental conditions by changes in its very nature.

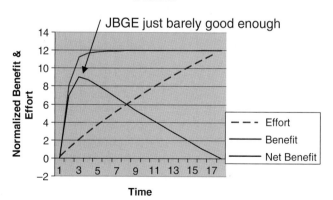

Relation Between Benefit, Effort and Net Benefit

FIGURE 19.2   Relationship between net benefit and total effort.

FIGURE 19.3 The relationship between realized value (net benefit) and cumulative traditional value (based on Ambler, 2002).

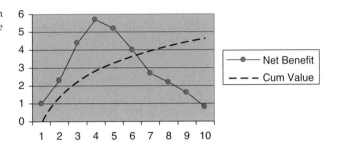

5. **The point of maximal benefit comes before you think it will:** In Figure 19.2, the point of maximum net benefit occurs at about the 4 level of effort. It may appear counterintuitive that additional effort is associated with a decline in benefit, but when additional costs are considered, the 4.0 level of effort is best.
6. **The realized value of a model may exceed the perceived value:** This statement may appear counterintuitive at first, but further consideration in the context of the business environment can clarify it. The traditional concept of value rises with additional effort. But much potential value can be masked by delays. The realized value of a timely model (even though it is not the most accurate possible) can far exceed that of an untimely model. For example, a medical diagnostic model of only moderate accuracy delivered in time to define a successful treatment may be much more valuable than a more accurate model delivered later, particularly if the patient dies in the meantime. We can see this dynamic expressed in Figure 19.3.

The cumulative value shown in Figure 19.3 represents the traditional view of defining value according to the accuracy of the model and the features included in it. Naturally, as effort increases throughout the development project, higher accuracy is achieved and more features are added. But the true utility value of the model may follow the curve of net benefit rather than the curve of the cumulative value.

POSTSCRIPT

This chapter (and indeed the whole book) is designed to present the case that many times, less is good enough. In the book as a whole, we present very few equations. Rather, we present intuitive explanations of the concepts presented which are sufficient to enable you to understand enough of the theory and mathematics underlying the practice of data mining to create acceptable models. Naturally, we would like to have models that are as accurate as possible. The definition of *possible*, though, must be composed of elements of time requirements, benefits relative to present methods in a time domain. As in the medical diagnosis example discussed earlier, it may be better to shoot for a model of lower accuracy finished sooner than to wait for a more predictive model later. In this context, our greatest challenge in data mining is not finding ways to analyze data, but deciding when less performance is good enough.

In the next chapter, we will discuss some caveats in data mining practice, which will help you to implement your appropriate elegant model designs. Commonly attributed to Robert Burns, the phrase "The best-laid plans of mice and men often go astray" really does apply to data mining practice. We can design the very best modeling system, and it may fail miserably when we try to apply it. Chapter 20 presents 11 common mistakes to avoid so that you are able to prevent this failure.

References

Ambler, S. (2002). *Agile Modeling*. San Francisco, CA: John Wiley & Sons.

Berry, M., & Linoff, G. (1997). *Data Mining Techniques for Marketing, Sales and Customer Relationship Management* (2nd ed.). San Francisco, CA: John Wiley & Sons.

Falvey, L. (2004). *Sustainability: Elusive or Illusory? Wise Environmental Intervention*. Institute for International Development.

Gates, B. (1999). *Business at the Speed of Thought*. New York, NY: Grand Central Publishing.

Imhoff, C., & Sousa, R. (1997). The information ecosystem. Part 1. *DM-Review*, January 27.

Inmon, W. H., Imhoff, C., & Sousa, R. (1998). *Corporate Information Factory*. New York, NY: Wiley Computer Publishing.

Juran, J. (1951). *Quality Control Handbook*. McGraw-Hill: New York.

Peters, T. (1987). *Thriving on Chaos: Handbook for a Management Revolution*. New York: HarperCollins.

Rothschild, M. (1991). *Bionomics: Economy as Ecosystem*. New York, NY: Henry Holt and Company, Inc.

Top 10 Data Mining Mistakes

PREAMBLE

Mining data to extract useful and enduring patterns remains a skill arguably more art than science. Pressure enhances the appeal of early apparent results, but it is all too easy to fool yourself. How can you resist the siren songs of the data and maintain an analysis discipline that will lead to robust results? It is essential to *not* lack (proper) data, focus on training, rely on one technique, ask the wrong question, listen (only) to the data, accept leaks from the future, discount pesky cases, extrapolate (practically and theoretically), answer every inquiry, sample casually, or believe the best model.

INTRODUCTION

It has been said that good judgment comes through experience, but experience seems to come through bad judgment! In two decades of mining data from diverse fields, we have made many mistakes, which may yet lead to wisdom. In the following sections, we briefly describe, and illustrate from examples, what we believe are the "Top 10" mistakes of data mining, in terms of frequency and seriousness. Most are basic, though a few are subtle. All have, when undetected, left analysts worse off than if they'd never looked at their data.

After compiling the list, we realized that an even more basic problem—mining without (proper) data—must be addressed as well. So, numbering like a computer scientist (with an overflow problem), here are mistakes 0 to 10.[1]

0. LACK DATA

To really make advances with an analysis, you must have labeled cases, i.e., an output variable. With input variables only, all you can do is look for subsets with similar character-istics (cluster) or find the dimensions that best capture the data variation (principal compo-nents). These unsupervised techniques are much less useful than a good (supervised) prediction or classification model. Even with an output variable, though, the most inter-esting class or type of observation is usually the most rare by orders of magnitude. For instance, roughly 1/10 of "risky" individuals given credit will default within 2 years, 1/100 people mailed a catalog will respond with a purchase, and perhaps 1/10,000 banking transactions of a certain size require auditing. The less probable the interesting events, the more data it takes to obtain enough to generalize a model to unseen cases. Some projects probably should not proceed until enough critical data are gathered to make them worthwhile.

For example, on a project to discover fraud in government contracting, known fraud cases were so rare that strenuous effort could initially only reduce the size of the haystack in which the needles were hiding.[2] That is, modeling served to assure that the great major-ity of contracts were almost surely not fraudulent, which did enable auditors to focus their effort. But more known fraud cases—good for data miners, but bad for taxpayers—could have provided the modeling traction needed to automatically flag suspicious new cases much sooner. This was certainly the situation on another project, which sought to discover collusion on tax fraud. Unfortunately (for honest taxpayers), there were plenty of training

---

[1] Most examples are from our own and our colleagues' experiences, but some identifying details are mercifully withheld.

[2] Virtually all known cases were government workers who had, out of guilt, turned themselves in. Most, it seems, meant to pay back what they had fraudulently obtained (but how?). One audacious fraudster was discovered, however, after coworkers realized that the clerk had been driving a different sports car to work every day of the week!

examples, but their presence did lead to stronger, immediate modeling results, which was ultimately beneficial to taxpayers.

You can't mine without data, but not just any data will work. Many data mining projects have to make do with "found" data, not the results of an experiment designed to illuminate the question studied. It's like making a salad out of weeds found in the yard.

One sophisticated credit-issuing company realized this when seeking to determine if there was a market for its products in the class of applicants previously routinely dismissed as being too risky. Perhaps a low-limit card would be profitable, and even help a deserving subset of applicants pull themselves up in their credit rating?[3] But the company had no data on such applicants by which to distinguish the truly risky from those worth a try; its traditional filters excluded such individuals from even initial consideration. So the company essentially gave (small amounts of) credit almost randomly to thousands of risky applicants and monitored their repayments for 2 years. Then it built models to forecast defaulters (those late on payments by 90+ days) trained only on initial application information. This large investment in *creating* relevant data paid off in allowing the company to rationally expand its customer base.

1. FOCUS ON TRAINING

Only out-of-sample results matter; otherwise, a lookup table would always be the best model. Researchers at the MD Anderson medical center in Houston a decade ago used neural networks to detect cancer. Their out-of-sample results were reasonably good, though worse than training, which is typical. They supposed that longer training of the network would improve it—after all, that's the way it works with doctors—and were astonished to find that running the neural network for a week (rather than a day) led to only slightly better training results and much worse evaluation results. This was a classic case of overfit,

---

[3] For a decade now, the credit industry has mailed over a billion offers a year to American households; the high-risk market was one of the few places not saturated a few years ago. Credit profits are nonlinear with risk, and remind us of the triage system established during the Napoleonic wars, when the *levee en masse* swelled the battlefields and, combined with the new technology of cannons, etc., led to an army's medical resources being completely overwhelmed. Battlefield wounds were classified into three levels: the most minor to be passed by and treated later (if at all), more serious to receive immediate attention, but the most serious were judged not likely to be worth a physician's time. (We can envision a combatant, aware of hovering between the latter two classes insisting, like the Black Knight in the *Monty Python* movie, "What? The leg gone? It's just a flesh wound!") Likewise, credit companies make the most profit on individuals in the middle category of "woundedness"—those who can't pay off their balance, but keep trying. But they lose 5–10 times as much on clients just a little worse off, who eventually give up trying altogether. So, for models to be profitable at this edge of the return cliff, they have to forecast very fine distinctions. Recent downturns in the economy have severely punished the stocks of companies that aggressively sought that customer niche—especially if they did not give obsessive attention to model quality.

IV. TRUE COMPLEXITY, THE "RIGHT MODEL," TOP MISTAKES, AND THE FUTURE

where obsession with getting as much as possible out of training cases focuses the model too much on the peculiarities of that data to the detriment of inducing general lessons that will apply to similar, but unseen, data. Early machine learning work often sought, in fact, to continue "learning" (refining and adding to the model) until achieving exact results on known data—which, at the least, insufficiently respects the incompleteness of our knowledge of a situation.

The most important way to avoid overfit is to reserve data. But since data—especially cases of interest—are precious, you must use resampling tools, such as bootstrap, cross-validation, jackknife, or leave-one-out. Traditional statistical significance tests are a flimsy defense when the model structure is part of the search process, though the strongest of penalty-based metrics, such as Bayesian Information Criterion or Minimum Description Length, can be useful in practice.

With resampling, multiple modeling experiments are performed, with different samples of the data, to illuminate the distribution of results. If you were to split the data into training and evaluation subsets a single time, the evaluation accuracy result might largely be due to luck (either good or bad). By splitting it, say, 10 different ways and training on the 90% sets and evaluating on the out-of-sample 10% sets, you have 10 different accuracy estimates. The mean of this distribution of evaluation results tends to be more accurate than a single experiment, and it also provides, in its standard deviation, a confidence measure.

Note that resampling evaluates whatever is held constant throughout its iterations, or "folds." That is, you can set the structure (terms) of a model and search for its parameter values over multiple data subsets; then the accuracy results would apply to that fixed model structure. Or you could automate multiple stages of the process—e.g., outlier detection, input selection, interaction discovery—and put that whole process inside a resampling loop. Then it's the accuracy distribution of that full process that is revealed.

In the end, you have multiple overlapping models; which is the model to use? One approach is to choose a single model (perhaps by its beauty rather than its accuracy). Another is to rerun the model-building process with all the data and assume that the resulting model inherits the accuracy properties measured by the cross-validation folds. A third is to use the variety of models in an ensemble, as discussed on Mistake 10, and in Chapter 13.

2. RELY ON ONE TECHNIQUE

"To a little boy with a hammer, all the world's a nail." All of us have had colleagues (at least) for whom the best solution for a problem happens to be the type of analysis in which they are most skilled! For many reasons, most researchers and practitioners focus too narrowly on one type of modeling technique. But, for best results, you need a whole toolkit. At the very least, be sure to compare any new and promising method against a stodgy conventional one, such as linear regression (LR) or linear discriminant analysis (LDA). In a study of articles in a neural network journal over a 3-year period (about a decade ago), only

17% of the articles avoided mistakes 1 and 2. That is, five of six referred articles either looked only at training data or didn't compare results against a baseline method, or made both of those mistakes. We can only assume that conference papers and unpublished experiments, subject to less scrutiny, are even less rigorous.

Using only one modeling method leads us to credit (or blame) it for the results. Most often, it is more accurate to blame the data. It is unusual for the particular modeling technique to make more difference than the expertise of the practitioner or the inherent difficulty of the data—and when the method will matter strongly is hard to predict. It is best to employ a handful of good tools. Once the data are made useful—which usually eats most of your time—running another algorithm with which you are familiar and analyzing its results adds only 5–10% more effort. (But to your client, boss, or research reviewers, it looks like twice the work!)

The true variety of modeling algorithms is much less than the apparent variety, as many devolve to variations on a handful of elemental forms. But there are real differences in how that handful builds surfaces to "connect the dots" of the training data, as illustrated in Figure 18.3 for five different methods—*decision tree, polynomial network, Delaunay triangles* (Elder, 1993), *adaptive kernels,* and *nearest neighbors*—on (different) two-dimensional input data. Surely some surfaces have characteristics more appropriate than others for a given problem.

Figure 18.1 (after Elder & Lee, 1997) reveals this performance issue graphically. The relative error of five different methods—*neural network, logistic regression, linear vector quantization, projection pursuit regression,* and *decision tree*—is plotted for six different problems from the Machine Learning Repository.[4] Note that "every dog has its day"; that is, that every method wins or nearly wins on at least one problem.[5] On this set of experiments, *neural networks* came out best, but how do you predict beforehand which technique will work best for your problem?[6] Best to try several and even use a combination (as covered in Mistake 10 and Chapter 13).

[4] The worst out-of-sample error for each problem is shown as a value near 1 and the best as near 0. The problems, along the x-axis, are arranged left-to-right by increasing proportion of error variance. So the methods differed least on the Pima Indians Diabetes data and most on the (toy) Investment data. The models were built by advocates of the techniques (using S implementations), reducing the "tender loving care" factor of performance differences. Still, the UCI ML repository data are likely overstudied; there are likely fewer cases in those data sets than there are papers employing them!

[5] When one of us used this colloquialism in a presentation in Santiago, Chile, the excellent translator employed a quite different Spanish phrase, roughly "Tell the pig Christmas is coming!" We had meant every method has a situation in which it celebrates; the translation conveyed the concept on the flip-side: "You think you're something, eh pig? Well, soon you'll be dinner!"

[6] An excellent comparative study examining nearly two dozen methods (though nine are variations on *decision trees*) against as many problems is (Michie et al., 1994) reviewed by (Elder, 1996). Armed with the matrix of results, the authors even built a decision tree to predict which method would work best on a problem with given data characteristics.

3. ASK THE WRONG QUESTION

It is first important to have the right project goal; that is, to aim at the right target. This was exemplified by a project at Shannon Labs, led by Daryl Pregibon, to detect fraud in international calls. Rather than use a conventional approach, which would have tried to build a model to distinguish (rare but expensive) fraud from (vast examples of) nonfraud, for any given call, the researchers characterized normal calling patterns for *each account* (customer) separately. When a call departed from what was the normal pattern for that account, an extra level of security, such as an operator becoming involved, was initiated. For instance, if one typically called a few particular countries each week, briefly, during weekdays, a call to a different region of the world on the weekend would bear scrutiny. Efficiently reducing historical billing information to its key features, creating a mechanism for the proper level of adaptation over time, and implementing the models in real time for vast streams of data provided interesting research challenges. Still, the key to success was asking the right question of the data. The ongoing "account signature" research won technical awards (at KDD, for example) but, more importantly, four researchers, part time in a year, were able to save their company enough money to pay the costs of the entire Shannon Labs (of 400 people) for the next year[7]—an impressive example of data mining return on investment (ROI).

Even with the right project goal, it is essential to also have an appropriate model goal. You want the computer to "feel" about the problem like you do—to share your multifactor score function, just as stock grants or options are supposed to give key employees a similar stake as owners in the fortunes of a company.[8] But analysts and tool vendors almost always use squared error as the criterion, rather than one tailored to the problem. Lured by the incredible speed and ease of using squared error in algorithms, we are like drunks looking for lost keys under the lamppost—where the light is better—rather than at the bar where they were likely dropped.

For instance, imagine that we're trying to decide whether to invest in our company's stock as a pension option, and we build a model using squared error. Say it forecasts that the price will rise from $10 to $11 in the next quarter, and it goes on to actually rise to $14. We've enjoyed a positive surprise; we expected a 10% gain but got 40%.[9] But when we're entering in that data for the next go-round, the computer has a different response; it sees an error of $3, between the truth and the estimate, and squares that to a penalty of 9. It would have more

---

[7] They were rewarded, as we techno-nerds like, with bigger toys. The group got a "Data Wall"—a $10' \times 20'$ computer screen, complete with couch, with which to visualize data. As it was often commandeered by management for demonstrations, the research group was eventually provided a second one to actually use.

[8] Unfortunately, the holder of an option has a different score function from the owner of the stock. The option is very valuable if the company thrives, but only worthless if it doesn't. Yet, the owner can be seriously hurt by a downturn. Thus, a manager's rational response to having options is to take on increased risk—to "shoot the moon" for the potential up-side reward. To better align owner and management interests, it is better to grant stock outright, rather than (cheaper, but more inflammatory) options.

[9] Of course, our joy is short-lived, as we kick ourselves for not mortgaging the house and betting even more! Fear and greed are always at war when dealing with the markets.

than twice "preferred" it if the price had dropped –$1 to $9; then its squared error would only have been 4. A criterion that instead punishes negative errors much more than positive errors would better reflect our preferences.

Though conventional squared error can often put a model into a serviceable region of performance, the function being optimized has a thorough effect on the suitability of the final model. "Inspect what you expect," a retired IBM friend often says about managing projects. Similarly, you won't produce the best-spelling students if your grading has focused on penmanship. When performance is critical, have the computer do not what's easiest for it (and thereby, us) but what's most useful. To best handle custom metrics, analysts need a strong multidimensional (and preferably, multimodal) optimization algorithm; still, use of even simple random search with a custom score function is usually better than not customizing.[10]

4. LISTEN (ONLY) TO THE DATA

Inducing models from data has the virtue of looking at the data afresh, not constrained by old hypotheses. But, while "letting the data speak," don't tune out received wisdom. Experience has taught these once brash analysts that those familiar with the domain are usually more vital to the solution of the problem than the technology we bring to bear.

Often, nothing *inside* the data will protect you from significant, but wrong, conclusions. Table 20.1 contains two variables about high school, averaged by state: cost and average SAT score (from about 1994). Our task, say, is to model their relationship to advise the legislature of the costs of improving our educational standing relative to nearby states. Figure 20.1 illustrates how the relationship between the two is significant: the LR *t*-statistic is over 4, for example, suggesting that such a strong relationship occurs randomly only 1/10,000 times. However, the sign of the relationship is the opposite of what was expected. That is, to improve our standing (lower our SAT ranking), the graph suggests we need to reduce school funding!

Observers of this example will often suggest adding further data—perhaps, for example, local living costs, or percent of the population in urban or rural settings—to help explain what is happening. But the real problem is one of self-selection. The high-SAT/low-cost states are clustered mainly in the Midwest, where the test required for state universities (the best deal for one's dollar) is not the SAT but the ACT. Only those students aspiring to attend (presumably more prestigious) out-of-state schools go to the trouble of taking an extra standardized test, and their resulting average score is certainly higher than the larger population's would be. Additional variables in the database, in fact (other than proportion of students taking the SAT), would make the model more complex and might obscure the fact that information external to the data is vital.

---

[10] (Elder, 1993) introduced a global search algorithm for multimodal surfaces that updates a piecewise planar model of the score surface as information is gathered. It is very efficient, in terms of function evaluations, but its required overhead restricts it to a handful of dimensions (simultaneous factors) in practice. The need remains for efficient, higher-capacity global search methods.

TABLE 20.1   Spending and Rank of Average SAT Score by State

USA State	SAT Rank	$ Spent
AK	31	7877
AL	14	3648
AR	17	3334
AZ	25	4231
CA	34	4826
CO	23	4809
CT	35	7914
DC	49	8210
DE	37	6016
FL	40	5154
GA	50	4860
HI	44	5008
IA	1	4839
ID	22	3200
IL	10	5062
IN	47	5051
KS	6	5009
KY	18	4390
LA	16	4012
MA	33	6351
MD	32	6184
ME	41	5894
MI	20	5257
MN	3	5260
MO	13	4415
MS	12	3322
MT	19	5184
NB	8	4381
NC	48	4802
ND	2	3685
NH	28	5504
NJ	39	9159
NM	15	4446
NV	29	4564
NY	42	8500
OH	24	5639
OK	11	3742
OR	26	5291
PA	45	6534
RI	43	6989
SC	51	4327
SD	5	3730
TN	9	3707
TX	46	4238
UT	4	2993
VA	38	5360
VT	36	5740
WA	30	5045
WI	7	5946
WV	27	5046
WY	21	5255

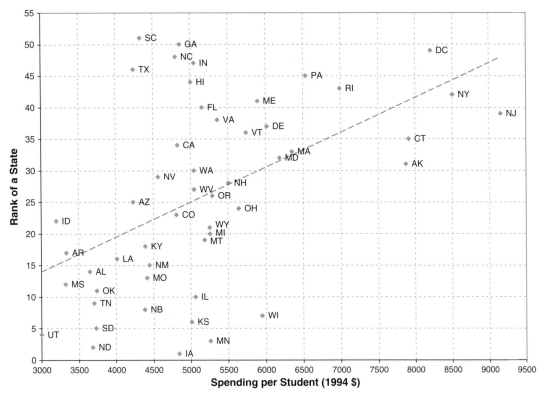

FIGURE 20.1 Rank of a state (in average SAT score) versus its spending per student (circa 1994) and the least-squares regression estimate of their relationship.

The preceding example employed typical "opportunistic," or found, data. But even data generated by a designed experiment need external information. A DoD project from the early days of neural networks attempted to distinguish aerial images of forests with and without tanks in them. Perfect performance was achieved on the training set, and then on an out-of-sample set of data that had been gathered at the same time but not used for training. This was celebrated but, wisely, a confirming study was performed. New images were collected on which the models performed extremely poorly. This drove investigation into the features driving the models and revealed them to be magnitude readings from specific locations of the images; i.e., background pixels. It turns out that the day the tanks had been photographed was sunny, and that for nontanks, cloudy![11] Even resampling the original data wouldn't have protected against this error, as the flaw was inherent in the generating experiment.

[11] PBS featured this project in a 1991 documentary series *The Machine That Changed the World*: Episode IV, "The Thinking Machine."

A second tanks and networks example: A colleague had worked at a San Diego defense contractor, where researchers sought to distinguish tanks and trucks from any aspect angle. Radars and mechanized vehicles are bulky and expensive to move around, so they fixed the radar installation and rotated a tank and a truck on separate large, rectangular platforms. Signals were beamed at different angles, and the returns were extensively processed—using polynomial network models of subsets of principal components of Fourier transforms of the signals—and great accuracy in classification was achieved. However, seeking transparency (not easy for complex, multistage models), the colleague discovered, much to his chagrin, that the source of the key distinguishing features determining vehicle type turned out to be the bushes beside one platform and not another![12] Further, it is suspected that the angle estimation accuracy came from the signal reflecting from the platform corners—not a feature you will encounter in the field. Again, no modeling technology alone could correct for flaws in the data, and it took careful study of how the model worked to discover its weakness.

5. ACCEPT LEAKS FROM THE FUTURE

One of us often evaluates promising investment systems for possible implementation. A Ph.D. consultant, with a couple of books under his belt, had prepared a neural network model for a Chicago bank to forecast interest rate changes. The model was 95% accurate—astonishing given the importance of such rates for much of the economy. The bank board was cautiously ecstatic and sought a second opinion. My colleagues found that a version of the output variable had accidentally been made a candidate input. Thus, the output could be thought of as only losing 5% of its information as it traversed the network.

One investment system we were called in to examine was 70% accurate in forecasting the direction a market index would move the next day. Its developers were quite secretive, but after a great deal of work on behalf of the client considering investing, we eventually duplicated its actions exactly with a simple moving average of 3 days of prices. This simplicity was disappointing, but much worse was that the 3 days were centered on *today*. That is, tomorrow's price was one of the inputs! (They'd have had 100% accuracy if they had have just dropped one of the input variables.) Another trading system, developed with monumental effort over several years and involving the latest research in Genetic Algorithms (GA), focused on commodities. Eventually, it was 99.9% matched by, essentially, two lines of code, which made obvious that its use was impractical. That is, the complex GA devolved to a simple model (the flaws of which then became quite clear), in a manner impossible to discern by examining the extremely complex modeling machinery. In all these cases, the model's author was the chief one deceived.

---

[12] This excellent practice of trying to break your own work is so hard to do even if you are convinced of its need that managers should perhaps pit teams with opposite reward metrics against one another in order to proof-test solutions.

One trick is to look hardest at any input variable that works too well. For instance, on a cross-sell project—trying to identify clients of an auto club who would be good prospects for a more profitable insurance product—we found a code that was present about 25% of the time but was always associated with insurance purchasers. After extended inquiry (as the meaning of data fields are often lost to the mists of time), we found that the code was the type of insurance cancellation; that is, that it really represented the fact that about a quarter of purchasers canceled their insurance each year. Dorian Pyle, author of a thorough book on *Data Preparation for Data Mining*, has recounted privately that he's encountered problems that required seven such "decapitation" passes, where the best variable turns out to be a leak from the future.

In general, data warehouses are built to hold the best information known to date on each customer; they are not naturally able to pull out what was known at the time that you wish to study. So, when you are storing data for future mining, it's important to date-stamp records and to archive the full collection at regular intervals. Otherwise, re-creating realistic information states will be extremely difficult and will lead to wrong conclusions. For instance, imagine you wished to study whether dot-com companies were, in aggregate, really a bad bet. Using a price-quoting service, you pull down all the histories of current such companies and study their returns. Quite likely, they would have been a great bet, despite the horrible shakeout in that market sector that started in roughly March 2000. Why? Were their early gains so great as to absorb later massive losses? Actually, you would have made a study error—"survivor bias"—by looking back from *current* companies, which is all most data services carry. A re-creation of the set of companies that existed at the earlier time, including the doomed ones, would provide a much more realistic (i.e., negative) result.

6. DISCOUNT PESKY CASES

Outliers and leverage points can greatly affect summary results and cloud general trends. Yet you must not routinely dismiss them; they could *be* the result. The statistician John Aitchison recalled how a spike in radiation levels over the Antarctic was thrown out for years, as an assumed error in measurement, when in fact it revealed a hole in the ozone layer that proved to be an impressive finding. To the degree possible, visualize your data to help decide whether outliers are mistakes to be purged or findings to be explored.

We find the most exciting phrase in research not to be a triumphal (and rare) "Aha!" of discovery, but the muttering of puzzlement, "That's odd ..." To be surprised, though, you must have expectations. So we urge colleagues to make hypotheses of how results will turn out from their upcoming experiments. After the fact, virtually everything can and will be plausibly interpreted. One master's engineering student at the University of Virginia was working with medical data (often an extremely tough domain) and presented some interim findings as a graph on an unlabeled transparency to the nurse and doctor leading the research. They were happily interpreting the results when he realized, to his horror, that the foil was upside-face-down, that is, that the relationship between the variables was reversed. He sheepishly set it right and in only seconds the medical experts exclaimed, "That makes sense too!" and continued interpreting its (new and completely opposite) nuances.[13]

Humans are, and likely will remain, the best pattern-recognizers in existence—for the low dimensions in which we operate. But we are perhaps too good; we tend to see patterns even when they don't exist. Two colleagues (and expert data miners, Dustin Hux and Steve Gawtry) worked at the Virginia State Climatology Office when a citizen sent in a videotape of purported cloud phenomena: "Could the weather experts explain the astonishing phenomena?" The un-narrated 3-hour tape contained nothing but typical summer (cumulus humulus) clouds. The citizen had seen "dragons in the clouds," where there (almost certainly) weren't any.

A valuable step early in analysis is to seek to validate your data internally: do the variables agree with one another? Finding, as we did on one data set, that "95% of the husbands are male" is useless in itself, but reveals something about the data's quality, and provides audit questions and flags observations. Reliable analysis depends so strongly on the quality of the data that internal inconsistencies can hobble your work, or they can be clues to problems with the flow of information within the company and reveal a key process obstacle. We worked closely with a direct mail client and dove deeply into the data, looking for relationships between what was known about a potential customer and resulting orders. We actually endangered our appearance of competence to the client by persisting in questioning about unexpectedly low numbers of catalogs being sent to some customers. Eventually, it was found that the "Merge/Purge house" was treating overseas purchasers the opposite of how they were instructed, and erroneously deleting from the mailing lists some of the best prospects. This finding was probably more helpful to the client's bottom line than most of our high-tech modeling work.[14]

7. EXTRAPOLATE

Modeling "connects the dots" between known cases to build up a plausible estimate of what will happen in related, but unseen, locations in data space. Obviously, models—and especially nonlinear ones—are very unreliable outside the bounds of any known data. (Boundary checks are the very minimum protection against "overanswering," as discussed in the next section.)

But there are other types of extrapolations that are equally dangerous. We tend to learn too much from our first few experiences with a technique or problem. The hypotheses we form—which our brains are desperate to do to simplify our world—are irrationally hard

---

[13] Those who don't regularly research with computers seem to give more credence to their output, we've noticed. Perhaps like sausage being enjoyed most by those least familiar with how it's made.

[14] The analysis work, though, combined with operational changes such as higher-quality catalog paper, did result in a doubling of the client's average sales per catalog within a year.

to dethrone when conflicting data accumulate. Similarly, it is very difficult to "unlearn" things we've come to believe after an upstream error in our process is discovered. (This is not a problem for our obedient and blindingly fast assistant: the computer. It blissfully forgets everything except what it's presented at the moment.) The only antidote to retaining outdated stereotypes about our data seems to be regular communication with colleagues and clients about our work, to uncover and organize the unconscious hypotheses guiding our explorations.[15]

Extrapolating also from small dimensions, d, to large is fraught with danger, as intuition gained on low-d is useless, if not counterproductive, in high-d. (That is, an idea may make sense on a white board, and not work on a many-columned database.) For instance, take the intuitive *nearest neighbor* algorithm, where the output value of the closest known point is taken as the answer for a new point. In high-d, no point is typically actually close to another; that is, the distances are all very similar and, by a univariate scale, not small. "If the space is close, it's empty; it it's not empty; it's not close" is how Scott (1992) describes this aspect of the "curse of dimensionality."

Friedman (1994) illustrates four properties of high-d space:

1. Sample sizes yielding the same density increase exponentially with d.
2. Radiuses enclosing a given fraction of data are disproportionately large.
3. Almost every point is closer to an edge of the sample space than to even the nearest other point.
4. Almost every point is an outlier in its own projection.[16]

As our most powerful technique—visualization—and our deep intuition about spatial relationships (in low-d) are rendered powerless in high-d, researchers are forced to employ much more simplistic tools at the early stages of a problem until the key variables can be identified and the dimensions thereby reduced.

The last extrapolation is philosophical. Most researchers in data mining, machine learning, artificial intelligence, etc., hold the theory of evolution as an inspiration, if not motivating faith. The idea that the awesome complexity observed of life might have self-organized through randomization and indirect optimization can bolster the belief that something similar might be accomplished in software (and many orders of magnitude faster). This deep belief can easily survive evidence to the contrary. We have heard many early users of *neural networks*, for instance, justify their belief that their technique will eventually provide the answer since "that's how the brain works."[17] Others have such faith in their mining algorithm that they concentrate only on obtaining all the raw materials that collectively contain the information about a problem and don't focus sufficiently on

---

[15] This is so critical that, if you don't have a colleague, rent one! A tape-recorder or a dog will even be preferable to keeping all of your dialog internal.

[16] That is, each point, when projecting itself onto the distribution of other points, thinks of itself as weird ... kind of like junior high.

[17] Though research from even a decade ago argues instead that each human neuron (of which there are billions) is more like a supercomputer than a simple potentiometer.

creating higher-order features of the raw data. They feed, say, the intensity values of each pixel of an image into an algorithm, in hopes of classifying the image—which is almost surely doomed to fail—instead of calculating higher-order features—such as edges, regions of low variance, or matches to templates—which might give the algorithms a chance.

A better mental model of the power and limitations of data mining is small-scale, rather than large-scale, evolution. We can observe, for instance, that one can take a population of mutts and, through selective breeding over several generations, create a specialized breed such as a greyhound. But it is a bold and unproven hypothesis that one could do so, even with infinite time, beginning instead with pond scum. Likewise, the features you extract from raw data strongly impact the success of your model. As a rule, use all the domain knowledge and creativity your team can muster to generate a rich set of candidate data features. Data mining algorithms are strong at sifting through alternative building blocks, but not at coming up with them in the first place.

The March 25, 1996, cover of *Time* magazine, provocatively asks: "Can Machines Think? They already do, say scientists. So what (if anything) is special about the human mind?"[18] Magazine covers can perhaps be forgiven for hyperbole; they're crafted to sell copies. But inside, someone who should know better (an MIT computer science professor) was quoted as saying, "Of course machines can think. After all, humans are just machines made of meat." This is an extreme version of the "high-AI (artificial intelligence)" view (or perhaps,

---

[18] *Time* magazine was reporting on the previous month's first chess match between Gary Kasparov (often called the best chess player in history) and "Deep Blue," a specialized IBM chess computer. Kasparov lost the first game—the first time a Grand Master had been beaten by a program—but handily won the full match. Still, that was to be the high-water mark of human chess achievement. A year later, "Deeper Blue" won the re-match, and it's likely humans will never reign again. (IBM enjoyed the publicity and didn't risk a requested third match and, of course, computer power has grown by over two orders of magnitude since then.) Unlike checkers, chess is still nearly infinite enough that computers can't play it perfectly, but they can simply march through a decision tree of possibilities as deep as time allows (routinely to a dozen or more plies, or paired move combinations). Though it seems like a good test of intelligence, the game of chess actually plays well to the strengths of a finite state machine: the world of possibilities is vast, but bounded, the pieces have precise properties, and there is close consensus on many of the game trade-offs (i.e., a bishop is worth about three times as much as an unadvanced pawn). There are also vast libraries of carefully worked special situations, such as openings, and end-game scenarios, where a computer can play precisely and not err from the known best path. The automation component with the greatest uncertainty is the precise trade-off to employ between the multiple objectives—such as attack position (strong forward center?), defense strength (take time to castle?), and the pursuit of materiel (capture that pawn?)—that vie for control of the next move. To define this score function by which to sort the leaf nodes of the decision tree, the Deep Blue team employed supervised learning. They took the best role models available (human Grand Masters) and trained on the choices made by the GMs over many thousands of recorded games to discover what parameter values for the move optimizer would best replicate this "gold standard" collection of choices. Lastly, the designers had the luxury of studying many of Kasparov's games and purportedly devised special anti-Kasparov moves. (Incidentally, Kasparov was refused the chance to study prior Deep Blue games.) Given how "computational" chess is then, it's a wonder any human does well against a machine! But our major point is that chess skill is a poor metric for "thinking."

the "low-human" view). But anyone who's worked hard with computers knows that the analytic strengths of computers and humans are more complementary than alike. Humans are vastly superior at tasks like image recognition and speech understanding, which require context and "common sense" or background knowledge to interpret the data, but computers can operate in vast numbers of dimensions—very simply, but with great precision. It's clear to us that the great promise being fulfilled by data mining is to vastly augment the productivity of—but not to replace—skilled human analysts. To believe otherwise—at the extreme, in an eventual "singularity event" in time where humans and machines will merge to create a type of immortal consciousness—is an extrapolation more like faith than science.

8. ANSWER EVERY INQUIRY

Early in our careers, one of us demonstrated a model estimating rocket thrust that used engine temperature, T, as an input. A technical gate-keeper for the potential client suggested we vary some inputs and tell what ensued. "Try $T = 98.6$ degrees." Naively, we complied—making mistake 7, as that was far outside its training bounds. The output (of a nonlinear polynomial network) was ridiculous, as expected, but no amount of calm technical explanation around that nonsurprising (to us) result could erase, in the decision-maker's mind, the negative impact of the breathtaking result that had briefly flashed by. We never heard from that company again. Obviously, a model should answer "don't know" for situations in which its training has no standing!

But how do we know where the model is valid, that is, has enough data close to the query by which to make a useful decision? The simplest approach is to note whether the new point is outside the bounds, on any dimension, of the training data. Yet, especially in high-d, the volume of the populated space is only a small fraction of the volume of the rectangle defined by the univariate bounds. With most real data, inputs are very far from mutually independent, so the occupied fraction of space is very small, even in low-d. (The data often look like an umbrella packed corner-to-corner diagonally in a box.) A second approach, more difficult and rare, is to calculate the convex hull of the sample—essentially, a "shrink wrap" of the data points. Yet even this does not always work to define the populated space. Figure 20.2 illustrates a 2-d problem similar to one we encountered in practice (in higher-d) in an aeronautical application. There, fundamental constraints on joint values of physical variables (e.g., height, velocity, pitch, and yaw) caused the data to be far from i.i.d. (independent and identically distributed). We found that even the sample mean of the data, μ, was outside the true region of populated space.

One approach that has helped the few times we've tried it, is to fit a very responsive, nonlinear model to the data, for instance through a polynomial network (Elder and Brown, 2000). High-order polynomials quickly go toward infinity outside the bounds of the training data. If the output estimate resulting from an unbounded, nonlinear (and even overfit) model is beyond the output bounds, then it is very likely the input point is outside the

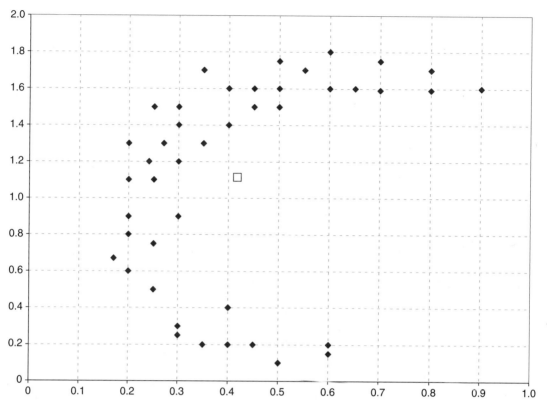

FIGURE 20.2 Sample two-dimensional problem for which the data mean (open box symbol) is outside the bounds of the (crescent-shaped) valid space.

training data. If a training data point had been near that input point, it would have better constrained the model's estimate.

Just as it is essential to know where a model has standing—i.e., in what regions of input space its estimates might be valid—it is also useful to know the uncertainty of estimates. Most techniques provide some measure of spread, such as σ, for the overall accuracy result (e.g., $+/- 3\%$ for a political survey), but it is rare indeed to have a conditional standard deviation, $\sigma(x)$, to go with the conditional $\mu(x)$. A great area of research, in our opinion, would be to develop robust methods of estimating certainty for estimates conditioned on where in input space you are inquiring.

One estimation algorithm, Delaunay Triangles, which does depend strongly on $\sigma(x)$ was developed to make optimal use of experimental information for global optimization (Elder, 1993). For experiments where results are expensive to obtain (core samples of soil, for instance), the challenge is to find, as efficiently as possible, the input location with the best result. If several samples and their results are known, you can model the score surface (relationship between input vector and output score) and rapidly ask the model for the best location to next probe (i.e., experimental settings to employ). If that result isn't yet good enough (and budget remains to keep going), its information could be used to update the model for the next probe location. The overall estimation surface consists of piecewise planes, as shown in Figure 20.3, where each region's plane has a quadratic variance "canopy" over it, as in Figure 20.4, revealing how the uncertainty of the estimation grows as you depart from the known points (the corners).[19] This approach worked very well, for low (fewer than about 10) dimensions, and the resulting multimodal search

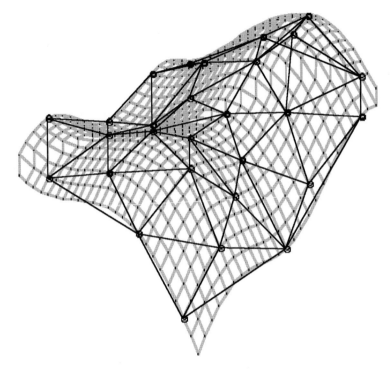

FIGURE 20.3 Estimation surface of Delaunay Triangle method (Elder, 1993) is piecewise planar. (The underlying functional surface is represented here by a mesh.)

[19] The modeling technique developed for GROPE was driven by the special requirements of optimizing an unknown function—especially that the response surface model had to agree exactly with the known samples. If you assume the least about the response surface—that there is Brownian motion (or a random walk) between the known points—then the ideal estimator turns out to be a plane. So, $\mu(x)$ is a piecewise planar collection of simplices (e.g., triangles when there are two input dimensions). The tiling or tessellation of the input space is done in such a way as to create the most uniform simplices (those with the greatest minimum angle), which is performed by Delaunay triangulation (a dual of nearest neighbor mapping). The key, though, was to pair this with an estimate of the standard deviation of $\mu(x)$, conditioned on x, $\sigma(x)$. (The Brownian motion assumption drives this to be the square root of a quadratic function of distance from the known corners.) Now, with both parts, $\mu(x)$ and $\sigma(x)$, you can rapidly calculate the location, x, where the probability of exceeding your result goal is the greatest. So, the model would suggest a probe location, you would perform the experiment, and the result would update the model, with greater clarity on the mean estimates (piecewise planes) and reduced variance (piecewise quadratic "bubbles" over each plane) with each iteration.

FIGURE 20.4   Each simplex (e.g., triangle in two dimensions) of the Delaunay method (Elder, 1993) pairs a planar estimation of $\mu(x)$ with a quadratic estimation of $\sigma^2(x)$.

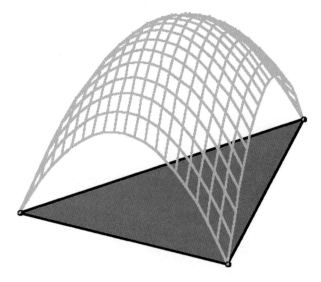

algorithm, GROPE (Global R^d Optimization when Probes are Expensive), took the fewest probes of all then-existing algorithms to converge close to the answer on a standard suite of test problems. By having, for every location, x, an estimate of the mean, $u(x)$, along with its uncertainty, $\sigma(x)$, the algorithm could, with every new result, refine its estimates and reduce its uncertainty, and thereby zero in on the locations with the greatest potential.

9. SAMPLE CASUALLY

The interesting cases for many data mining problems are rare, and the analytic challenge is akin to "finding needles in a haystack." However, many algorithms don't perform well in practice, if the ratio of hay to needles is greater than about 10 to 1. To obtain a near-enough balance, you must either *undersample*—removing most common cases—or *oversample*—duplicating rare cases. Yet it is a mistake to do either casually.

A direct marketing firm in Maryland had too many (99%) nonresponders (NR) for a decision tree model to properly predict who would give to a client charity. To better balance the data, the firm kept all responders and every tenth NR from the data set of over a million cases until it had exactly 100K cases. The firm's decision tree model then predicted that everyone in certain Alaskan cities (such as Ketchikan, Wrangell, and Ward Cove) would respond. And, it was right, on the training data. What had happened? It turns out the original data had been sorted by ZIP code, and since it was over 1M cases, the decimating (sampling every tenth) had stopped (at 100K) before reaching the end. Thus, only responders (all of whom had been taken at first) were sampled from the bottom of the file, where the highest (Alaskan) ZIP codes were.

Clearly, a good strategy is to "shake before baking," that is, to randomize the order of a file before sampling. Actually, first record the original case number, so you can reverse the process, but also because there may be useful information in the original order. One credit scoring problem we tackled, for instance, had very useful information in the case number. Perhaps the first cases in the file were the first to respond (most desperate for credit?) or were the top candidates to be given offers—in any case, the file position was a very significant predictor of creditworthiness. Unfortunately, no one could remember how the file was formed, so we lost the opportunity to use that clue to focus on related features that would improve the edge of the model.

Since case number is not always random, we recommend appending some truly random variables to the data to act as "canaries in the mine"[20] to indicate when your variable selection has pushed too far. If the model starts using those variables, you know that overfit is very likely. (Be sure to label them *random1*, *random2*, etc., instead of *R1*, *R2*, or someone will call you months later and ask where he or she might find the *R1* values, needed for the "improved" model!)

It is possible to make mistakes even more readily with up-sampling. Again, on a credit scoring problem, we had too few known defaulters to properly train several algorithms. (Some work well with case weights and variable misclassification costs, but most don't, and we wanted to make sure their training data were exactly equivalent for comparison.) We ran several experiments with cross-validation and many modeling cycles. Oddly, for several experiments, it was much harder than anticipated to obtain overfit; that is, results tended to improve with model complexity far beyond where we expected. Eventually, we noticed that when we separated the performance on new data of the rare cases of default from the common cases of creditworthiness, performance on the former (rare cases) kept improving with complexity, while the latter curve turned upward (higher error) as expected. It turns out that we had duplicated the defaults *before* splitting the data into cross-validation sets. This meant that copies of each rare case appeared in most of the data subsets, and thus no rare case was truly ever out-of-sample. The lesson learned is to split into sets first and then up-sample the rare cases in training only.

Note that you may need to define the unit of sampling at a higher granularity than the case. That is, some cases may need to stay bundled together and never be separated across data subsets. For instance, with medical data, each case often represents a doctor visit. In predicting outcomes, you must almost certainly keep all the records from one patient together and not use some for training and some for evaluation.

Lastly, remember that a stratified sample will almost always save you from trouble. Carefully consider which variables need to be represented in each data subset and sample them separately; e.g., take 10% of the red and 10% of the blue cases for each sample, instead of just a simple 10% of the whole. Good stratification is how fairly accurate political projections, for instance, can be made with very few (~1,000) interviews.

---

[20] Canaries are particularly susceptible to carbon monoxide and methane, so if the caged canary stopped singing and keeled over, it was time to evacuate quickly!

10. BELIEVE THE BEST MODEL

As George Box said, "All models are wrong, but some are useful." Unfortunately, we seem to have a great need to believe in our models. We want them to reveal deep underlying truth, rather than to just be useful leading indicators.[21] However, reading too much into models may do more harm than good. Usually, too much attention is paid to the particular variables used by the "best" data mining model—which likely barely won out over hundreds of others of the millions (to billions) tried, using a score function only approximating your goals, and on finite data scarcely representing the underlying data-generating mechanism. It is better to build several models and interpret the resulting distribution of variables rather than accept the set chosen by the single best model.[22]

Usually, many very similar variables are available, and the particular structure and variables selected by the best model can vary chaotically, that is, change greatly due to very small changes in the input data or algorithm settings. A decision tree, for instance, can change its root node due to one case value change, and that difference cascades through the rest of the tree. One polynomial network we built two decades ago changed drastically in appearance when only 999 of the 1,000 cases sent with it were used to test it. This structural variability is troubling to most researchers. But that is perhaps due to their reading too much into the "best" model. In cases we have examined, the functional similarity of competing models—that is, their vector of estimate values—is often much more similar than their structural form. That is, competing models often *look* more different than they *act*, and it's the latter we believe that matters.

Lastly, as argued in Chapter 13, the best model is likely to be an ensemble of competing, distinct, and individually good models. In Chapter 18, Figure 18.1 showed the relative performance of five algorithms on six test problems. Figure 18.2 went on to reveal that any of the four different ways of ensembling those models greatly reduced the error on out-of-sample data compared to the individual models.

Bundling models reduces the clarity of a model's details; yet, as Leo Breiman argued, only somewhat tongue-in-cheek:

$$\text{Interpretability} * \text{Accuracy} < \beta \text{ (Breiman's constant)}$$

meaning increased interpretability of a model comes, inevitably, at the cost of reduced accuracy. Some problems do require strict interpretability. For instance, insurance models have to be approved by state authorities. Also, credit applicants have to, by law, be able to be told the top five factors under their influence that hurt their credit score. But for the applications we're most familiar with, such as investment prediction or fraud detection, there is

---

[21] The ancient Egyptians believed the dog star/god, Sirius, was responsible for the seasonal flooding of the Nile, so essential to their survival. While not causal, the star's rise was a useful leading indicator, accurate enough for important decision making.

[22] In a similar vein, Box also said, "Statisticians are like artists; they fall in love with their models." To be great data miners, we need to work as hard to break our models as we did to build them. The real world will certainly break them if we don't!

a huge return to small but significant increases in accuracy, and we are content to worry about interpretation only after the model has proven it is worthy of the attention.

HOW SHALL WE THEN SUCCEED?

Fancier modern tools and harder analytic challenges mean we can now "shoot ourselves in the foot" with greater accuracy and power than ever before! Success is improved by learning, which best comes from experience, which seems to be most memorable due to mistakes. So go out and make mistakes early in your career!

We've found a useful PATH to success to be

- **P**ersistence: Attack a data mining problem repeatedly, from different angles. Automate the essential steps, especially so you can perform resampling tests. Externally check your work. A great idea is to hire someone to break your model, since you often won't have the heart.
- **A**ttitude: An optimistic, "can-do" attitude can work wonders for results, especially in a team setting.
- **T**eamwork: Business and statistical experts must cooperate closely to make the best progress. Does everyone want the project to succeed? Sometimes, passive-aggressive partners, such as the purported providers of data, can secretly see only danger in a project delving into their domain, so be sure that each partner can advance his or her career through the project's success.
- **H**umility: Learning from others requires vulnerability. When we data miners visit a client, we know the least about the subject at hand. However, we do know a lot about analysis and the mistakes thereof. Also be humble about the powers of technology. As shown in this chapter especially, data mining is no "silver bullet." It still requires a human expert who can step back, see the big picture, and ask the right questions. We think of data mining as something of an "Iron Man" suit, tremendously augmenting, for good or ill, the powers of the wearer.

POSTSCRIPT

So go out and do well, while doing good, with these powerful modern tools!

References

Elder, J. F., IV (1993). *Efficient optimization through response surface modeling: A GROPE algorithm*. Dissertation. Charlottesville: School of Engineering and Applied Science, University of Virginia.

Elder, J. F., IV (1996). A review of *Machine Learning, Neural and Statistical Classification* (Michie, Spiegelhalter, & Taylor, Eds., 1994). *Journal of the American Statistical Association, 91*(433), 436–437.

Elder, J. F., IV & Lee, S. S. (1997). *Bundling Heterogeneous Classifiers with Advisor Perceptrons*. University of Idaho Technical Report, October, 14.

Elder, J. F., IV, & Brown, D. (2000). Induction and Polynomial Networks. In M. Fraser (Ed.), *Network Models for Control and Processing* (Chapter 6). Bristol, UK: Intellect Press.

Friedman, J. H. (1994). An Overview of Computational Learning and Function Approximation. In V. Cherkassky, J. H. Friedman, & H. Wechsler (Eds.), *From Statistics to Neural Networks: Theory and Pattern Recognition Applications* (Chapter 1). New York: Springer-Verlag.

Michie, D., Spiegelhalter, D. J., & Taylor, C. C. (1994). *Machine Learning, Neural and Statistical Classification*. New York: Ellis Horwood.

Scott, D. W. (1992). *Multivariate Density Estimation: Theory, Practice, and Visualization*. New York: John Wiley.

Prospects for the Future of Data Mining and Text Mining as Part of Our Everyday Lives

PREAMBLE

Like most other areas of technology, data mining exists on a shifting landscape. Not only is the old part of the landscape being redefined continually, but new areas of interest always loom ahead. In this chapter, we will describe several of those opportunities for data mining on the road ahead.

From the perspectives of the coauthors, there are minimally four new or currently emerging areas in data mining. Some of them have been under development but are difficult areas just now being refined to a stage of high accuracy (e.g., 90% or better), whereas others are just emerging These four areas are

1. Radio frequency identification (RFID) technologies
2. Social networks
3. Image and object (or visual) data mining including object identification, 3D medical scanning, and photo–3D motion analysis
4. "Cloud computing" and the "elastic cloud": Software as a Service (SaaS)

RFID

The acronym *RFID* stands for *radio frequency identification*. RFID technologies simply put a radio frequency identification tag on anything from a kidney being transported to a medical center for a transplant to every box of corn flakes coming off a conveyor belt. One of the first businesses to make use of RFID in a large way is Wal-Mart. A few years ago, Wal-Mart mandated that its top 100 suppliers have all items RFID tagged, with the idea of eventually having all its vendors supplying goods with RFID tags. Why? Wal-Mart can keep track of where every box of any item is at any one moment—from when it enters the warehouse to when it leaves the warehouse via a conveyor belt and is routed to *this* shipping semi truck or *that* semi truck. As the items are being shipped, Wal-Mart knows the whereabouts of all items, whether on a particular interstate highway traveling west, east, south, or north; to how many boxes of any item were left at any store; to how many items remain for sale in any particular store.

RFID does not collect a lot of variables but does collect long data files with billions of cases. What to do with these data? Data miners are now trying to figure out the answer. These data must be the source of unfound patterns of information, i.e., just awaiting knowledge discovery that can be used for good purposes.

Following are other examples where RFID technology is being used:

- **Security:** Examples include access control to a building, a particular laboratory, etc., by either a card, like a credit card, that the user has to swipe to gain access, or RFID tags, like those being put in car ignition keys by some automobile manufacturers as a deterrent to theft.
- **Tracking:** Examples include tracking of merchandise, as in the previous Wal-Mart example, but also tags being used for human tissue transplants. At many marathon races in the United States, organizers now provide RFID tags, placed in the laces of the running shoes, that handle the logistics of thousands of runners, recording the start time as a runner passes the start line and at various mileposts and the finish line. RFID is also well suited to tagging/tracking cattle and other livestock, especially in this age of food products possibly carrying disease to humans, such as mad cow disease. Even pets can be tagged with an injected RFID behind the ear or neck, making it much easier to get a lost dog or cat back to its owner (the owner's name and address are part of the RFID tag). Airline luggage tags now provide a combination of RFID, barcodes, and printed information about the owner. Watermarks on bank notes and artists' signatures on paintings have been used in the past for authenticity, but they can be forged; RFID tags can provide added authenticity.
- **Authenticity:** RFID tags are being put in the front cover of U.S. passports issued since January 2007. These tags have all the information about the passport holders and enables faster movement through customs. There are even separate lines for those with the new RFID passports because these people can be processed faster, with greater accuracy and authenticity.
- **Electronic payments:** Examples include auto tolls on many U.S. toll roads, where the RFID tag in the car windshield is "read" as a person drives under the reader,

automatically debiting the driver's toll account. Transportation tickets (bus, train, airplane) provide a greater advantage than paper tickets because they can be read and validated while in a person's pocket and even "stamped" as "used."

- **Entertainment:** Some toy manufacturers are using RFID tags to make "smart toys" that will do things or make sound effects or speak, the exact output being dependent on the environmental circumstances (for example, the Hasbro Star Wars characters). RFID "ticket tags" are being used by ski resorts, so people can rapidly gain access to chair lifts multiple times over a predefined period of hours or days, a procedure that is clearly more efficient than users having to reach in their pockets for either money, a token, or a pass each time they need the chair lift.

From the preceding examples, you can see that RFID is rapidly becoming an integral part of our lives.

SOCIAL NETWORKING AND DATA MINING

A Google search on "social networking and data mining," done in late September 2008, brought up 1,770,000 sites; the first 10 of these are reproduced as a PC screen shot on the accompanying DVD (see "Social Networking Web Sites/PDFs" on the DVD). These 10 sites are as follows:

1. Data Mining in Social Networks (www.cs.purdue.edu/homes/neville/papers/ jensen-neville-nas2002.pdf; see the PDF file on the accompanying DVD)
2. Blog on Social Networks (www.resourceshelf.com/2008/08/25/ collabio-game-explores-social-network-data-mining . . . and-social-psychology/)
3. Data Mining, Text Mining, Visualization, and Social Media (http://datamining.typepad .com/data_mining/2007/04/twitter_social_.html)
4. Pentagon Sets Its Sights on Social Networking Websites: Data Mining and Homeland Security (www.newscientist.com/article/mg19025556.200)
5. Social Networking Sites: Data Mining and Investigative Techniques (https://www .blackhat.com/presentations/bh-usa-07/Patton/Presentation/bh-usa-07-patton.pdf)
6. Social-Network Data Mining: What's the Most Powerful, Untapped Information Repository on the Web Today? (research.microsoft.com/displayArticle.aspx?id=2075)
7. Discovering Social Networks and Communities in Email Flows (www.orgnet.com/ email.html)
8. Social Networks, Data Mining, and Intelligence: Trends (http://rossdawsonblog.com/ weblog/archives/2006/05/social_networks_2.html)
9. The 2nd ACM Workshop on Social Network Mining and Analysis at the KDD–August 2008 Annual Meeting (http://workshops.socialnetworkanalysis.info/SNAKDD2008/)
10. MySpace Has Data-Mining Plans: DMNews (www.dmnews.com/MySpace-has-data-mining-plans/article/98564/)

From the preceding 10 sites, you can see that social networking analyzed with data mining technology is invading many areas of our lives. These areas include social networks of people; networks of web pages; complex relational databases; and data on interrelated people, places, things, and events extracted from text documents.

Example 1

An example of a social network is Twitter. Twitter was started a couple of years ago as a means of friends communicating with fewer than 140 characters in a text-like message. As a communicator, you did not choose who would receive your message, but other people chose you, as "followers." Other people log onto Twitter and select that they want to "follow" you. Thus, a person could develop quite of network of followers, partly dependent on the content of a "tweet" (the message sent on Twitter), and also partly dependent on the "stardom" of the person (e.g., people like to follow movie stars or other celebrities).

During 2008, it was uncertain as to whether Twitter was going to succeed as a major communication method, or if it was a "flash in the pan." However, with the US Airbus 320 crash-landing into the Hudson River to the west of Manhattan, New York on January 15, 2009, this all changed. The plane, crashing into the river minutes after taking off from LaGuardia Airport, an event happened involving Twitter networks that dramatically illustrated that Twitter was the "news medium of the world." The first photo of the crash was taken on an I-Phone by a passenger on a Hudson River ferry; it was uploaded to "Twit-pix," and immediately was relayed by others on Twitter.com around the world. People knew about this US Airways crash while CNN and other news media helicopters were still searching for the plane on the river and before any tugboats and ferry boats reached the aircraft.

Photos similar to the one in Figure 21.1 were uploaded to Twitter immediately after this Airbus crash, circling the world in minutes. All passengers were picked up by ferry boats and all were in a warm building on the east shore of the Hudson River within 45 minutes of this crash landing. Only afterwards did the regular news media cameras

FIGURE 21.1   Photo of US Airways Flight 1549 after crashing into the Hudson River in New York City, United States, January 15, 2009. The photo was taken and uploaded to Flicker.com as a Creative Common License, within minutes of crash landing. Traditional news media did not arrive until about 45 minutes later (http://www.flickr.com/photos/22608787@N00/3200086900;   http://upload.wikimedia.org/wikipedia/commons/c/c8/Plane_crash_into_Hudson_River_muchcropped.jpg)

arrive at the scene. For the first time in history, a major world news event had been sent via pictures and words around the world by Twitter prior to availability of regular traditional news media methods.

Gary Miner caught the CNN News e-mail alert at about 3:30 p.m. CST on this day and immediately clicked on the streaming video reports. At this point, CNN was showing the plane sinking lower into the water with several tugboats and a ferry around the craft, but no passengers were visible on the plane's wings and front slides/rafts. CNN reporters were stating things like "We are not sure if anyone has gotten out, if the passengers are still in the plane; it appears the plane is drifting with the current down the river; we have no reports whether anyone has been saved, or what has happened to people on the aircraft . . . . "

Twitter had taken over as a major source of news for the world. What is important about this Twitter image of the jet liner floating in the Hudson River is that without Twitter, this would have been experienced by the world in an entirely different fashion. Twitter is considered a "microblog" service. During 2008, Twitter was in use by some people, but most of the world had not yet heard of it, or if they had, they had not joined the Twitter network.

However, this Hudson River airplane crash landing on January 15, 2009, brought Twitter into perspective, and shortly after, people signed up for Twitter at astronomical numbers. As of March, 2009, as this is written, people join the social network at a rate of about 10,000/hour.

This phenomenon with Twitter is important to data mining and knowledge discovery because it will be one of the next sources of data available with which to submit it to data analysis in order to gain new knowledge, and thus, make better decisions in the world of our future. So, Twitter has "come of age" (Ulanoff).

All of which is to say: The world is moving fast, and the predictions we co-authors made in this book last September are already taking place, including "Cloud Elastic Net Computing." Cloud Computing is discussed in this chapter as "something of the future," when written in September of 2008, but by November of 2008, companies like Google, IBM, Amazon.com, Zementis, and Microsoft were developing "Commercial Clouds."

Example 2

An email social network is shown in Figure 21.2.

Data Mining Email to Discover Social Networks and Emergent Communities

The social email map in Figure 21.2 illustrates the following according to Krebs (2008):

- This social network map shows email flows among a very large project team.
- In one sense, you might think of this map as an "X-ray" of the way the project team works.
- The nodes are color-coded as follows:
  - Team members department: Red, Blue, Green
  - Consultants outsourced/hired for the project: Yellow
  - External experts consulted: Grey

FIGURE 21.2 Email network among a project team with a deadline to produce a deliverable product.

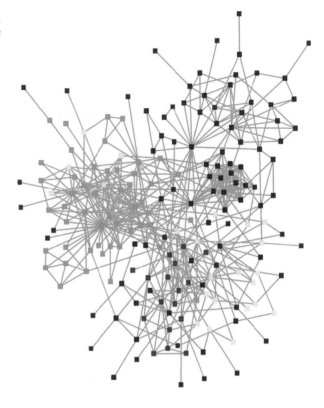

Grey links are drawn between two nodes if two persons sent email to each other at a weekly or higher frequency.

The above diagram shows the project network soon after one deadline was missed. Notice the clustering around formal departments—blues interacting with blues, greens interacting with greens. Several of the hubs in this network were under-performing and often came across as bottlenecks. Project managers saw the need for more direct integration between the departments. One of the solutions was very simple, yet effective—co-location of more project team members. A surprising solution in the age of the Internet! (*Source*: Figure 21.2 and parts of the explanation are abstracted from http://www.orgnet.com/email.html.)

Example 3

Another example of social networking and data mining is the MySpace web site initiating the Social Network Data Mining Initiative; see http://www.dmnews.com/MySpace-has-data-mining-plans/article/98564/.

The article mentions these key points:

- MySpace has data mining plans for a "social network solution."
- MySpace expects this "social network solution" to double its income.

- The plan is to capture personal information from profile pages and then use that information to target ads.
- Such a process could change the way online advertising in general is conducted.

Example 4

The 2nd ACM Workshop on Social Network Mining and Analysis was held in conjunction with the 13th ACM SIGKDD International Conference on Knowledge Discovery and Data Mining (KDD, 2008) in Las Vegas, Nevada, August 23–27, 2008.

"Social Network Mining" topics of interest at this session included the following:

- Community discovery and analysis in large-scale online and offline social networks;
- Personalization for search and for social interaction;
- Recommendations for product purchase, information acquisition, and establishment of social relations;
- Data protection inside communities;
- Misbehavior detection in communities;
- Web mining algorithms for clickstreams, documents, and search streams;
- Preparing data for web mining;
- Pattern presentation for end users and experts;
- Evolution of patterns in the Web;
- Evolution of communities in the Web;
- Dynamics and evolution patterns of social networks and trend prediction;
- Contextual social network analysis;
- Temporal analysis on social networks topologies;
- Search algorithms on social networks;
- Multiagent-based social network modeling and analysis;
- Application of social network analysis;
- Anomaly detection in social network evolution.

Never before had such a large number of social network papers been presented at KDD meetings.

From the preceding few selected examples that fall under the domain of social network/ data mining, you can see that this method of understanding how humans interact is invading almost all areas of life.

IMAGE AND OBJECT DATA MINING

Image and object data mining include visualization, 3D medical scanning and visual–photo movement analysis for development of better physical therapy procedures, security threat identifications, and other areas.

As we pointed out in Chapter 8, this is an area of current research involving development of new and modified algorithms that can better deal with the complexities of three-dimensional object identification. What is now called "visual object analysis" offers a better solution. Visual object analysis started in the late 1990s but has really developed during the first few years of the twenty-first century because:

- Machine learning methods offer greater accuracy in object identification; and
- Large amounts of training data are now available (computer storage space is now not a problem).

The problem in this area of object category recognition is simply: humans can recognize objects—whether people, animals, rocks, or stars in the sky—better than computers; in fact, this has been a very difficult problem for computers. But in the past few years, researchers at several places, notably the University of California–Berkeley, Yahoo!, Google, and California Institute of Technology, have developed newer modified data mining algorithms that have gone from 16% recognition of objects in their 2004 work to 90% correct object category recognition in 2008, with a selected number of objects, using the following algorithms:

- Nonlinear kernelized SVM (slower but more accurate)
- Boosted trees (works slowly)
- Linear SVM (fast but not accurate)
- Intersection kernels in SVM (improved to 90% correct object category recognition in 2008)

Today, 100 objects can be categorized with the methods listed here, but we need to go to several levels of magnitude larger for these methods to be fully successful in accurately recognizing images for such things as national security surveillance and precise medical identification of conditions when using three-dimensional imaging procedures.

Let's examine several areas and ways of looking at the problem of visualization by computer technology and data mining analysis for fields needing high levels of accuracy by actually looking at several visual scenarios (Figure 21.3).

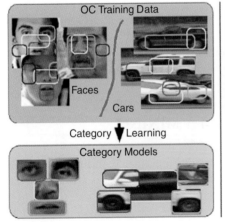 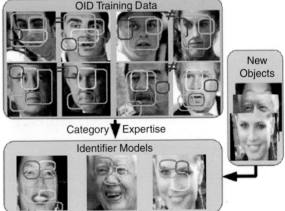

FIGURE 21.3   Object categorization (OC) versus objection identification (OID).

According to Ferencz et al. (2008):

This figure highlights the different learning involved in categorization and identification. The training sets for object categorization, shown on the left side, typically contain many examples of each category (e.g. faces and cars), which are then turned into a fixed model for each in a generative system, or a decision boundary in a discriminative system. A training set for object identification, on the other hand, contains pairs of images from a known category, with a label of "same" or "different" (denoted by = and ≠ in the figure) for each pair. From these labeled pairs, the system must learn how to generate an object instance identifier given a single image of a new object (e.g. Mr. Carter) from the category. For these identifiers to work well, they should highlight distinctive regions of the object. That is, the identifiers should be different for each object.

The statistical methods used in the study of object categorization versus object identification shown in Figure 21.3 used discriminate analysis methods, not the Fast Intersection Support Vector Machine methods that have been shown in 2008 to be more accurate in object identification. Nevertheless, the figure shows some of the types of problems involved in accurately getting computers and software to identify visual objects, including medical objects both for diagnosis and medical delivery.

Briefly, the four illustrations in Figure 21.4, each a visual object dataset, are defined as follows:

A. The MINIST dataset is a set of handwritten digits from two populations, one U.S. Census Bureau employees and the other high school students.
B. The USPS dataset contains handwritten digits collected from mail envelopes in Buffalo, NY.
C. The CURTeT dataset contains images of 61 real-world textures, like leather, rabbit fur, sponge, etc.

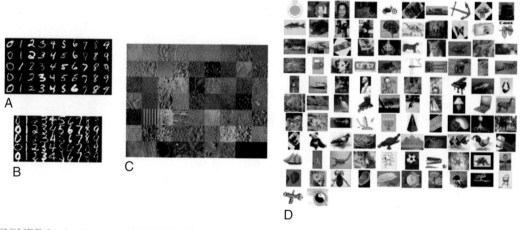

FIGURE 21.4 Data sets: (A) MNIST, (B) USPS, (C) CUReT, (D) Caltech-101. (See the DVD bound with this book for the complete pdf document where these four illustrations, A, B, C, and D are fully explained in the following source: Figure 2 in http://www.eecs.berkeley.edu/Research/Projects/CS/vision/shape/nhz-cvpr06.pdf)

D. The Caltech-101 dataset consists of images from 101 object categories and an additional "background class" for a total of 102 classes. There are significant differences in color and lighting which make this dataset quite challenging.

The next illustration for this object identification–visual data mining, shown in Figure 21.5, comes from work at the University of California–Berkeley and Yahoo! on what is called a *fast intersection kernel Support Vector Machine algorithm*. This is an SVM algorithm that has been modified to give about 90% accuracy in object identification of 100 objects, obtained during the summer of 2008 (Maji et al., 2008). We won't go into details here, but for those interested, this entire paper with further information can be found on the DVD bound with the book.

The fast interaction SVM program, both as C++ and as source code, is available at the following web site: http://www.cs.berkeley.edu/~smaji/projects/fiksvm/ (on the DVD that accompanies this book, see the file named *berkeley - visual object recognition_fast intersection kernel svms_source code - illustrations.doc* in the folder named *pdfs – extra datasets – powerpoints etc.*).

Visual data mining—including its extension into movement modeling, e.g., slices of a moving object such as frames of a motion picture, successive slices of a body scan, or specialized frames/segments of movement in physical therapy—could lead to much more accurate identification of disease states or security threat detections, or more effective physical therapy or athletic training regimens, among other domains.

The possibilities to accomplish this will be available with *STATISTICA* Data Miner Version 9 combined 32 bit/64 bit software. Version 8 is bound as a DVD with this book; Version 9 will not be released until after this book is published.

FIGURE 21.5   As reported in Maji et al. (2008): "Top two rows are a sample of errors on the INRIA pedestrian data set and the bottom two are on the Daimler-Chrysler data set. For each set the first row is a sample of the false negatives, while the second row is a sample of the false positives." (*Source:* Maji et al., 2008.)

Visual Data Preparation for Data Mining: Taking Photos, Moving Pictures, and Objects into Spreadsheets Representing the Photos, Moving Pictures, and Objects

A photo, illustration, graph, or any kind of object can be imported into a data spreadsheet and subsequently analyzed with any of the algorithms available in either the traditional statistics arsenal or the data mining algorithm group. Each pixel point in the photo/object becomes a data point in the spreadsheet. Let's look at an example to see how this is accomplished.

The photo we will be using for an example is shown in Figure 21.6.

In *STATISTICA*, open the File pull-down menu and select Import Picture, as shown in Figure 21.7.

Then find a diagram, photo, graph, or other object that you'd like to import into a spreadsheet with each cell in the spreadsheet representing one pixel in the photograph; thus, if you have a 16-bit photo, you'll have a 16-columns by 16-cases spreadsheet, and so on.

Now, click on the photo file name in the Open dialog box, and then click Open in the lower right of the dialog, as shown in Figure 21.8.

FIGURE 21.6 This is the CD cover photograph from co-author Gary Miner's son, Matt Miner's GHOSTS Christmas album, 2008.

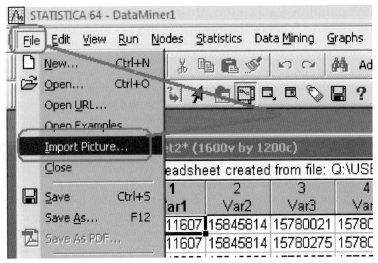

FIGURE 21.7   Importing the picture. Pull down the File menu, and choose Import Picture to open a dialog which can access picture files.

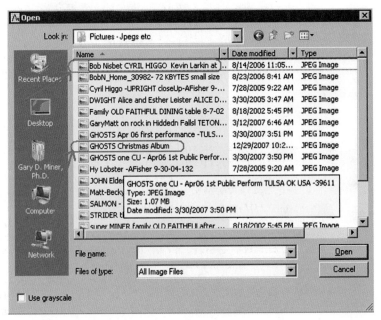

FIGURE 21.8   Locating the files to open. Notice the two JPEG files named Bob Nisbet and GHOSTS Christmas Album. We will be opening the GHOSTS Christmas Album to import into a pixel spreadsheet.

	1 Var1	2 Var2	3 Var3	4 Var4	5 Var5	6 Var6	7 Var7	8 Var8	9 Var9	10 Var10	11 Var11
1	15911607	15845814	15780021	15780021	15714482	15714482	15780275	15780275	15912116	15846323	1578078!
2	15911607	15845814	15780275	15780275	15714482	15714482	15780275	15780275	15846323	15846323	1584657(
3	15846068	15846068	15780275	15780275	15780275	15780275	15780530	15780530	15780530	15846323	1591237
4	15846323	15780530	15780530	15780530	15780530	15780530	15780530	15780530	15714737	15846323	1591237
5	15780785	15780785	15780785	15780785	15780785	15780785	15780785	15780785	15649199	15780785	1591237
6	15714990	15780783	15780783	15846576	15846576	15846576	15846578	15780785	15780785	15846578	1584657(
7	15715245	15715245	15781038	15846831	15846576	15846576	15846576	15780783	15978164	15846578	1578078!
8	15649452	15715245	15781038	15846831	15846831	15846831	15846576	15780783	16109750	15912371	1571499;
9	15780021	15780021	15780275	15846068	15846323	15846323	15846578	15912371	15780783	15846576	1584657(
10	15780021	15780021	15780275	15846068	15846323	15846323	15846578	15912371	15714990	15714990	1578078(
11	15780021	15780021	15780275	15846068	15846578	15846576	15912369	15583404	15649197	1571499(
12	15780275	15780275	15780530	15846323	15846578	15846578	15846576	15912369	15649452	15715245	1578078(
13	15780275	15780275	15780530	15846323	15846578	15846578	15846831	15912624	15846831	15846831	1591262
14	15780275	15780275	15780530	15846323	15846576	15846576	15846831	15912624	15912624	15912624	1591262
15	15780530	15780530	15780785	15846578	15846576	15846576	15846831	15912624	15847085	15847085	1584683
16	15780530	15780530	15780785	15846578	15846576	15846576	15846831	15912624	15715499	15715499	1578103(
17	15846323	15846323	15846323	15780530	15780785	15846578	15846578	15846578	15780785	15780785	1578078;
18	15846323	15846323	15846323	15846323	15846578	15846578	15846578	15846578	15780785	15780785	1578078;
19	15715251	15715251	15715251	15715251	15715506	15715506	15715506	15715506	15715506	15715506	1571576(

FIGURE 21.9 Spreadsheet showing pixel numbers representing the picture of the cover of the GHOSTS CD Album.

Then the spreadsheet shown in Figure 21.9 appears. It is a 1600-variables by 1200-cases spreadsheet; i.e., this was a 1600 by 1200 pixel photo.

Then open the Graphs pull-down menu and select 3D Sequential Graphs and then Raw Data Plots, as shown in Figure 21.10.

On the Advanced tab of the Raw Data Plots dialog, select Contour/Discrete in the Graph Type box, as shown in Figure 21.11.

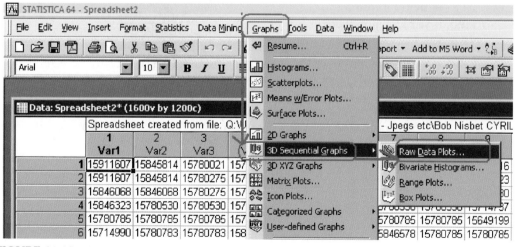

FIGURE 21.10 Opening the Raw Data Plots.

IV. TRUE COMPLEXITY, THE "RIGHT MODEL," TOP MISTAKES, AND THE FUTURE

FIGURE   21.11   Selecting graph type.

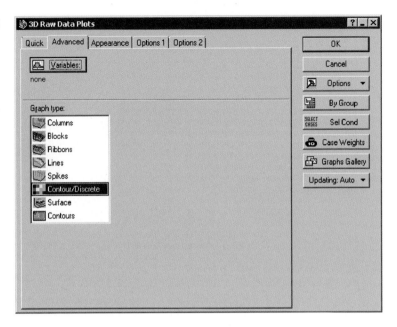

Then click on the Variables button at the top of the dialog to open the Select Variables dialog shown in Figure 21.12.

These examples illustrate just the beginning of working with this technology. Here is what is exciting: you can take things like the following and put each of them into a spreadsheet:

1. Sequential MRI slices or images through the brain or any body part;
2. Movies or individual frames;
3. Any kind of sequential movement.

Then you can analyze these combined spreadsheets to follow, for example, (1) proper physical therapy movements to rejuvenate an injured limb, (2) security screening, and (3) all kinds of object identification problems. The main use case would be to take the data matrices and do something important with them. For example, you might want to take a graph, reduce it to a more simple size graphically (shrink it perhaps to 300×300 or so) and then perform data reduction on it. In general, you now have a way to turn pictures into numbers; what you do with those numbers is a different question because almost any kind of analysis is possible. Perhaps the Kernel SVM algorithm mentioned previously from the University of California–Berkeley and California Institute of Technology might be one useful tool for specific applications.

In conclusion for this section on visual object recognition by data mining technology, we can say that shape-based object recognition is one key for the future in this area. It will not only involve visual object recognition, but will have to be extended to such things as audio waves, brain wave mapping, and similar three-dimensional phenomena.

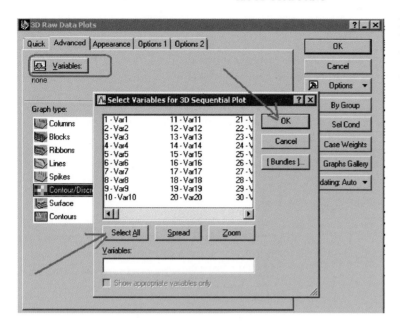

FIGURE 21.12 Selecting variables in the plot. Next, click the Select All button to select all the variables and click OK to get a graphical representation of the photo pixel spreadsheet.

CLOUD COMPUTING

"Cloud computing" is one of the most revolutionary concepts that has begun to turn into a reality in the past couple of years. In this section, we will look at information that began popping up during 2008 on the Internet and other sources. We'll start with the idea that the scientific method is becoming obsolete.

We might say that cloud computing is the final attempt at beginning to really understand the universe or get the next step closer to reality. We might also say that the year 2008 was the real beginning of "the Petabyte Age" because:

- We are now in the arena of "big data;" e.g., *more* is just not *more*, but *more* is *different*.
- Sixty years ago, digital computers made information readable; we might say that this information was stored in "paper manila folders."
- Twenty years ago, the Internet made information reachable; e.g., we stored it in a "file cabinet."
- Ten years ago, the first search engine crawlers made information into a "single database," e.g., a "library."
- Today, petabytes of information are stored in a "cloud."

Various web sites, from the National Science Foundation and other organizations, tell us about this phenomena of cloud computing that hit academia at the beginning of 2008, and later during the last few months of 2008, it "hit" the business world, including Google, IBM, Zementis, Amazon.com, and even Microsoft, as we will explain here.

The NSF made this announcement in February 2008:

National Science Foundation taps into IBM-Google computer cluster
Cluster created last year to support academic parallel computing initiative
By Brad Reed, Network World, 02/26/2008
The National Science Foundation announced this week that it had reached an agreement with Google and IBM that would let the academic research community conduct experiments through the companies' 1,600-processor computer cluster.

Google, IBM and the NSF's Computer Science and Engineering Directorate will be launching a joint initiative called the Cluster Exploratory (CluE) that will grant the academic research community access to the Google-IBM academic cluster. The NSF also says that the cluster will give researchers access to resources that would otherwise have [...] been prohibitively expensive.
(*Source:* http://www.networkworld.com/news/2008/022608-nsf-ibm-google.html)

The National Science Foundation is making the most out of this new cloud computing project, giving it public relations exposure in many ways, including the following:

NSF Head: All Hail the Cluster
By Alexis Madrigal March 14, 2008 | 4:00:11 PM
(For more information on this source, see http://blog.wired.com/wiredscience/2008/03/nsf-head-cluste.html.)

Here's a web story headline, from June 2008:

Google Cloud at Work for NSF, Academia
Google gives academics and students at some of the largest universities around the planet access to massive resources for academic quest and experiments. In February 2008, Google announced that it was working with National Science Foundation and IBM on the Cluster Exploratory (CluE) that would enable, "academic research community to conduct experiments and test new theories and ideas using a large-scale, massively distributed computing cluster."
(*Source:* http://gigaom.com/2008/06/28/how-google-is-taking-clouds-to-college/).

According to this story, the NSF-sponsored 16,000-processor cluster, with terabytes of memory and many hundreds of terabytes of storage, runs as follows and uses the following software sources:

- The clusters run an open source implementation of Google's published computing infrastructure (Map-Reduce and GFS from Apache's Hadoop project)
- Open source software designed by IBM to help students develop programs for clusters running Hadoop.
- The software works with Eclipse, an open source development platform.
(*Source:* http://gigaom.com/2008/06/28/how-google-is-taking-clouds-to-college/).

Academics and Google and IBM even refer to people in this field as "cloud people;" an entire new terminology is developing for this fifth phase of statistical and data mining analysis.

Additional characteristics of this Academic Cloud Computing Initiative of NSF are as follows:

- The NSF solicited proposals for research in the spring of 2008.
- Currently available large data collections include
  - Sloan Digital Sky Survey;
  - The Visible Human;
  - The IRIS Seismology Data Base;
  - The Protein Data Bank; and
  - The Linguistic Data Consortium.
- A term seen frequently with this cloud computing initiative is the *Cluster Exploratory (CluE)*.

As indicated in the National Science Foundation's solicitation for proposals:

In many fields, it is now possible to pose hypotheses and test them by looking in databases of already collected information. Further, the possibility of significant discovery by interconnecting different data sources is extraordinarily appealing. In data-intensive computing, the sheer volume of data is the dominant performance parameter. Storage and computation are co-located, enabling large-scale parallelism over terabytes of data. This scale of computing supports applications specified in high-level programming primitives, where the run-time system manages parallelism and data access. Supporting architectures must be extremely fault-tolerant and exhibit high degrees of reliability and availability.

The Cluster Exploratory (CluE) program has been designed to provide academic researchers with access to massively-scaled, highly-distributed computing resources supported by Google and IBM. While the main focus of the program is the stimulation of research advances in computing, the potential to stimulate simultaneous advances in other fields of science and engineering is also recognized and encouraged. (*Source:* http://blog.beagrie.com/archives/2008/04/)

A press release in April 2008 included a cartoon alluding to the unique relationship between Google, IBM, and NSF that allows the academic computing research community to access large-scale computer clusters.

Here is one way we can look at this cloud computing phenomenon: the scientific method is built around testable hypotheses. Scientists are trained to recognize that correlation is *not* causation, that no conclusions should be drawn simply on the basis of a correlation between X and Y. Instead, you must understand the underlying mechanisms that connect the two. Once you have a model, you can connect the data sets with confidence:

- Data without a model is just noise.
- But faced with massive data, this approach to science—hypothesize, model, test—is becoming obsolete.

So, now there is a better way. Petabytes allow us to say, "Correlation is enough." We can stop looking for models. We can analyze the data without hypotheses about what it might show. We can throw the numbers into the biggest computing clusters the world has ever seen and let statistical algorithms find patterns where science cannot.

In summary, the Cluster Exploratory (CluE) project, which might be called the "fifth phase of data analysis," is a program that funds research designed to run on a large-scale distributed computing platform developed by Google and IBM in conjunction with six pilot universities. The cluster will consist of 1,600 processors, several terabytes of memory, and hundreds of terabytes of storage, along with the software, including IBM's Tivoli and open source versions of Google File System and MapReduce.

IV. TRUE COMPLEXITY, THE "RIGHT MODEL," TOP MISTAKES, AND THE FUTURE

Early CluE projects will include simulations of the brain and the nervous system and other biological research that we might describe as lying somewhere between wetware and software.

We might say that there is no reason to cling to our old ways; the opportunity is great. The new availability of huge amounts of data, along with the statistical tools to crunch these numbers, offers a whole new way of understanding the world. Correlation supersedes causation, and science can advance even without coherent models, unified theories, or really without any mechanistic explanation at all. It may be time to ask this question:

What Can Science Learn from Google?

Modest quantities (a few hundred gigabytes) of data can be loaded on the cluster over the Internet. Some projects may require the mounting of large (terabytes +) quantities of data on the cluster. The process for loading data is expected to evolve over the life of the program and will be worked out with each project team on a case-by-case basis post award. It is anticipated that a growing number of public data sets will be available on the cluster for use by awardees. A catalog of these data sets will be accessible via http://www.nsf.gov/clue. Data created by awardees will be retained on the cluster and become public; i.e., it will be accessible by other authorized cluster users.

For more information regarding some of the preceding statements/concepts, see these web sites:

http://www.nsf.gov/funding/pgm_summ.jsp?pims_id=503270
http://www.nsf.gov/pubs/2008/nsf08560/nsf08560.htm
http://www.nsf.gov/div/index.jsp?div=IIS

But the story doesn't stop here. In the last few months of 2008, cloud computing invaded the commercial world as well. We'll tell this story in the following section.

The Next Generation of Data Mining

Let's review again the five phases of data analysis, introduced in Chapter 1 of this book:

1. Classical Bayesian Statistics
2. Classical Parametric Statistics
3. Machine Learning
   Data Mining with Neural Nets and Decision Trees
4. Statistical Learning Theory
   Data Mining with Support Vector Machines and Related Algorithms
5. Distributed Analytical Computing
   Grids and Clouds

Analytical modeling began (in large part) with the work of R. A. Fisher in 1921. Fisher devised the Parametric Model to correct the problem with Bayesian statistics; let's look at this topic in outline format:

- Bayesians brought "subjective prior" information to the table of decision making (what happened in the past).

- Two Bayesian scientists could start with the same data and come to very different conclusions, depending on their set of subjective priors.
- Fisher did away with the subjective priors and defined the probability of an event's occurrence, using only data from one controlled experiment.
- The focus of statistical analysis was on the complex mathematical operations necessary to analyze complicated and diverse data sets.
- Machine learning methods were devised to analyze strongly *nonlinear* problems.
  - Fisher's first methods could be applied well only to *linear* problems (straight-line relationships).
  - Many relationships in the real world outside Fisher's laboratories followed very "curvy" nonlinear lines. (*Note:* Many parametric methods were created in parallel for analyzing nonlinear relationships.)
- Statistical learning methods (Vapnik, 1995) enabled analysis of more complex problems by "mapping" input data points to higher dimensional spaces with "kernels" (complex mathematical functions used to transform data).

Grid and Cloud Computing

Grid and cloud computing are defined as follows:

- **Grid**: A dedicated group of networked computers, which divide and conquer very large and complex data processing tasks.
- **Cloud:** A nondedicated group of computers connected to the Internet, which can be drawn upon as needed to process tiny to extremely huge processing tasks.

The Way Data Mining Used to Be

Traditional data mining

- Starts with data;
- Trains a model;
- Publishes the model;
- Hands it off to the systems people for deployment.

The traditional solution landscape is characterized by the following:

- Data miners are focused very largely on the technology of building models.
- Training good models can be *very* time-consuming and artistic!
- When models are done, data miners just want to hand them off to systems people for deployment and go back to building other models (see Figure 21.13).

What are the fruits of this traditional style of data mining?

- The de-emphasis of deployment of most data mining tools causes
  Design of requirements for analytical data preparation steps that are hard to implement in deployment;
  Unnecessary model complexity, when a simpler "good-enough" model would have served and permitted much easier deployment (see Taylor and Raden, 2007).

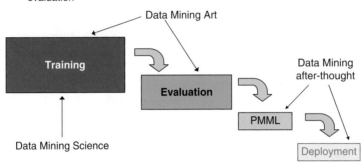

Data Mining Tools Follow Suit

FIGURE 21.13   Data mining tools follow suit.

What we need for the future is an "action orientation." Data mining is like training some-one who runs hurdles. It involves

- Painstaking practice of
  How many steps to run to the first hurdle;
  How high he must jump to clear the hurdle;
  How he must move his legs to clear the hurdle;
  How many steps he must run to the next hurdle.
- Practice over and over and over…
- This is "model" training for the race.

Likewise, businesses need to make decisions and act:

- Business decision making is like the track meet, *not the practice field.*
  The runner must draw upon the training of the "model" for winning the race.
  Decision making in the race must be nearly *automatic!*
- Winning *actions* are needed at race time, not model refinements.

So, for the future, the authors of this book are saying, "Shift up the gears!" (See Figure 21.14).
Data mining solutions should be

- Focused on deployment of winning models;
- Designed for ease of deployment;

Shifting Up Another Gear…

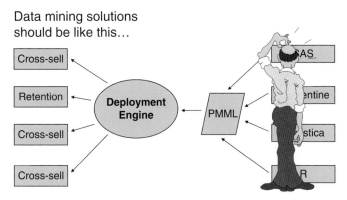

Adaptive Decision Technology

FIGURE 21.14   Shifting up another gear.

- Designed for running the *race* on any "track";
- Incorporated into enterprise decision systems:
  Tool agnostic (any data mining tool that outputs PMML will do);
  Business application agnostic (solutions should be vended for easy use by any business application).

A New Perspective on Data Mining

The Enterprise Decision Management (EDM) tool interface should be focused on *deployment,* not model training (see Figure 21.15).

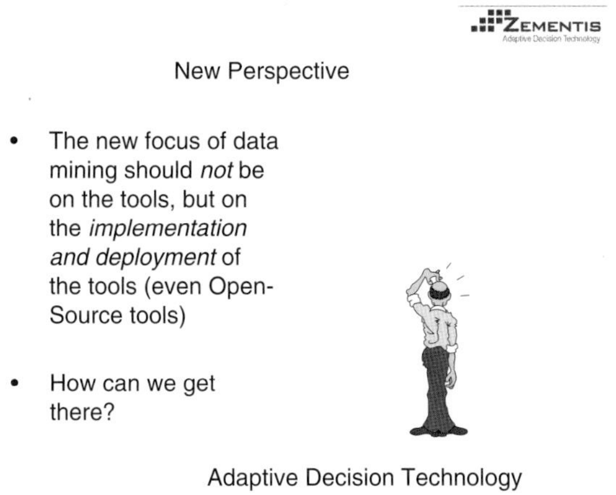

          FIGURE 21.15   Getting a new perspective.

New Perspective

- The new focus of data mining should *not* be on the tools, but on the *implementation and deployment* of the tools (even Open-Source tools)

- How can we get there?

Adaptive Decision Technology

EDM deployment should be

- Adaptive to any form of analytical model;
- Adaptive to any decision environment;
- Architected around a common model output;
- Linearly scalable to the highest data volumes.

Current data mining tools are *expensive!* A new generation of open source (free) data mining tools is being developed:

- R
- Weka
- Orange
- Yale
- RapidMiner (an implementation of Yale)

We Can "Take to the Clouds"

When you implement your models with ADAPA from Zementis (see Figure 21.16), you can

- Use your *existing* models and tools;
- Use your *existing* business rules;
- Seamlessly integrate them; and
- Fly to the clouds (like the Amazon.com Elastic Cloud) and deploy almost *infinite* scoring.

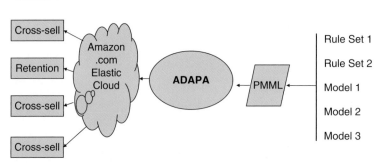

FIGURE 21.16   How ADAPA does it.

How ADAPA does it:

- It uses models:
  ADAPA can call the PMML from a model as a function within a rule set in PMML. ADAPA can call the PMML from a rule set as a function within the deployment of a model.
- Previously, neural nets and rule induction engines (for example) could be combined only in modeling ensembles.
- A model ensemble permits scoring with multiple models, which "vote" for the final score.
- ADAPA combines diverse model inputs at a much lower level to create an adaptive meta-model.

Data Mining Is Maturing

- Electronic spreadsheets (a la Lotus 1-2-3) began as a high-tech application.
- Excel moved them into the business mainstream.
- Now, Excel is a *component* of larger business operations.

Data Mining as a Utility

- No longer do you have to invest in an expensive software and hardware infrastructure to deploy models.
- Zementis deploys models seamlessly on the Amazon.com EC2 Elastic Cloud.
- You pay for only the processing time.
- Several cloud-based "utilities" are available now (IBM, Google, Amazon.com).

Benefits of Cloud Computing

Benefits of cloud computing include the following:

1. It has open standards.
   - It is based on PMML.
     PMML is like the IBM PC disk format standard, which replaced a multitude of vendor formats in the early 1980s.
     Standardization permits companies to focus on the ends rather than the means of computing.
   - It encourages development of open source data mining packages (R, RapidMiner, Weka).
   - Using it allows you to avoid vendor "lock-in."
2. It provides utility-based computing.
   - Electric utilities provide electricity, transmission lines, and billing systems. You pay only for usage.
   - You pay only for the usage of cloud computing resources, not the infrastructure that provides them.
   - Deployment usage happens in minutes!
   - Cost scales linearly with the usage.

3. It provides Software as a Service (SaaS), not as part of your infrastructure.
   - Again, you pay only for the usage, not huge software licenses fees.
   - You have no long-term commitment to a software vendor.
   - You pay operational costs only, not capital expenditures.
4. It provides secured computing instances (https and WS-Security).
   - You can keep your company and customer data safe.
5. Again, you pay only for the usage, not huge software licenses fees.
   - You have no long-term commitment to a software vendor.
   - You pay operational costs only, not capital expenditures.

Additional interesting developments in cloud computing and "elastic clouds" can be found at the following web sites:

1. Zementis (http://www.zementis.com) markets itself as "the first predictive analytics cloud computing solution."
2. Microsoft's SQL Server Data Mining (http://www.sqlserverdatamining.com/cloud/) provides cloud computing free. You can access it via the Web or via a simple Excel add-in.

From the Desktop to the Clouds . . .

Will cloud computing and elastic clouds, like Amazon.com is now doing in collaboration with Zemantis.com, take over data mining?

No, we don't think so. The cloud computing and elastic nets will serve a need for certain industries, businesses, and academic institutions. But there will always be corporations and organizations that want to keep their data and computing in house and/or outsource it to data mining consultants. With the needs of this ever-increasing global world, there is plenty to be done to keep everyone busy.

We are on the cutting edge of a new phase in data analysis. It will be interesting to see what develops during the next 5 to 10 years.

Note: The DVD that accompanies this book provides some interesting URLs and PDF files on cloud computing.

POSTSCRIPT

The entire next chapter is the postscript for this book.

References

Anderson, C. (2008). The end of theory: The data deluge makes the scientific method obsolete. *WIRED Magazine*, 16, 7. www.wired.com/print/science/discoveries/magazine/16-07/pb_theory.

Berg, A. C., Berg, T. L., & Malik, J. (2005). Shape matching and object recognition using low distortion correspondence. *CVPR*.

Bromley and Sackinger. (1991). Neural-network and K-nearest-neighbor classifiers. Technical Report 11359-910819-16TM, AT&T.

Ferencz, A., Learned-Miller, E. G., & Malik, J. (2008). Learning to locate informative features for visual identification. *International Journal of Computer Vision*, 77, 3–24.

Grauman, K., & Darrell, T. (2006). Pyramid match kernels: Discriminative classification with sets of image features (version 2). Technical Report CSAIL-TR-2006-020, MIT.

Holub, A. D., Welling, M., & Perona, P. (2005). Combining generative models and fisher kernels for object recognition. *ICCV*.

KDD-08. (2008). *The 2nd ACM Workshop on Social Network Mining and Analysis*. Las Vegas: Nevada, August 23–27.

Krebs, V. (2008). Data mining email to discover social networks and emergent communities. http://www.orgnet.com/email.html

LeCun, Y., Boser, B., Denker, J. S., Henderson, D., Howard, R. E., Hubbard, W., et al. (1989). Backpropagation applied to handwritten zip code recognition. *Neural Computation*, 1(4), 541–551.

Maji, S., Berg, A. C., & Malik, J. (2008). Classification using intersection kernel support vector machines is efficient. *IEEE Computer Vision and Pattern Recognition*, Anchorage. http://www.eecs.berkeley.edu/Research/Projects/CS/vision/shape/papers/mbm08cvpr_tag.pdf

Taylor, J., & Raden, N. (2007). *Smart (enough) systems*. Upper Saddle River, NJ: Prentice Hall.

Ulanoff, Lance. "Twitter Takes Off on the Wings of Flight 1549." *PCMag.com 21 January 2009*. http://www.pcmag.com/article2/0,2817,2339286,00.asp

Vapnik, V. N. (1995). *The Nature of Statistical Learning Theory*. New York: Springer-Verlag.

Vincent, P., & Bengio, Y. (2001). K-local hyperplane and convex distance nearest neighbor algorithms. *NIPS*, 16, 985–992.

Zhang, H., Berg, A. C., Maire, M., & Malik, J. (2006). SVM-KNN: Discriminative nearest neighbor classification for visual category recognition. *02*, 2126–2136, (2006) Abstract. http://www.eecs.berkeley.edu/Research/Projects/CS/vision/shape/nhz-cvpr06.pdf

Summary: Our Design

PREAMBLE

How do you summarize a book like this, when the whole book is a summary by design? One way is to synthesize recommended effects of this design. To begin, we can revisit the goals laid down in the Preface of this book. These goals were to

- Conduct you through a relatively thin slice across a wide practice of data mining in many industries and disciplines;

- Show you how to create powerful predictive models in your own organization in a relatively short period of time;
- Serve as a springboard to launch you into higher-level studies of the theory and practice of data mining.

Only you can answer whether or not we reached any of those goals, as you follow the path to implementing them in any of the many tutorials in this book. Rather than go back over the major areas of this book, we would like to use this chapter to underscore the major themes and "take-aways" we want you to absorb from this book:

1. Beware of overtrained models.
2. A diversity of methods and techniques is best.
3. The process is more important than the tool.
4. Mining of unstructured data (e.g., text) is becoming as important as data mining of structured data in databases.
5. Practice thinking about your organization as *organism* rather than as *machine.*
6. Good solutions *evolve* rather than just appear after initial efforts.
7. What you *don't do* is just as important as what you *do*: avoid common mistakes; they are the "plague" of data mining.
8. Very intuitive graphical interfaces are replacing procedural programming.
9. Data mining is no longer a boutique operation; it is firmly established in the mainstream of our society.
10. "Smart" systems are the direction in which data mining technology is going.

BEWARE OF OVERTRAINED MODELS

This theme is listed first because it is the single-most overperformed and underappreciated mistake in the field of data mining. Machine learning algorithms are notoriously prone to overtraining. The first line of defense against overtraining is to test the model as a part of the iterative training exercise. Most algorithms in data mining packages provide facilities to do this. Determine if the implementation of an algorithm in your data mining package does this automatically; if not, look for an option to set the algorithm to do it. After the model is trained, do not believe the first model. Test it against another (validation) data set, which was *not* used in the training in any way. Otherwise, your model results may not present an objective reflection of the true systematic error in your modeling algorithm or the random error in the data used to build it. The second line of defense against overtraining is to choose an appropriate form of the final solution. For example, make sure that you select an appropriate minimum number of observations to define a terminal leaf node of a decision tree. If you let the tree go without constraint, the algorithm might create a perfect model for that data set but could fail miserably upon validation with other data sets. Experiment with different stopping functions in decision trees and neural networks. See how the generalization ability of the model is affected by different model configurations. Use data mining algorithms to model a data pattern as an artist uses different colors of paint to define an image. Data mining is a very artistic endeavor!

A DIVERSITY OF MODELS AND TECHNIQUES IS BEST

A statistical or data mining algorithm is a mathematical expression of certain aspects of the patterns they find in data. Different algorithms provide different perspectives (or "colors") on the complete nature of the pattern. No one algorithm can see it all. It is true that for a given data set, there exists one algorithm that does it best. But don't be satisfied with just the best *single* view of the pattern. The best perspective among other views nearly as poor is not sufficient to define the pattern properly. A number of different "weak" perspectives of the pattern can be combined to create a relatively good definition of the pattern, even if it is incomplete. Most data mining tool packages have the capability to create ensemble models, which can "vote" for the best overall prediction (review the use of ensembles in Chapters 13 and 18). This approach can be applied also in the scoring of new data sets, where the target variable values are not known. Data mining can draw upon the best concepts in a democratic process to produce the best results.

THE PROCESS IS MORE IMPORTANT THAN THE TOOL

To some degree, data mining tools lead you through the data mining process. SAS-Enterprise Miner organizes its top toolbar to present groups of operations performed in each of the major phases of Sample, Explore, Modify, Model, and Assess (SEMMA). This organization keeps the correct sequence of operations central in the mind of the data miner. *STATISTICA* Data Miner divides the modeling screen into four general phases of data mining: (1) data acquisition; (2) data cleaning, preparation, and transformation; (3) data analysis, modeling, classification, and forecasting; and (4) reports. This group of activities expands somewhat on the six phases in the SEMMA process flow. The *STATISTICA* Data Miner approach constrains you to do the appropriate group of operations in sequence when building a model in the visual programming diagram of a workspace. The CRISP-DM process presented in Chapter 3 is tool-independent and industry (or discipline)-independent. You can follow this process with any data mining tool. Whichever approach you take, it is wise to follow the Marine drill sergeant's appeal to "Get with the program!"

Many algorithms will perform similarly on the same data set, although one may be best. Your choice of algorithm will have much less impact on the quality of your model than the process steps you went through to get it. Focus on following the process correctly, and your model will be at least acceptable. Refinements can come later.

TEXT MINING OF UNSTRUCTURED DATA IS BECOMING VERY IMPORTANT

When your data are contained in data structures like databases, analyzing it with standard statistical or data mining algorithms is relatively easy (in principle). It is not so easy when the data consists of a bunch of words in a document, email, or web site. Yet, much of the information crucially important to proper decision making in a company may be present *only* in text form. In an insurance company, underwriter notes contain very valuable items of information for quantifying the risk of a potential customer. The notes are a distillation of the mental process the underwriter goes through to assign the proper risk category to the customer, but much of it is not converted to database elements. Text mining of insurance underwriter notes can represent a treasure trove of information for risk modeling. The Internet is another source of very valuable information for many companies and organizations. Even generic office tools like Microsoft Excel can be configured to load text from web sites into a spreadsheet format prior to analysis. Excel macros can be created to automatically load text strings from specific locations in web site displays to monitor the occurrence of target words through time. Most data mining tool packages provide text mining modules or add-ons that can be combined with standard statistical and data mining algorithms for creating models based on textual input data.

PRACTICE THINKING ABOUT YOUR ORGANIZATION AS *ORGANISM* RATHER THAN AS *MACHINE*

This shift in thinking is hard to do, but it is very profound in its effects. One of the byproducts of the Industrial Revolution was to present a picture of ideal processes in a very mechanistic way (review Chapter 1). This approach helped to convert the previous *ad hoc* and customized approaches to manufacturing to become standardized and much more efficient. But soon, the complexity of the systems outgrew the machine metaphor. Interactions, feedbacks, and synergy became dominant factors rather than just the directional workflow (e.g., the assembly line). Businesses became complex systems of suppliers, subcontractors, distributors, vendors, and customers, which defied the constraints of the machine metaphor. A much better metaphor is that of the organism and ecosystem, which gained prominence in the 1990s. Today, the concept of the *business ecosystem* has become firmly established as the best structure in which to organize and manage the complexities of modern business and management.

Practice thinking of your business as a social ecological system rather than viewing it as just an economic engine. It is not without reason that two of the authors of this book are former biologists (an ecologist and a geneticist). The third author was originally an engineer. All three authors have a great deal of experience in teaching highly technical subjects to nontechnical audiences. This book is an attempt to merge the analytical perspectives of biology, engineering, and teaching pedagogy in a business environment to structure systemically (from biology) and optimize (from engineering) various approaches to search for and teach Truth in the business world. A mentor of the senior author, Dr. Daniel Botkin

(1990) observes in his book *Discordant Harmonies* that the true harmony of nature is not static, but rather dynamic and discordant. It is created from

> ... the simultaneous movements of many tones, the combination of many processes flowing at the same time along various scales, leading not to a simple melody but to a symphony at times harsh and at times pleasing. (25)

Dr. Botkin is not the first person to think this way. He cites Plotinus (the famous Neo-Platonist of third century Greece) for the concept of discordant harmony and H. E. Clements (an ecologist in the early part of the twentieth century) for the concept of a forest as a "super-organism." Botkin's call for an interdisciplinary study of the environment (from the perspective of a global ecologist) is also appropriate for businesses in the global economy of the twenty-first century. His view of discordant harmony is the way the natural world works, and businesses are embedded in it, whether they recognize it or not. Our call to you is to recognize it and to start practicing data mining from that perspective.

GOOD SOLUTIONS EVOLVE RATHER THAN JUST APPEAR AFTER INITIAL EFFORTS

The biological metaphor of business processes as "super-organism" or as "ecosystem" contains the concept of evolution. Directed change in nature seldom follows a straight line. There are many dead ends (errors from an evolutionary standpoint). Peters (1987) promotes the concept of "fast failures" as a recommended tactic in business management. Managing to "create" failures was a novel idea in business 20 years ago. But he argues that we must go through failures to identify the true course; otherwise, we will fail anyway and never find it. This concept is a just a form of the oldest scientific method: trial-and-error. If we make our failures happen very fast, we can learn from them quickly and *evolve* the best solution with minimal cost. The best business structures are shaped by evolution rather than revolution.

Data mining provides one of the best environments for evolving business solutions. Many models can be created and validated by experience (like survival of the fittest species). If we try to induce the final model by our initial efforts, we will almost surely fail. A series of "fast failure" models created in a data mining "sandbox" can serve to evolve the best solution for a given business problem.

WHAT YOU *DON'T DO* IS JUST AS IMPORTANT AS WHAT YOU *DO*

Serious analytical mistakes are the "plague" of data mining. Chapter 20 discusses 11 of the most serious of these mistakes. Think on them! The old saying that those who ignore the mistakes of history are bound to repeat them is certainly true in the practice of data mining. Immerse yourself in the study of the mistakes of others; they are like a road map of the routes not to take. You may not know exactly where you are going in a data mining project, but you can know where *not* to go. If you avoid the most common (and serious) mistakes, you have a much higher probability of finding the right solution than just blundering ahead.

VERY INTUITIVE GRAPHICAL INTERFACES ARE REPLACING PROCEDURAL PROGRAMMING

The program code of the final solutions will not go away, but the means of generating it is changing. No longer do data miners have to write long, complicated data mining programs in a programming language. Many powerful visual graphical programming interfaces have been built on top of statistical analysis and data mining algorithms to permit users to leverage their power without a deep understanding of the underlying technology. The "automobile interface" for data mining is here in the form of semi-automated data mining "dashboards," like those highlighted in the tutorials of this book (*STATISTICA* Data Miner, SAS-Enterprise Miner, SPSS Clementine) and others like KXEN.

The combination of these graphical interfaces and the step-by-step tutorials provided in this book (with more on the CD-DVD) will permit you to navigate through the complexity of statistical and data mining techniques to create powerful models (and you don't have to have a doctorate to do it).

DATA MINING IS NO LONGER A BOUTIQUE OPERATION; IT IS FIRMLY ESTABLISHED IN THE MAINSTREAM OF OUR SOCIETY

Fifteen years ago, data mining was the realm of the mathematical "gurus." Most of the analysis was performed with statistical analysis algorithms, which were associated with a lot of potential "pilot error." To get good results, you had to really know what you were doing from a mathematical standpoint. Most of the data mining in business was performed by highly paid consultants. Today, early successes of data mining, combined with the refinement of data mining tool packages, have thrust data mining into the mainstream of business. The bar has been raised. Businesses that do not leverage data mining technology will be outcompeted by those who do. The recognition and fear of this business reality has led most large- and medium-sized companies to bring data mining in-house. Analytical groups have been formed, sometimes in a central organization, and sometimes in a distributed design. The concept of the importance of data in a company has been trumped by the importance of information. Chief information officers (CIOs) have arisen in many companies to oversee not only the old Management Information Systems (MIS) processes, but also guide the transformation of data into information and knowledge useful to drive intelligent decision making.

Large companies can hire a hoard of doctors of philosophy to pump out the information and knowledge from analytical models, but smaller companies must rely on business data analysts without specialized training in statistical analysis or data mining. It is to that resource base that this book is targeted. This book is designed to get these users up and running with data mining to create acceptable models in a very short period of time.

"SMART" SYSTEMS ARE THE DIRECTION IN WHICH DATA MINING TECHNOLOGY IS GOING

These smart systems are composed of the union of many individual techniques designed to work together to create powerful data mining models. These models are embedded within many business systems, like the recommender engines used by Amazon.com to suggest additional products that might be of interest to you. This union of solution elements has expanded beyond the data mining process to the entire data processing flow. Groups of computers can be networked together in grid computing systems, which can tackle computing problems that were impossible just a few years ago. The Internet can host a "cloud" of computers available to be linked together upon demand, which can form an almost unlimited virtual grid system.

Visual data mining is not new. Powerful graphical techniques have been incorporated into statistical and data mining packages for the past 20 years. Recently, however, resurgence in powerful new graphical systems may move data mining away from the traditional "left-brained" statisticians toward "right-brained" people who comprise the majority of people in business management. The same thing happened when Microsoft Excel moved into the mainstream of business practice. Accountants used paper spreadsheets for years and combined them into vast spreadsheet models such as cash flow plans. Excel provided business analysts without any accounting background the ability to do very similar things with their data. Excel is now a standard office tool. Visual data mining technology will permit business managers to do similar things with data mining technology. Soon, managers will have the capability to do what-if analyses, which draw automatically on their corporate data and utilize the same technology that has driven their marketing direct mail operations for many years (for example). Then business managers will be able to guide and direct their departments (business organisms) and their entire enterprise (business ecosystem) with true business intelligence rather than just gut-level responses.

POSTSCRIPT

We are heading into a new and exciting age in the twenty-first century. The atom may be the limit for nanotechnology development of the hardware, but the globe (and even the universe) is the limit for data mining development. If we follow Dan Botkin's advice, we will be able to understand and build optimal business systems, with a dynamic stability harmonious with sustained profitability, however discordant some of the elements may appear to be.

References

Botkin, D. B. (1990). *Discordant Harmonies: A New Ecology for the Twenty-First Century.* New York: Oxford University Press.

Peters, T. (1987). *Thriving on Chaos: Handbook for a Management Revolution.* New York: HarperCollins.

Glossary

TEXT MINING TERMINOLOGY

categorization of text material: Assignment of documents to one or more categories based on content.

clustering in text analysis: Document clustering and text clustering are closely related in text mining. It can be an unsupervised clustering, involving trees or other algorithms like SVM for the organization of either documents or text in documents, sometimes referred to as filtering.

collocations: Text units that cover more than one word, such as *United States of America*.

concept mining: A text mining approach based on artificial intelligence and traditional statistics. Thesauri have been used to look up words to convert them into "concepts," and similar computer programs have been developed to do the same, such as Princeton's WordNet.

co-occurrence: Occurrence of similar or the same words/concepts in the same area of text—typically a sentence or fixed window of tokens (words) surrounding the focus token, but sometimes as large as an abstract—suggesting a relationship between them.

document classification: See *categorization of text material*.

GATE: General Architecture for Text Engineering (Cunningham, 2002). A system of looking at the structure of text material.

gene mention: The act of keeping tabs on mention of "genes" in text, for example; this has been a major focus of biomedical text mining research. For identifying a wider range of biomedical categories like diseases, drugs, chemicals, and methods of treatment, a specific tool is the National Library of Medicine's MetaMap

grammar: The structure of a language; different languages have different "associations" of word forms, so this can be a very difficult pattern to understand.

Hidden Markov Model (HMM): A process to determine hidden parameters from observable parameters where the system being modeled is assumed to be a "Markov process." HMMs were first applied to speech recognition problems in the mid-1970s; and in the 1980s HMMs were applied to biological processes, particularly DNA.

hybrid methods for text analysis: Hybrid n-gram/lexical analysis tokenization is one type of "hybrid text analysis" system; it comprises a lexicon and a hybrid tokenizer that perform both N-gram tokenization of a text and lexical analysis tokenization of a text. Another is HybGFS, which is a hybrid method for genome-fingerprint scanning.

information extraction (IE) from text: Automatic extraction of structured information from unstructured text data in natural language processing; often by rule induction and generalization.

inverse document frequency: A transformation of raw word frequency counts computed in text mining. This process is used to express simultaneously the frequencies with which specific terms or words are used in a collection of documents, as well as the extent to which particular words are used only in specific documents in the collection. For more information, see Manning and Schütze (2002); and the online help in the *STATISTICA* Data Miner that is on the DVD that comes with this book.

LingPipe: A program composed of a set of modules for biomedical (and general) language processing tasks, ranging from low-level preprocessing, like sentence segmentation, to part of speech tagging (http://www.alias-i.com/lingpipe/).

natural language: A language that has evolved naturally, such as English, German, Russian, Chinese, etc.

noise: Extraneous text that is not relevant to the task at hand, necessitating filtering it out from important text.

PubMed/MEDLINE: A database of publications in medicine and biomedicine/bioscience that can be reached and searched online; particularly useful for genomics-related publications.

relationship extraction: The process of finding relationships between entities in text.

rule-based or knowledge-based system: One of the three approaches to text mining, the other two being (1) co-occurrence and (2) statistical or machine-learning-based systems. Rule-based systems make use of some sort of knowledge. This might take the form of general knowledge about how language is structured, specific knowledge about how a discipline's relevant facts are stated in the literature, knowledge about the sets of things that this discipline talks about and the kinds of relationships that they can have with one another, and the variant forms by which they might be mentioned in the literature, or any subset or combination of these.

semantic specificity: See *inverse document frequency.*

sentence segmentation: Splitting a multisentence input into its individual sentences. Performance varies widely, with some tools doing as poorly as 40% on this task. High-performing systems include the rule-based Perl sentence segmenter that is packaged with the KeX gene name recognizer and the statistical Java sentence segmentation module that is distributed with LingPipe; both perform in the mid to high 90% range (Baumgartner et al., in progress).

sentiment analysis: An examination that aims to determine the attitude of a speaker or a writer on a given topic, which may be his or her "judgment" or "evaluation" of that topic and/or "emotional state" connected to that topic.

statistical or machine-learning-based systems: One of the three approaches to text mining, the other two being 1) co-occurrence and (2) rule-based or knowledge-based system. Statistical or machine-learning-based systems operate by building classifiers that may operate on any level, from labeling parts of speech to choosing syntactic parse trees to classifying full sentences or documents.

text analytics: A set of linguistic, lexical, pattern recognition, extraction, tagging/structuring, visualization, and predictive techniques. Text analytics also describes processes that apply these techniques to solve business problems from facts, business rules, and relationships.

text mining: The process of automatically extracting "meaning" from a collection of documents. A first step might be calculating statistics about the words, terms, and structure of the documents.

tokenization: The task in text mining of splitting an input into individual words (as well as other units, such as punctuation marks).

UIMA: Unstructured Information Management Architecture (Ferrucci and Lally, 2004; Mack et al., 2004).

universal grammar: A theory proposing that all natural languages have certain underlying rules which constrain the structure of the language. It tries to explain language acquisition in general and proposes rules to explain language acquisition during child development.

DATA MINING, DATABASE, AND STATISTICAL TERMINOLOGY

algorithm: Sets of steps, operations, or procedures that will produce a particular outcome; like a recipe.

assessment: The process of determining how well a model estimates data that are not used during training, usually by using a completely new set of data of the same variables.

association rule: A data mining technique used to describe relationships among items. The A-priori algorithm (see Witten and Frank, 2000) is a popular and efficient algorithm for deriving such association rules.

bagging: Bootstrapped aggregating—combining the results from more than one model as your final model. Bagging can give a more accurate model, especially when data sets are small; this is done by repeated random sampling of the data set, with replacement and model fitting, and then averaging or voting the outputs of the separate models together.

basis functions: Functions involved in the estimation of Multiple Adaptive Regression Splines (MARS). These basis functions approximate the relationships between the response and predictor variables.

Bayes' theorem: The theorem that uses new information to update the probability (or prior probability) of a phenomenon or target variable.

Bayesian networks: Neural networks based on Bayes' theorem.

Bayesian statistics: An approach based on Bayes' law. The current probability of a factor is proportional to the prior probability of that factor multiplied by the likelihood of that factor as reflected in your current data. In other words, as more and more data are collected, you can refine the probability of a factor until you get closer and closer to reality about that factor or phenomenon.

binary variable: A variable that contains two discrete values (for example, Sex: male or female).

bootstrapping: In statistics, a resampling process, of which there is more than one type; the Jackknife and *V*-fold cross-validation are names of two bootstrapping methods.

BPMD: Business Process Modeling Notation.

branch: Part of a classification tree that is rooted in one of the initial divisions of a segment of a tree. For example, if a rule splits a segment into five subsets, then five branches grow from the parent segment.

C-SVM classification: A support vector machine method for solving multiclassification prediction problems.

CART: Classification and Regression Tree algorithm developed by Breiman et al. (1984). Generic versions are often named C&RT.

categorical dependent variable: A variable measured on a nominal scale, identifying class or group membership (e.g., male and female). For example, a good or bad credit risk score; see the Credit Scoring Tutorial, or the Credit Risk Tutorial, one in the pages of this book and the other on the DVD that accompanies this book.

categorical predictor variable: A variable, measured on a nominal scale, whose categories identify class or group membership, like good or bad credit risk score.

CHAID: A classification trees algorithm developed by Kass (1980) that performs multilevel splits in classification trees. CHAID represents "Chi-square Automatic Interaction Detection." The CHAID technique specifies a significance level of a chi-square test to stop tree growth.

champion model: The best predictive model developed from a "competitive evaluation" of data mining algorithms to which you submit a data set.

classification trees: One of the main "workhorse" techniques in data mining; used to predict membership of cases in the classes of a categorical dependent variable from their measurements predictor variables. Classification trees typically split the sample on simple rules and then resplit the subsamples, etc., until the data can't sustain further complexity. The following illustration (from the *STATISTICA* manual) shows how to "read" a classification tree diagram:

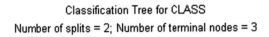

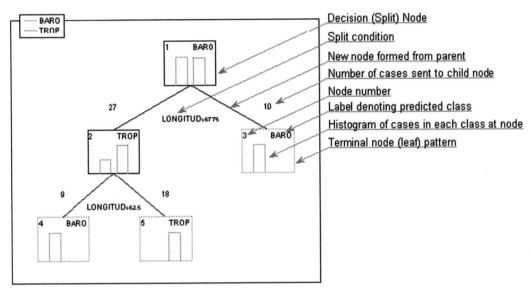

Classification Tree for CLASS
Number of splits = 2; Number of terminal nodes = 3

Decision (Split) Node
Split condition
New node formed from parent
Number of cases sent to child node
Node number
Label denoting predicted class
Histogram of cases in each class at node
Terminal node (leaf) pattern

clustering: A type of analysis that divides data (cases or variables, depending on how specified) into groups such that members of each group are as close as possible to each other, while different groups are as far apart from each other as possible.

CRISP: Cross-Industry Standard Process for data mining; proposed in the mid-1990s by a European consortium of companies to serve as a nonproprietary standard process model for data mining. Other models for data mining include DMAIC—the Six Sigma methodology, involving the steps of Define→Measure→Analyze→Improve→Control. Another model is SEMMA, proposed by the SAS Institute, involving the steps of Sample→Explore→Modify→Model→Assess.

CRM: Customer Relationship Management; processes a company uses to handle its contact with its customers. CRM software is used to maintain records of customer addresses, quotes, sales, and future needs so that customers can be easily and effectively supported.

cross-validation: The process of assessing the predictive accuracy of a model in a test sample compared to its predictive accuracy in the learning or training sample that was used to make the model. Cross-validation is a primary way to assure that over learning does not take place in the final model, and thus that the model approximates reality as well as can be obtained from the data available.

data mining: A process that minimally has four stages: (1) data preparation that may involve "data cleaning" and even "data transformation," (2) initial exploration of the data, (3) model building or pattern identification, and (4) deployment, which means subjecting new data to the "model" to predict outcomes of cases found in the new data.

epoch in neural networks: A single pass through the entire training data set, followed by scoring of the verification or "testing" data set.

FACT: A classification tree algorithm developed by Loh and Vanichestakul (1988).

feature extraction: A technique that attempts to combine or transform predictors to make clear the information contained within them. Feature extraction methods include factor analysis, principal components analysis, correspondence analysis, multidimensional scaling, partial least square methods, and singular value decomposition.

feature selection: A method by which to decide on which features (columns) to keep in the analysis that will be done by the data mining algorithms. One of the first things to be done in a data mining project; this uncovers the most important variables among the set of predictor variables. Many of the predictor variables in a data set may not really be important for making an accurate predictive model, and only dilute/reduce the accuracy score of the model if included.

gains chart: A summary graph showing which data mining algorithm provides the best model for predicting a binomial (categorical) outcome variable (dependent variable), as compared to a baseline level. This gains chart, or its analog the lift chart, is the single most important output result to examine at the completion of a data mining project.

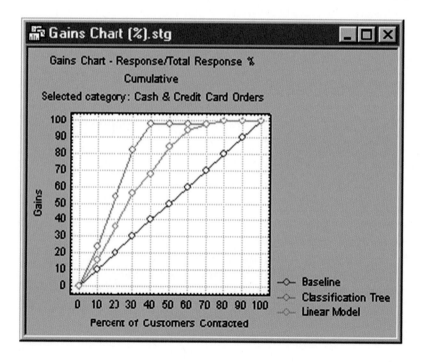

The baseline is the straight line from coordinate points 0,0 to 100,100; the algorithm curve that "bends" the most to the upper left and is consistent in that "bend" is the best model for your data. In the preceding example, the "classification tree" is the best model. [Gains chart taken from online help of *STATISTICA*: StatSoft, Inc. (2008). *STATISTICA* (data analysis software system), version 8. www.statsoft.com.]

generalization: The process of creating a model based on specific instances that is an acceptable predictor of other instances.

genetic algorithm: The type of algorithm that locates optimal binary strings by processing an initially random population of strings using artificial mutation, crossover, and selection operators, in an analogy with the process of natural selection (Goldberg, 1989).

Gini Measure of node impurity: A goodness-of-fit measure in classification problems, used in C&RT and interactive trees in data mining.

GRNN (Generalized Regression Neural Network): A type of neural network using a kernel-based method to perform regression. One of the Bayesian networks (Speckt, 1991; Bishop, 1995; Patterson, 1996).

hazard rate: The probability per time unit that a case that has survived to the beginning of an interval will fail in that interval.

hidden layers in neural networks: The layers between the "input" and the "output" in a neural network model; hidden layers provide the neural network's nonlinear modeling capabilities.

hold-out data: The observations removed from the data set and set aside to be used as test data to benchmark the fit and accuracy of the predictive model produced from the training data set.

imputation: The process in which methods are used to compute replacement values for missing values; for example, when case values for certain variables are missing in a data set, but it is determined important enough to keep these cases, various methods can be used to compute replacement values for these missing values.

incremental algorithms: Algorithms that derive information from the data to predict new observations, which require only one or two complete passes through the input data. Nonincremental learning algorithms are those that need to process all observations in each iteration of an iterative procedure for refining a final solution. Incremental learning algorithms are usually much faster.

input variable: A variable that is used to predict the value of one or more target variables.

interval variable: A continuous variable over a range of values; for example, height of humans with values like 5 feet, 5.5 feet, 6 feet, 6.3 feet, etc.

isotropic deviation assignment: An algorithm in neural networks for assigning radial unit deviations; it selects a single deviation value using a heuristic calculation based on the number of units and the volume of pattern space they occupy, with the objective of ensuring "a reasonable overlap" (Haykin, 1994).

jogging weights: Adding a small random amount to the weights in a neural network, in an attempt to escape a local minima in error space.

K-means algorithm: An algorithm used to assign K centers to represent the clustering of N points (K< N). The points are iteratively adjusted so that each of the N points is assigned to one of the K clusters, and each of the K clusters is the mean of its assigned points (Bishop, 1995).

K-nearest algorithm: An algorithm used to assign deviations to radial units. Each deviation is the mean distance to the K nearest neighbors of the point.

KDD: Originally, Knowledge Discovery in Databases; later, Knowledge Discovery and Data Mining.

KDM: Knowledge Discovery Metamodel.

kernels: A function with two vectors as input that returns a scalar representing the inner product of the vectors in some alternate dimension.

knowledge discovery: The process of automatically searching large volumes of data for patterns that can be described as "knowledge" about the data.

Kohonen networks: Neural networks based on hypothesized topological properties of the human brain. Also known as self-organizing feature maps (SOFMs) (Kohonen, 1982; Fausett, 1994; Haykin, 1994; Patterson, 1996).

leaf: In a classification tree diagram, any node that is not further segmented. The final leaves in a tree are called terminal nodes or leaves.

lift chart: A visual summary of the usefulness of the statistical or data mining models for predicting a binomial (categorical) outcome variable (dependent variable). For multinomial (multiple-category) outputs, the chart shows how useful the predictive models may be, compared to baseline.

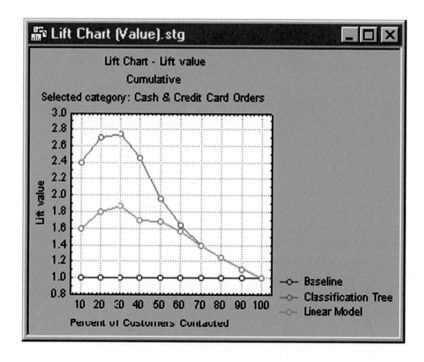

Lift may also be defined as the calculation equal to the "confidence factor" divided by the "expected confidence." The model having the highest value at the left side of the chart and continuing more or less as highest as the model's curve goes across to the right side of the chart is the "best" model for the data.

likelihood: The probability of an event based on current observations.

logistic regression: A form of regression analysis in which the target variable (response variable) is a binary-level or ordinal-level response and the target estimate is bounded at the extremes.

lookahead: For time series analysis, the number of time steps ahead of the last input variables that the output variable is to be predicted.

machine learning: A term often used to denote the application of generic model-fitting or classification algorithms for predictive data mining. This differs from traditional statistical data analysis, which is usually concerned with the estimation of population parameters by statistical inference and p-values. The emphasis in data mining machine learning algorithms is usually on the accuracy of the prediction as opposed to discovering the relationship and influences of different variables.

market basket analysis: See *association rule*.

metadata: A description or definition of data or information.

Multilayer Perceptron (MLP): A neural network that has one or more hidden layers, each of which has a linear combination function and executes a nonlinear activation function on the input to that layer.

Naïve Bayes: A statistical method based on Bayesian theorem that is primarily used for classification tasks.

neural networks: Techniques modeled after the (hypothesized) processes of learning in the cognitive system and the neurological functions of the brain and capable of predicting new observations (on specific variables) from other observations (on the same or other variables) after inducing a model from existing data. These techniques are also sometimes described as flexible nonlinear regression models, discriminant models, data reduction models, and multilayer nonlinear models.

node: In SAS-EM, SPSS Clementine, and *STATISTICA* Data Miner, a graphical icon that represents a data mining task in a process flow diagram. These nodes can perform data cleaning, transformation, sorting, and other such tasks; or they can represent algorithms from traditional statistics, machine learning, or statistical learning theory.

noise addition in neural networks: The addition of random noise to input patterns during training to prevent overfitting during back propagation training (and so "blurring" the position of the training data).

one-off in neural networks: A single case submitted to the neural network to be estimated that is not part of a data set and not used in training.

overfitting: The situation that occurs when an algorithm has too many parameters or is run for too long and fits the noise as well as the signal. Overfit models become too complex for the problem or the available quantity of data.

predicted value: The estimate of a dependent variable calculated by running the model on (or scoring) a set of values of the predictor variables.

Probabilistic Neural Network (PNN): A neural network using kernel-based approximation to form an estimate of the probability density functions of classes in a classification problem. One of the Bayesian networks (Speckt, 1990; Bishop, 1995; Patterson, 1996).

profit matrix: A table of expected revenues and expected costs for each level of a target variable.

QUEST: A classification tree program developed by Loh and Shih (1997).

R programming language: A programming language for statistics and graphics. Originally created by Ross Ihaka and Robert Gentleman at the University of Auckland, it is now further kept in development by the R Development Core Team. R developed out of the S language of Bell Labs. See the R Tutorial and associated white papers on the DVD that is packaged with this book, for further details, and web sites from which you can download the R-language and a multitude of statistical programs as freeware. (Main web site: http://www.r-project.org/)

radial basis function neural network: A neural network using a hidden layer of radial units and an output layer of linear units, and characterized by fast training and compact networks. Introduced by Broomhead and Lowe (1988) and Moody and Darkin (1989), they are described in many neural network textbooks (e.g., Bishop, 1995; Haykin, 1994).

regularization in neural networks: A modification to training algorithms that attempts to prevent over- or underfitting of training data by building in a penalty factor for network complexity, usually by penalizing large weights, which correspond to networks modeling functions of high curvature (Bishop, 1995).

RDF: Resource Description Framework.

root node: The beginning of a decision tree; the root node holds the entire data set submitted to the tree, before any splits are made.

scoring: Computing the values for new cases, based on the model developed from a "training data set." This can be rapidly done with PMML code of the saved model by applying the new data to this code.

self-organizing feature maps (SOFMs): Same as Kohonen networks; an SOFM classifies the parameter space into multiple clusters, while at the same time organizing the clusters into a map that is based on the relative distances between clusters. A competitive learning neural network that is also used for visualization.

statistical significance (p-level type in "traditional"/"frequentist" statistics): An estimated measure of the degree to which a result is "true." The higher the p-level, the less we can believe that there is a significant difference in what has been measured compared to the population as a whole. For example, the p-level of 0.05 indicates that there is a 5% probability that the relation between the variables found in our sample is a fluke. This means that there is a 95% chance that our result represents reality, but this also means that if we did 10 t-tests on the same data, we would get one that was significant at the $p < 0.05$ level just by chance, and thus would have to doubt that this represented reality.

In many areas of research, the *p*-level of 0.05 is treated as a borderline acceptable error level. In medical diagnoses, for example, we would like our *p*-level to be <0.001 or more—e.g., we want a medical Dx to be as accurate as possible.

Support Vector Machine (SVM): A classification method based on the maximum margin hyperplane.

target variable: The dependent variable for which the data mining independent variables (or predictor variables) make a model; it is what we want to score for in new cases of data.

transformation: The application of a function to a variable to adjust its range, variability, shape, etc.

unsupervised learning in neural networks: The type of learning that occurs when algorithms adjust the weights in a neural network by reference to a training data set that includes input variables only. Unsupervised learning algorithms attempt to locate clusters in the input data.

V-fold cross-validation: The process of drawing repeated random samples (V replicates) from the data for analysis so that the algorithm is then applied to compute predicted values, classifications, etc. Instead of one accuracy estimate, you get V, providing a more accurate estimate of the mean accuracy, as well as its uncertainty (e.g., standard deviation). This method is most often used with small data sets. The stability of the training and evaluation accuracies points to the quality of the model This method is used in tree classification and regression methods and is very useful in Support Vector Machines as an added check on the SVM Training Accuracy, Test Accuracy, and Overall Accuracy; e.g., in SVM, if the training and test accuracy are almost identical, then we have a good model; but if the training and text accuracy are a little off, then the V-fold cross-validation accuracy can be checked, and if it is identical or almost identical to the TEST accuracy, then we can accept the model on the basis of test accuracy.

validation data: A subset of a data set held out to run through the model produced by the training and testing (or evaluation) data sets, to see if the model holds up or generalizes.

variable: A column in an SAS, SPSS, or *STATISTICA* data set.

REFERENCES

Breiman, L., Friedman, J. H., Olshen, R. A., & Stone, C. J. (1984). *Classification and Regression Trees.* Monterey, CA: Wadsworth & Brooks/Cole Advanced Books & Software.

Broomhead, D. S., & Lowe, D. (1988). Multivariable functional interpolation and adaptive networks. *Complex Systems 2*, 321–355.

Cunningham, H. (2002). GATE, a general architecture for text engineering. *Computers and the Humanities 36*(2), 223–254.

Fausset, L. (1994). *Fundamentals of Neural Networks.* Upper Saddle River, NJ: Prentice Hall.

Ferrucci, D., & Lally, A. (2004) UIMA: An architectural approach to unstructured information processing in the corporate research environment. *Natural Language Engineering 10*(3/4), 327–348.

Goldberg, D. E. (1989). *Genetic Algorithms.* Reading, MA: Addison Wesley.

Haykin, S. (1994). *Neural Networks: A Comprehensive Foundation.* New York: Macmillan Publishing.

Kass, G. V. (1980). An exploratory technique for investigating large quantities of categorical data. *Applied Statistics 29,* 119–127.

Kohonen, T. (1982). Self-organized formation of topologically correct feature maps. *Biological Cybernetics, 43,* 59–69.

Loh, W.-Y., & Shih, Y.-S. (1997). Split selection methods for classification trees. *Statistica Sinica, 7,* 815–840.

Loh, W.-Y., & Vanichestakul, N. (1988). Tree-structured classification via generalized discriminant analysis (with discussion). *Journal of the American Statistical Association, 83,* 715–728.

Mack, R., Mukherjea, S., Soffer, A., Uramoto, N., Brown, E., Coden, A., et al. (2004). Text analytics for life science using the unstructured information management architecture. *IBM Systems Journal 43*(3), 490–515.

Manning, C. D., & Schütze, H. (2002). *Foundations of Statistical Natural Language Processing.* Cambridge, MA: MIT Press.

Moody, J., & Darkin, C. J. (1989). Fast learning in networks of locally-tuned processing units. *Neural Computation 1*(2), 281–294.

Speckt, D. F. (1990). Probabilistic neural networks. *Neural Networks 3*(1), 109–118.

Speckt, D. F. (1991). A generalized regression neural network. *IEEE Transactions on Neural Networks 2*(6), 568–576.

Witten, I. H., & Frank, E. (2000). *Data Mining: Practical Machine Learning Tools and Techniques.* New York: Morgan Kaufmann.

Index

Note: Page numbers followed by *f* or *t* indicate figures or tables, respectively.

DVD Install Instructions

1. Put the DVD in the CD-DVD read drive of your computer.
2. Open MY COMPUTER [from START → My Computer; or if you have a MY COMPUTER ICON placed on your desktop, click on this]
3. Click on the D-DRIVE [or whatever letter you have for your CD-DVD drive] to open the contents of this CD-DVD
4. There will be 2 primary folders on the HANDBOOK DVD:
    a. STATISTICA Data Miner Ver 8 [Note: this is Version 8/SERIES 0608c]
    b. TUTORIALS_etc_for CD_ELSEVIER [Note: It was unknown at the time of creating the DVD if it would be a CD or DVD. When there is a reference to CD, DVD, or CD-DVD, it means the DVD.]

5. **If you want to look at the TUTORIALS, open the "TUTORIALS_etc_for CD_ELSEVIER folder by clicking on it;** from there you can click on the sub-folders and examine each to see what is available, and pick the folder of interest.

6. **If you want to INSTALL and RUN the STATISTICA software:** Click on the "STATISTICA Data Miner Ver 8" folder; there will be several files inside:
    a. ENGLISH [a folder]
    b. MUILTIMED [a folder; containing videos/statistical learning instructions]
    c. Autorun.inf [a setup information text file]
    d. CDSTART.exe

7. To START installing STATISTICA software do either of the following:
    a. Click on the CDSTART.exe → a BLUE DIALOG will appear on the screen
    OR: to accomplish the same thing:
    b. Click on ENGLISH folder and then click on either the "setup.exe" or the "autorun.exe", which will also bring up the BLUE INSTALL dialog.
    c. Then proceed through the following set of numbered instructions [1 – 14] immediately below to install STATISTICA Data Mining software and/or if you prefer "visual instructions", jump down to Section II, below.

INSTALLING STATISTICA

1. The *STATISTICA* installation screen will appear. Click on *Install STATISTICA*.
2. The Welcome screen will appear. Click the *Next* button.

3. Read the software license agreement, and then select "I accept the terms of the license agreement," and click *Next* if you agree with the terms and wish to continue the installation process.
4. Select *Typical Setup* then click *Next*. *Typical Setup* will install *STATISTICA* with the most common options; this is the recommended selection. *Custom Setup* options are not covered in these instructions. If you have questions about the custom installation, please contact StatSoft technical support
5. On the *Register with StatSoft* dialog, enter the requested information in the appropriate boxes. Note: It is important that you enter a valid email address, otherwise registration cannot complete. Click *Next* to continue.
6. A dialog will prompt you to enable your wireless network adaptor. If your computer has a wireless network adaptor, please enable it until installation is complete in order to ensure proper licensing of the software. Once it is enabled, click *OK*.
7. On the following dialog, you will be informed that your license registration is pending and that a registration email has been sent to you.
8. Open your email application. Go to your Inbox and open the registration email from license@statsoft.com. The email will ask you to verify your email address in order to continue the installation of *STATISTICA*. Click on the hyperlink in the email. Alternately, you can copy and paste the link, in its entirety, into the address bar of your web browser.

   Note: If you do not receive an email from license@statsoft.com, you may need to look in your Junk E-mail folder. Due to the hyperlink in the email, your email application may have flagged the email as spam. Alternately, there may be an issue with your internet connection or firewall.

9. In your web browser, the *StatSoft Email Address Confirmation* webpage appears. Your email address has been confirmed.
10. You may now return to the installer and click the *Continue* button to finish the installation of *STATISTICA*. If you have closed the installer, restart it and continue as normal. A message will state that registration for this license is complete. Your license has been successfully registered. Click *OK*. If the registration process fails, a different dialog will open, indicating the failure. See notes below for additional details of failed registration.
11. You will be asked if you want to install the Multimedia files to your hard drive. These are movies that provide overviews of various aspects of the *STATISTICA* system. We recommend that you install them if you have sufficient disk space but they can also be viewed from the CD at any time
12. If you would like to create a Desktop shortcut to *STATISTICA*, press *Yes*. If you do not, press *No*.
13. *STATISTICA* is ready to install. Click *Install*.
14. You should receive a message stating that the installation is complete. You may be asked if you wish to reboot now or reboot later, depending on the components that were previously installed on your machine. If you are asked, it will be necessary to reboot before you run *STATISTICA*. Click *Finish* to complete the installation process.